ESO ASTROPHYSICS SYMPOSIA
European Southern Observatory

Series Editor: Bruno Leibundgut

Alvio Renzini Ralf Bender (Eds.)

Multiwavelength Mapping of Galaxy Formation and Evolution

Proceedings
of the ESO Workshop
Held at Venice, Italy, 13-16 October 2003

Springer

Volume Editors

Alvio Renzini

European Southern Observatory
Karl-Schwarzschild-Str. 2
85748 Garching, Germany

Ralf Bender

Universitäts-Sternwarte
der Ludwig-Maximilians-Universität München
Scheinerstr. 1
81679 München, Germany

and

Max-Planck-Institut für extraterrestrische Physik
Giessenbachstr.
85741 Garching, Germany

Series Editor

Bruno Leibundgut

European Southern Observatory
Karl-Schwarzschild-Str. 2
85748 Garching, Germany

ISBN 978-3-642-06509-5 **e-ISBN 978-3-540-31641-1**

Springer is a part of Springer Science+Business Media
springeronline.com

Softcover reprint of the hardcover 1st edition 2005

Cover-Design: Erich Kirchner, Heidelberg

Printed on acid-free paper 55/3141/Yu - 5 4 3 2 1 0

Preface

There are times in astronomy when new ideas and paradigms appear on the scene, and theory blooms and expands far beyond the reach of observations. There are other times when the opposite situation arises, and the sudden availability of new observational capacities reveals unanticipated aspects of nature. In the case of galaxy formation and evolution in a cosmological context, the '80s and early 90's were years clearly dominated by the theoretical developments, with simulations predicting far more than could be observationally tested at the time. However, starting in the mid '90's a new trend has progressively reversed this situation, to the extent that we now witness a dramatic, unprecedented expansion of observational studies in this field.

There is no doubt that this is due to the coming into play of major new observational facilities: First HST, then the 8–10 m class telescopes associated with tremendous progress in efficiency and multiplexing of their instrumentation, then a series of other space observatories such as Chandra and XMM-Newton, Spitzer, Galex, ... Now, together with radio and sub-mm facilities we have access to virtually the whole electromagnetic spectrum of galaxies, all the way from the local to the highest redshift universe. And we are just at the beginning.

To check at which stage is the mapping of galaxy formation and evolution through cosmic times, from October 13 to 16, 2003 over 170 astronomers from 15 countries met for a 4-day workshop hosted by the Venice International University (VIU). All major ongoing efforts were extensively illustrated and discussed at the meeting, including those just about to start and which promise quantum jumps in area and spectral coverage, depth, angular resolution, etc., thus making possible really unprecedented cosmological explorations.

While the meeting was primarily designed to overview the various observational projects, ongoing theoretical efforts to model galaxy formation and evolution were also presented at the meeting. It is no surprise that it is now harder for such models to match the plethora of data that are provided by observations, compared to ten or so years ago, when very little was known about high redshift galaxies, just a little beyond our local universe. Yet, understanding will of course emerge from the interplay of the empirical mapping of galaxy populations with those provided by theoretical simulations, with the general appreciation of the complexity of the baryonic physical processes one is trying to reproduce.

These massive observational efforts are inevitably changing our ways of doing astronomy. Multiwavelength, multi-facility coordinated projects are growing to sizes directly involving dozens of astronomers just for the data taking and processing. In several instances data soon are becoming public worldwide, then attracting more and more scientists in their scientific exploitation. But this is by no means a linear, one way process. Indeed, others feel overwhelmed by the difficulty of just keeping track of the pace with which data become available, or wonder whether it is worth investing time in a project while somewhere else quite the same project may already have been completed. We are at the brink of a sociological change in astronomy. The meeting has ostensibly shown that large new groups and collaborations have emerged that involve unprecedented numbers of astronomers. At the same time, others feel disaffected with such gigantism, and prefer finding a niche they can fully control and continue to work in smaller, human-size groups. All this, for the good and the bad, with coexisting elation and frustration, is at this stage unavoidable. We have constructed the machines, and we are bound to exploit them thoroughly, as mapping the whole cosmic evolution is now within reach.

The venue of the meeting was instrumental in facilitating discussions and interaction among the participants. The VIU is singularly located on the S. Servolo Island in the Venetian Lagoon, offering a special view of Venice itself. Substantial time was indeed devoted to discussions at the end of each session, and discussions continued during breaks taking advantage of the isolated environment, as well as during the morning and evening boat trips to and from the island. Sunny weather beautifully cooperated to the success of the meeting, along with the excellent Venetian dishes catered to the island at lunch time.

The workshop was organized jointly by the European Southern Observatory, the Max-Planck-Institut für Extraterrestrische Physik, and the Observatory of the Ludwig-Maximilians-Universität.

The Scientific Organizing Committee included R. Bender (Co-Chair), A. Cimatti, M. Colless, M. Dickinson, S. Guilloteau, G. Hasinger, D. Koo, O. LeFevre, M. Pettini, A. Renzini (Co-Chair) and I. Smail.

The Local Organizing Committe included P. Bristow, G. Hubert, C. Maraston, D. Pierini, M. Salvato, C. Stoffer, and S. Teupke, who we warmly thank for their smooth and efficient organization, before, during and after the meeting.

Finally, we would like to thank F. Nisii and the staff of the Venice International University for their most helpful cooperation.

Munich, Garching, *R. Bender*
December 2004 *A. Renzini*

Contents

List of Participants

AFONSO, Jose
CAAUL,
Observatorio Astronomico de Lisboa
jafonso@oal.ul.pt

ALEXANDER, David
Institute of Astronomy, Cambridge
dma@ast.cam.ac.uk

ALMAINI, Omar
Royal Observatory Edinburgh
omar@roe.ac.uk

ARIMOTO, Nobuo
NAOJ, Tokyo
arimoto@optik.mtk.nao.ac.jp

BALCELLS, Marc
Instituto de Astrofisica de Canarias
balcells@ll.iac.es

BALDRY, Ivan
Johns Hopkins University
baldry@pha.jhu.edu

BARDELLI, Sandro
INAF –
Osservatorio Astronomico di Bologna
bardelli@bo.astro.it

BELL, Eric
MPI für Astronomie, Heidelberg
bell@mpia.de

BENDER, Ralf
University Observatory Munich /
MPE Garching
bender@usm.uni-muenchen.de

BERGVALL, Nils
Uppsala Astronomical Observatory
nils.bergvall@astro.uu.se

BERNARDI, Mariangela
Carnegie Mellon University,
Pittsburgh
bernardi@cmu.edu

BERTOLA, Francesco
Università di Padova,
Dip. di Astronomia
bertola@pd.astro.it

BERTOLDI, Frank
MPI für Radioastronomie, Bonn
bertoldi@mpifr-bonn.mpg.de

BETTONI, Daniela
Osservatorio Astronomico di Padova
bettoni@pd.astro.it

BLAIN, Andrew
Caltech
awb@astro.caltech.edu

BORCH, Andrea
MPI für Astronomie, Heidelberg
borch@mpia-hd.mpg.de

BORYS, Colin
Caltech
borys@caltech.edu

BOUWENS, Rychard
University of California, Santa Cruz
bouwens@ucolick.org

BURGARELLA, Denis
Observatoire Astronomique Marseille – LAM
denis.burgarella@oamp.fr

BURKERT, Andreas
MPI für Astronomie, Heidelberg
burkert@mpia-hd.mpg.de

BUZZONI, Alberto
Telescopio Nazionale Galileo,
S. Cruz de La Palma
buzzoni@tng.iac.es

CAPAK, Peter
Institute for Astronomy,
Univ. of Hawaii
capak@ifa.hawaii.edu

CAPUTI, Karina
Royal Observatory Edinburgh
kic@roe.ac.uk

CASERTANO, Stefano
STScI
stefano@stsci.edu

CHAPMAN, Scott
Caltech
schapman@astro.caltech.edu

CHEN, Hsiao-Wen
MIT Center for Space Research
hchen@space.mit.edu

CIMATTI, Andrea
INAF –
Osservatorio Astrofisico di Arcetri
cimatti@arcetri.astro.it

CIOTTI, Luca
Università di Bologna,
Dip. di Astronomia
ciotti@bo.astro.it

COELHO, Paula
Universidade de Sao Paulo
pcoelho@usp.br

COLBERT, James
UCLA Astronomy
colbert@astro.ucla.edu

CONSELICE, Christopher
Caltech
cc@astro.caltech.edu

CORSINI, Enrico Maria
Università di Padova,
Dip. di Astronomia
corsini@pd.astro.it

COWIE, Len
Institute for Astronomy,
Univ. of Hawaii
cowie@mach.ifa.hawaii.edu

COWIE SONGAILA, A.
Institute for Astronomy,
Univ. of Hawaii
acowie@ifa.hawaii.edu

CRAMPTON, David
National Research Council of Canada,
Victoria
David.Crampton@nrc.ca

CRISTIANI, Stefano
INAF – Trieste
cristiani@ts.astro.it

DADDI, Emanuele
ESO, Garching
edaddi@eso.org

DALLA BONTÀ, Elena
Università di Padova,
Dip. di Astronomia
dallabonta@pd.astro.it

DAVE, Romeel
University of Arizona
rad@as.arizona.edu

DE BREUCK, Carlos
ESO, Garching
cdebreuc@eso.org

DE MELLO, Duilia
NASA Goddard Space Flight Center
duilia@ipanema.gsfc.nasa.gov

DE PROPRIS, Roberto
RSAA, ANU
propris@mso.anu.edu.au

DEKEL, Avishai
The Hebrew University of Jerusalem
dekel@phys.huji.ac.il

DICKINSON, Mark
STScI
med@stsci.edu

DRORY, Niv
University of Texas at Austin
drory@astro.as.utexas.edu

ELBAZ, David
Service d'Astrophysique/CEA
elbaz@ucolick.org

ERB, Dawn
Caltech
dke@astro.caltech.edu

FALL, Michael
STScI
fall@stsci.edu

FALOMO, Renato
INAF – Osservatorio di Padova
falomo@pd.astro.it

FERGUSON, Harry
STScI
ferguson@stsci.edu

FERRARA, Andrea
SISSA, Trieste
ferrara@sissa.it

FONTANA, Adriano
INAF –
Osservatorio Astronomico di Roma
fontana@mporzio.astro.it

FONTANOT, Fabio
Università di Trieste,
Dip. di Astronomia
fontanot@ts.astro.it

FORSTER SCHREIBER, Natascha
Leiden Observatory
forster@strw.leidenuniv.nl

FOUCAUD, Sebastien
IASF Sezione di Milano
foucaud@mi.iasf.cnr.it

FRANCESCHINI, Alberto
Università di Padova
afrances@mostro.pd.astro.it

GALLEGO, Jesus
Universidad Complutense de Madrid
jgm@astrax.fis.ucm.es

GARDNER, Jonathan
NASA Goddard Space Flight Center
jonathan.p.gardner@nasa.gov

GAVAZZI, Giuseppe
Università di Milano – Bicocca
giuseppe.gavazzi@mib.infn.it

GENZEL, Reinhard
MPI für extraterrestrische Physik, Garching
genzel@mpe.mpg.de

GIAVALISCO, Mauro
STScI
mauro@stsci.edu

GIOIA, Isabella
Istituto di Radioastronomia, Bologna
gioia@ira.cnr.it

GRAY, Meghan
Royal Observatory Edinburgh
meg@roe.ac.uk

GREVE, Thomas
Royal Observatory Edinburgh
tgreve@roe.ac.uk

GRONWALL, Caryl
Penn State University
caryl@astro.psu.edu

GUZMAN, Rafael
University of Florida
guzman@astro.ufl.edu

HAMMER, Francois
GEPI, Observatoire de Paris
francois.hammer@obspm.fr

HASINGER, Guenther
MPI für extraterrestrische Physik, Garching
ghasinger@mpe.mpg.de

HEAP, Sara
NASA Goddard Space Flight Center
Sara.R.Heap@nasa.gov

HEMPEL, Angela
MPI für Astronomie, Heidelberg
hempel@mpia-hd.mpg.de

HILL, Gary
McDonald Observatory,
University of Texas at Austin
hill@astro.as.utexas.edu

HOLDEN, Bradford
UCO/Lick Observatory
holden@ucolick.org

HOPKINS, Andrew
University of Pittsburgh
ahopkins@phyast.pitt.edu

HUANG, Jiasheng
Harvard-Smithsonian Center
for Astrophysics
jhuang@cfa.harvard.edu

ILLINGWORTH, Garth
University of California
gdi@ucolick.org

IVISON, Rob
Royal Observatory Edinburgh
rji@roe.ac.uk

JAHNKE, Knud
Astrophysical Institute Potsdam
kjahnke@aip.de

JARVIS, Matt
University of Oxford
mjj@astro.ox.ac.uk

KAUFFMANN, Guinevere
MPI für Astrophysik, Garching
gamk@Mpa-Garching.MPG.DE

KELM, Birgit
Università di Bologna,
Dip. di Astronomia
birgit@orion.bo.astro.it

KODAMA, Tadayuki
NAOJ, Tokyo
kodama@th.nao.ac.jp

KOEKEMOER, Anton
STScI
koekemoe@stsci.edu

KOO, David
Lick Observatory
koo@ucolick.org

KRETCHMER, Claudia
Johns Hopkins University
claudia@pha.jhu.edu

KURK, Jaron
INAF, Arcetri
kurk@arcetri.astro.it

LABBE, Ivo
Leiden Observatory
ivo@strw.leidenuniv.nl

LACEY, Cedric
Dept. of Physics,
University of Durham
cedric.lacey@obs.univ-lyon1.fr

LANZONI, Barbara
INAF –
Osservatorio Astronomico di Bologna
lanzoni@bo.astro.it

LE FÈVRE, Olivier
Laboratoire d'Astronomie Spatiale,
Marseille
olivier.lefevre@astrsp-mrs.fr

LEDLOW, Michael
Gemini Observatory, La Serena
mledlow@gemini.edu

LEHNERT, Matt
MPI für extraterrestrische Physik,
Garching
mlehnert@mpe.mpg.de

LOTZ, Jennifer
University of California, Santa Cruz
jlotz@scipp.ucsc.edu

LUCAS, Ray
STScI
lucas@stsci.edu

MACARTHUR, Lauren
UBC / Physics and Astronomy,
Vancouver
lauren@astro.ubc.ca

MACCAGNI, Dario
IASF Sezione di Milano
dario@mi.iasf.cnr.it

MADAU, Piero
University of California, Santa Cruz
pmadau@ucolick.org

MAIOLINO, Roberto
INAF, Arcetri
maiolino@arcetri.astro.it

MALLER, Ariyeh
University of Massachusetts
ari@astro.umass.edu

MARASTON, Claudia
MPI für extraterrestrische Physik,
Garching
maraston@mpe.mpg.de

MARINONI, Christian
Laboratoire d'Astrophysique
de Marseille
christian.marinoni@oamp.fr

MARTIN, Chris
Caltech
cmartin@srl.caltech.edu

MASEGOSA, Pepa
Instituto de Astrofisica de Andalucia,
Granada
pepa@iaa.es

MAZZARELLA, Joe
Caltech / JPL
mazz@ipac.caltech.edu

MAZZEI, Paola
INAF –
Osservatorio Astronomico Padova
mazzei@pd.astro.it

MCLURE, Ross
Institute for Astronomy,
University of Edinburgh
rjm@roe.ac.uk

MENDES DE OLIVEIRA, Claudia
IAG, Sao Paulo
oliveira@astro.iag.usp.br

METEVIER, Anne
University of California, Santa Cruz
anne@ucolick.org

MILVANG-JENSEN, Bo
MPI für extraterrestrische Physik,
Garching
milvang@mpe.mpg.de

MONACO, Pierluigi
Università di Trieste,
Dip. di Astronomia
monaco@ts.astro.it

MOORWOOD, Alan
ESO, Garching
amoor@eso.org

MORELLI, Lorenzo
ESO Chile, and
Univ. Padova, Dip. di Astronomia
lmorelli@eso.org

MORGANTI, Raffaella
ASTRON, Dwingeloo
morganti@astron.nl

MOUSTAKAS, Leonidas
STScI
leonidas@stsci.edu

NEWMAN, Jeffrey
U.C. Berkeley – Dept. of Astronomy
jnewman@astro.berkeley.edu

NIPOTI, Carlo
Università di Bologna,
Dip. di Astronomia
nipoti@bo.astro.it

NOLL, Stefan
Landessternwarte Heidelberg
snoll@lsw.uni-heidelberg.de

OLIVER, Sebastian
Astronomy Centre,
University of Sussex
S.Oliver@sussex.ac.uk

OOSTERLOO, Thomas
ASTRON, Dwingeloo
oosterloo@astron.nl

PERCIVAL, William
Institute for Astronomy,
University of Edinburgh
wjp@roe.ac.uk

PIERINI, Daniele
MPI für extraterrestrische Physik,
Garching
dpierini@mpe.mpg.de

PIZZELLA, Alessandro
Università di Padova,
Dip. di Astronomia
pizzella@pd.astro.it

POPESSO, Paola
MPI für extraterrestrische Physik,
Garching
popesso@mpe.mpg.de

POZZETTI, Lucia
Osservatorio Astronomico di Bologna
lucia@bo.astro.it

PRANDONI, Isabella
Istituto di Radioastronomia, Bologna
prandoni@ira.cnr.it

PREETHI, Nair
University of Toronto
nair@astro.utoronto.ca

RADOVICH, Mario
INAF –
Oss. Astronomico Capodimonte
radovich@na.astro.it

RENZINI, Alvio
ESO, Garching
arenzini@eso.org

RETTURA, Alessandro
ST-ECF, Garching
arettura@eso.org

RICH, Michael
UCLA
rmr@astro.ucla.edu

RIDGWAY, Susan
Johns Hopkins University
ridgway@pha.jhu.edu

RIX, Hans-Walter
MPI für Astronomie, Heidelberg
rix@mpia.de

ROCCA-VOLMERANGE, Brigitte
Inst. d'Astrophysique Paris
rocca@iap.fr

RODIGHIERO, Giulia
Università di Padova
rodighiero@pd.astro.it

ROSATI, Piero
ESO, Garching
prosati@eso.org

RÖTTGERING, Huub
Leiden Observatory
rottgeri@strw.leidenuniv.nl

RUDNICK, Gregory
MPI für Astrophysik, Garching
grudnick@mpa-garching.mpg.de

SALVATO, Mara
MPI für extraterrestrische Physik, Garching
mara@mpe.mpg.de

SANCISI, Renzo
Osservatorio Astronomico di Bologna
sancisi@bo.astro.it

SANDERS, David
Institute for Astronomy,
Univ. of Hawaii
sanders@galileo.ifa.hawaii.edu

SARACCO, Paolo
INAF –
Osservatorio Astronomico di Brera
saracco@brera.mi.astro.it

SCARAMELLA, Roberto
Oss. Astronomico di Roma,
Monteporzio
kosmobob@mporzio.astro.it

SCHADE, David
National Research Council of Canada,
Victoria
David.Schade@nrc.gc.ca

SCHIMINOVICH, David
Caltech
ds@srl.caltech.edu

SCHULTE-LADBECK, Regina
Univ. of Pittsburgh
rsl@phyast.pitt.edu

SCOVILLE, Nick
Caltech
nzs@astro.caltech.edu

SEITZ, Stella
University Observatory Munich
stella@usm.uni-muenchen.de

SEKIGUCHI, Kazuhiro
Subaru Telescope, NAOJ
kaz@naoj.org

SHAVER, Peter
ESO, Garching
pshaver@eso.org

SIMPSON, Chris
Dept. of Physics,
University of Durham
chris.simpson@durham.ac.uk

SMITH, Joanna
Institute of Astronomy, Cambridge
joanna@ast.cam.ac.uk

SOMERVILLE, Rachel
STScI
somerville@stsci.edu

SQUIRES, Gordon
SIRTF Science Center – Caltech
squires@ipac.caltech.edu

STEIDEL, Chuck
Caltech
ccs@astro.caltech.edu

STERN, Daniel
JPL/Caltech
stern@zwolfkinder.jpl.nasa.gov

TAKAGI, Toshinobu
Imperial College London
t.takagi@imperial.ac.uk

TANIGUCHI, Yoshiaki
Astronomical Institute,
Tohoku University
tani@terra.astr.tohoku.ac.jp

TASCA, Lidia
MPI für Astrophysik, Garching
tasca@mpa-garching.mpg.de

THOMAS, Daniel
MPI für extraterrestrische Physik,
Garching
dthomas@mpe.mpg.de

TRAGER, Scott
Kapteyn Institute, Groningen
sctrager@astro.rug.nl

VACCARI, Mattia
Università di Padova,
Dip. di Astronomia
vaccari@pd.astro.it

VAZDEKIS, Alexander
Instituto de Astrofisica de Canarias
vazdekis@ll.iac.es

VENEMANS, Bram
Leiden Observatory
venemans@strw.leidenuniv.nl

VERDE, Licia
Dep. Physics and Astronomy,
Univ. of Pennsylvania
lverde@astro.princeton.edu

VERNET, Joël
INAF –
Osservatorio Astrofisico di Arcetri
vernet@arcetri.astro.it

WEBB, Tracy
Leiden Observatory
webb@strw.leidenuniv.nl

WECHSLER, Risa
University of Chicago
wechsler@umich.edu

WHITE, Simon
MPI für Astrophysik, Garching
swhite@mpa-garching.mpg.de

WIKLIND, Tommy
ESA/STScI
wiklind@stsci.edu

WOLF, Marsha
University of Texas at Austin
mwolf@astro.as.utexas.edu

WOLF, Christian
University of Oxford
cwolf@mpia.de

ZAMORANO, Jaime
Astrofisica –
Universidad Complutense, Madrid
jaz@astrax.fis.ucm.es

ZIRM, Andrew
Leiden Observatory
azirm@strw.leidenuniv.nl

ZUCCA, Elena
INAF –
Osservatorio Astronomico di Bologna
zucca@bo.astro.it

The Dawn of Galaxies

P. Madau

Department of Astronomy and Astrophysics, University of California,
1156 High Street, Santa Cruz, CA 95064, USA

Abstract. The development of primordial inhomogeneities into the non-linear regime and the formation of the first astrophysical objects within dark matter halos mark the transition from a simple, neutral, cooling universe – described by just a few parameters – to a messy ionized one – the realm of radiative, hydrodynamic, and star formation processes. The recent measurement by the *WMAP* satellite of a large optical depth to electron scattering implies that this transition must have begun very early, and that the universe was reionized at redshift $z_{\rm ion} = 17 \pm 5$. It is an early generation of extremely metal-poor massive stars and/or 'seed' accreting black holes in subgalactic halos that may have generated the ultraviolet radiation and mechanical energy that reheated and reionized most of the hydrogen in the cosmos. The detailed thermal, ionization, and chemical enrichment history of the universe during the crucial formative stages around $z = 10 - 20$ depends on the power-spectrum of density fluctuations on small scales, the stellar initial mass function and star formation efficiency, a complex network of poorly understood 'feedback' mechanisms, and remains one of the crucial missing links in galaxy formation and evolution studies.

1 Introduction

The last decade has witnessed great advances in our understanding of the high redshift universe. The pace of observational cosmology and extragalactic astronomy has never been faster, and progress has been equally significant on the theoretical side. The key idea of currently popular cosmological scenarios, that primordial density fluctuations grow by gravitational instability driven by cold, collisionless dark matter (CDM), has been elaborated upon and explored in detail through large-scale numerical simulations on supercomputers, leading to a hierarchical ('bottom-up') scenario of structure formation. In this model, the first objects to form are on subgalactic scales, and merge to make progressively bigger structures ('hierarchical clustering'). Ordinary matter in the universe follows the dynamics dictated by the dark matter until radiative, hydrodynamic, and star formation processes take over. Perhaps the most remarkable success of this theory has been the prediction of anisotropies in the temperature of the cosmic microwave background (CMB) radiation at about the level subsequently measured by the *COBE* satellite and most recently by the *BOOMERANG*, *MAXIMA*, *DASI*, *CBI*, *Archeops*, and *WMAP* experiments.

In spite of some significant achievements in our understanding of the formation of cosmic structures, there are still many challenges facing hierarchical clustering theories, and many fundamental questions remain, at best, only partially answered. While quite successful in matching the observed large-scale density distribution (like, e.g., the properties of galaxy clusters, galaxy clustering, and the statistics of the Lyman-α forest), CDM simulations appear to produce halos that are too centrally concentrated compared to the mass distribution inferred from the rotation curves of (dark matter-dominated) dwarf galaxies, and to predict too many dark matter subhalos compared to the number of dwarf satellites observed within the Local Group.[38,53,37,27] Another perceived problem (possibly connected with the 'missing satellites'[8]) is our inability to predict when, how, and to what temperature the universe was reheated and reionized, i.e. to understand the initial conditions of the galaxy formation process. While N-body+hydrodynamical simulations have convincingly shown that the intergalactic medium (IGM) – the main repository of baryons at high redshift – is expected to fragment into structures at early times in CDM cosmogonies, the same simulations are much less able to predict the efficiency with which the first gravitationally collapsed objects lit up the universe at the end of the 'dark ages'. The crucial processes of star formation, preheating and feedback (e.g. the effect of the heat input from the first generation of sources on later ones), and assembly of massive black holes in the nuclei of galaxies are poorly understood.[30] We know that at least some galaxies and quasars were already shining when the universe was less than 10^9 yr old. But when did the first luminous objects form, what was their nature, and what impact did they have on their environment and on the formation of more massive galaxies? While the excess H I absorption measured in the spectra of $z \sim 6$ quasars in the Sloan Digital Sky Survey (SDSS) has been interpreted as the signature of the trailing edge of the cosmic reionization epoch[3,17,15], the recent detection by the *Wilkinson Microwave Anisotropy Probe* (*WMAP*) of a large optical depth to Thomson scattering, $\tau_e = 0.17 \pm 0.04$ suggests that the universe was reionized at higher redshifts, $z_{\rm ion} = 17 \pm 5$.[28,51] This is of course an indication of significant star-formation activity at very early times.

In this talk I will summarize some recent developments in our understanding of the dawn of galaxies and the impact that some of the earliest cosmic structure may have had on the baryonic universe.

2 The Dark Ages

In the era of precision cosmology we know that, at a redshift $z_{\rm dec} = 1088 \pm 1$, exactly $t_{\rm dec} = (372 \pm 14) \times 10^3$ years after the big bang, the universe became optically thin to Thomson scattering[51], and entered a 'dark age'.[42] At this epoch the electron fraction dropped below 13%, and the primordial radiation cooled below 3000 K, shifting first into the infrared and then into the radio.

We understand the microphysics of the post-recombination universe well. The fractional ionization froze out to the value $\sim 10^{-4.8}\Omega_M/(h\Omega_b)$: these residual electrons were enough to keep the matter in thermal equilibrium with the radiation via Compton scattering until a thermalization redshift $z_t \simeq 800(\Omega_b h^2)^{2/5} \simeq 150$, i.e. well after the universe became transparent.[40] Thereafter, the matter

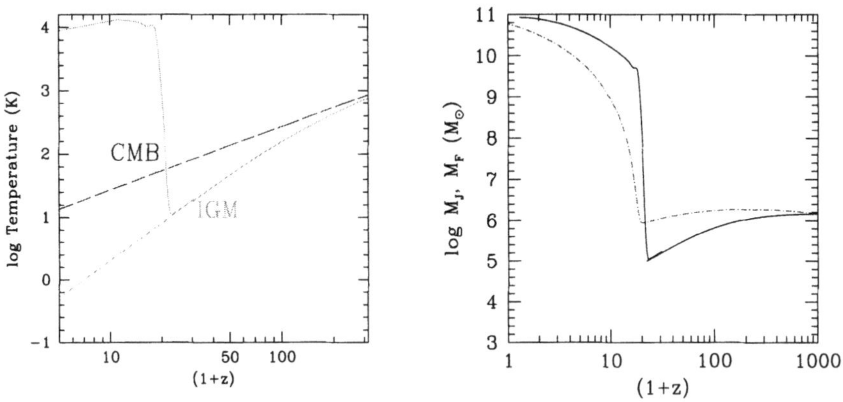

Fig. 1. *Left:* Evolution of the radiation (*long-dashed line*, labeled CMB) and gas (*solid line*, labeled IGM) temperatures after recombination. The universe is assumed to be reionized by ultraviolet radiation at $z \simeq 20$. The *short-dashed line* is the extrapolated gas temperature in the absence of any reheating mechanism. *Right:* Cosmological (gas + dark matter) Jeans (*solid line*) and filtering (*dot-dashed line*) mass.

temperature decreased as $(1+z)^2$ due to adiabatic expansion (Figure 1) until primordial inhomogeneities in the density field evolved into the non-linear regime. The minimum mass scale for the gravitational aggregation of cold dark matter particles is negligibly small. One of the most popular CDM candidates is the neutralino: in neutralino CDM, collisional damping and free streaming smear out all power of primordial density inhomogeneities only below $\sim 10^{-7}$ $M_\odot$.[26] Baryons, however, respond to pressure gradients and do not fall into dark matter clumps below the cosmological Jeans mass (in linear theory this is the minimum mass-scale of a perturbation where gravity overcomes pressure),

$$M_J = \frac{4\pi\bar{\rho}}{3}\left(\frac{5\pi k_B T}{3G\bar{\rho}m_p\mu}\right)^{3/2} \approx 2.5\times 10^5\, h^{-1}\ M_\odot (aT/\mu)^{3/2}\Omega_M^{-1/2}. \quad (1)$$

Here $a = (1+z)^{-1}$ is the scale factor, $\bar{\rho}$ the total mass density including dark matter, μ the mean molecular weight, and T the gas temperature. The evolution of M_J is shown in Figure 1. In the post-recombination universe, the baryon-electron gas is thermally coupled to the CMB, $T \propto a^{-1}$, and the Jeans mass is independent of redshift and comparable to the mass of globular clusters, $M_J \approx 10^6$ $M_\odot$. For $z < z_t$, the temperature of the baryons drops as $T \propto a^{-2}$, and the Jeans mass decreases with time, $M_J \propto a^{-3/2}$. This trend is reversed by the reheating of the IGM. The energy released by the first collapsed objects drives the Jeans mass up to galaxy scales (Figure 1): previously growing density perturbations decay as their mass drops below the new Jeans mass. In particular, photoionization by the ultraviolet radiation from the first stars and quasars would heat the IGM to temperatures of $\approx 10^4$ K (corresponding to a Jeans mass $M_J \lesssim 10^{10}$ $M_\odot$ at $z \simeq 20$), suppressing gas infall into low mass halos and preventing new (dwarf) galaxies from forming.

3 Linear Theory

When the Jeans mass itself varies with time, linear gas fluctuations tend to be smoothed on a (filtering) scale that depends on the full thermal history of the gas instead of the instantaneous value of the sound speed.[23] From linear perturbation analysis, and for a flat universe at high redshift, the growth of density fluctuations in the gas is suppressed for comoving wavenumbers $k > k_F$, where the filtering scale k_F is related to the Jeans wavenumber k_J by[22]

$$\frac{1}{k_F^2(a)} = \frac{3}{a} \int_0^a \frac{da'}{k_J^2(a')} [1 - (a'/a)^{1/2}]. \tag{2}$$

Here $k_J \equiv (a/c_s)\sqrt{4\pi G\bar{\rho}}$, and c_s is the sound speed. This expression for k_F accounts for an arbitrary thermal evolution of the IGM through $k_J(a)$. Corresponding to the critical wavenumber k_F there is a critical (filtering) mass M_F, defined as the mass enclosed in the sphere with comoving radius equal to k_F,

$$M_F = (4\pi/3)\bar{\rho}(2\pi a/k_F)^3. \tag{3}$$

The Jeans mass M_J is defined analogously in terms of k_J. It is the filtering mass that is central to calculations of the effects of reheating and reionization on galaxy formation. The filtering mass for a toy model with early photoionization is shown in Figure 1: after reheating, the filtering scale is actually smaller than the Jeans scale. Numerical simulations of cosmological reionization confirm that the characteristic suppression mass is typically lower than the linear-theory Jeans mass.[22]

4 The Emergence of Cosmic Structure

As mentioned in the introduction, some shortcomings on galactic and sub-galactic scales of the currently favored model of hierarchical galaxy formation in a universe dominated by CDM have recently appeared. The significance of these discrepancies is still debated, and 'gastrophysical' solutions involving feedback mechanisms may offer a possible way out. Other models have attempted to solve the apparent small-scale problems of CDM at a more fundamental level, i.e. by reducing small-scale power. Although the 'standard' ΛCDM model for structure formation assumes a scale-invariant initial power spectrum of density fluctuations, $P(k) \propto k^n$ with $n = 1$, the recent *WMAP* data favor (but don't require) a slowly varying spectral index, $dn/d\ln k = -0.031^{+0.016}_{-0.018}$, i.e. a model in which the spectral index varies as a function of wavenumber k.[51] This running spectral index model predicts a significantly lower amplitude of fluctuations on small scales than standard ΛCDM. The suppression of small-scale power has

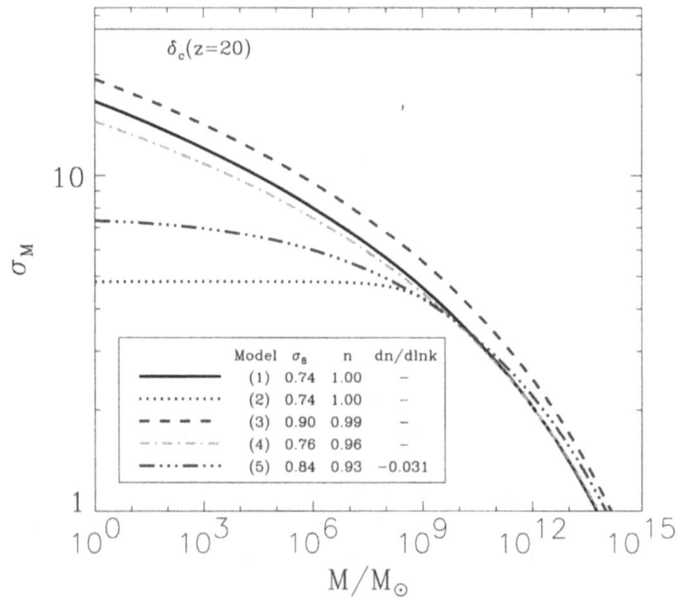

Fig. 2. The variance of the matter-density field vs. mass M, for different power spectra. All models assume a 'concordance' cosmology with parameters $(\Omega_M, \Omega_\Lambda, \Omega_b, h) = (0.29, 0.71, 0.045, 0.7)$. *Solid curve:* standard ΛCDM with no tilt, cluster normalized. *Dotted curve:* ΛWDM with a particle mass $m_X = 2\,\text{keV}$, cluster normalized, no tilt. *Dashed curve:* tilted *WMAP* model, *WMAP* data only. *Dash-dotted curve:* tilted *WMAP* model, including 2dFGRS and Lyman-α data. *Dash-triple dotted curve:* running spectral index *WMAP* model, including 2dFGRS and Lyman-α data. Here n refers to the spectral index at $k = 0.05\,\text{Mpc}^{-1}$. The horizontal line at the top of the figure shows the value of the extrapolated collapse overdensity $\delta_c(z)$ at $z = 20$.

the advantage of reducing the amount of substructure in galactic halos and makes small halos form later (when the universe was less dense) hence less concentrated,[38,57] relieving some of the problems of ΛCDM. But it makes early reionization a challenge.

Figure 2 shows the linearly extrapolated (to $z = 0$) variance of the mass-density field smoothed on a scale of comoving radius R,

$$\sigma_M^2 = \langle(\delta M/M)^2\rangle = \frac{1}{2\pi^2}\int_0^\infty dk\, k^2 P(k) T^2(k) W^2(kR), \tag{4}$$

for different power spectra. Here $M = H_0^2 \Omega_M R^3/2G$ is the mass inside R, $T(k)$ is the transfer function for the matter density field (which accounts for all modifications of the primordial power-law spectrum due to the effects of pressure and dissipative processes), and $W(kR)$ is the Fourier transform of the spherical top-hat window function, $W(x) = (3/x^2)(\sin x/x - \cos x)$. The value of the rms mass fluctuation in a $8\,h^{-1}$ Mpc sphere, $\sigma_8 \equiv \sigma(z=0, R=8\,h^{-1}\,\mathrm{Mpc})$, has been fixed for the $n=1$ models to $\sigma_8 = 0.74$, consistent with recent normalization by the $z=0$ X-ray cluster abundance constraint.[43]

In the CDM paradigm, structure formation proceeds 'bottom-up', i.e., the smallest objects collapse first, and subsequently merge together to form larger objects. It then follows that the loss of small-scale power modifies structure formation most severely at the highest redshifts, significantly reducing the number of self-gravitating objects then. This, of course, will make it more difficult to reionize the universe early enough.

It has been argued, for example, that one popular modification of the CDM paradigm, warm dark matter (WDM), has so little structure at high redshift that it is unable to explain the *WMAP* observations of an early epoch of reionization.[51,2] And yet the *WMAP* running-index model may suffer from a similar problem.[49] A look at Figure 2 shows that $10^6\,\mathrm{M}_\odot$ halos will collapse at $z=20$ from $2.9\,\sigma$ fluctuations in a tilted ΛCDM model with $n=0.99$ and $\sigma_8 = 0.9$, from $4.6\,\sigma$ fluctuations in a running-index model, and from $5.7\,\sigma$ fluctuations in a WDM cosmology. The problem is that scenarios with increasingly rarer halos at early times require even more extreme assumptions (i.e. higher star formation efficiencies and UV photon production rates) in order to be able to reionize the universe by $z \sim 17$ as favored by *WMAP*.[49,56,24,11,9]

5 The Epoch of Reionization

Since hierarchical clustering theories provide a well-defined framework in which the history of baryonic material can be tracked through cosmic time, probing the reionization epoch may then help constrain competing models for the formation of cosmic structures. Quite apart from uncertainties in the primordial power spectrum on small scales, however, it is the astrophysics of baryons that makes us unable to predict when reionization actually occurred. Consider the following illustrative example:

Hydrogen photoionization requires more than one photon above 13.6 eV per hydrogen atom: of order $t/\bar{t}_{\rm rec} \sim 10$ (where $\bar{t}_{\rm rec}$ is the volume-averaged hydrogen recombination timescale) extra photons appear to be needed to keep the gas in overdense regions and filaments ionized against radiative recombinations.[21,33]. A 'typical' stellar population produces during its lifetime about 4000 Lyman continuum (ionizing) photons per stellar proton. A fraction $f \sim 0.25\%$ of cosmic baryons must then condense into stars to supply the requisite ultraviolet flux. This estimate assumes a standard (Salpeter) initial mass function (IMF), which determines the relative abundances of hot, high mass stars versus cold, low mass ones.

The very first generation of stars ('Population III') must have formed, however, out of unmagnetized metal-free gas: numerical simulations of the fragmentation of pure H and He molecular clouds[6,1] have shown that these characteristics likely led to a 'top-heavy' IMF biased towards very massive stars (VMSs, i.e.

stars a few hundred times more massive than the Sun), quite different from the present-day Galactic case. Metal-free VMSs emit about 10^5 Lyman continuum photons per stellar baryon[7], approximately 25 times more than a standard stellar population. A corresponding smaller fraction of cosmic baryons would have to collapse then into VMSs to reionize the universe, $f \sim 10^{-4}$. There are of course further complications. Since, at zero metallicity, mass loss through radiatively-driven stellar winds is expected to be negligible[29], Population III stars may actually die losing only a small fraction of their mass. If they retain their large mass until death, VMSs with masses $100 \lesssim m \lesssim 250$ M$_\odot$ will encounter the electron-positron pair instability and disappear in a giant nuclear-powered explosion[18], leaving no compact remnants and polluting the universe with the first heavy elements. In still heavier stars, however, oxygen and silicon burning is unable to drive an explosion, and complete collapse to a black hole will occur instead.[5] Thin disk accretion onto a Schwarzschild black hole releases about 50 MeV per baryon. The conversion of a trace amount of the total baryonic mass into early black holes, $f \sim 3 \times 10^{-6}$, would then suffice to reionize the universe.

6 Preheating and Galaxy Formation

Even if the IMF at early times were known, we still would remain uncertain about the fraction of cold gas that gets retained in protogalaxies after the formation of the first stars (this quantity affects the global efficiency of star formation at these epochs) and whether – in addition to ultraviolet radiation – an early input of mechanical energy may also play a role in determining the thermal and ionization state of the IGM on large scales. The same massive stars that emit ultraviolet light also explode as supernovae (SNe), returning most of the metals to the interstellar medium of pregalactic systems and injecting about 10^{51} ergs per event in kinetic energy. A complex network of feedback mechanisms is likely at work in these systems, as the gas in shallow potential is more easily blown away,[13] thereby quenching star formation. Furthermore, as the blastwaves produced by supernova explosions – and possibly also by winds from 'miniquasars' – sweep the surrounding intergalactic gas, they may inhibit the formation of nearby low-mass galaxies due to 'baryonic stripping'[45], and drive vast portions of the IGM to a significantly higher temperature than expected from photoionization,[54,31,32,52,10] so as to 'choke off' the collapse of further galaxy-scale systems. Note that this type of global feedback is fundamentally different from the 'in situ' heat deposition commonly adopted in galaxy formation models, in which hot gas is produced by supernovae within the parent galaxy. We refer here to this global early energy input as 'preheating'.[4] Note that a large scale feedback mechanism may also be operating in the intracluster medium: studies of X-ray emitting gas in clusters show evidence for some form of non-gravitational entropy input [41]. The energy required there is at a level of $\sim 1\,\mathrm{keV}$ per particle, and must be injected either in a more localized fashion or at late epochs in order not to violate observational constraints on the temperature of the Lyman-α forest at $z \sim 3$. The thermal and ionization history of a preheated universe may be very different from one where hydrogen is photoionized. The gas will be heated up to a higher adiabat, and collisions with hot electrons will be the dominant ionization mechanism. The higher energies associated with preheating may doubly ionize helium at high-z, well before the 'quasar epoch' at $z \sim 3$. Galaxy formation and evolution will also be different, as preheating will drive the filtering mass above $10^{10} - 10^{11}$ $M_\odot$ and will tend to flatten the faint-end slope of the present-epoch galaxy luminosity function, in excellent agreement with the data and without the need for SN feedback at late times.[4]

It is interesting to set some general constraints on the early star-formation episode and stellar populations that may be responsible for an early preheating of the IGM at the levels consistent with the temperature of intergalactic gas inferred at $z \approx 3$. Let us characterize the energy input due to preheating by the energy per baryon, E_p, deposited in the IGM at redshift z_p. We examine a homogenous energy deposition since the filling factor of pregalactic outflows is expected to be large.[32,19] Let Ω_* be the mass density of stars formed at z_p in units of the critical density, $E_{\rm SN}$ the mechanical energy injected per SN event, and f_w the fraction of that energy that is eventually deposited into the IGM. Denoting with η the number of SN explosions per mass of stars formed, one can write

$$\frac{\Omega_*}{\Omega_b} = \frac{E_p}{f_w \eta E_{\rm SN} m_p}, \tag{5}$$

where m_p is the proton mass. For a Salpeter IMF between 0.1 and 100 $M_\odot$, the number of Type II SN explosions per mass of stars formed is $\eta = 0.0074$ $M_\odot^{-1}$, assuming all stars above 8 $M_\odot$ result in SNe II. Numerical simulations of the dynamics of SN-driven bubbles from subgalactic halos have shown that up to 40% of the available SN mechanical luminosity can be converted into kinetic energy of the blown away material, $f_w \approx 0.4$, the remainder being radiated away.[36] With $E_{\rm SN} = 1.2 \times 10^{51}$ ergs, equation (5) then implies

$$\left(\frac{\Omega_*}{\Omega_b}\right)_{\rm sp} = 0.05\ (E_p/0.1\ {\rm keV}). \tag{6}$$

SN-driven pregalactic outflows efficiently carry metals into intergalactic space.[32] For a normal IMF, the total amount of metals expelled in winds and final ejecta (in SNe or planetary nebulae) is about 1% of the input mass. Assuming a large fraction, $f_Z = 0.5$, of the metal-rich SN ejecta escape the shallow potential wells of subgalactic systems, the star-formation episode responsible for early preheating will enrich the IGM to a mean level

$$\langle Z\rangle_{\rm sp} = \frac{0.01\,\Omega_*\, f_Z}{\Omega_b} = 0.014\, Z_\odot\ (E_p/0.1\ {\rm keV}). \tag{7}$$

The weak C IV absorption lines observed in the Lyman-α forest at $z = 3 - 3.5$ imply a minimum universal metallicity relative to solar in the range [−3.2] to [−2.5].[50]. Preheating energies in excess of 0.1 keV appear then to require values of Ω_* and $\langle Z\rangle$ that are too high, comparable to the total mass fraction in stars seen today[20] and in excess of the enrichment of the IGM inferred at intermediate redshifts, respectively.

The astrophysics of first light may not be as simple, however. The metal constraint assumes that metals escaping from protogalaxies are evenly mixed into the IGM and the Lyman-α clouds.[54] Inefficient mixing could instead produce a large variance in intergalactic metallicities. The metal abundances of the Lyman-α clouds may underestimate the average metallicity of the IGM if there existed a significant warm-hot gas phase component with a higher level of enrichment, as detected for example in O VI.[48] Today, the metallicity of the IGM may be closer to $\sim 1/3$ of solar if the metal productivity of galaxies within clusters is to be taken as representative of the universe as a whole.[44] Uncertainties in the early IMF make other preheating scenarios possible and perhaps even more likely. Population III stars with main-sequence masses of approximately $140-260$ $\mathrm{M}_\odot$ will encounter the electron-positron pair instability and be completely disrupted by a giant nuclear-powered explosion.[25] A fiducial 200 $\mathrm{M}_\odot$ Population III star will explode with a kinetic energy at infinity of $E_{\rm SN} = 4 \times 10^{52}$ ergs, injecting about 90 $\mathrm{M}_\odot$ of metals. For a very 'top-heavy' IMF with $\eta = 0.005\ \mathrm{M}_\odot^{-1}$, equation (5) now yields

$$\left(\frac{\Omega_*}{\Omega_b}\right)_{\rm III} = 0.001\ (E_p/0.1\,\mathrm{keV}), \tag{8}$$

and a mean IGM metallicity

$$\langle Z\rangle_{\rm III} = \frac{0.45\,\Omega_*\,f_Z}{\Omega_b} = 0.02\,Z_\odot\ (E_p/0.1\,\mathrm{keV}) \tag{9}$$

(in both expressions above we have assumed $f_w = f_Z = 1$). This scenario can yield large preheating energies by converting only a small fraction of the comic baryons into Population III stars, but tends to produce too many metals for $E_p \gtrsim 0.1\,\mathrm{keV}$. The metallicity constraint, of course, does not bound preheating from winds produced by an early, numerous population of faint 'mini-quasars'.[1] Accretion onto black holes releases 50 MeV per baryon, and if a fraction f_w of this energy is used to drive an outflow and is ultimately deposited into the IGM, the accretion of a trace amount of the total baryonic mass onto early black holes,

$$\frac{\Omega_{\rm BH}}{\Omega_b} = \frac{E_p}{f_w\,50\,\mathrm{MeV}} = 2 \times 10^{-6}\ f_w^{-1}\,(E_p/0.1\,\mathrm{keV}), \tag{10}$$

may then suffice to preheat the whole universe. Note that this value is about $50\,f_w$ times smaller than the density parameter of the supermassive variety found today in the nuclei of most nearby galaxies, $\Omega_{\rm SMBH} \approx 2 \times 10^{-6}\,h^{-1}$.[35]

[1] Because the number density of *bright* quasi-stellar objects at $z > 3$ is low[16], the thermal and kinetic energy they expel into intergalactic space must be very large to have a global effect, i.e. for their blastwaves to fill and preheat the universe as a whole. The energy density needed for rare, luminous quasars to shock-heat the entire IGM would in this case violate the *COBE* limit on y-distortion.[54]

7 Conclusions

The above discussion should make it clear that, despite much recent progress in our understanding of the formation of early cosmic structure and the high-redshift universe, the astrophysics of first light remains one of the missing links in galaxy formation and evolution studies. We are left very uncertain about the whole era from 10^8 to 10^9 yr – the epoch of the first galaxies, stars, supernovae, and massive black holes. Some of the issues discussed above are likely to remain a topic of lively controversy until the launch of the *James Webb Space Telescope* (*JWST*), ideally suited to image the earliest generation of stars in the universe. If the first massive black holes form in pregalactic systems at very high redshifts, they will be incorporated through a series of mergers into larger and larger halos, sink to the center owing to dynamical friction, accrete a fraction of the gas in the merger remnant to become supermassive, and form binary systems.[55] Their coalescence would be signalled by the emission of low-frequency gravitational waves detectable by the planned *Laser Interferometer Space Antenna* (*LISA*). An alternative way to probe the end of the dark ages and discriminate between different reionization histories is through 21 cm tomography.[34] Prior to the epoch of full reionization, 21 cm spectral features will display angular structure as well as structure in redshift space due to inhomogeneities in the gas density field, hydrogen ionized fraction, and spin temperature. Radio maps will show a patchwork (both in angle and in frequency) of emission signals from H I zones modulated by H II regions where no signal is detectable against the CMB.[12] The search at 21 cm for the epoch of first light has become one of the main science drivers of the *LOw Frequency ARray* (*LOFAR*). While remaining an extremely challenging project due to foreground contamination from unresolved extragalactic radio sources[14] and free-free emission from the same halos that reionize the universe[39], the detection and imaging of large-scale structure prior to reionization breakthrough remains a tantalizing possibility within range of the next generation of radio arrays.

Acknowledgements

Support for this work was provided by NSF grant AST-0205738 and by NASA grant NAG5-11513.

References

1. Abel, T., Bryan, G., & Norman, M. 2000, ApJ, 540, 39
2. Barkana, R., Haiman, Z., & Ostriker, J. P. 2001, ApJ, 558, 482
3. Becker, R. H., et al. 2001, AJ, 122, 2850
4. Benson, A. J., & Madau, P. 2003, MNRAS, 344, 835
5. Bond, J. R., Arnett, W. D., & Carr, B. J. 1984, ApJ, 280, 825
6. Bromm, V., Coppi, P. S., & Larson, R. B. 2002, ApJ, 564, 23
7. Bromm, V., Kudritzki, R. P., & Loeb, A. 2001, ApJ, 552, 464

8. Bullock, J. S., Kravtsov, A. V., & Weinberg, D. H. 2001, ApJ, 548, 33
9. Cen, R. 2003, ApJ, 591, L5
10. Cen, R., & Bryan, G. L. 2001, ApJ, 546, L81
11. Ciardi, B., Ferrara, A., & White, S. D. M. 2003, MNRAS, 344, L7
12. Ciardi, B., & Madau, P. 2003, ApJ, 596, 1
13. Dekel, A., & Silk, J. 1986, ApJ, 303, 39
14. Di Matteo, T., Perna, R., Abel, T., & Rees, M. J. 2002, ApJ, 564, 576
15. Djorgovski, S. G., Castro, S. M., Stern, D., & Mahabal, A. A. 2001, ApJ, 560, L5
16. Fan, X., et al. 2001, AJ, 121, 54
17. Fan, X., et al. 2002, AJ, 123, 1247
18. Fryer, C. L., Woosley, S. E., & Heger, A. 2001, ApJ, 550, 372
19. Furlanetto S., & Loeb A. 2003, ApJ, 588, 18
20. Glazebrook, K., et al. 2003, ApJ, 587, 55
21. Gnedin, N. Y. 2000, ApJ, 535, 530
22. Gnedin, N. Y. 2000, ApJ, 542, 535
23. Gnedin, N. Y., & Hui, L. 1998, MNRAS, 296, 44
24. Haiman, Z., & Holder, G. P. 2003, ApJ, 595, 1
25. Heger, A., & Woosley, S. E. 2002, ApJ, 567, 532
26. Hofmann, S., Schwarz, D. J., & Stocker, H. 2001, PhRvD, 64, 083507
27. Klypin, A., Kravtsov, A. V., Valenzuela, O., & Prada, F. 1999, ApJ, 522, 82
28. Kogut, A., et al. 2003, ApJS, 148, 161
29. Kudritzki, R. P. 2000, in The First Stars, ed. A. Weiss, T. Abel, & V. Hill (Heidelberg: Springer), 127
30. Loeb, A., & Barkana, R. 2001, ARA&A, 39, 19
31. Madau, P. 2000, RSPTA, 358, 2021
32. Madau, P., Ferrara, A., & Rees, M. J. 2001, ApJ, 555, 92
33. Madau, P., Haardt, F., & Rees, M. J. 1999, ApJ, 514, 648
34. Madau, P., Meiksin, A., & Rees, M. J. 1997, ApJ, 475, 492
35. Merritt D., & Ferrarese, L. 2001, ApJ, 547, 140
36. Mori, M., Ferrara, A., & Madau, P. 2002, ApJ, 571, 40
37. Moore, B., et al. 1999, ApJ, 524, L19
38. Navarro, J. F., Frenk, C. S., & White, S. D. M. 1997, ApJ, 490, 493
39. Oh, S. P., & Mack, K. J. 2003, MNRAS, 346, 871
40. Peebles, P. J. E. 1993, Principles of Physical Cosmology (Princeton: Princeton University Press)
41. Ponman, T. J., Cannon, D. B., & Navarro, J. F. 1999, Nature, 397, 135
42. Rees, M. J. 2000, RSPTA, 358, 1989
43. Reiprich, T. H., & Böhringer, H. 2002, ApJ, 567, 716
44. Renzini, A. 1997, ApJ, 488, 35
45. Scannapieco, E., Ferrara, A., & Madau, P. 2002, ApJ, 574, 590
46. Schneider, R., Ferrara, A., Natarajan, P., & Omukai, K. 2002, ApJ, 571, 30
47. Seager, S., Sasselov, D. D., & Scott, D. 1999, ApJ, 523, L1
48. Simcoe, R. A., Sargent, W. L. W., & Rauch, M. 2002, ApJ, 578, 737
49. Somerville, R. S., Bullock, J. S., & Livio, M. 2003, ApJ, 593, 616
50. Songaila, A. 1997, ApJ, 490, L1
51. Spergel, D.N., et al. 2003, ApJS, 148, 175
52. Theuns, T., Mo, H.J., & Schaye, J. 2001, MNRAS, 321, 450
53. van den Bosch, F. C., Robertson, B. E., Dalcanton, J., & de Block, W. J. G. 2000, AJ, 119, 1579
54. Voit, G. M. 1996, ApJ, 465, 548
55. Volonteri, M., Haardt, F., & Madau, P. 2003, ApJ, 582, 559
56. Wyithe, J. S. B., & Loeb, A. 2003, ApJ, 588, L69
57. Zentner, A. R., & Bullock, J. S. 2002, PhRvD, 66, 43003

The Cosmic Infrared Background (CIRB) and the Role of the "Local Environment of Galaxies" in the Origin of Present-Day Stars

David Elbaz, Delphine Marcillac, and Emmanuel Moy

CEA Saclay/DSM/DAPNIA/Service d'Astrophysique, Orme des Merisiers, F-91191 Gif-sur-Yvette Cedex, France

Abstract. A combination of evidence is presented suggesting that the majority of the stars in today's galaxies were born during a luminous infrared phase (LIRP) triggered by the local environment of galaxies. The CIRB is a fossil record of these LIRPs and therefore reflects the influence of triggered star formation through galaxy-galaxy interactions, including non merging tidal encounters. This scenario, in which galaxies experienced several LIRPs in their history, is consistent with the measured redshift evolution of the cosmic density of star formation rates and of stellar masses of galaxies.

1 Introduction

Stars represent only 15 % of the cosmic baryonic density, itself only equal to about 4% of the critical density ($\Omega_b \simeq 0.04$), and are unequally distributed into spheroids (10% of Ω_b, including spiral bulges) and disks (5% of Ω_b, Fukugita, Hogan & Peebles). It is usually assumed that disk stars formed quiescently while bulge stars formed more efficiently and rapidly, as suggested by their redder colors and overabundance in α-element over iron ratio, typical of a SNII origin. However recent studies of the history of the star formation of the disk of the Milky Way indicate that during the last 2 Gyr it has experienced about five major events of star formation, starbursts, whose signatures can be found in the peaked ages of these open clusters (de la Fuente Marcos & de la Fuente Marcos 2004, and references therein). As a result we may wonder whether quiescent star formation did play a major role in the formation of present-day stars at all. There are some evidence that star formation mostly takes place in globular clusters and is rarely isolated. These clusters are thought to evolve into unbound stellar associations, which evolve and dissolve in a time-scale of about 50 Myr. This timescale is longer than the lifetime of massive stars which dominate the luminosity of starbursting galaxies or regions of galaxies. Hence, it is logical to expect that if star formation occurred mainly in starburst episodes within galaxies, then the bulk of galaxies luminosity will be absorbed by dust in the giant molecular clouds, while if star formation is quiescent then only a minor fraction of a galaxy's luminosity will be affected by dust extinction. We will argue in the following that there is presently a solid collection of evidence suggesting that most stars that we see in the local universe formed during starburst episodes.

A major peace of evidence for that comes from the detection of a strong diffuse cosmic infrared background (CIRB, Puget et al. 1996, Hauser & Dwek and references therein) which is majoritarily produced by luminous infrared phases (LIRP) within galaxies located around $z \sim 0.7$, for the peak of the CIRB at $\lambda \sim 140\,\mu$m, while the $\lambda \geq 240$ μm is due to galaxies at $z \sim 2$ and above (Elbaz et al. 2002, Chary & Elbaz 2001). We have introduced the term LIRP instead of the classical one, luminous and ultra-luminous infrared galaxies, i.e. LIRGs and ULIRGs, because we wish to emphasize the idea that the scenario that is emerging from the study of distant galaxies is that LIRGs do not represent a population of galaxies that would require to be studied independently in order to determine which present-day galaxies are the remnants of these LIRGs, but what is suggested instead is that they illustrate the omnipresence of rapid and efficient star formation as a leading process in shaping the present-day universe, i.e. that any galaxy that we see today must have experienced a phase when it radiated the bulk of its light in the infrared (see also Elbaz & Cesarsky 2003). This phase should not be restricted to LIRGs and ULIRGs, i.e. galaxies with infrared (IR) luminosities larger than 10^{11} $L_\odot$ or star formation rates (SFR) larger than ~ 20 $M_\odot$ yr^{-1}, since the closest starburst M82, for example, presents a spectral energy distribution typical of most LIRGs, with the bulk of its luminosity radiated in the IR although its luminosity is only 4×10^{10} $L_\odot$.

We present evidence suggesting that the bulk of local stars formed during a LIRP. A spectroscopic diagnostic is used to quantify the typical duration of this phase and the amount of stars that formed during it. From the combination of both we will advocate that not only did all galaxies experience a LIRP but that they must have experienced several of them. Finally we will discuss the physical origin of the LIRP and present evidence that a major event in the lifetime of galaxies was probably underestimated: the effect of the "local environment of galaxies" (LEG) and its impact in terms of driving the conversion of gas into stars through passing-by galaxies.

2 Luminous IR Phases and the Origin of Present-Day Stars

The detection of a CIRB came as a surprise since local galaxies only radiate $\sim 30\,\%$ of their bolometric luminosity in the mid to far IR range, i.e. sharing as a common origin stellar photons reprocessed by dust above $\lambda \sim 3\,\mu$m. With about half of the diffuse background light radiated above and below this wavelength cutoff, the extragalactic background light tells us that in the past, major star formation events were strongly affected by dust extinction even when galaxies were less metal rich. Deep surveys in the mid infrared ($\lambda \sim 15\,\mu$m with ISOCAM onboard ISO, Elbaz et al. 1999) brought independent evidence that infrared was more ubiquitous in the past. These surveys detected ten times more objects at faint flux levels than expected from the extrapolation of the local universe (no evolution models). These galaxies turned out to belong to the class of LIRGs and ULIRGs discovered by IRAS in the local universe but located at a median redshift of z$\sim$0.7. They do not exhibit any optical signature of such strong SFRs neither in their optical colors nor in their emission lines, except if careful correction for extinction is applied using the Balmer decrement (Cardiel et al. 2003, Flores et al. 2004). Only less than 20 % of them were found to harbor or be dominated by an active galactic nucleus (AGN; see Fadda et al. 2002). Unexpectedly, because of the complex and multiple physical origins of the mid and far IR photons (PAHs, Very Small Grains, Big Grains), local galaxies do exhibit a strong correlation between their mid and far IR luminosities over three decades in luminosity, including the extreme LIRGs and ULIRGs (Elbaz et al. 2002). When applied to galaxies up to z∼1 these correlations can be used to derive far IR luminosities, i.e. SFRs, which are consistent with those derived from the radio, using the radio-far IR correlation, suggesting that these correlations are still valid at these redshifts (Elbaz et al. 2002).

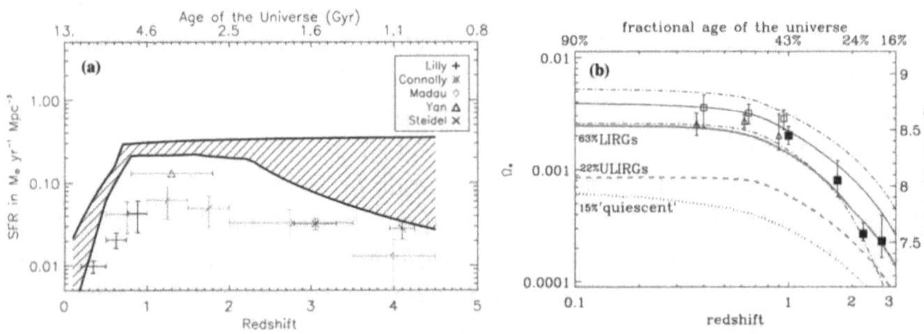

Fig. 1. a) cosmic star formation rate (CSFR) as a function of redshift and universe age, in a $H_0 = 50$ km s^{-1} Mpc^{-1} and $q_0 = 0.5$ cosmology (Fig.15 of Chary & Elbaz 2001). The data represent the SFR density derived from UV or Hα uncorrected for extinction (the various authors are quoted on the plot and the references can be found in Chary & Elbaz 2001). **b)** cosmic stellar mass history (CSMH) or redshift evolution of $\Omega_\star$, the cosmic stellar mass density over critical density (cosmology: Λ=0.7, $\Omega_{\rm m}$=0.3, $H_0 = 70$ km s^{-1} Mpc^{-1}). The data are from Dickinson et al. (2003). See text for description.

The median SFR of the galaxies responsible for the evolution of the mid IR counts is ~ 50 M$_\odot$ yr^{-1} and their contribution to the CIRB from 100 to 1000 μm is derived to be as large as two thirds of its measured intensity. Hence the bulk of the CIRB results from LIRPs at redshifts of the order of z∼1, but due to cosmological dimming the influence of more distant galaxies is more limited to the large wavelength tail of the CIRB. Some models have been designed to reproduce number counts in the mid IR (ISOCAM, 15 μm), far IR (ISOPHOT-90, 175 μm) and sub-mm (SCUBA, 850 μm) together with the CIRB which can be used to derive the cosmic star formation history of the universe (CSFH) unaffected by dust extinction and to disentangle the relative roles of rapid (LIRPs) and quiescent star formation. In Chary & Elbaz (2001), we suggested to define a region delimiting all possible histories of star formation (see Fig. 1a), instead of a single line for any favorite model that would not represent the uncertainties of the model and observational constraints. One way to check the robustness of such models consists in comparing to direct measurements the resulting cosmic stellar masses history (CSMH), i.e. the redshift evolution of $\Omega_{\star}$, the mass density of stars per comoving volume over the critical density of the universe. The hatched area of Fig. 1a which represents the range of possible CSFH was converted into a range of possible CSMH in Fig. 1b, with the thin plain line showing the mean value and the range of possible histories delimited by the two dot-dashed thin lines. The model fits the data collected in Dickinson et al. (2003), where stellar masses were directly measured from optical-near IR magnitudes.

Before deriving any conclusion, we wish to remind our assumptions: the CSFH was derived assuming mid to far IR correlations similar to locally (in agreement with the radio), the AGN fraction was supposed to make a minor contribution (see above), we assumed a universal IMF (we used the one of Gould et al. 1996, which combines a Salpeter IMF with the now standard inflexion of the IMF below 1 $M_{\odot}$). If these assumptions are indeed justified, then the fit of the data in Fig. 1b illustrates the fact that the photons emitted by star forming regions do reflect the stellar mass building of galaxies and as a consequence, it is now possible to derive which fraction of present-day stars were formed in a LIRP. The model CSFH was separated into three components shown as thick grey lines in the Fig. 1b, with 63 % of present-day stars born during an IR phase of the LIRG type, which would be the dominant mode of star formation in the universe. The shape of the redshift evolution of $\Omega_{\star}$ implies that ~80 % of present-day stars were born below z=2, and ~50 % below z=1, most of which during a LIRP, leaving little room for quiescent star formation.

If most of today's stars formed during a LIRP and if this phase reflects efficient and rapid star formation then this suggests that some "positive feedback", i.e. triggering, might be at play. This idea is comforted by the morphology and local environment (LEG) of distant LIRGs. It is well-known that galaxies lie in large-scale structures made of walls, filaments and clusters but LIRGs tend to appear exclusively in high density environments around z~0.7. The deepest ISOCAM survey was performed in a region of $27'^2$ centered on the Hubble Deep Field North (HDFN), detecting 95 galaxies among which 47 lie above the completeness limit of ~ 0.1 mJy.

The histogram of field galaxies presents several redshift peaks. The location and the extent of these peaks can be quantified by adopting a thresholding S/N ratio of 3, where the "background" is simply the redshift distribution smoothed with a gaussian with σ = 15,000 km s^{-1}. Monte-Carlo simulations performed by extracting 100 random samples from the real distribution of field galaxies show that the probability of reproducing by chance the level of clustering of the ISOCAM galaxies is much less than one chance over ten (largest error bars). These results are illustrated on Fig. 2a, where we plot the fraction of ISOCAM galaxies included in redshift peaks above a given S/N ratio as a function of this S/N. The corresponding curves for the field galaxies and the simulated samples are also indicated. The thin (resp. thick) error bars show the 68 % (resp. 90 %) confidence level. The strong clustering of ISOCAM galaxies illustrates that at z~0.7, galaxies experiencing a LIRP are more clustered on average than field

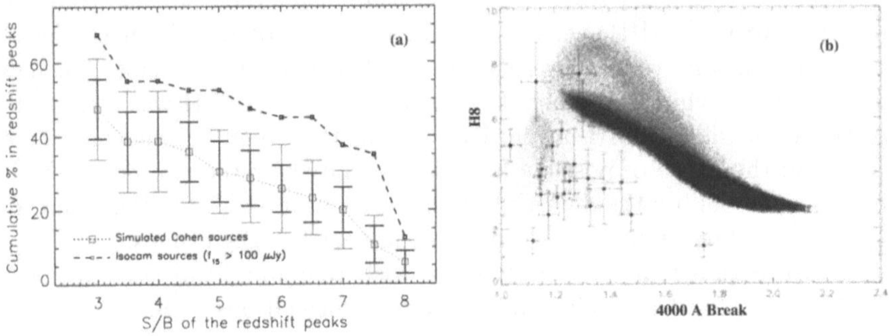

Fig. 2. a) Cumulative fraction of galaxies located in a redshift peak of a given S/N, i.e. number of galaxies in all redshift peaks above a given S/N (see text). **b)** Equivalent width of the high-order Balmer absorption line, H8, as a function of the 4000 Åbreak. Data with error bars are a sub-sample of $z \sim 0.7$ LIRGs. Dots were generated by a Monte Carlo simulation of 200,000 galaxies. The darkest region is populated by galaxies dominated by continuous star formation, or bursts older than $\sim$ 2 Gyr.

galaxies. On the contrary, the study of LIRPs in the local universe using the shallower ELAIS survey (Oliver et al., in these proceedings) shows that they are less clustered than field galaxies locally. The natural explanation for this behavior is that galaxies might experience a LIRP when located in a region which is collapsing over a large scale, i.e. star formation would be triggered by large-scale structures in the process of formation.

How much stellar mass does a LIRG form ? We addressed this question using an optical spectroscopic diagnostic combining the equivalent width of the high order Balmer absorption line H8 to the 4000 Åbreak to characterize these starburst events (Marcillac et al., in prep.). The advantage of using H8 instead of Hδ is that it is in a bluer side of the spectrum, hence less affected by sky lines, and that its underlying nebular emission line can be neglected. Fig. 2b shows a sub-sample of 22 LIRGs, at z$\sim$0.7, selected in three different locations of the sky compared to a Monte Carlo simulation of 200,000 galaxies using the code of Bruzual & Charlot (2004) such as those used by Kauffmann et al. (2003) to reproduce the behavior of local galaxies in the Sloane survey. This simulation can be used to determine which histories of star formation would reproduce these galaxies. It is found that only galaxies presently experiencing a burst of star formation can fall in this region of the diagram and that this burst lasts approximately 10^8 years and produce about 10 % of the stars of the galaxies. These numbers were both derived by the simulation but they perfectly agree with the measured median mass of the ISOCAM galaxies of $\sim 5\times10^{10}$ $M_\odot$ (from Dickinson et al. 2003). Indeed in 10^8 years and with their median SFR$\sim$ 50 $M_\odot yr^{-1}$, they produce $\sim 5\times10^9$ $M_\odot$ of stars, i.e. 10 % of the galaxy mass. The mass of newly formed stars is also consistent with the typical mass of molecular gas found in local LIRGs. The model used in the Fig. 1 predicts that nearly 50 % of present-day stars were formed in a LIRP below z∼1, hence if this phase only makes up 10 % of a galaxy's stars then the majority if not all of today's galaxies experienced up to five or even more LIRPs. This suggests that the dense surrounding of galaxies can trigger successive luminous IR phases, LIRPs, in galaxies. Finally, the optical morphology of $z \sim 0.7$ LIRGs derived from HST-ACS observations (Elbaz et al., in prep, see also Hammer et al. in these proceedings) shows that less than half of them look like the major mergers that we see locally in LIRGs. Major mergers might not be numerous enough to explain such a behavior and passing-by galaxies might play an important role by triggering strong star formation events through tidal effects. Hence the local environment of galaxies, or LEG, might be considered as a better candidate to understand the origin of distant LIRGs. The cosmic star formation history is therefore probably strongly dependent on the local density of galaxies which will also possibly determine their final morphology, i.e. spirals versus ellipticals, instead of the standard picture of the merger of two massive disks.

The recently launched Spitzer satellite will provide ideal observations to check the robustness of this scenario by observing larger patches of the sky (in particular the SWIRE legacy program), lowering the effect of cosmic variance, and to extend the study of luminous IR phases to higher redshifts and lower luminosities (with the MIPS GTO and GOODS legacy program).

References

1. G. Bruzual, S. Charlot: MNRAS **344**, 1000 (2003)
2. N. Cardiel et al.: ApJ **584**, 76 (2003)
3. R.R. Chary, D. Elbaz: ApJ **556**, 562 (2001)
4. R. de la Fuente Marcos, C. de la Fuente Marcos: to appear in New Astronomy (astro–ph/0401360)
5. M. Dickinson et al.: ApJ **587**, 25 (2003)
6. D. Elbaz et al.: A&A, **351**, L37 (1999)
7. D. Elbaz et al.: A&A, **384**, 848 (2002)
8. D. Elbaz, C.J. Cesarsky: Science **300**, 270 (2003)
9. D. Fadda et al.: A&A **383**, 838 (2002)
10. H. Flores et al.: A&A **415**, 885 (2004)
11. M. Fukugita, C.J. Hogan, P.J.E. Peebles: ApJ **503**, 518 (21998)
12. A. Gould, J.N. Bahcall, C. Flynn: ApJ **465**, 759 (2003)
13. M. Hauser, E. Dwek: ARAA **39**, 249 (2001)
14. G. Kauffmann et al.: MNRAS **341**, 33 (2003)
15. J.-L. Puget et al.: A&A **308**, L5 (1996)

High Redshift Lyman Break Galaxies

M.D. Lehnert[1] and M.N. Bremer[2]

[1] Max-Planck-Institut für extraterrestrische Physik
[2] Department of Physics, University of Bristol

Abstract. Two of the most outstanding issues in modern astrophysics are what reionized the Universe and how did the first objects form. Observations of galaxies selected through the Lyman-Break technique indicate that UV photon output at the end of reionization was dominated by relatively faint low mass galaxies and not AGN.

1 Background

Galaxies at the highest redshifts, z>5, are the key to our developing understanding of how galaxies have formed and evolved, and how the Universe was reionized. The recent results from the Wilkinson Microwave Anisotropy Probe (Kogut et al. 2003) combined with the fact that the Universe appears to be opaque to Lyman continuum photons at z≈6 (Becker et al. 2001; but see Songaila 2004), suggests that the Universe had a very complex reionization history.

Probing the re-ionization epoch directly will be exceedingly difficult with the current generation of telescopes and instrumentation. On physical grounds, observing Ly-α emission may be difficult. An optical depth of a few in the Lyman continuum is reached at very low neutral fraction in the IGM ($\approx 10^{-5}$). Thus damping wings of the neutral material will effectively eat into the galaxies Lyα emission even at relatively low neutral fractions. Although sources will create their own HII regions, observability of Lyα will depend on the star-formation rate, lifetime, local galaxy and halo gas density, and kinematics of neutral halo gas. Although lines like HeIIλ1640 or CIVλ1549 could be used, they are likely to be much weaker except at low metallicities. On technical and functional grounds, obtaining robust compete samples of z- or J-band drop outs (6.5<z<10.3) will be difficult. Current near-infrared imagers on 8m class telescopes have small fields of view (<20 arcmin^2) and the source density of (even bright) sources is likely to be very small (few tenth per arcmin^2 at most). Moreover, the "spectral filling factor" of the night sky lines is 40-50% even at moderate resolutions (R=4000) and Nyquist sampling implying that many line emitters will be missed. Perhaps more importantly, it will be challenging and time consuming to detect these faint objects in two band longward of Lyman continuum break. Having a two band detection is important to eliminating spurious or contaminating sources like stars or lower redshift red galaxies.

2 Survey for High Redshift LBGs

However, large samples of high redshift galaxies are required in order to carry out detailed studies of their properties and to ensure that conclusions drawn are not subject to cosmic variance. In addition, the selection techniques must be varied so that biases do not hamper us in obtaining a broad understanding of early galaxy evolution. Thus large areas need to be surveyed and both continuum break techniques and narrow-band imaging should be used. The later of these is particularly important. Galaxy selected using the Lyman break technique are sensitive to the brightness of the rest-frame UV continuum and would be generally expected to be sensitive to galaxies which have low equivalent width Lyα lines. On the other hand, narrow-band selection favors galaxies with high equivalent width emission lines – typically greater than a significant fraction of the filter width and will discover galaxies with faint rest-frame continua. In addition, narrow band selection allows one to pick a region with few night sky lines and thus make spectroscopic follow-up that much easier (see contributions by Tanaguichi and Cowie). However, the disadvantage of narrow-band selection is that wide areas need to be surveyed since the volume contributed by the redshift coverage is small.

Until now, we have been in the exciting "discovery phase", showing that it is indeed possible to detect high redshift galaxies with current instrumentation on 8-m class telescopes. To create such large samples, we have an on-going program to obtain deep R-, I-, and z-band images with FORS2 on the VLT. This 3 color technique relies on the fact that the continuum opacity in the Lyman continuum is high at redshifts above about 5 giving galaxies at these redshifts very red R-I or I-z colors. Galaxies selected in this manner have redshifts between 4.8 and 6.4 (using R-I and I-z color selection technique with the FORS2 filters and CCD response).

We identified a field of about 200 arcmin^2 with extremely low galactic extinction and infrared cirrus emission that was well-placed in RA for ease of service observing. The field was chosen to have a declination of -35, so that it went roughly overhead at Paranal but meant that the telescope faced south, out of the prevailing wind, minimising the time the field was unobservable due to weather conditions. In 2002 we imaged 40 arcmin^2 to a depth of R_{AB}=27.8, I_{AB}=26.5 and z_{AB}=26. In 2003 we deepened this field in z to z_{AB}=26.5 and imaged a further 40 arcmin^2 to the same depths in the 3 filters. Two more similarly-sized regions should be imaged in 2004, leading to complete imaging of a 160 arcmin^2 region of sky.

Sources that appear to have spectral breaks can be selected from the imaging data. Starting with a flux-limited sample in the I-band (to I_{AB}=26.3) we can identify such sources by requiring R-I>1.5. To-date we have followed up spectroscopically the sources identified in the 80 arcmin^2 of imaging data we have obtained so far (see Lehnert & Bremer 2003 for details of the first 40 arcmin^2). All of these sources meeting the flux and colour criteria, except one, have been observed with FORS2 using the MXU mode.

For the objects that were either not detected, or were marginal detections in the R-band, we found that only those sources exhibited signatures suggesting they have redshifts between $4.8<z<5.8$ as expected from our simulations. These signatures were either or both a strong emission line with an asymmetric profile and a break in continuum light. Sources with large breaks but bright magnitudes in R or I were found to be either very red cool stars or intermediate redshift red galaxies. The number of bright objects meeting the colour criteria in our fields that turned out to be stars or lower redshift galaxies was not large (~5-10). However, compared to the number of high redshift sources with confirmed redshifts, only thirteen out of a total 26 sources, implies that the "contamination rate" is at most 1/3. Thus spectroscopy to confirm the nature of the colour selected sources is key in determining important parameters such as the co-moving space density of high redshift star-forming galaxies. Without the spectroscopy, lower redshift galaxies and stars would make any such estimates very uncertain.

3 The Properties of LBGs at $z \approx 5.5$

In our total field of about 80 arcmin^2 investigated to date, we have determined redshifts for 13 galaxies so far through the R-band drop-out technique. The Lyα emission from these galaxies (used to determine the redshifts) has fluxes of about few $\times$ 10^{-18} to few $\times$ 10^{-17} ergs s^{-1} cm^{-2} and has high equivalent widths (>20-30 angstroms in the rest-frame). The fluxes imply Lyα luminosities of about 10^{42-43} ergs s^{-1} and their high emission line equivalent widths suggest very young ages (about to less than 10^8 yrs). However, it is worth noting that inferring such young ages does not rule out the existence of an older population of stars – it simply indicates that the UV continuum relative to the number of ionizing photons is relatively weak but there could be older populations of stars which do not contribute significantly to the UV continuum.

Interestingly, the widths of the Lyα emission line are relatively modest, being at most several hundred km s^{-1}. Since we did not detect any broad or high ionization emission lines in any of our spectra, it appears that none of the high redshift sources were AGN. This is in agreement with a study we conducted on the CDF-S to investigate the X-ray properties of similarly high redshift galaxies and found no evidence for Compton thin AGN (Bremer et al. 2004).

The colour and magnitude distribution of the sources without spectroscopic redshifts is very similar to that of the sources with redshifts. This being the case, it is a reasonable assumption that those without spectroscopic redshift are also at similar redshifts. Making this assumption doubles the sample to 26 high redshift galaxies in an area of about 80 arcmin^2 down to an I-band flux limit of I_{AB}=26.3. The rest-frame UV flux densities as probed by the I and z-band fluxes implies that the star-formation rate of the high redshift galaxies is about a few tenths to almost 20 solar masses per year. This is about a factor of a few to 10 higher than the values estimated using the Lyα luminosities. This is not surprising since Lyα suffers from both IGM absorption which may remove a considerable amount of the intrinsic Lyα for the high redshift galaxies and, due to high optical depth of the line, radiative transfer effects allow for the destruction of Lyα by dust grains.

We find that the I-z colour correlate with redshift of the sources. Such a trend comes about due to the overwhelming influence of absorption by the IGM and the relatively small dispersion in the intrinsic colours of the sources. If the colours of the galaxies were intrinsic, then there is no logical reason for such a correlation. It would require something like age or reddening to correlate with redshift which would need to be carefully tuned and thus is ad hoc. Using a model for the IGM (Madau et al. 1996) and galaxy evolutionary synthesis models we can predict the colours of galaxies for a wide range of ages and extinctions but requiring that no galaxy is older than the age of the Universe at redshifts of between 4.8 and 5.8. Doing this we find that the colours of the galaxies as a function of redshift are consistent with them being both young (less than a 100 Myrs) and relatively lightly extincted (visual extinction of a few tenths of a magnitude at most). This result is consistent with the galaxies exhibiting relatively strong Lyα emission, because it is easily destroyed by dust, also suggests low extinction. The high equivalent width of Lyα also implies relatively young ages for the burst of star formation in the galaxies.

4 The Co-Moving Space Density of High Redshift Galaxies

Using the estimated number of high redshift galaxies, the area covered in the images, an estimate of the completeness as a function of magnitude, and assumptions about the cosmological parameters, it is possible to estimate the co-moving density of sources (see Lehnert & Bremer 2003). However, as discussed earlier, the colours and magnitudes of the sample galaxies are sensitive to the source redshift due to the strong influence of the IGM absorption. Therefore, it is not a simple matter to translate all of the parameters into a co-moving density as a function of magnitude. To mitigate against these effects, we chose a more conservative approach of comparing the co-moving number density of sources at lower redshift selected using a similar technique to the one outlined here. We used the co-moving density as a function of magnitude for a sample of similarly selected galaxies at z$\sim$3 and $\sim$4 from Steidel et al. (1999) and applied a offset to the magnitudes due to larger distance of the high redshift sources, put in the incompleteness as determined for the high redshift sample, and then used a linear relationship between I-z and redshift covering the range of values we determined for the high redshift sample. What we found is that the number of bright (luminous) galaxies declined significant from z$\sim$3 or 4 to z$\sim$5.3 (the mid-point of our sample). The decline was roughly a factor of 2 to 3.

Given that we have a luminosity function of galaxies at z$\sim$3 or 4 we can then adjust the fiducial co-moving density and luminosity until we get a good representation of the data. Doing this we find that we need to decrease L* by about a factor of 3 (about 1 magnitude) and increase ϕ^* by about a factor of 3 from z=3 to 5.3. The reason that this is uncertain is that we are only observing galaxies down to about L* and thus we are not detecting galaxies over a wide enough portion of the luminosity function to make this estimate robust (see contribution by Dickinson et al.). At any rate, there is clear evidence for a steepening of the bright end of the luminosity function as originally observed and suggested by Lehnert & Bremer 2003.

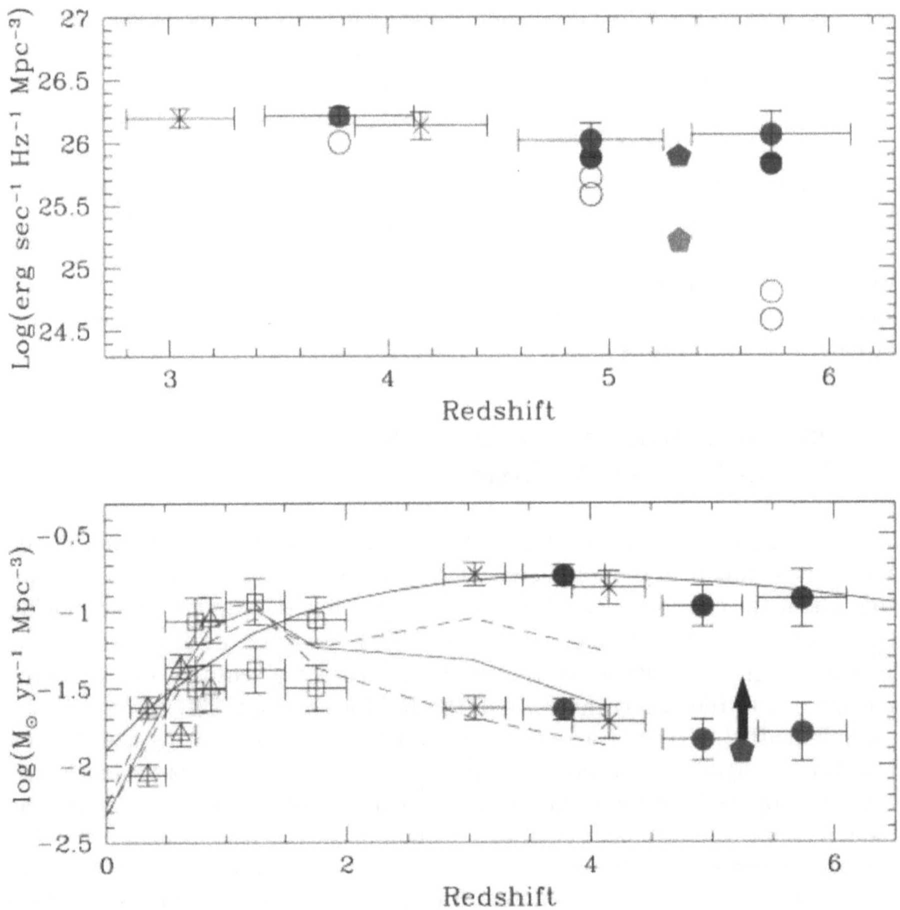

Fig. 1. A reproduction of a figure from Giavalisco et al. (2004) of the specific luminosity density at 1500Å versus redshift (*top*) and the average density of star-formation as a function of redshift (*bottom*). The hexagons represents the observed UV energy density and an extrapolation of our best fit luminosity function. In the bottom panel, the hexagon represents the total star-formation rate density estimated using the extrapolated luminosity density. Giavalisco et al. (2004) applied an extinction correction of 0.8 magnitudes in the visible to estimate the total star formation density (the bottom set of points versus those higher in the diagram). Our analysis suggests that something less is more likely and the arrow represents a conservative value of 0.4 magnitudes of visual extinction.

This analysis indicates that the observed rate of UV photon production per unit volume of observed sources is low compared to the results for lower redshift galaxies. However, by fitting and extrapolating the luminosity function, we find rough agreement. This estimate can then be used to derive a rate of star-formation per unit volume. Making such an estimate for the best fit luminosity function, we again find rough agreement with results for z=3 to 4 galaxies. However, to infer the the true rate of star-formation per unit volume, one must correct for extinction of the ensemble of galaxies. We have found evidence that the extinction in these sources is low, the optical extinction is probably less than

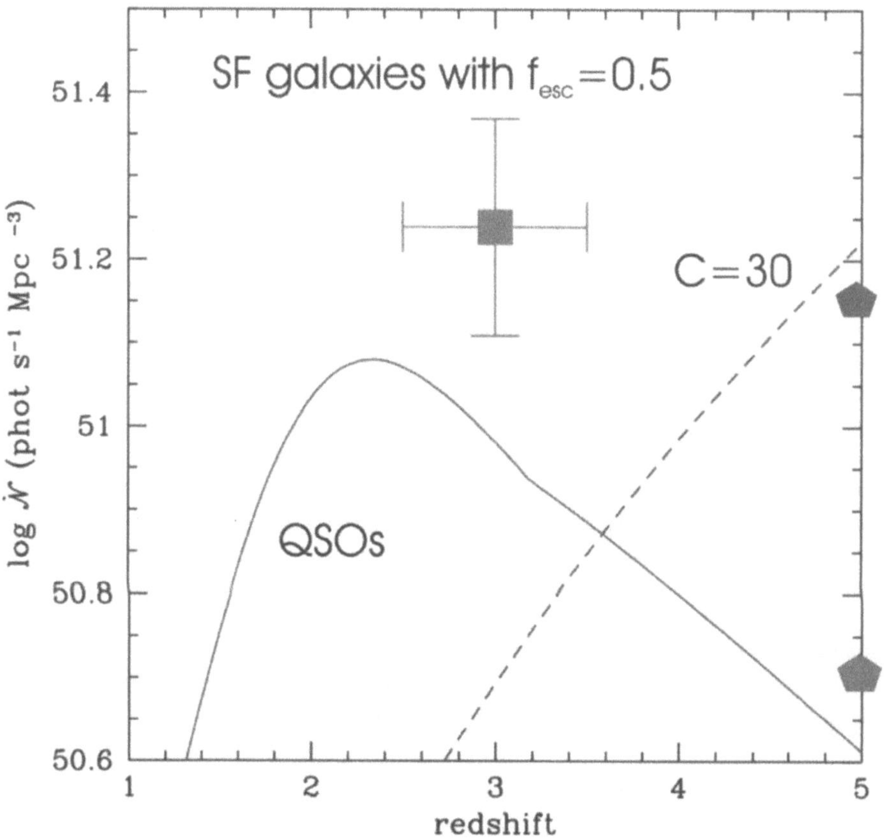

Fig. 2. A reproduction of a figure from Madau et al. (1999) which shows the number of ionizing photons per unit volume versus redshift. The solid line shows the contribution from optically selected QSOs while the dotted curve shows the number of photons needed assuming a clumpy distribution of Hydrogen. The large solid square represents star-forming galaxies at z∼3 assuming an escape fraction of 50%. Our results are shown as the two hexagons representing the UV ionizing photon density we have observed (lower hexagon) and based an extrapolation of our best fit luminosity function to 0.2 L* (upper hexagon). In both cases, the escape fraction is 100%.

a few tenths of a magnitude. If it is this low, then the star-formation rate per unit volume may have declined from z=3 or 4 to z∼5.5. However, our ability to determine this robustly is limited by our understanding of the extinction in a large sample which contributes most of the star-formation (i.e., around L^*), only sampling the luminous part of the luminosity function implying that that the luminosity function could be even steeper than we have estimated, and cosmic variance.

5 Keeping Their Local Volume Ionised and Re-Ionisation

From the UV luminosity function, with caveats, it is possible to estimate the number of ionising photons emitted per unit co-moving volume both by the sources we have directly observed and then by extrapolating to fainter sources using the best-fit luminosity function. Once we have done this we can compare the derived photon density to that required to keep the volume ionized. The UV photon density from our detected sources fell short by a factor of three relative to that produced by similar luminosity galaxies at z≈3 and 4. Ferguson et al. (2002) and others had previously shown that even this higher photon density is insufficient to ionize the high redshift Universe.

The clear implication of our analysis is that the objects we have detected emit insufficient ionizing photons to maintain ionization at z≈5.3, and so the bulk of ionizing photons must come from less luminous objects. Given that our sources are observed within 100-200 Myr of the end of reionization, this also implies that the bulk of the photons that reionized the Universe were emitted by relatively low luminosity sources. As we detect no quasars or AGN in our volume (see also Bremer et al. 2004), but many galaxies, it follows that unless the AGN luminosity function has a bizarre shape, these less luminous sources must be galaxies. Is there any way that we could have underestimated the ionizing impact of the more luminous detected sources? The photon density required to ionize a volume of IGM depends linearly on the clumping factor of the IGM. Only in the case where this factor is close to unity does the required photon density become comparable to our measured density. Given the results of simulations of structure formation for z>5 (e.g., Gnedin & Ostriker 1997), this is unlikely.

We thank the ESO OPC for their generous allocation of telescope time and the our friends on Paranal for their effort in efficiently executing these observations. And finally we thank Alvio, Ralf and the rest of the LOC/SOC for organizing a wonderful conference and their (immense) patience in waiting for this contribution.

References

1. Becker, R. H. et al. 2001, AJ, 122, 2850
2. Bremer, M., et al. 2004, MNRAS, 347, L7

3. Fan et al. 2001, AJ, 122, 2833
4. Ferguson, H. C., Dickinson, M., & Papovich, C. 2002, ApJ, 569, L65
5. Giavalisco, M. et al. 2004, ApJ, 600, 103
6. Gnedin, N. Y., & Ostriker, J. P. 1997, ApJ, 486,581
7. Kogut, A. et al. 2003, ApJS, 148, 161
8. Lehnert M., & Bremer M., 2003, ApJ, 593, 630
9. Madau, P., Ferguson, H. C., Dickinson, M. E., Giavalisco, M., Steidel, C. C., & Fruchter, A. 1996, MNRAS, 283, 1388
10. Madau, P., Haardt, F. R., & Rees, M. J. 1999, ApJ, 514, 648
11. Schmidt, M., Schneider, D. P., & Gunn, J. E. 1995, AJ, 110, 68
12. Songaila, A. 2004, astro-ph/0402347
13. Steidel, C. C., Adelberger, K. L., Giavalisco, M., Dickinson, M., & Pettini, M. 1999, ApJ, 519, 1

Subaru Surveys for High-z Galaxies

Yoshiaki Taniguchi

Astronomical Institute, Graduate School of Science, Tohoku University, Aramaki, Aoba, Sendai 980-8578, Japan

Abstract. We present a summary of optical/NIR deep surveys for very high-z galaxies using the 8.2m Subaru Telescope operated by National Astronomical Observatory of Japan. The prime focus mosaic CCD camera, Suprime-Cam, with a very wide field of view, $34' \times 27'$, allows us to carry out efficient optical deep surveys. In particular, the Subaru Deep Field project has provided us a number of Lymanα emitters beyond $z = 6$. We discuss the star formation history in the early universe based on this project.

1 Introduction

Since the discovery of Lyα emission from a galaxy at $z = 5.34$ [4], more than two dozen of Lyα emitters (LAEs) have identified spectroscopically; see for reviews, [27]; [23]. The most distant LAE known to date is SDF J132418.3+271455 at $z = 6.578$ [14]. Another very high-z LAE is HCM-6A at $z = 6.56$ [9]. These discoveries are actually thanks to the great observational capability of 8-10m class optical telescopes. Furthermore, the GOODS survey has provided a sample of very high-z Lyman break galaxies (LBGs) at $z \sim 6$, thanks to the high-quality imaging capability of the Advanced Camera for Surveys (ACS) on the Hubble Space Telescope (e.g., [8], [5], [24]). These exciting observations enable us to investigate the cosmic star formation history and mass assembly history in the early universe. In this review, we present a summary of recent deep surveys for very high-z (i.e., $z > 5$) galaxies based on the 8.2m Subaru Telescope.

2 Subaru Surveys for High-z Galaxies

2.1 The Subaru Deep Survey

The 8.2m Subaru Telescope [12] has seven instruments; see http://www. subarutelescope.org/Observing/Instruments/index.html. During the commissioning phase of three instruments (FOCAS, OHS/CISCO, and Suprime-Cam), these instruments team members organized a systematic deep survey using these three instruments to investigate high-z galaxies; the Subaru Deep Survey (SDS). All the observations were done during a period between 1999 and 2001. Their target fields are (1) the Subaru Deep Field (SDF) centered at RA(J2000) = $13^{\rm h}$ $24^{\rm m}$ $21.^{\rm s}38$ and DEC(J2000) = $+27°29'23''$, and (2) the Subaru XMM-Newton Deep Field (SXDF) centered at RA(J2000) = $2^{\rm h}$ $18^{\rm m}$ $00.^{\rm s}00$ and DEC(J2000) = $-5°12'00''$. The SDF is used to make a very deep imaging survey while the

SXDF is used to make a wide-field, medium deep one; see for the SXDF project, [20], & [17].

(1) NIR Deep Imaging Survey: Very deep J and K' images of the central $2' \times 2'$ field of the SDF were obtained with use of CISCO [15]. The integration times of the J and K' bands were 12.1 hr and 9.7 hr, resulting in 5σ limiting magnitudes of 25.1 and 23.5 mag (the Vega system), respectively. These data are used to investigate the NIR galaxy number count, colors, and size distribution; see also [30]. They also found a population of hyper extremely red objects (HEROs) with $J - K' > 3 - 4$ [29]. These deep NIR data were also utilized to investigate the diffuse extragalactic background light (EBL) [28]. They found that ~ 90 % of the EBL from galaxies were resolved in their deep NIR images.

This NIR data set was also used to construct a K'-selected galaxy sample, consisting of 439 galaxies for which both optical (B, V, R, I, and z') and NIR photometric data are available [13]. Comparing the star formation rate density (SFRD) at $z \sim 3$ for their K'-selected sample with those based on previous LBG surveys, they found that a large fraction of SFRD at $z > 1.5$ may come from a faint blue galaxy population.

(2) Optical Narrowband Deep Survey: One of narrowband filters, NB711 centered at $\lambda_C = 7126$ Å with $\Delta\lambda = 73$ Å was used to search for LAEs at $z \sim 4.9$ [16], [22]. They found 87 reliable LAE candidates at $z \sim 4.9$, and then analyzed their luminosity function and clustering properties [16]. They also found a large-scale clustering of LAEs with a scale of ~ 20 Mpc $\times$ 50 Mpc [22].

(3) Optical Broad Band Deep Survey: In order to investigate photometric and clustering properties of LBGs at $z \sim 4 - 5$, optical broad band data of both the SDF and the SXDF, covering 1200 sq. arcmin in total were carefully analyzed by [17], [18]. They obtained a large sample of LBGs (2600 objects) at $z \simeq 3.5 - 5.2$.

Their analysis shows that the correlation lengths are $\simeq 4.1\ h_{100}^{-1}$ Mpc and $5.9\ h_{100}^{-1}$ Mpc in co-moving units for all the detected LBGs at $z \simeq 4$ and $z \simeq 5$, respectively.They also found that a typical mass of dark matter halos hosting LBGs with $L > L^*$ amounts to $\sim 1 \times 10^{12} M_\odot$, being comparable to those of typical massive disk galaxies like our Milky Way.

Based on the CDM model, they also estimated the mass of dark matter halos which could form from such high-z objects. Since they obtained a mass range between $\sim 10^{13} - 10^{15} M_\odot$, they suggested that dark matter halos hosting high-z LBGs could evolve to groups and clusters in the local universe. On one hand, faint LBGs, LAEs, and K'-selected galaxies could evolve to present-day galaxies after experiencing a few merger events.

2.2 The Subaru Deep Field (SDF) Project

As outlined in the previous subsection, the SDS gave a number of important findings in the research field of galaxy evolution. This success seems to be attributed to the very wide-field of view of Suprime-Cam and excellent seeing conditions at Mauna Kea. In order to make the SDS much more fruitful, the

Subaru Telescope Office decided to promote big surveys using guaranteed observing time that each Subaru builder member has. Then they proposed two big surveys for the extragalactic research; (i) the Subaru Deep Field Project led by Nobunari Kashikawa, and (ii) the Subaru XMM-Newton Deep Survey Project led by Kaz Sekiguchi. As mentioned before, the latter project is aimed to carry out a wide-field (1 sq. degree), medium deep survey in collaboration with the XMM-Newton Observatory. Since the SDF is dedicated to a very deep search for high-z galaxies, we present a brief summary of the current status of the SDF. It is noted that a common-use, intensive program on "A Search for Lyα Emitters at $z = 5.7$ and $z = 6.6$" (Proposal ID = S02A-IP2; PI = Y. Taniguchi) joined to the SDF project.

(1) SDF2002: Thirteen nights were allocated to the SDF project in the semester S02A. In this semester, we performed a deep optical imaging survey using a narrowband filter ($NB921$) centered at $\lambda = 9196$ Å together with i' and z' broadband filters covering an 814 arcmin^2 area of the SDF. We obtained a sample of 73 strong $NB921$-excess objects based on the following two color criteria; $z' - NB921 > 1$ and $i' - z' > 1.3$. We then obtained optical spectroscopy of nine objects in our $NB921$-excess sample, and identified at least two Lyα emitters at$z = 6.541 \pm 0.002$ and $z = 6.578 \pm 0.002$, each of which shows the characteristic sharp cutoff together with the continuum depression at wavelengths shortward of the line peak.

These new data allow us to estimate the first meaningful lower limit of the star formation rate density beyond redshift 6 [14]. First, we estimate the total star formation rate of 73 LAEs in our photometric sample using the equivalent width of NB921 flux. Our follow-up optical spectroscopy found that two among the nine LAE candidates are real LAEs, it seems reasonable to assume that approximately 22% (=2/9) of 73 LAE candidates are real LAEs at $z \approx 6.5$ - 6.6; $f(\mathrm{LAE}) \simeq 22\%$. If we assume that all the 73 LAE candidates are true LAEs at $z \approx 6.5 - 6.6$, we obtain nominally a total star formation rate of $SFR_{\mathrm{total}}^{\mathrm{nominal}} = 475 h_{0.7}^{-2}\ M_{\odot}\ \mathrm{yr}^{-1}$. Adopting $f(\mathrm{LAE}) \simeq 22\%$, we can estimate the total star formation rate, $SFR_{\mathrm{total}} \simeq 0.22 \times SFR_{\mathrm{total}}^{\mathrm{nominal}} \simeq 105 h_{0.7}^{-2}\ M_{\odot}\ \mathrm{yr}^{-1}$. Given the survey volume, 202,000 $h_{0.7}^{-3}$ Mpc^3, we thus obtain a star formation rate density of $\rho_{\mathrm{SFR}} \simeq 5.2 \times 10^{-4} h_{0.7}\ M_{\odot}\ \mathrm{yr}^{-1}\ \mathrm{Mpc}^{-3}$. This observation reveals that a moderately high level of star formation activity already occurred at $z \sim 6.6$ (see also [9]).

(2) SDF2003: Fifteen nights were allocated for the SDF project in the semester S03A. We made optical deep imaging and spectroscopy again, and finished our optical imaging survey. We obtained optical spectra of additional 18 LAE candidates using FOCAS, and thus we obtained a spectroscopic sample of 27 LAE candidates including our spectroscopy made in 2002. From our spectroscopy, we identify nine LAEs at $z = 6.50 - 6.60$. The remaining 18 objects are; nine single-line emitters, one [O II] emitter at $z = 1.46$, two [O III] emitters at $z = 0.84$ and $z = 0.85$, and six unclassified objects. The single-line emitters are either [O II] emitters at $z \sim 1.46$ or LAEs at $z \sim 6.6$. Much higher-resolution spectroscopy will be necessary to identify them unambiguously.

Since our new spectroscopy leads to a new value of $f(\mathrm{LAE}) = 9/27 \simeq 33\%$, we obtain a star formation rate density of $\rho_{\mathrm{SFR}} \simeq 7.8 \times 10^{-4} h_{0.7}\ M_{\odot}\ \mathrm{yr}^{-1}\ \mathrm{Mpc}^{-3}$. It should be reminded that we apply neither any reddening correction nor integration by assuming a certain luminosity function for LAEs. Therefore, this value should be regarded as a lower limit.

We also made follow-up spectroscopy of a small sample of very red objects in $i' - z'$ color, and then identified a new bright LAE at $z = 6.33$. This suggests that a number of LAEs may be found in such very red objects. Finally, we remind you that the data reduction of SDF data taken in 2003 is still underway.

2.3 Deep Surveys Based on Common-Use Observations

(1) Lyman Break Galaxies at $z \sim 5$: [11] (Proposal ID = S00-017; PI = K. Ohta) made deep optical imaging of 618 arcmin^2 including the Hubble Deep Field-North to search for LBGs at $z \sim 5$. They found ~ 100 LBG candidates at $23.0 \leq I_{\mathrm{C}} \leq 24.5$ and ~ 300 LBG candidates at $23.0 \leq I_{\mathrm{C}} \leq 25.5$. These data were used to estimate the rest-frame UV luminosity function at $4.4 \leq z \leq 5.3$. They found that the UV luminosity density at this redshift range is lower by a factor of two than that at $z \sim 3$.

(2) Lymanα Emitters at $z > 5$: [2] (Proposal ID = S01B-051; PI = Y. Taniguchi) made a survey for Lyα emitters at $z \approx 5.7$ based on optical narrow-band ($\lambda_{\mathrm{c}} = 8150$ Å and $\Delta\lambda = 120$ Å), and broad-band (B, R_{C}, I_{C}, and z')

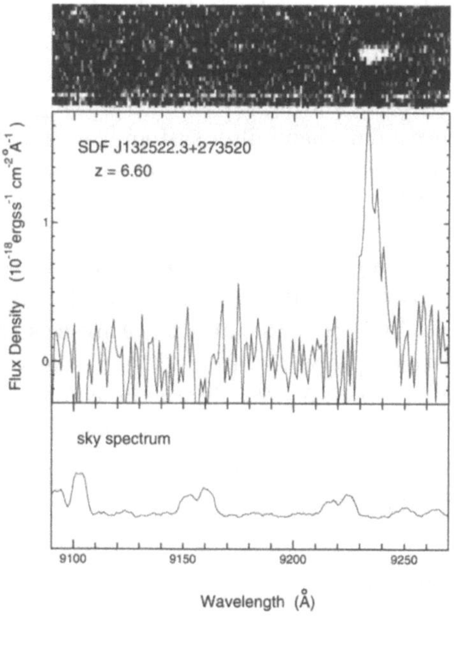

Fig. 1. The most distant LAE found in the SDF project; SDF J132522.3+273529 at $z = 6.60$.

observations of the field surrounding the high redshift quasar, SDSSp J104433.04 −012522.2 at $z = 5.74$. This survey covers a sky area of ≈ 720 arcmin^2 and a co-moving volume of $\simeq 2 \times 10^5\ h_{0.7}^{-3}$ Mpc^3. They found 20 LAE candidates at $z \approx 5.7$ with $\Delta z \approx 0.1$. This survey leads to a new estimate of the star formation rate density at $z \approx 5.7$, $\sim 1.2 \times 10^{-3} h_{0.7} M_\odot$ yr^{-1} Mpc^{-3}. It is also noted that this NB816 survey was used to investigate field Hα emitters at $z \approx 0.24$ [7].

Among their 20 LAE candidates, two objects were confirmed star-forming galaxies at $z = 5.655$ and $z = 5.687$ from their follow-up optical spectroscopy made with FOCAS on Subaru and/or ESI on Keck II. LAE J1044−0130 is identified as a probable superwind galaxy at $z = 5.687 \pm 0.002$ [2]. Its emission line profile is strongly truncated at wavelengths blueward shortward the line peak while shows red-wing emission. The observed broad line width, FWHM (full width at half maximum) $\simeq 340$ km s^{-1} as well as the red wing emission suggest that this object is experiencing the superwind activity. The emission-line morphology appears to show a triangle shape. This may be also interpreted in terms of the superwind activity.

LAE J1044−0123 is identified as a star forming galaxy at $z = 5.655 \pm 0.002$ with a star formation rate of $\sim 13\ h_{0.7}^{-2}\ M_\odot$ yr^{-1} [26]. Remarkably, the velocity dispersion of Lyα-emitting gas is only 22 km s^{-1}. Since a blue half of the Lyα emission could be absorbed by neutral hydrogen gas, perhaps in the system, a modest estimate of the velocity dispersion may be ~ 44 km s^{-1}. Together with a linear size of 7.7 $h_{0.7}^{-1}$ kpc, we estimate a lower limit of the dynamical mass of this object to be $\sim 2 \times 10^9 M_\odot$. Therefore, LAE J1044−0123 seems to be a star-forming dwarf galaxy (i.e., a subgalactic object or a building block).

[3] also made a unique deep survey for LAEs at $z \sim 5.8$ using an intermediate-band filter centered at $\lambda_c \approx 8270$ Å with $\Delta\lambda_{\rm FWHM} \approx 340$ Å (i.e., the spectroscopic resolution is $R \approx 23$) during the same observing run as that of [2]; see for details of this intermediate-band filter system [25], In this survey, they found four Lyα-emitter candidates from the intermediate-band image ($z \approx 5.8$ with $\Delta z \approx 0.3$); see also [6] for a similar survey for LAEs at $z \simeq 3.7$ using another intermediate-band filter IA 574.

In the above LAE survey, they observed a sky are surrounding the high redshift quasar, SDSSp J104433.04−012522.2 at $z = 5.74$. They found a foreground lensing galaxy with m_B(AB) ≈ 25, located at 1.9 arcsec southwest of the quasar [21]. Its broad band color properties from B to z' suggest that the galaxy is located at a redshift of $z \sim 1.5$ - 2.5. Since the counter image of the quasar cannot be seen in our deep optical images, the magnification factor seems not so high. Our modest estimate is that this quasar is gravitationally magnified by a factor of 2 ;see also [31].

3 Concluding Remarks

The Hubble Space Telescope and 8-10m class optical telescopes have been contributing to the progress in deep searches for high-z galaxies. Although the

Subaru Telescope came later to this research field, as we see above, it is also powerful to search for high-z galaxies as well as the other 8-10m class telescopes.

Up to date, approximately several tens of LAEs beyond $z = 5$ have already been identified spectroscopically. However, we still need any systematic deep surveys for such LAEs to understand the whole history of cosmic star formation in the early universe. In particular, one of important things related to LAEs is to construct reliable Lyα luminosity functions of LAEs as a function redshift and the compare them UV luminosity functions; see for recent progress, [2], [10], & [19].

We would like to thank all the SOC and LOC members, in particular, Alvio Renzini and Ralf Bender. We would like to thank Norio Kaifu, Hiroyasu Ando, Hiroshi Karoji, Hy Spinrad, Nobunari Kashikawa, Yutaka Komiyama, Sadanori Okamura, Kazuhiro Shimasaku, Masami Ouchi, Yasuhiro Shioya, Takashi Murayama, and Tohru Nagao for useful discussion and encouragement. We also thank all members of the SDF project.

References

1. M. Ajiki et al. ApJ, 576, L25 (2002)
2. M. Ajiki et al. AJ, 126, 2091 (2003)
3. M. Ajiki et al. AJ, submitted (2004)
4. A. Dey et al. ApJ, 498, L93 (1998)
5. M. Dickinson et al. astro-ph/0309070 (2003)
6. S. S. Fujita et al. AJ, 125, 13 (2003a)
7. S. S. Fujita et al. ApJ, 586, L115 (2003b)
8. M. Giavalisco et al. astro-ph/0309150 (2003)
9. E. M. Hu et al. ApJ, 568, L75 (2002)
10. E. M. Hu et al. astro-ph/0311528 (2003)
11. I. Iwata et al. PASJ, 55, 415 (2003)
12. N. Kaifu et al. PASJ, 52, 1 (2000)
13. N. Kashikawa et al. AJ, 125, 53 (2003)
14. K. Kodaira et al. PASJ, 55, L17 (2003)
15. T. Maihara et al. PASJ, 53, 25 (2001)
16. M Ouchi et al. ApJ, 582, 60 (2003a)
17. M. Ouchi et al. ApJ, submitted (2003b)
18. M. Ouchi et al. ApJ, submitted (2003c)
19. M. R. Santos et al. astro-ph/0310478 (2003)
20. K. Sekiguchi, in this volume (2003)
21. Y. Shioya et al. PASJ, 54, 975 (2002)
22. K. Shimasaku et al. ApJ, 586, L11 (2003)
23. H. Spinrad Astrophysics Update, in press (astro-ph/0308411)
24. E. Stanway et al. astro-ph/0308124 (2003)
25. Y. Taniguchi 2001, the Japan-Germany Workshop on Studies of Galaxies in the Young Universe with New Generation Telescopes (astro-ph/0301097)
26. Y. Taniguchi et al. ApJ, 585, L97 (2003a)
27. Y. Taniguchi et al. JKAS, 36, 123 (2003b)
28. T. Totani et al. ApJ, 559, 552 (2001a)
29. T. Totani et al. ApJ, 558, L87 (2001b)
30. T. Totani et al. ApJ, 550, L137 (2001c)
31. S. F. Yamada et al. PASJ, 55, 733 (2003)

The First 1–2 Gyrs of Galaxy Formation: Dropout Galaxies from $z \sim 3 - 6$

Garth Illingworth and Rychard Bouwens

UCO/Lick Observatory, Astronomy and Astrophysics Department, University of California, Santa Cruz, CA 95064

Abstract. The unique high–resolution wide–field imaging capabilities of HST with ACS have allowed the characterization of galaxies at redshift 6, less than 1 Gyr from recombination. The dropout technique, applied to deep ACS i, z images in the RCDS 1252–2927, GOODS and UDF–Parallel fields has yielded large samples, allowing determination of their properties (e.g., size, color) and meaningful comparisons against lower redshift dropout samples. The use of cloning techniques has enabled us to control for many of the strong selection biases that affect the study of high redshift populations. A clear trend of size with redshift has been identified, and its impact on the luminosity density and star formation rate can be estimated. There is a significant, though modest, decrease in the star formation rate from redshifts $z \sim 2.5$ out through $z \sim 6$. The latest data also allow for the first robust determination of the luminosity function at $z \sim 6$.

1 Introduction

The advent of the HST Advanced Camera, the ACS (Ford et al 2003) has greatly increased our ability to "watch galaxies form". The sensitivity, resolution and excellent filter set have provided us with images from which large samples of high redshift galaxies can be derived. Of particular interest are those galaxies with red enough i-z colors to qualify as i-dropouts – galaxies at redshifts $z \sim 6$. Such objects have been the focus of a number of papers over the last year (e.g., Bouwens et al 2003b, Stanway et al 2003, Yan et al 2003, Dickinson et al 2004). Spectroscopic confirmation is beginning to appear (e.g., Bunker et al 2003, Dickinson et al 2004) but is challenging as Weymann et al (1998) demonstrated with their $z = 5.6$ object, which took over 6 hours on Keck.

The current frontier for high redshift objects is at $z \sim 6$ (the ACS UDF and NICMOS UDF–IR images together will likely extend the dropout sources to redshifts 7 and beyond, but the samples will be small). Rapid changes in the properties of high redshift galaxies must occur beyond $z \sim 6$ and so careful characterization of objects even those separated by small intervals of time, is a well-justified goal – especially given that only 650 million years separates $z \sim 15$ from $z \sim 6$. In this context there is great value in having large samples of $z \sim 3$–5 objects which are more amenable to thorough, quantitative study.

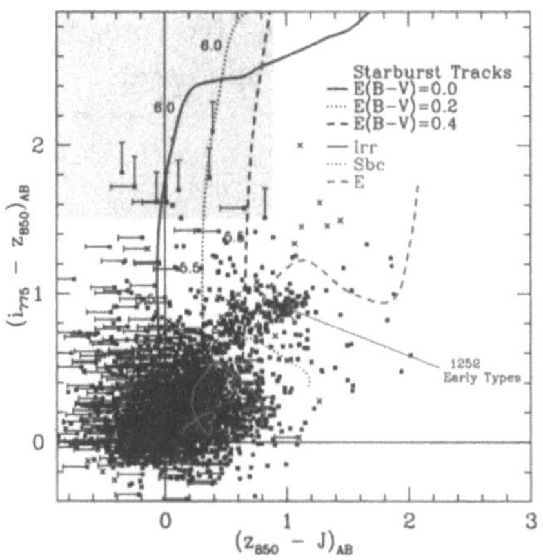

Fig. 1. Selection of i-dropouts in the i-z z-J two–color plane. This example is for the HST ACS from the RCDS 1252–2927 field (Rosati et al 1998). The selection limits (particularly the $(i-z) > 1.5$ cut – see Bouwens et al 2003b) returns $z \sim 6$ galaxies with little contamination (an estimated 11% contamination rate).

2 Fields and Object Selection

While data from several fields have been used to identify high redshift dropouts, three fields have stood out for their value for dropout studies over the last year – namely the RCDS 1252–2927 field, the GOODS fields, and the UDF–parallels (UDF–Ps). All have excellent HST ACS i_{775} and z_{850} data, while the GOODS and the UDF-Ps fields also have deep B_{435} and V_{606} data. The excellent IR data (Lidman et al 2004) in RCDS 1252–2927 also makes a substantial contribution to the selection of i–dropouts, helping to establish the degree of contamination in the samples.

The selection of i–dropouts is shown in Fig 1 for the RCDS 1252–2927 field (from Bouwens et al 2003b). The ACS data reaches typically to $z_{850,AB} \sim 27.3$ mag (6σ), while the ground-based IR data goes impressively deep, down to $J_{AB} = 25.7$ and $Ks_{AB} = 25.0$ mag (5σ). The fraction of $z \sim 6$ objects in the IR coverage in RCDS 1252–2927 is impressively small (0.3%), only 12 out of ~3000 galaxies. Even so the estimated contamination is only about 11%. A number of these candidates have been observed with Keck and the VLT and confirmed to be at $z \sim 6$. A total of 23 $z \sim 6$ galaxies are found in four ACS pointings of the RCDS 1252–2927 field, giving a surface density of 0.5 ± 0.2 i–dropouts per square arcmin to $z_{AB} = 26.5$ mag. The objects are very small, though all are resolved, with typical half-light radii of 0.15″ or ~ 0.9 kpc. The $z \sim 6$ objects reach down to $\sim 0.3 L_{*,z=3}$ (Steidel et al 1999).

Two of the brighter i–dropouts from the RCDS 1252–2927 field are shown in Fig 2, along with their location in the two–color plane, and SED fits that are used to establish the redshifts. The ACS i and z data from the HDF–N also allowed for a search for i–dropouts. A reassuring result was that the Weymann et al (1998) object in the HDF–N, spectroscopically verified to be at $z = 5.60$, was very close to meeting our i–dropout criterion (its $i - z = 1.2$ color was just a little too blue). While not a true i–dropout, it suggested that our selection was yielding bona–fide high redshift objects. Other spectroscopic results from Bunker et al (2003) and Dickinson et al (2004), and our own ongoing Keck programs, have only served to strengthen our confidence in the dropout approach.

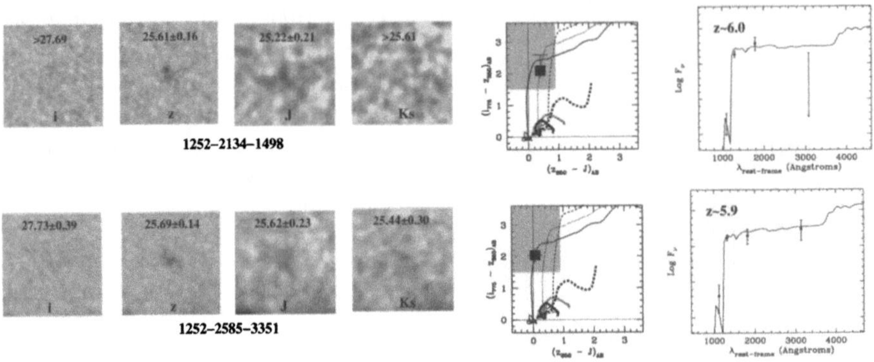

Fig. 2. Images ($3'' \times 3''$) in *i, z, J, Ks* of $z \sim 6$ objects, along with two–color schematics (showing starburst tracks as a function of redshift for different reddenings - see Fig 1), and starburst galaxy SEDs (10^8 Gyr), with the best fit redshift. The sources are all in RCDS 1252–2927. The magnitudes given for the sources are AB magnitudes.

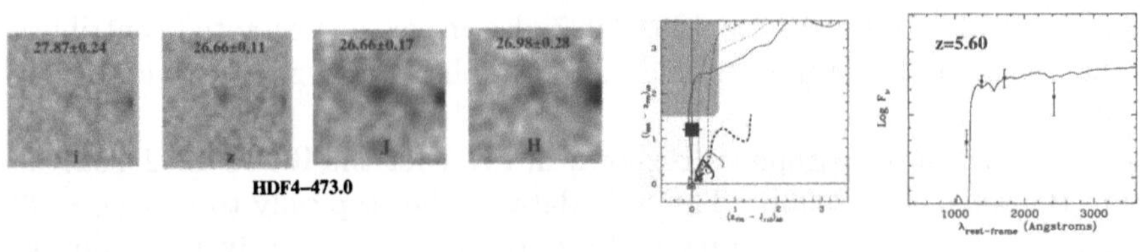

Fig. 3. As in Fig 2, but for the Weymann et al (1998) galaxy in the HDF–N whose redshift was measured to be $z = 5.6$ from over 6 hours of integration with the LRIS spectrograph on Keck. The redshift determined from the photometric data was also $z = 5.6$.

While the RCDS 1252–2927 field provided a significant sample of i–dropouts (with a good assessment of the contamination from the deep VLT IR data), the best samples of brighter dropouts come from the two HST ACS GOODS fields, CDF–S and HDF–N (see Giavalisco et al 2004). From these fields, Bouwens et al (2004b) derived a large number of B-, V- and i–dropouts, augmenting them with a smaller but very useful sample of U–dropouts from the HDF–N and HDF–S fields so that a self–consistent differential analysis could be applied across a large redshift range, $z \sim 3$ to $z \sim 6$. Even with relatively conservative selection criteria, Bouwens et al (2004b) derive 1235 $z \sim 4$ B–dropouts, 407 $z \sim 5$ V–dropouts, and 59 $z \sim 6$ i–dropouts. These samples go as faint as 0.2, 0.3, 0.5

$L_{*,z=3}$ (using the Steidel et al 1999 value for $L_{*,z=3}$), respectively, with 10σ limiting magnitudes of 27.4 in the $i_{775,AB}$ band and 27.1 in the $z_{850,AB}$ band.

The large samples and wide areal coverage of the GOODS fields are nicely complemented by the two UDF–parallel fields (UDF–Ps) obtained in parallel with the deep NICMOS images of the UDF. These fields have overlapping ACS images on a 45″ grid with 9 orbits each in B and V, 18 orbits in i and 27 orbits in z (as well as 9 orbits with the grism). They reach impressively faint, to 28.8, 29.0, 28.5 and 27.8 mag (10σ) in $B_{435}, V_{606}, i_{775}$, and z_{850} AB-mag, respectively – or to 0.1–0.2$L_{*,z=3}$. The UDF itself will be an impressive addition to these fields, taking the limits to $< 0.1L_{*,z=3}$.

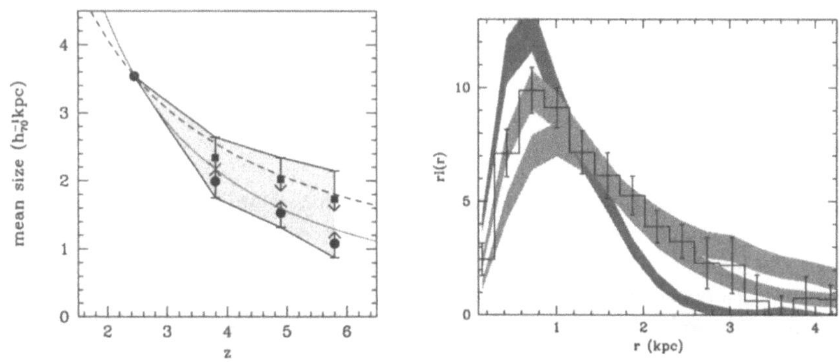

Fig. 4. (Left) Size evolution of $1-2L_{*,z=3}$ galaxies derived from composite radial flux profiles for objects with redshifts from $z \sim 2.5$ to $z \sim 6$ (Bouwens et al 2004b). The solid black circles (1σ errors) give the observed sizes with redshift, and should be a lower limit. The solid black squares are expected to be upper limits and show the evolution obtained by bootstrapping the sizes from $z \sim 2.5$, comparing each sample with the one adjacent to it in redshift. A clear evolution towards smaller size is observed with redshift, consistent with the scalings predicted from hierarchical models ($H(z)^{-1} \sim (1+z)^{-3/2}$ for fixed circular velocity [solid line] and $H(z)^{-2/3} \sim (1+z)^{-1}$ for fixed mass [dashed line]). (Right) The mean radial flux profile for the 10 brightest i–dropouts in the UDF–Ps (histogram) compared with "cloned" projections of the HDF–N and HDF–S U–dropout sample scaled in size as $(1+z)^m$, where $m = 0$, -1.5 and -3; $m = -1.5$ is the best fit (Bouwens et al 2004a).

3 Results

A major issue with deriving the evolution of galaxy properties at high redshift is systematic error – primarily through the many selection effects that can influence the nature of the samples, even when derived from very similar datasets. Of these the $(1+z)^4$ surface brightness dimming is the dominant effect, but many others affect the derived samples (e.g., size evolution, color evolution, definition of selection volumes, data properties as a function of redshift, filter band, and instrument, etc.). To treat these effects, we compare our highest redshift samples with "cloned" projections of our lower redshift samples (e.g., Bouwens et al. 1998; Bouwens et al. 2003a), allowing us to contrast intrinsic evolution from changes brought about by the selection process itself.

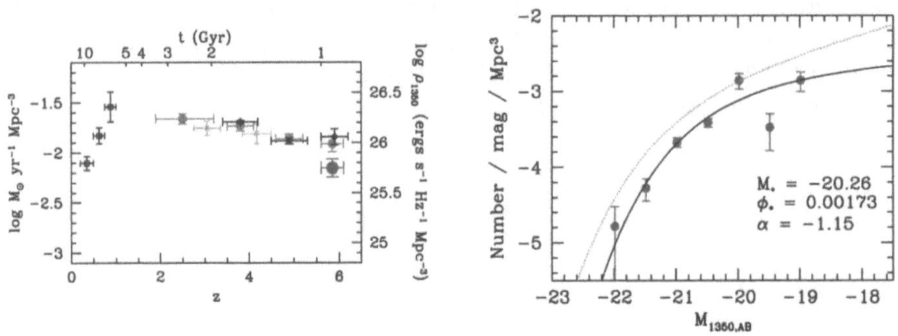

Fig. 5. Star formation rate evolution (dust–free) with redshift and age (top), integrated down to $0.2L_{*,z=3}$ (rest–frame UV continuum luminosity density on the right vertical axis for the high redshift values). Note the small Δt from $z \sim 6$ to $z \sim 3$. A Salpeter IMF is used to convert the luminosity density to SFR (see Madau et al 1998). (Left) The four solid circles from $z = 2.5$ to $z = 6$ are the values from the GOODS data (Bouwens et al 2004b). Other determinations are Lilly et al (1996 – open squares), Steidel et al (1999 – crosses), Bouwens et al (2004a – solid circles at $z = 6$) and Giavalisco et al (2004 – solid diamonds). The Thompson et al (2001) values are similar to those shown here. The low point at $z = 6$ includes the effect of size evolution on the $z \sim 6$ Bouwens et al (2004a) value (indicating how significant this effect can be). (right): The rest frame continuum UV (at 1350 Å) luminosity function at $z \sim 6$ from the GOODS field (for $M_{1350,AB} < -19.7$) and the UDF–Ps. The best fit values for a Schechter luminosity function are shown on the figure. The Steidel et al (1999) $z \sim 3$ luminosity function (dotted line) is also shown. The Steidel luminosity function had a best fit faint end slope $\alpha = -1.6$. Such a slope is also consistent with our $z \sim 6$ data.

One of the key results is that of size evolution. This is demonstrated in Fig 4a (and also discussed in the context of the GOODS datasets by Ferguson et al 2004). There is clearly size evolution with redshift. The deeper UDF–Ps provide stronger and even more conclusive evidence for this (see Bouwens et al 2004a), as well as indicating that the best fit appears to be with $(1 + z)^{-1.5}$ (Fig 4b).

A major goal of these studies is to extend the constraints on the luminosity density and the star formation rate with redshift to higher redshifts $z \sim 6$ and beyond. A related goal is to improve the constraints at lower redshifts ($z \sim 2-5$). These new datasets are proving to be of great value for these two goals. Fig 5a gives the most recent estimate of the (dust-free) star formation rate (including the effect of size evolution) out to $z \sim 6$.

Another major development over the last year is that the HST ACS data is now of sufficient quality and depth that a luminosity function can be derived to significantly fainter than L_*, as was recently done by Bouwens et al. (2004a) with the GOODS + UDF–Ps data (Fig 5b). The UDF will extend this luminosity function one magnitude fainter.

4 Summary

There has been a remarkable growth in the number of objects known at high redshifts ($z \sim 6$) since the HST Advanced Camera came into operation after servicing mission SM3B. Not only are large numbers of sources being detected at high redshift, but the development of new techniques for detecting, characterizing and comparing high redshift objects from photometric datasets has led to many quantitative results on the nature and evolution of galaxies in the first $\sim 1-3$ Gyrs.

Acknowledgements

We would like to thank the organizers for an excellent meeting in a great place. We acknowledge the remarkable advances that have come about because of HST and its amazing imagers, and regret the decision to cancel SM4 that will lead to the premature death of HST. We owe a lot to our team members on the ACS GTO team and the UDF–IR team, and particularly the PIs, Holland Ford and Rodger Thompson. Support from NASA grant NAG5–7697 and NASA/STScI grant HST–GO–09803.05–A is gratefully acknowledged.

References

1. Bouwens, R., Broadhurst, T. and Silk, J. 1998, ApJ., **506**, 557.
2. Bouwens, R., Broadhurst, T., & Illingworth, G. ApJ., **593**, 640 (2003a).
3. Bouwens, R. J., et al. ApJ., **595**, 589 (2003b).
4. Bouwens, R. J. et al. ApJ., in press (2004a).
5. Bouwens, R. J. et al. ApJ., in press (2004b).
6. Bunker, A. J., Stanway, E. R., Ellis, R. S., McMahon, R. G., & McCarthy, P. J. MNRAS, **342**, L47 (2003).
7. Dickinson, M., et al. ApJ., **600**, L99 (2004).
8. Ferguson, H.C., et al. ApJ., **600**, L107 (2004).
9. Ford, H. C. et al. Proc. SPIE, **4854**, 81 (2003).
10. Giavalisco, M., et al. ApJ., **600**, L93 (2004).
11. Lidman, C., et al. A.& A., in press (2004).
12. Lilly, S.J., Le Fevre, O., Hammer, F., & Crampton, D. ApJ., **460**, L1 (1996).
13. Madau, P., Pozzetti, L. & Dickinson, M. ApJ., **498**, 106 (1998).
14. Rosati, P., et al. ApJ., **492**, L21 (1998).
15. Stanway, E. R., Bunker, A. J., & McMahon, R. G. MNRAS, **342**, 439.
16. Steidel, C. C., Adelberger, K. L., Giavalisco, M., Dickinson, M. and Pettini, M. ApJ., **519**, 1 (1999).
17. Thompson, R. I., Weymann, R. J., & Storrie-Lombardi, L. J. ApJ., **546**, 694 (2001).
18. Weymann, R. J., et al. ApJ., **505**, L95 (1998).
19. Yan, H., Windhorst, R. A., & Cohen, S. H. ApJ., **585**, L93 (2003).

The Phoenix Deep Survey: Evolution of Star Forming Galaxies

A.M. Hopkins[1,2], J. Afonso[3], A. Georgakakis[4], M. Sullivan[5], B. Mobasher[6], and L.E. Cram[7]

[1] University of Pittsburgh, Department of Physics and Astronomy 3941 O'Hara St, Pittsburgh, PA 15206, USA
[2] Hubble Fellow; ahopkins@phyast.pitt.edu
[3] CAAUL, Observatory of Lisbon, Tapada da Ajuda, 1349-018 Lisbon, Portugal
[4] National Athens Observatory, I.Metaxa & Vas.Pavlou str., Athens 15236, Greece
[5] University of Toronto, 60 St. George St, Toronto, Ontario M5S 3H8, Canada
[6] STScI, 3700 San Martin Drive, Baltimore, MD 21218, USA
[7] Australian Research Council, GPO Box 2702, Canberra, ACT 2601, Australia

Abstract. The Phoenix Deep Survey (PDS) is a multiwavelength survey based on deep 1.4 GHz radio observations used to identify a large sample of star forming galaxies to $z = 1$. Photometric redshifts are estimated for the optical counterparts to the radio-detected galaxies, and their uncertainties quantified by comparison with spectroscopic redshift measurements. The photometric redshift estimates and associated best-fitting spectral energy distributions are used in a stacking analysis exploring the mean radio properties of U-band selected galaxies. Average flux densities of a few μJy are measured.

1 Introduction

The study of galaxy evolution in recent years has included a strong focus on the star formation properties of galaxies. Many of these studies are based primarily on selection at ultraviolet (UV) and optical wavelengths, known to be strongly affected by obscuration due to dust. It has been shown that selection at these wavelengths results in samples of star forming systems that miss a significant fraction of heavily obscured galaxies [20]. There have moreover been suggestions that the most vigorous star forming (SF) systems suffer the most obscuration [2,17,6,22,16]. Radio selection provides an efficient tool to construct a SF galaxy sample free from dust induced biases, and the average obscuration in such samples indeed appears significantly higher than in optically selected samples [2,17].

Motivated to construct a homogeneously selected sample of SF galaxies, unbiased by the effects of obscuration due to dust, the Phoenix Deep Survey (PDS, see http://www.atnf.csiro.au/people/ahopkins/phoenix/) is based on a deep (60 μJy), wide-area (4.5 square degree) 1.4 GHz survey with the Australia Telescope Compact Array. This provides one of the largest existing deep 1.4 GHz source catalogues [14] containing a large fraction of SF galaxies spanning the broad redshift range $0 < z < 1$. The PDS has already been highly successful in providing a basis for several investigations of the nature of SF galaxies and their evolution ([12,2,14] and references therein). Throughout the present investigation we assume a $(\Omega_M = 0.3, \Omega_\Lambda = 0.7, H_0 = 70)$ cosmology.

2 Photometric Redshift Analysis

Deep *UBVRI* observations of about one square degree within the PDS have recently been analysed (these are described in detail in [21]). These data achieve a 5σ level of $R_{\rm AB} \approx 24.5$ and optical catalogues have been constructed and cross-correlated with the 1.4 GHz catalogues. This multicolour data has been used to estimate photometric redshifts for all ≈ 40000 galaxies detected in each of the 5 bands. This includes about 800 optical counterparts of the radio detected galaxies in this area.

For our analysis we use the photometric redshift code of Connolly et al. [10] (see also [7]). We use SED templates based on those of Coleman et al. [8] (hereafter CWW), providing four standard SEDs (E/S0, Sbc, Scd, Im), extended in the UV and IR wavelength regions using the GISSEL98 code [5]. Rather than using these four SEDs directly, we use the method of optimal subspace filtering [7] to provide a large number (61) of smoothly interpolated SEDs based on the reference CWW SEDs. This supports more realistic type estimates for most of the galaxies. We allow the possible photometric redshifts to range from 0.0 to 1.3 and also apply a prior constraining the absolute magnitudes of the galaxies to the broad range $-29 < M_B < -16$ (having the effect of removing photometric redshift fits with unphysically high or low redshifts).

Figure 1 compares the photometric redshift estimate with spectroscopic redshift for 116 radio sources with an optical counterpart having both *UBVRI* detections and spectroscopic data. The filled symbols (including points) indicate the spectroscopic classification, while the open symbols give an estimate of the best-fitting SED template type (after binning the 61 subspace filtered templates into four bins, based approximately on the closest CWW-type template). The reliability of the photometric redshifts can be characterised in several ways. The rms of $|\Delta z| = |z_{\rm photo} - z_{\rm spec}|$ is 0.1 for these 116 galaxies. The rms of $|\log[(1+z_{\rm photo})/(1+z_{\rm spec})]|$ is 0.028 (implying a typical uncertainty of 7% in $1+z_{\rm photo}$). The rms of $|\Delta z|/(1+z_{\rm spec})$ is 0.065. This level of reliability in the photometric redshifts compares favourably with that of other analyses [11,19]. As well as providing reasonable redshift estimates, the best-fitting SED is also a good indicator of the galaxy type, in the sense that spectroscopic absorption-line systems are mostly well-fit by early-type SEDs, while SF systems are mostly well-fit by late-type SEDs.

3 Stacking Analysis of *U*-Band Galaxies

The technique of stacking small subregions of an image at the locations of a known population of objects that are not otherwise detected, in order to extract a rough estimate of the mean emission properties of a population, has been used with some success at X-ray wavelengths [4,18]. This technique has been applied to XMM observations of the PDS [12] to explore the X-ray properties of radio-detected SF galaxies. Following this success we extended the method to radio wavelengths, performing a stacking analysis using the 1.4 GHz mosaic

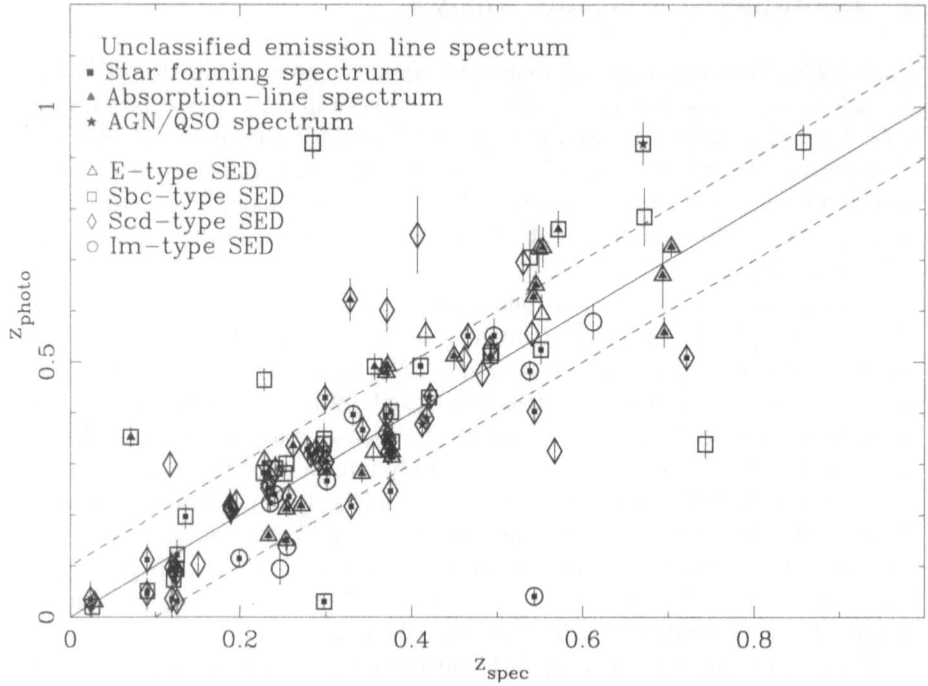

Fig. 1. A comparison of spectroscopic and photometric redshifts for the spectroscopically observed sub-sample. The points and filled symbols refer to the spectroscopic classification of the galaxies [13], while the open symbols refer to the best-fitting SED from the photometric redshift estimation. The open symbols are estimated by binning the 61 subspace filtered SEDs approximately into the closest CWW-type classification. Note that this figure shows redshifts in linear units, with the dashed lines indicating offsets of ±0.1 in $z_{\rm photo}$ about the one-to-one line.

image of the PDS to explore the mean radio properties of extremely red galaxies (ERGs) not otherwise detected at 1.4 GHz [15]. These results have implications for the expected average radio luminosities and inferred star formation rates of ERGs. Further investigation of the radio properties of ERGs through stacking analyses are explored elsewhere in this volume [1]. We now further extend this technique to explore the average 1.4 GHz properties of a population of U-band selected galaxies, which is expected to include a large fraction of SF galaxies. This will provide an estimate of the typical radio properties for the population of "normal" or quiescent SF galaxies, as opposed to the starbursts that often dominate studies of star formation in galaxies.

The available U-band data reaches a 5σ detection limit of $U_{\rm AB} \approx 25.0$. To explore radio flux density trends with both redshift and galaxy type, we take advantage of the photometric redshifts and best-fitting SED types to split the 40000 galaxies into three bins in redshift and three bins in SED type (described as "early," "mid" and "late," the CWW types Sbc and Scd being included in

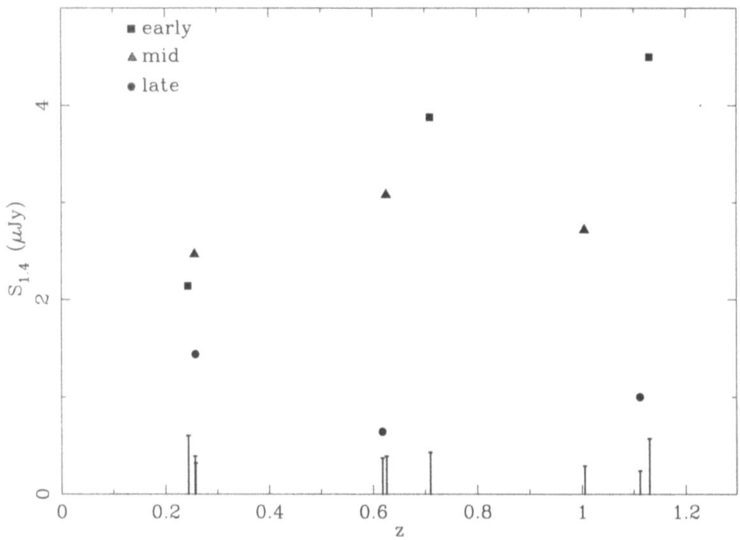

Fig. 2. Average 1.4 GHz flux density inferred from the stacking analysis for galaxies of early (squares), mid (triangles) and late (circles) SED type. The flux densities are shown as a function of the median photometric redshift of all the objects contributing to each of the nine stacked images. The vertical bars below each point indicate the rms noise level of the stacked image in which each measurement was made, with detections ranging from $3.6 - 9\sigma$. The only non-detection is the late-type SED (circle) in the middle redshift bin, and is shown to indicate the upper limit for this stacked image (an arrow was omitted to avoid clutter with the rms noise level bars).

the "mid" type). The numbers of galaxies in each bin are given in Table 1. For the stacking analysis a subregion of $2'$ square was extracted from the PDS 1.4 GHz image at the location of each of the U-band galaxies. To ensure that radio detections or uncatalogued low signal-to-noise (S/N) radio emission do not bias the stacking signal, subregions are excluded from the stacking analysis if the average 1.4 GHz emission in a $14'' \times 14''$ region centred at the location of the U-band galaxy is above some S/N threshold. The threshold chosen was a fairly conservative 1.5σ, although the results change only marginally if slightly higher thresholds ($2 - 3\sigma$) are used. Since the noise level is not uniform across the PDS 1.4 GHz image [14], the individual subregions are weighted by the inverse square of the rms noise background during the averaging step, in order to maximise the S/N of the resulting stacked image.

The results of the stacking analysis are shown in Figures 2 and 3. Of the nine stacked images constructed, eight show confident detections ($> 3.6\sigma$). The exception is the middle redshift bin for the late-type SED systems, where the peak flux at the expected location of the source is 1.7σ. The measured 1.4 GHz flux densities for the stacking results are shown in Figure 2 as a function of the median photometric redshift for the objects contributing to each final stacked

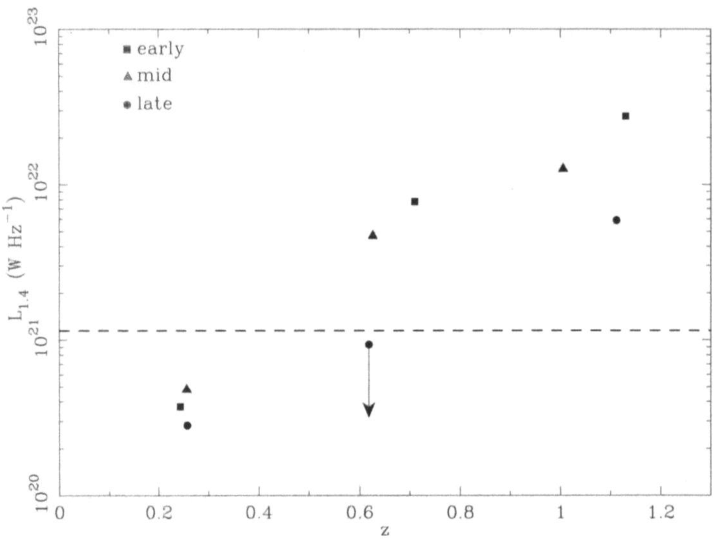

Fig. 3. Average 1.4 GHz luminosity inferred from the flux density measured in the stacking analysis, and the median photometric redshift. The non-detection is now indicated as an upper limit. The dashed line indicates the luminosity corresponding to an SFR of $1\,M_{\odot}\,yr^{-1}$ using the calibration of [3].

image. The measured flux densities are of order a few μJy for each of the stacked detections, in rms backgrounds around 0.3 to 0.5 μJy. The "mid"-type SED class seem to predominantly show flux densities between the "early" and "late" types, consistent with what might be expected if it was comprised of a combination of both active galactic nuclei (AGNs) and SF systems, or of systems driven by both processes.

The luminosities derived from these flux densities and the median photometric redshifts are shown as a function of photometric redshift in Figure 3. The trend to higher luminosities at higher redshifts for all SED types is likely to be a result of our magnitude-limited selection, since at high redshifts only the higher (U-band) luminosity systems are being sampled, and the radio luminosity is likely to be correlated at some level with the U-band luminosity [17]. The dashed line in this figure indicates a star formation rate (SFR) of $1\,M_{\odot}\,yr^{-1}$ using the calibration of [3] (this line is about 40% higher than if the calibration of [9] were used). This shows that through the use of the stacking technique it is possible to develop some insight into the 1.4 GHz properties of "normal" or quiescent SF galaxies. The fact that early type SED systems are detected at similar flux densities may suggest either that low luminosity AGN are present in significant numbers in normal galaxies, or that early type galaxies can support these low levels of star formation, or even (since the flux density in early types seems consistently higher than in late types) that both processes might be occurring in these systems.

Table 1. Numbers of galaxies in redshift and type bins for the U-band stacking

Redshift range	Early	Mid	Late
$0.00 < z < 0.43$	1358	3349	6377
$0.43 < z < 0.87$	3349	4009	3384
$0.87 < z < 1.30$	1562	6910	8663

AMH acknowledges support provided by NASA through Hubble Fellowship grant HST-HF-01140.01-A awarded by STScI. JA acknowledges the support from the Science and Technology Foundation (FCT, Portugal) through the fellowship BPD-5535-2001 and the research grant POCTI-FNU-43805-2001.

References

1. J. Afonso, A. Hopkins, B. Mobasher, L. Cram, A. Georgakakis, M. Sullivan: This Proceedings (2003)
2. J. Afonso, A. Hopkins, B. Mobasher, C. Almeida: ApJ **597**, 269 (2003)
3. E. F. Bell: ApJ **586**, 794 (2003)
4. W. N. Brandt, et al.: AJ **122**, 1 (2001)
5. A. G. Bruzual, S. Charlot: ApJ, 405, 538 (1993)
6. V. Buat, A. Boselli, G. Gavazzi, C. Bonfanti: A&A **383**, 801 (2002)
7. T. Budavári, A. S. Szalay, A. J. Connolly, I. Csabai, M. Dickinson: AJ **120**, 1588 (2000)
8. G. D. Coleman, C.-C. Wu, D. W. Weedman: ApJS **43**, 393 (1980)
9. J. J. Condon: ARA&A **30**, 575 (1992)
10. A. J. Connolly, I. Csabai, A. S. Szalay, D. C. Koo, R. G. Kron, J. A. Munn: AJ **110**, 2655 (1995)
11. A. Fernández-Soto, K. M. Lanzetta, A. Yahil: ApJ **513**, 34 (1999)
12. A. Georgakakis, A. M. Hopkins, M. Sullivan, J. Afonso, I. Georgantopoulos, B. Mobasher, L. E. Cram: MNRAS **345**, 939 (2003)
13. A. Georgakakis, B. Mobasher, L. Cram, A. Hopkins, C. Lidman, M. Rowan-Robinson: MNRAS **306**, 708 (1999)
14. A. M. Hopkins, J. Afonso, B. Chan, L. E. Cram, A. Georgakakis, B. Mobasher: AJ **125**, 465 (2003)
15. A. M. Hopkins, J. Afonso, A. Georgakakis, M. Sullivan, B. Mobasher, L. E. Cram: 'Extremely red galaxies in the Phoenix Deep Survey'. In: *Multiwavelength Cosmology,* (in press; astro-ph/0309147)
16. A. M. Hopkins, A. J. Connolly, D. B. Haarsma, L. E. Cram: AJ **122**, 288 (2001)
17. A. M. Hopkins, et al.: ApJ (2003) (in press; astro-ph/0306621)
18. K. Nandra, R. F. Mushotzky, K. Arnaud, C. C. Steidel, K. L. Adelberger, J. P. Gardner, H. I. Teplitz, R. A. Windhorst: ApJ **576**, 625 (2002)
19. M. Rowan-Robinson: MNRAS **345**, 819 (2003)
20. I. Smail, F. N. Owen, G. E. Morrison, W. C. Keel, R. J. Ivison, M. J. Ledlow: ApJ **581**, 844 (2002)
21. M. Sullivan, A. M. Hopkins, J. Afonso, A. Georgakakis, B. Chan, L. E. Cram, B. Mobasher: (in preparation)
22. M. Sullivan, B. Mobasher, B. Chan, L. Cram, R. Ellis, M. Treyer, A. Hopkins: ApJ **558**, 72 (2001)

A Study of Distant Lyα Emitters in Overdense Regions

Bram P. Venemans, Huub J. A. Röttgering, and George K. Miley

Leiden Observatory, Niels Bohrweg 2, 2333 CA Leiden, The Netherlands

Abstract. Recently, we conducted a Very Large Telescope (VLT) large program to search for forming clusters by looking for overdensities of Lyα emitters around high redshift radio galaxies. In total seven proto-clusters were discovered, including a proto-cluster around the radio galaxy MRC 0316–257 at $z \sim 3.13$. This structure has an excess of Lyα emitters by a factor of 3 as compared to the field, and the derived mass is 2–5 $\times 10^{14}$ $M_\odot$. The Lyα emitters in the proto-cluster are on average bluer than Lyman Break Galaxies (LBGs). Also, the galaxies are faint (sub L_*) and small (half light radii < 1.7 kpc, which is smaller than the average size of LBGs). This might indicate that, at least a fraction of, Lyα emitters could be young ($\sim 10^6$ yr), nearly dust-free, forming galaxies.

1 VLT Large Program: Overdense Regions in the Early Universe

Forming clusters of galaxies (proto-clusters) could provide information on the formation of large scale structure in the early Universe. Since a substantial number of galaxies can be detected at the same redshift and features in the spectral energy distribution can be studied systematically by choosing the appropriate filters, they are also ideal places to study galaxy evolution. By comparing proto-cluster galaxies to field galaxies, changes in in galaxy properties as a function of environment can be studied.

An important issue is where to search for these proto-clusters. During the last decade evidence has accumulated that high redshift radio galaxies (HzRGs, $z > 2$) are forming brightest cluster galaxies at the centers of clusters or proto-clusters (for an overview, see e.g. [21]). Supporting evidence includes: (i) HzRGs are amongst the most massive ellipticals in the early Universe [9,5], (ii) they can have extreme radio rotation measures, indicative of dense hot gas [3] and (iii) radio galaxies at $0.5 < z < 1.5$ lie in moderately rich clusters [8,1,2]. A pilot project conducted on the Very Large Telescope (VLT) to search for a proto-cluster around radio galaxy PKS 1138–262 at $z = 2.16$ resulted in the detection of 15 Lyα emitters within 1000 km s^{-1} of the central radio galaxy [11,15]. This provided direct evidence that distant luminous radio galaxies can be used as tracers of proto-clusters.

We initiated a VLT large program to search for Lyα emitters around HzRGs. To select the most likely progenitors of cD ellipticals, the radio galaxies in this program satisfied the following criteria: large radio luminosities, and bright optical and IR continua. To be able to find Lyα emitters, the redshift of the radio

Table 1. Overview of the observed radio galaxy fields. The Table shows the name of the radio galaxy, its redshift, the number of candidate Lyα emitters in the imaging ('IMG'), the number of spectroscopically confirmed Lyα emitters, excluding the radio galaxy ('SPC') and the velocity dispersion of the confirmed cluster members.

Name of RG	z	IMG	SPC	Δv (km s^{-1})	Notes
MRC 2048–272	2.06	16	2	N/A	High extinction
PKS 1138–262	2.16	70	14	~ 1000	See [11,15,10]
MRC 0052–241	2.86	73	37	~ 900	
MRC 0943–242	2.92	~ 70	29	~ 800	
MRC 0316–257	3.13	85	31	~ 625	
TN J2009–3040	3.15	20	11	~ 470	Radio loud quasar from [4]
TN J1338–1942	4.10	50	32	~ 350	See [20] for more details
TN J0924–2201	5.19	~ 20	6	N/A	

galaxies had to be suitable for Lyα imaging with the available VLT/FORS narrowband filters. Two objects at high redshift (radio galaxies at $z = 4.1$ and one at $z = 5.2$) were added and for which a custom narrowband filter was purchased.

In total, 20 nights on the VLT and 2 nights on Keck in 2001–2003 were used to observe 7 radio galaxy fields with redshifts up to 5.2. In these fields, roughly 400 candidate Lyα emitters were found, of which 162 were confirmed to be Lyα emitters near the redshift of the radio galaxy (Table 1). All six fields studied to sufficient depth turned out to be overdense in Lyα emitters, as compared to blank field surveys. The galaxy overdensities in the narrowband filters are 3–5. The structures have sizes of > 3 Mpc and masses of 10^{14-15} $M_\odot$ [10,20].

In Sect. 2, we will describe the observations of one of the radio galaxy fields, the field of 0316–257, in detail. In Sect. 3, we will discuss the properties of the Lyα emitters.

2 An Overdensity Around MRC 0316–257 at $z = 3.13$

One of the targets in our program was the radio galaxy MRC 0316–257. There were several additional reasons that this radio galaxy was chosen. First of all, it was already known that the object had two spectroscopically confirmed companions [12]. Secondly, the redshift of the radio galaxy ($z = 3.13$) allows for an efficient search for Lyman Break Galaxies (LBGs) and for [OIII] λ 5007 Å emitters using a K-band narrowband filter mounted on an infrared camera.

2.1 Imaging and Spectroscopy

The field surrounding 0316–257 was imaged in a narrowband for 6.5 hours, and 1.3 hours in both V and I with the VLT/FORS2 camera. As part of an imaging program with the *Hubble Space Telescope*, the field was observed for 3 orbits with the Advanced Camera for Surveys (ACS) in the F814W filter. For each object detected in the narrowband image, the line flux, continuum strength and continuum slope and their uncertainties were calculated using the measured narrowband and V and I magnitudes. Subsequently, with the derived line flux and continuum strength, the rest-frame equivalent width (EW_0) of the objects was computed. Following [20], galaxies with a rest-frame equivalent width $EW_0 > 15$ Å and a significance $\Sigma \equiv EW_0/\sigma_{EW_0} > 3$ were selected as candidate Lyα emitters. 85 objects were marked as candidate emitters. Follow-up FORS2 spectroscopy of 40 candidate Lyα emitters revealed 33 emission line galaxies of which 2 turned out to be [OII] λ 3727 Å emitters with $z \sim 0.35$ and 31 were confirmed to be Lyα emitters at a redshift of ~ 3.13.

2.2 A Proto-Cluster at $z = 3.13$

The velocity distribution of the 31 confirmed emitters has a FWHM of 1510 km s^{-1}, which is much smaller than the width of the filter (FWHM ~ 3500 km s^{-1}), and the peak of the velocity distribution is located within 200 km s^{-1} of the redshift of the radio galaxy. The volume density of Lyα emitters within our narrowband filter is a factor 3.0 ± 0.8 higher as compared to the volume density of field emitters (e.g. [7]). Because the structure is not virialized (taking the velocity dispersion as a typical velocity, it would take a galaxy at least 5 Gyr to cross the structure, while the age of the Universe at $z = 3.13$ is only 2.2 Gyr), the virial theorem cannot be used to calculate the mass of the proto-cluster. Therefore, following [17], we use the volume occupied by the proto-cluster, the mean density of the Universe at $z = 3.13$, the measured galaxy overdensity and the bias parameter, the computed mass is in the range 2–5 $\times 10^{14}$ $M_\odot$. Because the structure does not seem to be bound in the FORS images, this mass estimate is a lower limit.

3 Properties of the Lyα Emitters

Sizes Including the radio galaxy, 19 of the 32 confirmed emitters were located within the ACS image. Two of these emitters were not detected in the image to a depth of $I_{814} > 27.1$ arcsec^{-2} (3σ). The emitters that do have a counterpart in the ACS image show a range of morphologies. Four of them, including the radio galaxy, appear to consist of several small (half light radii < 1.8 kpc) clumps, six are single, resolved objects and the rest (seven) are unresolved, single objects. The half light radii of the single objects range between < 0.8 and ~ 1.7 kpc. Interestingly, this is smaller than the average size of LBGs at a redshift of ~ 3 [6], which is 2.3 kpc.

Spectra The line full width half maximum of the Lyα emission line ranges from 120 km s^{-1} to 800 km s^{-1}, with a median of ~ 260 km s^{-1}. One emitter has a FWHM of almost 2500 km s^{-1}, and is likely to harbor an AGN. Emitters

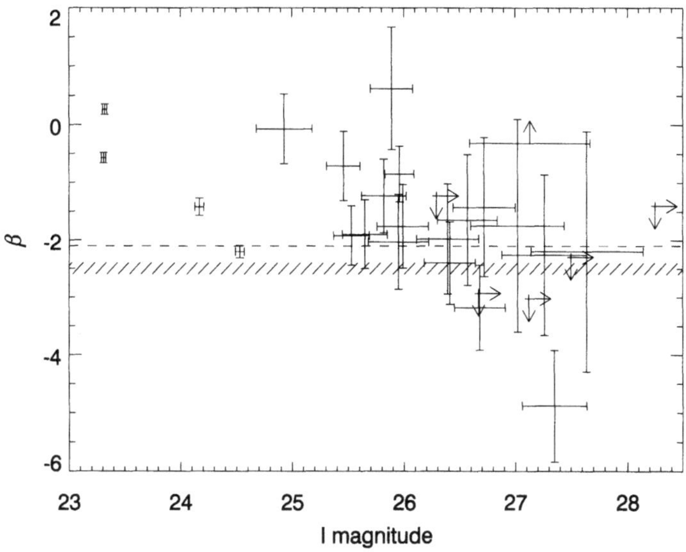

Fig. 1. I band magnitude plotted against UV continuum slope β. The dashed line represents the color of an unobscured, continuously star forming galaxy with an age $> 10^6$ yr [13]. The area with parallel lines indicate the color of a young, unobscured starburst.

with a high signal-to-noise spectrum show an asymmetric line profile with an absorbed blue wing. In those cases, the lines were fitted by a Gaussian emission line with one or more Voigt absorption profiles. The inferred column densities of the absorbers are in the range of 10^{14}–10^{15} cm^{-2}. Taking the half light radius as measured in the VLT narrowband image as the radius of the object, the amount of projected neutral HI is in the range of 5–400 $M_\odot$.

UV colors Characterizing the spectral energy distribution as $f_\lambda \propto \lambda^\beta$, the UV continuum slope β of the emitters was computed. In Fig. 1 the slope β is plotted against the I magnitude of the confirmed emitters. Excluding the radio galaxy and the emitter containing an AGN, the median UV continuum slope of the confirmed emitters $\beta = -1.65$. Also, the color of a flux limited sample was determined. This sample contained 23 candidate Lyα emitters with a Lyα flux $> 1.5 \times 10^{-17}$ erg s^{-1} cm^{-2}, of which 20 are confirmed. Again excluding the radio galaxy and AGN, the median slope of the remaining emitters is $\beta = -1.65$, the same as the median slope of the confirmed emitters. This is bluer than the average LBG with Lyα in emission, which has a slope of -1.09 ± 0.05 [16].

Models of galaxies with active star formation predict an UV continuum slope around $\beta = -2.1$ for an unobscured, continuously star forming galaxy with an age between a few $\times 10^6$ yr and over a Gyr [13]. 18 out of the 27 (67%) confirmed Lyα emitters for which the slope are measured, have, within their 1σ errors, colors consistent with being an unobscured star forming galaxy. Of those, 15

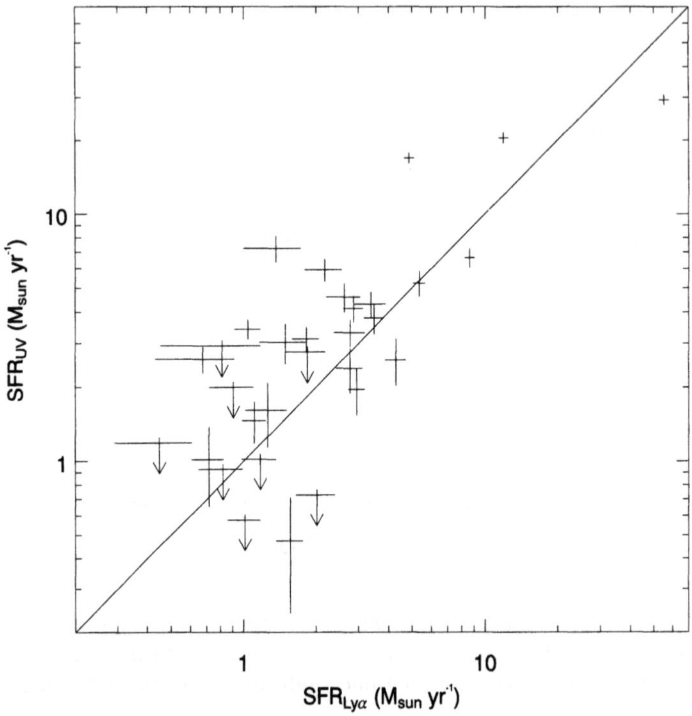

Fig. 2. Star formation rate as derived from the Lyα line luminosity, assuming Case B recombination, and the star formation rate derived from the UV continuum at 1500 Å.

(56%) have (within 1σ) such blue colors ($\beta \sim -2.5$) that they could be star-forming galaxies with very young ages ($\sim 10^6$ yr [13]).

Star formation rates The star formation rates (SFR) of the emitters were estimated in two ways. The first method was to derive a Hα luminosity from the Lyα luminosity (assuming Case B recombination) and convert the Hα luminosity to a SFR following [14]. The second method was to compute the SFR from the UV luminosity density at a rest-frame wavelength of 1500 Å [14]. The Lyα line inferred SFR of the confirmed emitters are 0.5–20 $M_\odot$ yr^{-1}, with an average of 2.6 $M_\odot$ yr^{-1}. In Fig. 2 the SFR derived from the Lyα flux is plotted against the UV continuum SFR rate. On average, both methods to calculate the SFR give roughly the same result.

4 Summary and Discussion

Last year we finished a large program to search for proto-clusters around HzRGs. Around all seven well studied radio galaxies overdensities of Lyα emitters were found. This includes the discovery of 31 confirmed Lyα emitters around the radio galaxy MRC 0316–257 at $z \sim 3.13$.

The Lyα emitters in this proto-cluster have blue continuum slopes. Furthermore, the emitters are small and faint compared to LBGs. Besides the radio galaxy, only one emitter is brighter than m_* at $z \sim 3$ [17], the other 30 confirmed emitters have a continuum luminosity fainter than L_*. Given these properties, a fraction of the Lyα emitters might be very young, forming galaxies in their first starburst phase, without significant dust absorption. Currently, we are investigating the other radio galaxy fields to confirm this.

At $z \sim 3$ the star formation rate density (SFRD) derived from observations of LBGs with $R \sim 27$ is 0.0184 ± 0.0034 $M_\odot$ yr^{-1} Mpc^{-3} [18]. With the same magnitude limit applied to the confirmed emitters, we find a SFRD of 0.0109 ± 0.0002 $M_\odot$ yr^{-1} Mpc^{-3} in the volume probed by the narrowband filter. Because no correction has been made for incompleteness and only confirmed emitters were considered, this is a lower limit on the SFRD. Studies of LBGs showed that only 20–25% of the star forming, UV bright galaxies at $z \sim 3$ have a Lyα line with an equivalent width > 20 Å [19,16]. Correcting for this incompleteness, then the SFRD around MRC 0316–257 is roughly 2.4–3.0 times higher than in the field, in agreement with the overdensity computed from volume density of the Lyα emitters.

References

1. P. Best: MNRAS **317**, 720 (2000)
2. P. Best, M. Lehnert, G. Miley, H. Röttgering: MNRAS **343**, 1 (2003)
3. C. Carilli, H. Röttgering, R. van Ojik, G. Miley, W. van Breugel: ApJS **109**, 1 (1997)
4. C. De Breuck, W. van Breugel, H. Röttgering, G. Miley: A&AS **143**, 303 (2000)
5. C. De Breuck, W. van Breugel et al.: AJ **123**, 637 (2002)
6. H. Ferguson, M. Dickinson et al.: ApJL (2003) (accepted; astro-ph/0309058)
7. J. Fynbo, C. Ledoux, P. Möller, B. Thomsen, I. Burud: A&A **407**, 147 (2003)
8. G. Hill, S. Lilly: ApJ **367**, 1 (1991)
9. M. Jarvis, S. Rawlings et al.: MNRAS **326**, 1585 (2001)
10. J. Kurk, L. Pentericci et al.: A&A (2004) (submitted)
11. J. Kurk, H. Röttgering et al.: A&A **358**, L1 (2000)
12. O. Le Févre, J. Deltorn, D. Crampton, M. Dickinson: ApJ **471**, L11 (1996)
13. C. Leitherer, D. Schaerer et al.: ApJS **123**, 3 (1999)
14. P. Madau, L. Pozzetti, M. Dickinson: ApJ **498**, 106 (1998)
15. L. Pentericci, J. Kurk et al.: A&A **361**, L25 (2000)
16. A. Shapley, C. Steidel, M. Pettini, K. Adelberger: ApJ **588**, 65 (2003)
17. C. Steidel, K. Adelberger et al.: ApJ **492**, 170 (1998)
18. C. Steidel, K. Adelberger, M. Giavalisco, M. Dickinson, M. Pettini: ApJ **519**, 1 (1999)
19. C. Steidel, K. Adelberger et al.: ApJ **532**, 170 (2000)
20. B. Venemans, J. Kurk et al.: ApJ **569**, L11 (2002)
21. B. Venemans, G. Miley, J. Kurk, H. Röttgering, L. Pentericci: The Messenger **111**, 36 (2003)

Clustering and Proto-Clusters in the Early Universe

Huub Röttgering[1], Carlos De Breuck[2], Emanuele Daddi[2], Jaron Kurk[3], George Miley[1], Laura Pentericci[4], Roderik Overzier[1], and Bram Venemans[1]

[1] Leiden Observatory, Niels Bohrweg 2, NL-2333 CA Leiden, The Netherlands
[2] European Southern Observatory, Garching bei München, Germany
[3] INAF, Osservatorio Astrofisico di Arcetri, Firenze, Italy
[4] Dipartimento di Fisica, Università degli Studi Roma Tre, Italy

Abstract. The clustering properties of clusters, galaxies and AGN as a function of redshift are briefly discussed. It appears that extremely red objects at $z \sim 1$, and objects with $J - K > 1.7$ and photometric redshifts $2 < z_{phot} < 4$ are highly clustered, indicating that a majority of these objects are the progenitors of nearby ellipticals. Similarly clustered seem luminous radio galaxies at $z \sim 1$, indicating that these objects comprise a short lived phase in the lifetime of these red objects. The high level of clustering furthermore suggests that distant powerful radio galaxies (e.g. $z > 2$) might be residing in the progenitors of nearby clusters – proto-clusters. A number of observational projects targetting fields with distant radio galaxies, including studies of Lyα and Hα emitters, Lyman break galaxies and (sub)mm and X-ray emitters, all confirm that such radio galaxies are indeed located in proto-clusters. Estimates of the total mass of the proto-clusters are similar to the masses of local clusters. If the total star formation rate which we estimate for the entire proto-clusters is sustained up to $z \sim 1$, the metals in the hot cluster gas of local clusters can easily be accounted for.

1 Introduction

An important theme in astrophysics is to understand the nature of clustering of galaxies, active galaxies and even clusters of galaxies. Observationally, two different methods have been employed to measure clustering and its evolution with redshift.

A first method is to measure the level of clustering in various distinct populations of objects. A number of statistical methods have been used to quantify clustering including counts in cells, nearest-neighbor statistics and the two-point correlation function. By introducing the "bias parameter", which basically measures the clustering of light compared to the underlying clustering of dark matter, the clustering of objects in the "local" universe can be directly linked to the measured fluctuations in the cosmic microwave background. For the art of understanding how galaxies and AGN form, an interesting way to rephrase aspects of this problem is to understand the evolution of this bias parameter for various classes of galaxies and AGN. In this contribution we will consider measurements of clustering of a number of different classes of galaxies and AGN and how these evolve as a function of redshift. Briefly, some implication for the understanding of the nature of these objects will be discussed.

A second method is to actually find and study ensembles of galaxies in the early Universe for which the case can be made that they will evolve into present day clusters. This will not only constrain the formation and evolution of clusters, but will also allow detailed studies of how the environment in which galaxies are born influences their subsequent evolution. To find such early clusters, "proto-clusters", we have a number of programmes to observe fields centered on extremely luminous radio galaxies with redshifts in the range $2 < z < 5$. In all the well studied fields, it seems that there are significant excesses of galaxies compared to the field, suggesting masses of the proto-clusters in the range of $10^{14} - 10^{15}$ $M_{\odot}$.

2 Clustering of Galaxies and AGN

A well established way of characterizing the clustering of objects is through measuring their spatial two-point correlation function (e.g. Peebles 1980). In the usual case of a limited number of objects with well determined distances for which clustering is being assessed, only the correlation length r_0 can be constrained. This parameter indicates the spatial scale at which the chance of detecting an object at a distance from a given object is a factor of two more than for random distribution. Often, as in the case of new imaging surveys, the distances to the objects are not known and only the angular clustering can be determined. If the redshift distribution of such a population is known, the spatial correlation length can be derived using the well known Limber's equation (e.g. Daddi et al. 2000 and references therein).

A consideration of the correlation length for a number of samples at different redshifts of galaxies and AGN can be found in Overzier et al. (2003) and Röttgering et al. (2003). At low redshift, the most clustered objects are clusters of galaxies, followed by elliptical galaxies and spirals. Detailed measurements of the local correlation functions have recently been carried out on the basis of the 2dF survey (Hawkins et al. 2003). In the last few years, a number of interesting results have been obtained which constrain the correlation length of various types of objects at $z \sim 1$.

A major achievement of the VLA was the production of the NVSS radio catalogue, which now contains 2 million radio sources observed at 1.4 GHz (Condon et al. 1998). Recently, two groups have measured the angular correlation function of NVSS sources (Overzier et al. 2003, Blake and Wall 2002), and determined the correlation lengths using the redshift distribution from Dunlop and Peacock (1990). In their analysis Overzier et al. concluded that the most luminous radio sources at $z \sim 1$ are clustered at about a level approaching that of local clusters. This may indicate that these sources are preferentially located in the progenitors of nearby massive clusters, while the less luminous sources would then be associated with a field population.

The availability of large area infrared surveys made it possible to study the clustering of extremely red objects (EROs). The definition of the color cut that distinguishes EROs from the general population is often taken to be $R - K > 5$.

The main two reasons that these objects are so red are that (i) they straddle the 4000 Å break of a $z \sim 1$ elliptical, or that (ii) the dust in a starburst galaxy has absorbed most of the optical light (see Cimatti et al. 2003 and references therein). Various clustering measurements for EROs gave correlation lengths of similar magnitude as found for the luminous radio sources (Daddi et al. 2000, Roche et al. 2002, Firth et al. 2002). Taking into account that the lifetime of radio sources is relatively small ($\sim 10^7$ years), the inferred space density of both EROs and radio galaxies are similar (Mohan et al. 2002; Willott et al. 2001). This all has the interesting implication that luminous radio sources and EROs might be similar objects at a different stage of their evolution.

Measurements of clustering of optical quasars at $z \sim 1$ has been done to an unprecedented accuracy by the 2dF team (Croom et al. 2001). Interestingly, the correlation lengths that are found ($\sim 4 - 5$ Mpc) are significantly smaller than that for the powerful radio galaxies. This has the immediate implication that these optical quasars can not evolve into or descent from luminous radio sources. Optical quasars are therefore probably associated with a field population of objects with modest mass black holes as supposed to the clustered radio galaxies with very massive black holes.

With the availability of wide and deep surveys conducted with 10-m class optical telescopes, large numbers of Lyman break galaxies (LBGs) have been detected. With these samples of LBGs it was found that r_0 was constant over the measured redshift range of $3 < z < 5$ (Hamana et al. 2004). This is in agreement with hierarchical models of galaxy formation that state that the highest redshift galaxies are the very biased tracers of the underlying dark matter distribution.

The FIRES survey (Franx et al. 2000) is a very deep survey with the VLT that is designed to obtain a sample of high redshift galaxies selected in the rest-frame optical rather than the restframe UV, like the LBGs. For this programme, in total 100 hours were spent on observing the Hubble Deep Field South, reaching in each of the infrared bands J, H and K, limiting magnitudes of 26.0, 24.9, and 24.5, respectively (Labbé et al. 2003). One of the interesting discoveries was the presence of a population of galaxies with $J - K > 2.3$. Although these objects had photometric redshifts $z > 2$, in general there was only limited overlap with the LBG population (Franx et al. 2003). This is because these objects are too faint in the optical to detect the Lyman break according to the usual criteria and/or are too red. Although the number of such galaxies is limited, an attempt to measure the clustering was presented by Daddi et al. 2003. Down to $K > 24$, it was found that the galaxies with a $2 < z_{phot} < 4$ and $J - K > 1.7$ had a correlation length a factor of 2-3 larger than LBG galaxies. Although the statistics are clearly somewhat limited, the objects with $J - K > 2.3$ seem to be even more clustered. Models of the evolution of the clustering of ellipticals by for example Kauffmann et al. (1999) predicted that the correlation length for local elliptical and their progenitors should be at the same high level. Since this is what we observe, this is an important indication that these red objects are indeed the progenitors of local ellipticals. Further evidence for this comes from considerations related to their number density, masses and sizes (van Dokkum et al. 2004, submitted).

3 Radio Galaxies at the Centers of Proto-Clusters

As already noted, the powerful radio galaxies at $z \sim 1$ seems to have large correlation lengths, suggesting that they are located in progenitors of local clusters. This suggests that proto-clusters could be found in fields containing distant and powerful radio galaxies. There is significant additional supporting evidence that radio galaxies can be associated with proto-clusters, including (i) they are amongst the most massive objects at high redshifts, (ii) they have large Lyα halo's whose outer regions are consistent with originating from a cooling flow, (iii) 20 % have high rotation measures as measured with the VLA, indicating the presence of a dense medium. We therefore started a project to study fields centered at distant powerful radio galaxies with the aim of finding proto-clusters at a range of redshifts. A second aim was to study the various classes of galaxies in these proto-clusters.

4 The Proto-Cluster Associated with 1138-262 at $z = 2.2$

The first radio galaxy for which we studied its associated proto-cluster in great detail was 1138-262 at $z = 2.2$ (Kurk et al. 2003 and references therein) For a number of reasons this was among the very best objects to study. It is among the brightest object both at K-band and at radio wavelength. It has the highest rotation measure (6200 rad/m^2) of a well studied sample of 80 $z > 2$ radio galaxies. The morphology both in the radio, optical and near infrared is very clumpy, consistent with simulation of forming brightest cluster galaxy in which many star forming regions are merging together to ultimately form a massive galaxy. With a size of almost 200 kpc, the very luminous Lyα halo represents a significant reservoir of gas from which part of the stars that will make up the final system can be formed. The first programme was to carry out deep narrow band imaging to obtain candidate Lyα emitting galaxies. This resulted in 70 candidates. As compared to the field the overdensity was derived to be at least a factor of two. Subsequent spectroscopy resulted in 14 confirmed galaxies at the redshift of the radio galaxy. Deep narrow band imaging in the infrared combined with infrared spectroscopy resulted in the detection of 7 confirmed Hα emitting galaxies. The K-band imaging data combined with the optical contained 44 galaxies with $I - K > 4.3$. Unfortunately, these objects have no measured redshift – this is notoriously difficult –, but their surface density peaks at the location of the radio sources. Finally, we looked at the X-ray emitting galaxies using a deep Chandra image. On the basis of either spectroscopy or colour information, for 6 objects the case could be made that they are at the redshift of the proto-cluster. The spatial distribution of the various classes is given in Figure 1. In Figure 2, the surface density of these classes as a function of distance from the radio galaxy is given. An important point apparent from these two figures, is that

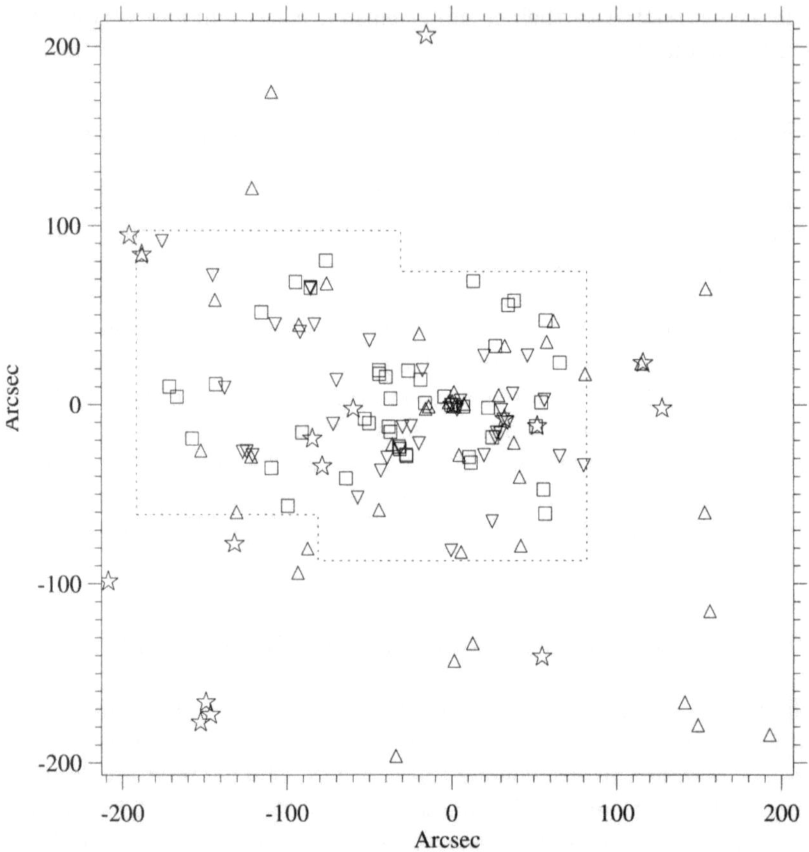

Fig. 1. The spatial distribution of candidate cluster galaxies in the field of the powerful radio galaxy 1138-262 ($z = 2.2$). Indicated are the candidate Hα emitters with downward pointing triangles, candidate Lyα emitters with upward pointing triangles and extremely red objects with squares. The X-ray emitters are indicated with stars. The radio galaxy is located at the origin. The dotted line indicates the boundaries of the two fields of the IR observations with ISAAC/VLT within which Hα emitters and EROs have been searched for. Note that spectroscopy yielded 15 confirmed Lya and 7 confirmed Ha emitters.

the distribution of Hα galaxies seems much more concentrated than the Lyα emitting galaxies. Further differences are that the population of Hα emitters on average have brighter K-band magnitudes and as a whole have a lower velocity dispersion. We interpreted this as the population of Hα emitting galaxies being older and dustier and further in the process of having their orbits virialized in the "proto-cluster" potential. It seems that the seeds of the morphology-density relation are already in place at $z = 2.2$. Following the analysis of Steidel et al. (1998) as they carried out for the high "redshift spike" at $z = 3.1$, the mass associated with the over density is in the range $10^{14} - 10^{15} M_{\odot}$. And indeed this is what is expected for a progenitor of a nearby massive cluster.

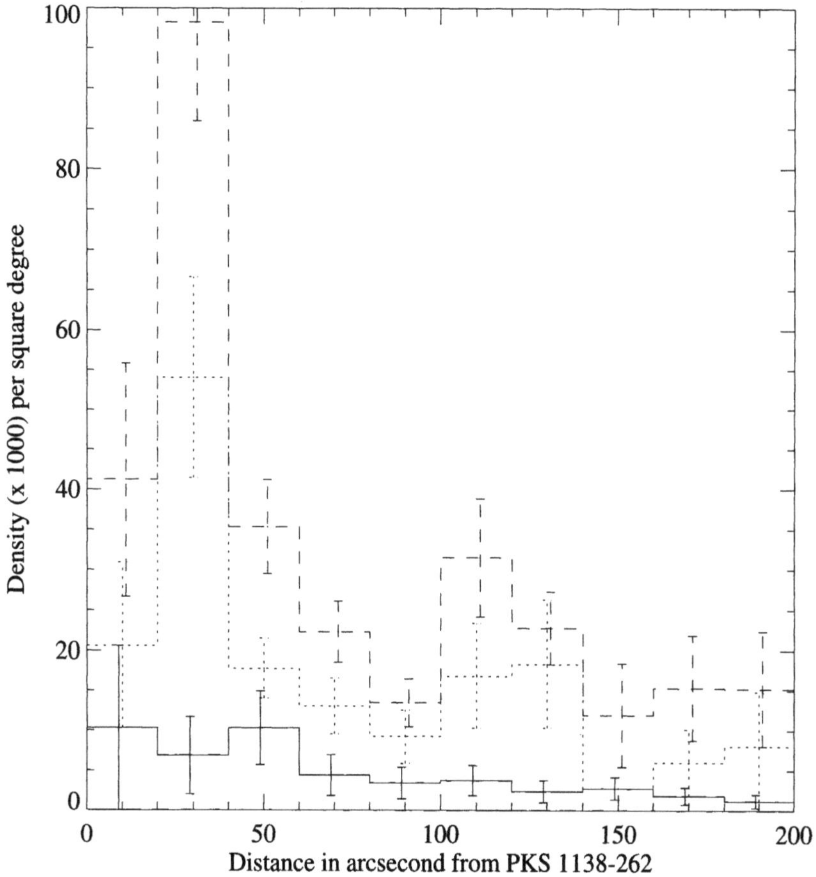

Fig. 2. The surface density of candidate cluster galaxies as a function of distance from the powerful radio galaxy 1138-262 ($z = 2.2$). Indicated are candidate Lyα emitters with black lines, candidate Hα emitters with dotted lines, and extremely red objects with dashed lines.

5 Towards a Sample of Proto-Clusters

The results on the proto-cluster of 1138-262 prompted us to extend this study to more objects. We initiated an ESO large programme to search for Lyα emitting galaxies in potential proto-clusters around radio galaxies with redshifts in the range $2 < z < 5$. An account of the results of this programme has been presented in this conference by Venemans et al. The main conclusion is that all of the 7 radio galaxies that have been studied in sufficient detail show over densities and resulting masses similar to that found for 1138-262.

6 Hubble Space Telescope ACS Imaging

In the course of our ESO large programme, we studied the region of the powerful radio galaxy 1338-193 at $z = 4.1$ and subsequently found the highest redshift group of galaxies. In total we have 50 candidate Lyα emitters distributed over two partly overlapping 7×7 arcminute field of views of the FORS instrument of the VLT. Of these, 32 are spectroscopically confirmed. This field was studied in detail using data from the new Advanced Camera for Surveys on the Hubble space telescope (Miley et al. 2004). It was observed in three filters (F475m, F625m, F775m), for optimum sensitivity for detecting LBG galaxies at $z \sim 4$. Comparisons with the surface densities in the Hubble Deep Field and the Subaru deep field led us to conclude that there are a factor of 2 more $z \sim 4$ LBGs than in a random field. Given that the three filters in principle select objects over the entire redshift range of $3.5 < z < 4.5$, the actual spatial over density compared to the field is likely to be well in excess of ten.

7 mm/submm Imaging with JCMT and IRAM

Both the LB galaxies and the Lyα emitters are discovered at optical wavelengths, which naturally biases the obtained samples to galaxies that are relatively unobscured. With (sub)mm telescopes, a number of galaxy fields have been observed with the aim of detecting very dusty and potentially very obscured galaxies located in the proto-cluster. Using the SCUBA array mounted on the JCMT, 7 radio galaxy fields have been observed. An excess of dusty galaxies of about a factor of two compared to the field has been found (Stevens et al. 2003). De Breuck et al. (2004, submitted) carried out a detailed investigation of the field of the radio galaxy 1338−192 at a redshift of $z = 4.1$ using MAMBO at the IRAM 30m. Combined with deep optical and radio data 5 dusty objects were found with properties consistent with being located at the distance of the proto-cluster. This is further evidence for the reality of an excess of starburst galaxies in distant radio galaxy fields.

8 Discussion

Both from the analysis of the correlation function of galaxies and AGN and the observational data, it seems that the evidence is now very convincing that distant powerful radio galaxies are located in "proto-clusters". It is even plausible that every proto-cluster has gone through a radio-active phase as it turns out that the space density of powerful radio galaxies is comparable to the space density of local clusters, taking into account the short lifetime of the radio activity ($10^6 - 10^7$ years). Again following the analysis of Steidel et al., the associated masses of the structures are in the range of $10^{14} - 10^{15}$ $M_\odot$. A simple estimate of the total star formation rate of all the galaxies combined in the proto-clusters is in excess of 20,000 $M_\odot$ yr^{-1}. If such a high rate is sustained, then the high metal content in the hot gas in $z < 1$ clusters can be easily accounted for.

References

1. Blake, C. & Wall, J. 2002, MNRAS, 329, L37
2. Cimatti, A., Daddi, E., Cassata, P., et al. 2003, A&A, 412, L1
3. Condon, J. J., Cotton, W. D., Greisen, E. W., et al. 1998, AJ, 115, 1693
4. Croom, S. M., Shanks, T., Boyle, B. J., et al. 2001, MNRAS, 325, 483
5. Daddi, E., Cimatti, A., Pozzetti, L., et al. 2000, A&A, 361, 535
6. Daddi, E., Röttgering, H. J. A., Labbé, I., et al. 2003, ApJ, 588, 50
7. Dunlop, J. & Peacock, J. 1990, MNRAS, 247, 19
8. Firth, A. E., Somerville, R. S., McMahon, R. G., et al. 2002, MNRAS, 332, 617
9. Franx, M., Labbé, I., Rudnick, G., et al. 2003, ApJ, 587, L79
10. Franx, M., Moorwood, A., Rix, H., et al. 2000, The Messenger, 99, 20
11. Hamana, T., Ouchi, M., Shimasaku, K., Kayo, I., & Suto, Y. 2004, MNRAS, 347, 813
12. Hawkins, E., Maddox, S., Cole, S., et al. 2003, MNRAS, 346, 78
13. Kauffmann, G., Colberg, J. . M., Diaferio, A., & White, S. D. M. 1999, MNRAS, 307, 529
14. Kurk, J., Röttgering, H., Pentericci, L., Miley, G., & Overzier, R. 2003, New Astronomy Review, 47, 339
15. Labbé, I., Franx, M., Rudnick, G., et al. 2003, AJ, 125, 1107
16. Miley, G. K., Overzier, R. A., Tsvetanov, Z. I., et al. 2004, Nature, 427, 47
17. Mohan, N. R., Cimatti, A., Röttgering, H. J. A., et al. 2002, A&A, 383, 440
18. Overzier, R. A., Röttgering, H. J. A., Rengelink, R. B., & Wilman, R. J. 2003, A&A, 405, 53
19. Peebles, P. J. E. 1980, The large-scale structure of the universe (Research supported by the National Science Foundation. Princeton, N.J., Princeton University Press, 1980. 435 p.)
20. Röttgering, H., Daddi, E., Overzier, R., & Wilman, R. 2003, New Astronomy Review, 47, 309
21. Roche, N. D., Almaini, O., Dunlop, J., Ivison, R. J., & Willott, C. J. 2002, MNRAS, 337, 1282
22. Stevens, J. A., Ivison, R. J., Dunlop, J. S., et al. 2003, Nature, 425, 264
23. Willott, C. J., Rawlings, S., & Blundell, K. M. 2001, MNRAS, 324, 1

Fact: Many SCUBA Galaxies Harbour AGNs

D.M. Alexander[1], F.E. Bauer[1], S.C. Chapman[2], I. Smail[3], A.W. Blain[2], W.N. Brandt[4], and R.J. Ivison[5]

[1] Institute of Astronomy, Madingley Road, Cambridge CB3 0HA, UK
[2] California Institute of Technology, Pasadena, California 91125, USA
[3] Institute for Computational Cosmology, University of Durham, South Road, Durham DH1 3LE, UK
[4] Department of Astronomy and Astrophysics, Pennsylvania State University, 525 Davey Laboratory, University Park, PA 16802, USA
[5] Astronomy Technology Centre, Royal Observatory, Blackford Hill, Edinburgh EH9 3HJ, UK

Abstract. Deep SCUBA surveys have uncovered a large population of ultra-luminous galaxies at $z > 1$. These sources are often assumed to be starburst galaxies, but there is growing evidence that a substantial fraction host an AGN (i.e., an accreting super-massive black hole). We present here possibly the strongest evidence for this viewpoint to date: the combination of ultra-deep X-ray observations (the 2 Ms Chandra Deep Field-North) and deep optical spectroscopic data. We argue that upward of 38% of bright ($f_{850\mu m} \geq 5$ mJy) SCUBA galaxies host an AGN, a fraction of which are obscured QSOs (i.e., $L_X > 3 \times 10^{44}$ erg s^{-1}). However, using evidence from a variety of analyses, we argue that in almost all cases the AGNs are not bolometrically important (i.e., < 20%). Thus, star formation appears to dominate their bolometric output. A substantial fraction of bright SCUBA galaxies show evidence for binary AGN activity. Since these systems appear to be interacting and merging at optical/near-IR wavelengths, their super-massive black holes will eventually coalesce.

1 Introduction

Blank-field SCUBA surveys have uncovered a large population of submillimetre (submm; $\lambda = 300$–$1000\ \mu$m) emitting galaxies (≈ 1000–10000 sources deg^{-2} at $f_{850\mu m} \approx 1$–5 mJy; e.g., [5,12,13,22,27,39]). The majority of these sources are faint at all other wavelengths, hindering source identification studies. However, due to a considerable amount of intensive multi-wavelength follow-up effort, it is becoming clear that almost all are dust-enshrouded galaxies at $z > 1$ (e.g., [17,30,41,47]). With estimated bolometric luminosities of $\approx 10^{12}$–10^{13} L$_\odot$, these galaxies outnumber comparably luminous local galaxies by several orders of magnitude.

Central to the study of submm galaxies is the physical origin of their extreme luminosities (i.e., starburst or AGN activity). If these sources are shown to be ultra-luminous starburst galaxies then their derived star-formation rates suggest a huge increase in star-formation activity at $z > 1$. Conversely, if these sources are shown to be ultra-luminous AGNs then they will outnumber comparably luminous optical QSOs by ≈ 1–2 orders of magnitude. Both of these scenarios provide challenges to models of galaxy formation and evolution.

2 Searching for AGN Activity in SCUBA Galaxies

It was initially expected that deep SCUBA surveys would identify large numbers of high-redshift ultra-luminous starburst galaxies. However, the first SCUBA source to be unambiguously identified via optical/near-IR spectroscopy was actually found to contain an obscured AGN, probably a BALQSO (i.e., the $z = 2.81$ source SMMJ02399–0136; [28,44]); subsequent moderately deep *Chandra* observations showed that this AGN is luminous at X-ray energies ($L_X \approx 10^{45}$ erg s^{-1}; [11]). Although clearly containing a powerful AGN, sensitive CO (3–2) observations of SMMJ02399–0136 suggested that star formation contributes $\approx 50\%$ of the total bolometric luminosity [25]. The lesson learnt from SMMJ02399–0136 is that the identification of an AGN at optical/near-IR wavelengths does not necessarily imply that the AGN is bolometrically dominant.

Extensive optical follow-up observations of other SCUBA sources have revealed further AGNs, a few starburst galaxies, and many sources with uncertain classifications (e.g., [17,29,41]). Although clearly successful in identifying the presence of an AGN in some SCUBA galaxies, optical spectroscopy has a number of limitations for AGN identification in all SCUBA galaxies:

- The large positional uncertainty of SCUBA sources often makes it challenging to identify an optical counterpart securely.
- A large fraction of the SCUBA sources with secure optical counterparts are optically faint ($I \geq 24$), making optical source classification challenging despite success in redshift identification (e.g., [17]).
- AGNs do not always show clear AGN signatures at optical wavelengths due to dust absorption and dilution by host-galaxy light (e.g., [20]).

Arguably the best discriminator of AGN activity is the detection of luminous hard X-ray emission (i.e., > 2 keV).[1] Hard X-ray emission is relatively insensitive to obscuration (at least for sources that are Compton thin; i.e., $N_{\rm H} < 1.5 \times 10^{24}$ cm^{-2}), and any hard X-ray emission from star formation in the host galaxy is often insignificant when compared to that produced by the AGN. Hard X-ray observations can even provide a secure AGN identification in sources where the optical signatures and counterparts are weak or even non existent (e.g., [1]). Due to their high-energy X-ray coverage, high X-ray sensitivity, and excellent positional accuracy, the *Chandra* and *XMM-Newton* observatories offer the best opportunities for the X-ray investigation of SCUBA galaxies.

3 The Promise of *Chandra* and *XMM-Newton*

The first cross-correlation studies of moderately deep X-ray surveys with deep SCUBA surveys yielded no overlap between the X-ray and submm detected

[1] Radio observations can also be useful in identifying AGNs in SCUBA galaxies; however, not all AGNs are bright at radio wavelengths, and radio emission from strong star formation can easily mask the radio emission from a radio-quiet AGN.

source populations (<10–20%; [24,26,38]).[2] This was somewhat contrary to expectations given that some submm galaxies clearly host an AGN. However, in retrospect, considering that the two most obvious AGNs identified via optical spectroscopy were both detected by *Chandra* (i.e., SMMJ02399–0134 and SMMJ02399–0136; [11,41]), AGN identification with moderately-deep X-ray observations had not performed any worse than that achieved via optical spectroscopy. These early studies concluded that a reasonable fraction of bolometrically dominant AGNs can only be present in the submm galaxy population if they are obscured by Compton-thick material. In this scenario, almost all of the direct emission from the AGN is obscured at X-ray wavelengths.

Later studies with the Chandra Deep Field-North (CDF-N; 1 Ms exposure: [15], 2 Ms exposure: [2]) survey showed that a significant fraction of submm galaxies are detected in ultra-deep X-ray observations [3,7]. In the 2 Ms *Chandra* exposure, seven (> 36%) bright SCUBA galaxies ($f_{850\mu m} \geq 5$ mJy; S/N>4) had an X-ray counterpart in a 70.3 arcmin^2 region centred around the CDF-N aimpoint [3].[3] At the time of this study, complete SCUBA observations across the 70.3 arcmin^2 region were not available and only a lower limit on the X-ray-submm fraction could be placed. The X-ray detected submm galaxy fraction is 54% when the most complete SCUBA observations of [14] are used.

Figure 1 shows the X-ray detected submm galaxy source density versus X-ray flux for the X-ray-submm cross-correlation studies to date. Clearly, the detection of significant numbers of submm galaxies at X-ray energies requires deep or wide-area observations. For example, ≈ 15 ($\approx$ 10–20%) SCUBA sources should have X-ray counterparts in the forthcoming 0.25 deg^2 SHADES survey (P.I. J. Dunlop) of the *XMM-Newton* Subaru Deep Survey (XMM-SDS) region, even though the X-ray coverage is only moderately deep (50–100 ks *XMM-Newton* exposures; P.I. M. Watson).

4 The Fraction of Bright SCUBA Galaxies Hosting an AGN

On the basis of radio constraints, the expected X-ray luminosity from star formation in bright submm galaxies is $L_X = 10^{42}$–10^{43} erg s^{-1} [3,9]. Since it is possible to detect sources of this luminosity out to $z \approx 1$–3 with the 2 Ms CDF-N survey, the detection of X-ray emission from a submm galaxy does not necessarily imply that it hosts an AGN. Indeed, two of the seven X-ray detected submm galaxies in [3] have X-ray properties consistent with those expected from star formation activity (i.e., soft and comparatively low luminosity X-ray emission

[2] Further cross-correlation studies with moderately deep X-ray observations have revealed some overlap [4,6,30,46]. This is probably due to the larger areal coverage of the SCUBA observations.

[3] The X-ray sources were matched to radio-detected SCUBA and radio-undetected SCUBA sources using 1″ and 4″ search radii, respectively; the respective probabilities of projected chance associations are <1% and 4% per SCUBA source.

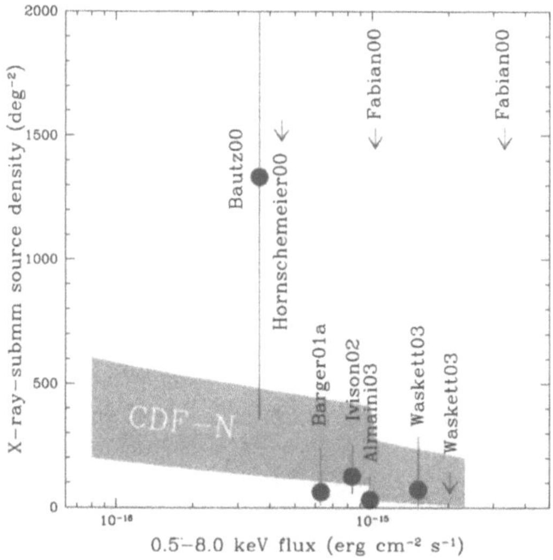

Fig. 1. X-ray detected submm galaxy source density versus X-ray flux. The shaded region corresponds to the constraints from the CDF-N studies. The data points and error bars correspond to the constraints placed by different X-ray-submm cross-correlation studies (see references); the arrows indicate upper limits. The flux limits are calculated assuming $\Gamma = 1.0$, and a 5 (*Chandra*) or 20 (*XMM-Newton*) count detection threshold for each X-ray survey; the flux limits are corrected for gravitational lensing amplification where necessary.

that is correlated with the radio emission; see §3.2 in [3]). With X-ray luminosities of $\approx 4 \times 10^{42}$ erg s^{-1}, these would be the most X-ray luminous starburst galaxies known. The other five X-ray detected submm galaxies have X-ray properties consistent with those of obscured AGNs [i.e., hard ($\Gamma < 1.0$), luminous ($L_X > 10^{43}$ erg s^{-1}) X-ray emission that is in excess of that expected from star formation]. Only one of these sources would have been classified as an AGN based on its optical properties, underlining the potency of ultra-deep X-ray observations in AGN identification. On the basis of these classifications at least 38% of bright ($f_{850\mu m} \geq 5$ mJy) SCUBA galaxies host AGNs.

Interestingly, all of the radio-detected submm galaxies are X-ray detected while relatively few of the radio-undetected submm galaxies are X-ray detected. This X-ray-radio dichotomy is unlikely to be due to AGN activity since in almost all cases the radio emission appears to be dominated by star formation. Possible explanations for the X-ray-radio dichotomy are

1. Some of the radio-undetected submm galaxies may be spurious. Due to the steep number counts at submm wavelengths most SCUBA sources are detected at a comparatively low significance.
2. The radio emission from the radio-undetected submm galaxies may be extended and therefore "resolved out" in the radio observations.

3. The radio-undetected submm galaxies might contain a cooler dust component, detected at submm wavelengths but corresponding to lower bolometric and radio luminosities when compared to hotter star-formation regions (e.g., cirrus emission; [13,23]).
4. The radio-undetected submm galaxies may be essentially the same as the radio-detected submm galaxies but lie at higher redshifts.

A combination of these effects might explain the X-ray-radio dichotomy of submm galaxies. Explanation 1 is unlikely to affect significantly the results of [3] since only SCUBA sources with S/N>4 were investigated. Explanations 2 and 3 would imply that the radio-undetected submm galaxies are different beasts to the radio-detected submm galaxies and might not typically host AGNs. Explanation 4 justifies the non detection of the radio-undetected submm galaxies at X-ray and radio wavelengths as due to a lack of sensitivity. Under this assumption in particular, the fraction of bright SCUBA galaxies hosting AGNs could be considerably higher than 38% (i.e., $\approx$ 75% if explanation 4 is correct and half of all bright submm galaxies are radio detected).

The closest local analogs to SCUBA galaxies are ultra-luminous infrared galaxies (ULIRGs: $L_{8-1000\mu m} > 10^{12}$ L$_\odot$; [37]). Based on spectroscopic classifications (optical: [43], mid-IR: [32]), it has been found that the fraction of ULIRGs hosting an AGN increases with luminosity [from $\approx$ 25% ($L_{8-1000\mu m} = 10^{12}$–$10^{12.3}$ L$_\odot$) to $\approx$ 50% ($L_{8-1000\mu m} > 10^{12.3}$ L$_\odot$)]. These AGN fractions are generally consistent with that found here for bright SCUBA galaxies.

5 What Powers Bright SCUBA Galaxies?

The ultra-deep CDF-N studies showed that a significant fraction of submm galaxies host obscured AGNs. As discussed in §2 and §3, in order to determine if these AGNs are bolometrically dominant it is crucial to determine if the obscuration toward the AGN is Compton thick or Compton thin. The most direct discrimination between Compton-thick and Compton-thin absorption is made with X-ray spectral analyses. The X-ray spectrum of a Compton-thick AGN is generally characterised by a large equivalent width Iron Kα emission line ($EW \geq 1$ keV; e.g., [8,34]) and a flat or inverted ($\Gamma < 0$) X-ray spectral slope, due to pure reflection.[4] By contrast, the X-ray spectrum of a Compton-thin AGN is usually well fitted by an absorbed power-law model and a smaller equivalent width Iron Kα emission line (generally $EW \approx$ 0.1–0.5 keV; e.g., [8,35]).

Basic X-ray spectral analyses were performed on the five submm galaxies hosting AGNs in the 2 Ms CDF-N study of [3]. Three of the sources showed the characteristics of Compton-thin absorption, one source was likely to be Compton thick, and the constraints for the other source were poor. A comparison of the X-ray-to-submm spectral slopes of these submm galaxies to those of three nearby luminous galaxies (Arp 220, a starburst galaxy; NGC 6240, an obscured AGN; 3C273, a quasar) suggested that the AGNs contributed only a small fraction

[4] A minority of Compton-thick AGNs have $EW \approx$ 0.5–1.0 keV [8].

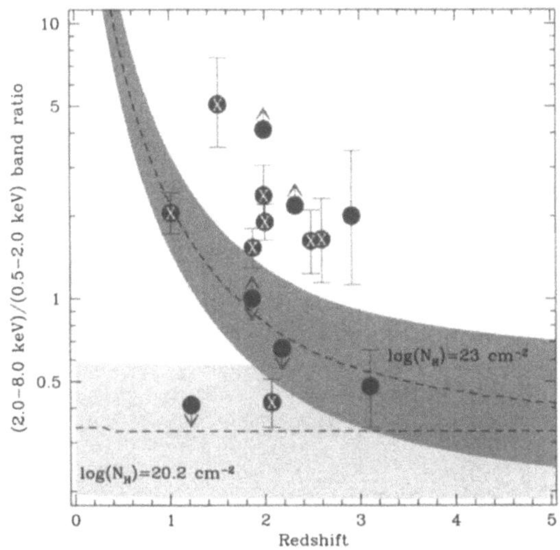

Fig. 2. X-ray band ratio versus spectroscopic redshift for the X-ray detected submm galaxies. The light and dark shaded regions show the range of expected band ratios for an unabsorbed and absorbed AGN, respectively. These regions were calculated assuming a $\Gamma = 1.7 \pm 0.5$ power law with differing amounts of absorption (as shown). This simple figure has limited diagnostic utility; however, it suggests that almost all of the sources are absorbed. The sources indicated by an "X" are further investigated using X-ray spectral analyses; see Figure 3.

of the bolometric luminosity (i.e., a few percent). However, although the study of [3] provided the tightest X-ray constraints on submm galaxies to date, only one source had a spectroscopic redshift [the rest of the sources had redshifts determined with the considerably less certain radio-submm photometric redshift technique (e.g., [16]), restricting more accurate and quantitative conclusions].

5.1 The X-Ray Properties of AGNS in Bright SCUBA Galaxies

Considerable progress in the optical identification of SCUBA galaxies has been recently made due to the pioneering deep optical spectroscopic work of [17,19]. By targeting radio and/or X-ray detected SCUBA galaxies, source redshifts for a sizable fraction of the submm galaxy population have now been obtained. CO emission line observations have confirmed that both the redshift and counterpart are correct in many cases (e.g., [36]; R. Genzel et al., these proceedings). The CDF-N field was one of the fields targeted for this intensive spectroscopic follow-up: optical spectroscopic redshifts have been obtained for 24 SCUBA galaxies in the CDF-N. The combination of this deep spectroscopic data with the 2 Ms CDF-N observations provides powerful constraints on AGNs in submm galaxies. In particular, reliable spectroscopic redshifts improve the accuracy of the X-

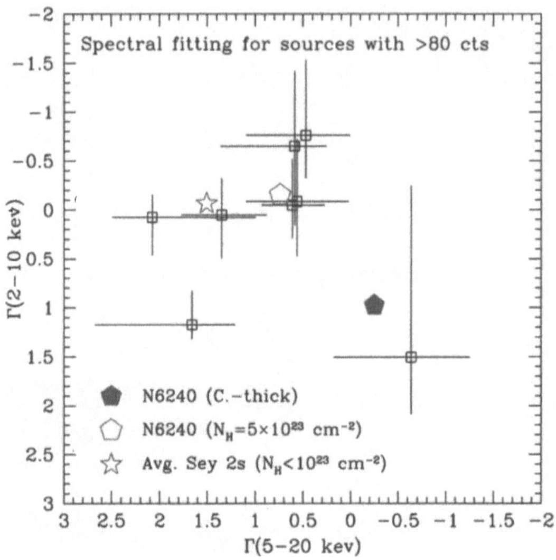

Fig. 3. Rest-frame 2–10 keV versus 5–20 keV spectral slopes for the X-ray brightest (>80 counts) submm galaxies compared to nearby absorbed AGNs. The X-ray spectral properties for the nearby absorbed AGNs consist of the average for a sample of Seyfert 2 galaxies (with $N_{\rm H} < 10^{23}$ cm^{-2}, open star; [33]), NGC 6240 adjusted as though it was Compton thin (with $N_{\rm H} = 5 \times 10^{23}$ cm^{-2}, open pentagon) and NGC 6240 (i.e., Compton thick, filled pentagon; [45]). The X-ray-submm galaxies generally reside in the region found for Compton-thin AGNs.

ray spectral analyses through the identification of subtle X-ray spectral features and making comparisons in rest-frame energy bands. In these analyses we will focus on the $z > 1$ sources, which are more typical of the general submm galaxy population [17].

Fifteen of the 20 $z > 1$ submm galaxies have X-ray counterparts: 12 ($\approx 75\%$) of the 16 $z > 1$ radio-detected submm galaxies have X-ray counterparts, continuing the X-ray-radio trend (see §4). Since some of the X-ray detected submm galaxies do not have enough counts for X-ray spectral analyses, we will first compare their X-ray band ratios [defined as the ratio of the hard-band (2–8 keV) to soft-band (0.5–2 keV) count rate] to those expected from a simple AGN model; see Figure 2. This figure has limited diagnostic utility because AGNs often have more complex spectra than that of power-law emission with differing amounts of absorption; however, it suggests that few of the sources are unabsorbed.

We have performed more detailed X-ray spectral analyses for the eight submm galaxies with > 80 X-ray counts; the X-ray spectra were extracted following the procedure outlined in [10]. In Figure 3 we compare their fitted rest-frame spectral slopes (in the 2–10 keV and 5–20 keV bands) to those found for nearby AGNs with differing amounts of absorption. Although this is still a relatively crude diagnostic, there are distinctions between Compton-thin and Compton-

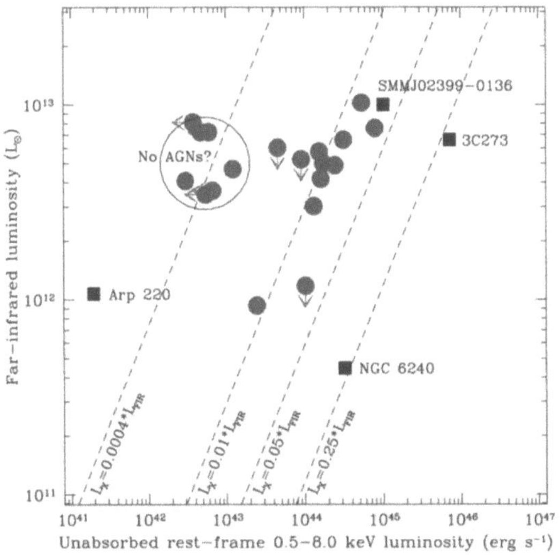

Fig. 4. Rest-frame far-IR versus unabsorbed 0.5–8.0 keV X-ray luminosities for the X-ray detected submm galaxies with spectroscopic redshifts. The far-IR luminosities have been calculated from the 1.4 GHz radio luminosity (assuming the far-IR-radio correlation) and the X-ray luminosities have been corrected for the effect of absorption. The submm galaxy SMMJ02399–0136, and three luminous nearby galaxies (Arp 220, NGC 6240, and 3C273) are shown for comparison. The line indicating the smallest L_X/L_{FIR} ratio corresponds to the average found for starburst galaxies (e.g., [9]).

thick sources. Based on this analysis, the majority of the sources appear to be heavily obscured but still only Compton thin (i.e., $N_{\rm H} \approx$ few $\times 10^{23}$ cm^{-2}). We also searched for the presence of Iron Kα emission lines. The rest-frame equivalent-width constraints are generally quite weak: Iron Kα emission lines are possibly detected in two sources (with $EW \approx 0.7$ keV and $EW \approx 1.7$ keV) while the other six only have upper limits (all have $EW < 1.8$ keV and five have $EW < 1.0$ keV). From these analyses it appears unlikely that more than $\simeq 3$ of these 8 submm galaxies contain a Compton-thick AGN. We cannot say much about the individual X-ray properties of the seven X-ray-submm galaxies with < 80 X-ray counts; however, we note that since their band ratios are consistent with the sources for which we have performed X-ray spectral analyses, they probably have similar amounts of absorption (see Figure 2).

5.2 The Bolometric AGN Contribution to Bright SCUBA Galaxies

Based on these analyses, the full range of unabsorbed X-ray luminosities is $L_X \approx 10^{42}$–10^{45} erg s^{-1}; see Figure 4.[5] While a number of the sources have X-ray luminosities consistent with those of X-ray luminous starburst galaxies (i.e.,

[5] The corrections for absorption are generally a factor of ≈ 3.

no AGNs), the majority clearly host AGNs. The AGNs generally have X-ray luminosities consistent with those of Seyfert galaxies; however, three could be considered obscured QSOs (i.e., $L_X > 3 \times 10^{44}$ erg s^{-1}).

We calculated rest-frame far-IR luminosities for all of the sources, following Equation 2 in [3] and assuming the local radio-far-IR correlation. The comparison of rest-frame far-IR luminosity with unabsorbed X-ray luminosity is shown in Figure 4. This figure provides an indicator of the AGN contribution to the bolometric luminosity.[6] Assuming that the far-IR emission from NGC 6240 and 3C273 is dominated by AGN activity, the AGNs in these submm galaxies contribute at most 20% of the bolometric luminosity and more typically a few percent. If instead we determine the AGN bolometric contributions based on the spectral energy distribution of SMMJ02399–0136 (i.e., ≈ 50%: [11,25]; see §2) then the AGN contributions increase by a factor of ≈ 2.5. Clearly, there is a range of X-ray to bolometric luminosity conversions for AGNs; however, on average the AGNs are unlikely to contribute more than ≈ 10–20% of the bolometric luminosity. Hence, although a large fraction of bright SCUBA galaxies host an AGN (i.e., at least ≈ 38%), in general, star formation is likely to dominate their bolometric output.

6 Evidence for Binary AGN Activity

An unexpected result in the 2 Ms study of [3] was that two (≈ 30%) of the seven X-ray detected submm galaxies were individually associated with X-ray pairs. The small angular separations of these pairs (≈ 2–3″) correspond to just ≈ 20 kpc at $z = 2$ (approximately one galactic diameter); the probability of a projected chance association is <1%. From *HST* imaging, it has been shown that the majority of SCUBA sources appear to be galaxies involved in major mergers (e.g., [18,21,29,40]). At X-ray energies we are presumably witnessing binary AGN activity fuelled by galaxy mergers that will ultimately lead to the coalescence of the super-massive black holes. Recent X-ray studies are only just beginning to show the late stages of binary AGN activity in AGNs at much lower redshifts (e.g., NGC 6240; [31]); however, in this ultra-deep X-ray observation we have good evidence that this is occurring at $z \approx 2$! Since the smallest linear separation we can resolve with *Chandra* at $z \approx 2$ is ≈ 10 kpc, many of the other X-ray detected submm galaxies could be binary AGNs with smaller separations (e.g., the linear separation of the two AGNs in NGC 6240 is ≈ 1 kpc). Five (≈ 3%) of the 193 X-ray sources in this region are close X-ray pairs, showing that binary AGN behaviour appears to be closely associated with submm galaxies (see also [42]). Qualitatively, this picture is consistent with that expected for major merger activity.

[6] When the radio emission has a large AGN component the far-IR luminosity will be overestimated and the AGN bolometric contribution will be underestimated; however, in general the radio emission appears to be star-formation dominated.

References

1. Alexander, D. M., Brandt, W. N., Hornschemeier, A. E., et al. 2001, AJ, 122, 2156
2. Alexander, D. M., Bauer, F. E., Brandt, W. N., et al. 2003a, AJ, 126, 539
3. Alexander, D. M., Bauer, F. E., Brandt, W. N., et al. 2003b, AJ, 125, 383
4. Almaini, O., Scott, S. E., Dunlop, J. S., et al. 2003, MNRAS, 338 303
5. Barger, A. J., Cowie, L. L., Sanders, D. B. 1999, ApJ, 518, L5
6. Barger, A. J., Cowie, L. L., Mushotzky, R. F., et al. 2001a, AJ, 121, 662
7. Barger, A. J., Cowie, L. L., Steffen, A. T., et al. 2001b, ApJ, 560, L23
8. Bassani, L., Dadina, M., Maiolino, R., et al. 1999, ApJS, 121, 473
9. Bauer, F. E., Alexander, D. M., Brandt, W. N., et al. 2002, AJ, 124, 2351
10. Bauer, F. E., Vignali, C., Alexander, D. M., et al. 2003, Adv. Space Res., in press (astro-ph/0209415)
11. Bautz, M. W., Malm, M. R., Baganoff, F. K., et al. 2000, ApJ, 543, L119
12. Blain, A. W., Smail, I., Ivison, R. J., et al. 1999, MNRAS, 302, 632
13. Blain, A. W., Smail, I., Ivison, R. J., et al. 2002, Physics Reports, 369, 111
14. Borys, C., Chapman, S., Halpern, M., et al. 2003, MNRAS, 344, 385
15. Brandt, W. N., Alexander, D. M., Hornschemeier, A. E., et al. 2001, AJ, 122, 2810
16. Carilli, C. L., Yun, M. S. 1999, ApJ, 513, L13
17. Chapman, S. C., Blain, A. W., Ivison, R. J., Smail, I. R. 2003a, Nature, 422, 695
18. Chapman, S. C., Windhorst, R., Odewahn, S., et al. 2003b, ApJ, 599, 92
19. Chapman, S. C., et al. 2004, ApJ, submitted
20. Comastri, A., Mignoli, M., Ciliegi, P., et al. 2002, ApJ, 571, 771
21. Conselice, C. J., Chapman, S. C., Windhorst, R. A. 2003, ApJ, 596, L5
22. Cowie, L. L., Barger, A. J., Kneib, J.-P. 2002, AJ, 123, 2197
23. Efstathiou, A., Rowan-Robinson, M. 2003, MNRAS, 343, 322
24. Fabian, A. C., Smail, I., Iwasawa, K., et al. 2000, MNRAS, 315, L8
25. Frayer, D. T., Ivison, R. J., Scoville, N. Z., et al. 1998, ApJ, 506, L7
26. Hornschemeier, A. E., Brandt, W. N., Garmire, G. P., et al. 2000, ApJ, 541, 49
27. Hughes, D. H., Serjeant, S., Dunlop, J., et al. 1998, Nature, 394, 241
28. Ivison, R. J., Smail, I., Le Borgne, J.-F., et al. 1998, MNRAS, 298, 583
29. Ivison, R. J., Smail, I., Barger, A. J., et al. 2000, MNRAS, 315, 209
30. Ivison, R. J., Greve, T. R., Smail, I., et al. 2002, MNRAS, 337, 1
31. Komossa, S., Burwitz, V., Hasinger, G., et al. 2003, ApJ, 582, L15
32. Lutz, D., Spoon, H. W. W., Rigopoulou, D., et al. 1998, ApJ, 505, L103
33. Malizia, A., Bassani, L., Stephen, J. B., et al. 2003, ApJ, 589, L17
34. Matt, G., Brandt, W. N., Fabian, A. C. 1996, MNRAS, 280, 823
35. Nandra, K., George, I. M., Mushotzky, R. F., et al. 1997, ApJ, 477, 602
36. Neri, R. et al. 2003, ApJ, 597, L113
37. Sanders, D. B. & Mirabel, I. F. 1996, ARA&A, 34, 749
38. Severgnini, P., Maiolino, R., Salvati, M., et al. 2000, A&A, 360, 457
39. Smail, I., Ivison, R. J., Blain, A. W. 1997, ApJ, 490, L5
40. Smail, I., Ivison, R. J., Blain, A. W., Kneib, J.-P. 1998, ApJ, 507, L21
41. Smail, I., Ivison, R. J., Blain, A. W., et al. 2002, MNRAS, 331, 495
42. Smail, I., Scharf, C. A., Ivison, R. J., et al. 2003, ApJ, 599, 86
43. Veilleux, S., Kim, D.-C., Sanders, D. B. 1999, ApJ, 522, 113
44. Vernet, J., Cimatti, A. 2001, A&A, 380, 409
45. Vignati, P., Molendi, S., Matt, G., et al. 1999, A&A, 349, L57
46. Waskett, T. J., Eales, S. A., Gear, W. K., et al. 2003, MNRAS, 341, 1217
47. Webb, T. M. A., Lilly, S. J., Clements, D. L., et al. 2003, ApJ, 597, 680

Evolution of X-Ray Selected AGN

Günther Hasinger

Max-Planck-Institut für extraterrestrische Physik, 85748 Garching, Germany

Abstract. Deep X–ray surveys have shown that the cosmic X–ray background (XRB) is largely due to the accretion onto supermassive black holes, integrated over the cosmic time. These surveys have resolved more than 90% of the X–ray background at ~ 1 keV and about 50% at 10 keV into discrete sources. Optical spectroscopic identifications show that the sources producing the bulk of the X–ray background are a mixture of unobscured (type-1) and obscured (type-2) AGNs, as predicted by the XRB population synthesis models. A class of highly luminous type-2 AGN, so called QSO-2s, has been detected in the deepest Chandra and XMM-Newton surveys. The fraction of type-2 AGN among all AGN, however, decreases significantly with luminosity. The new Chandra AGN redshift distribution peaks at much lower redshifts ($z \sim 0.7$) than that based on ROSAT data. The low redshift peak applies both to absorbed and unabsorbed AGN and is also seen in the 0.5-2 keV band alone. The new, preliminary X-ray luminosity function changes shape between low and high redshifts, confirming the luminosity-dependent density evolution model. The space density of Seyfert galaxies evolves much slower than that of QSOs.

1 Introduction

In recent years the bulk of the extragalactic X–ray background in the 0.1-10 keV band has been resolved into discrete sources with the deepest *ROSAT*, *Chandra* and *XMM-Newton* observations [1–4]. Optical identification programmes with Keck [5–8]) and VLT [9,10] find predominantly unobscured AGN-1 at X-ray fluxes $S_X > 10^{-14}$ erg cm^{-2} s^{-1}, and a mixture of unobscured and obscured AGN-2 at fluxes $10^{-14} > S_X > 10^{-15.5}$ erg cm^{-2} s^{-1} with ever fainter and redder optical counterparts, while at even lower X-ray fluxes a new population of star forming galaxies emerges [11–13]. At optical magnitudes R>24 all these surveys suffer, however, from spectroscopic incompleteness, so that photometric redshift techniques have to be applied [14].

After having understood the basic contributions to the X-ray background, the interest is now focusing on understanding the physical nature of these sources, the cosmological evolution of their properties, and their role in models of galaxy evolution. The X-ray observations have been roughly consistent with X-ray background population synthesis assuming a mixture of absorbed and unabsorbed AGN, folded with the corresponding luminosity function and cosmological evolution, e.g.[15–17]. However, inputs to these models are still rather uncertain, like e.g. the cosmological evolution of low-luminosity AGN or the fraction of type-1 to type-2 AGN as a function of redshift and intrinsic luminosity, and a wide range of different assumptions has been invoked for these parameters [18,19], see also [20]. Finally, the source statistics and optical incompleteness are rather poor at high redshifts.

The deep Chandra and XMM surveys, but also wider ASCA surveys have already provided important new constraints. Several examples of the long-sought class of high redshift, radio quiet type-2 QSO have been detected in deep fields [21,22,8,9]. These allow for the first time to constrain the fraction of type-2 AGN as a function of X–ray luminosity. At low luminosities a type-2 fraction of 75-80% is found, consistent with local optically selected Seyfert galaxies, while at high luminosities the type-2 fraction is significantly smaller [23,9]. The redshift distribution of Chandra deep survey sources peaks at $z \approx 0.7$. This is related to the finding of a much slower cosmic evolution for Seyferts compared to QSOs [23–26,10]. And finally, significant spikes are found in the redshift distributions [27,28], indicating that AGN prefer to live in sheets of large-scale structure.

2 Deep XMM-Newton Survey in the Lockman Hole

The Lockman Hole is the region on the sky having the lowest interstellar hydrogen column density and thus provides an excellent window to the distant Universe. It had been chosen as the location of the deepest ROSAT survey, which resolved the majority of the diffuse soft X-ray background into discrete sources [1]. It has also been selected for the first XMM-Newton deep survey in the PV phase [3]. Recently we have been awarded a very long XMM-Newton survey exposure in this field, bringing the net exposure to about 800 ksec, corresponding to almost 20 days of XMM-Newton exposure [29,30]. Figure 1 shows an X-ray image of this observation. The spacecraft pointing directions have been dithered between successive exposures in order to smooth out the effects of the gaps between the CCD chips.

This data has been used to determine the redshift and temperature of one of the most distant clusters of galaxies known [31,32], as well as to determine the distribution of spectral properties of the X-ray sources [33,34]. The fraction of the background resolved into discrete sources decreases from more than 90% at energies below 2 keV to about 50% at energies of 10 keV [35], leaving a significant population of hard sources still to be resolved. Spectroscopic follow-up observations have been performed at the Keck telescope in spring 2003 and 2004 in collaboration with Pat Henry and Maarten Schmidt, using the Deimos wide field spectrograph. Together with the already existing ROSAT sources we now have 125 spectroscopic identifications in this field. For a subsample of sources with X-ray fluxes $> 10^{-15}$ erg cm^{-2} s^{-1} we reach a spectroscopic completeness of $\sim 80\%$ in the inner field with 10 arcmin radius.

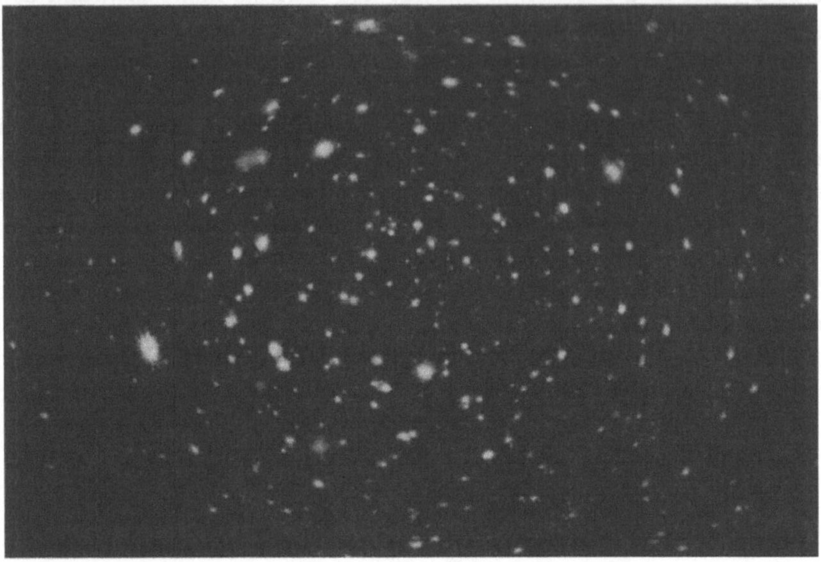

Fig. 1. Color composite image of the ~800 ksec XMM-Newton image of the Lockman Hole. The image was obtained combining three energy bands: 0.5-2 keV, 2-4.5 keV, 4.5-10 keV (respectively red, green and blue). The image has a size of $\sim 43 \times 30$ arcmin^2.

3 AGN Obscuration Dependence on X-Ray Luminosity

The optical identifications in the deepest Chandra survey fields, the 2 Msec Hubble Deep Field North (HDF-N) [7,8], the 1 Msec Chandra Deep Field South (CDF-S) [9,14], and the XMM-Newton survey in the Lockman Hole (see above) have now progressed to a state, where a statistical analysis of the samples is possible. Spectroscopic identifications using the VLT and Keck telescopes reach a limiting magnitude of R 24-25 and spectroscopic catalogues have been published for the HDF-N [28,8] and CDF-S [9]. In the CDF-S there is in addition the COMBO-17 survey, a 17-channel imaging survey providing high-quality photometric redshifts across the whole field to R=24 [36]. However, about 25% of the X-ray sources have counterparts too faint for optical spectroscopy or COMBO-17. For these sources it is very fortunate that both the HDF-N and the CDF-S are covered by the Great Observatories Origins Deep Survey (GOODS) providing HST-ACS imaging in four optical filters over a significant part of the field [37]. The center of the CDF-S is in addition covered by extremely deep NIR imaging with ISAAC on the VLT to K 25. Using a combination of GOODS and COMBO-17 data, photometric redshifts could be determined for almost all of of the optically faint sources[14], so that redshifts are now available for 80 and 90% of the sources in the HDF-N and CDF-S, respectively.

Using the X-ray luminosity and hardness ratio we can classify the X-ray sources in both fields into unabsorbed (type-1) and absorbed (type-2) AGN and

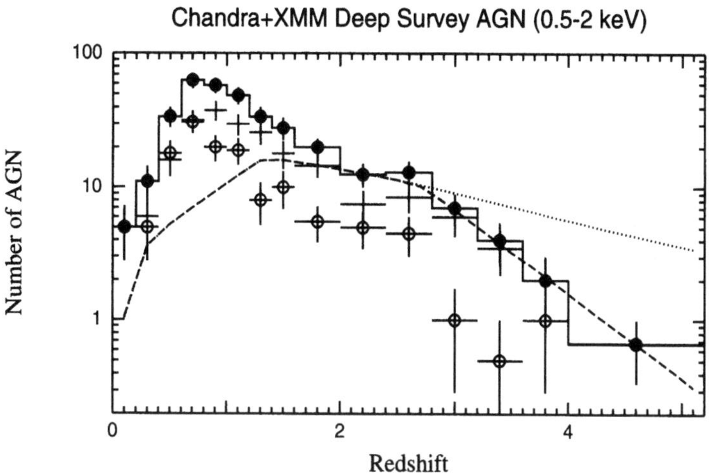

Fig. 2. Redshift distribution of 403 AGN selected in the 0.5-2 keV band from the Chandra CDF-S and HDF-N and the XMM-Newton Lockman Hole survey samples (solid circles and histogram), compared to model predictions from population synthesis models from [17]. The dashed line shows the prediction for a model, where the comoving space density of high-redshift QSO follows the decline above z=2.7 observed in optical samples [38,39]. The dotted line shows a prediction with a constant space density for $z > 1.5$. The two model curves have been normalized to fit the observed distribution in the redshift range z=2-2.4. Simple crosses and open circles give the redshift distribution separately for 252 AGN-1 and 151 AGN-2, respectively. The peak of the distribution is at redshifts below one for all three samples.

a set of normal or starforming galaxies. Figure 2 shows the redshift distribution measured for a total of 403 AGN selected in the 0.5-2 keV band in the Chandra and XMM-Newton deep fields. Contrary to some theoretical predictions [18,19] both the unabsorbed and the absorbed AGN peak at redshifts below one, indicating that the evolution of the total AGN population saturates at relatively low redshifts.

An important achievement in the Chandra Deep Fields is the establishment of a class of radio-quiet, high-luminosity, highly absorbed type-2 AGN, which are the type-2 equivalent of the unobscured QSO and which we therefore call QSO-2 [21,22]. Using the total identified Chandra samples we can estimate the fraction of absorbed AGN as a function of X-ray luminosity. In order to minimise selection effects due to obscuration, we have restricted this analysis to AGN detected in the hard (2-10 keV) band in the HDF-N and CDF-S. Figure 3 shows the ratio of type-2 to all AGN in comparison to data derived independently from brighter ASCA and HEAO-1 samples by Ueda et al. [23]. There is a significant decrease of the type-2 fraction with increasing luminosities: while at low L_X the ratio is consistent with that of local optically selected Seyfert galaxies 75-80% [40], at high luminosities type-2 QSOs account for only 30-40% of all AGN. This finding is an important new ingredient for the next generation of population synthesis models of the X-ray background.

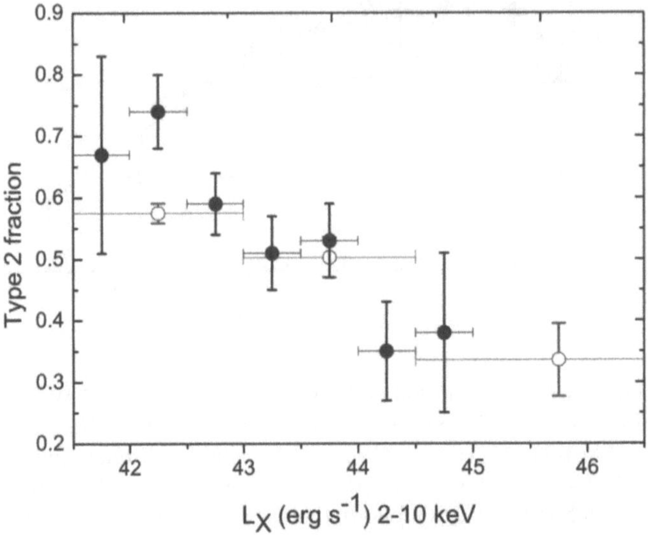

Fig. 3. Fraction of absorbed (type-2) AGN to all AGN selected in the 2-10 keV band in the deep Chandra surveys (HDF-N and CDF-S) as a function of X-ray luminosity (data points with filled circles) in comparison with similar data derived from shallower ASCA and HEAO-1 surveys (open circles; from [23]).

4 Cosmological Evolution of Seyfert Galaxies

For the first time we are now able to merge the Chandra and XMM-Newton deep survey data with the whole body of previously identified ROSAT AGN samples. We have selected only the type-1 AGN in all samples and treated only the detections and X-ray fluxes in the 0.5-2 keV band. A total of 1023 X-ray selected type-1 AGN have been used. The different samples cover an unprecedented six orders of magnitude in flux limit and seven orders of magnitude in survey solid angle between the ROSAT Bright and serendipitous surveys [41,42], the XMM-Newton Lockman Hole survey (see above) and the Chandra Deep Surveys. The new Chandra and XMM-Newton sources are predominantly Seyfert galaxies at a median luminosity of $\sim 10^{43}$ erg s^{-1} and a median redshift around 0.7. A total number of 75 sources in all samples together is still unidentified.

The X-ray luminosity function was calculated using the V/Va method [26] and is shown in two redshift shells (z=0.015-0.2 and z=1.6-2.3) in Figure 4. The shape of the two luminosity functions is significantly different, so that the cosmological evolution can be described neither by pure luminosity nor pure density evolution. The surprising result is, however, that the high-redshift luminosity function is almost horizontal at luminosities below 10^{44} erg s^{-1} and approaches the local space density in the Seyfert range. The strong positive density evolution, well known from previous AGN samples in the optical, radio and X-ray range, therefore only holds for relatively luminous AGN (i.e. QSOs), while the lower luminosity AGN (Seyfert galaxies) show much less or even negative density evolution. These results are similar to those derived in the 2-10 keV band by Ueda et al. [23], but the spectroscopic incompleteness at faint optical magnitudes still leaves substantial uncertainties in the derived space densities, which are discussed and quantified in [26].

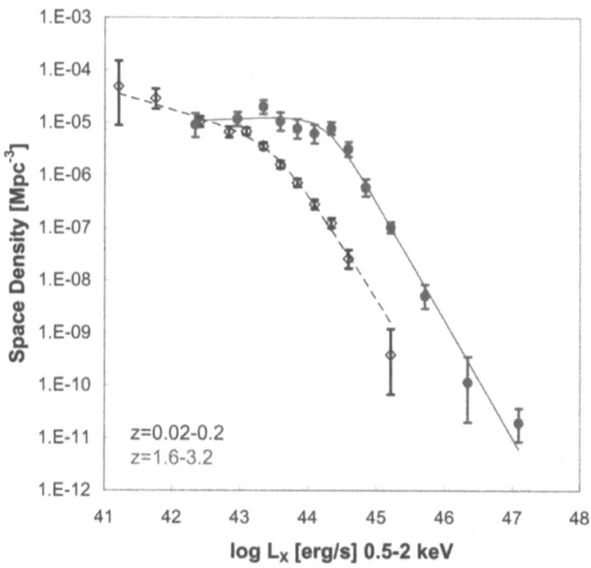

Fig. 4. Luminosity function of type-1 AGN selected in the 0.5-2 keV band in two redshift shells: z=0.02-0.2 (open symbols, dashed line) and z=1.6-2.3 (solid circles, solid line).

To illustrate the different evolutionary behaviour for different luminosity classes in more detail we show the space densities as a function of redshift for different luminosity classes in Figure 5. While the evolution of the highest luminosity class (45-48), the QSOs, follows very well the strong positive evolution with an increase of more than two orders of magnitude in space density, saturating at z∼2 known from optical and radio samples of QSOs, the evolution of lower luminosity classes is weaker and saturates at significantly later redshifts. The highest space density is achieved for the Seyferts of luminosity class 41-42 at redshifts around 0.5 at a space density about a factor of 1000 higher than that of the most luminous QSOs. Beyond z=0.7 there is a significant decline of the Seyfert space density. This is the reason, why the Chandra deep surveys are dominated by Seyfert galaxies at low redshifts and not, as originally expected by objects at higher redshifts.

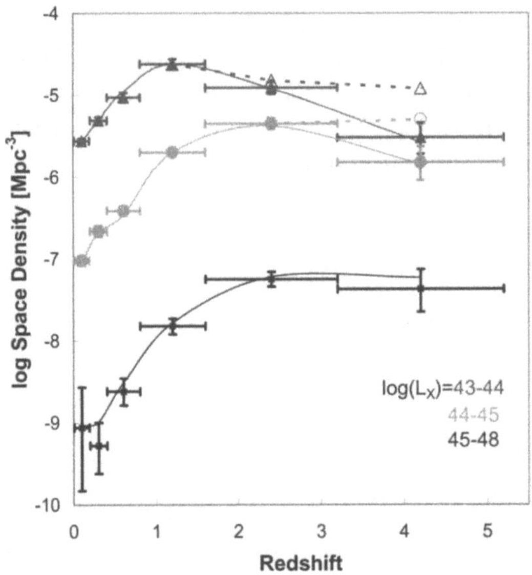

Fig. 5. Space density of type-1 AGN selected in the 0.5-2 keV band for different luminosity classes. The dashed lines give an upper bound for the space densities, assuming that all unidentified sources are lying at the corresponding redshift.

These new results paint a dramatically different evolutionary picture for low-luminosity AGN compared to the high-luminosity QSOs. While the rare, high-luminosity objects can form and feed very efficiently rather early in the universe, with their space density declining more than two orders of magnitude below z=2, the bulk of the AGN has to wait much longer to grow with a decline of space density by less than a factor of 10 below a redshift of one. This could indicate two modes of accretion and black hole growth with different accretion efficiency (see e.g. [43]). The late evolution of the low-luminosity Seyfert population is very similar to that which is required to fit the Mid- infrared source counts and background [18] and also the bulk of the star formation in the Universe [44], while the rapid evolution of powerful QSOs traces more the merging history of spheroid formation [45].

References

1. Hasinger G., Burg R., Giacconi R., et al., 1998, A&A 329, 482
2. Giacconi R., Rosati P., Tozzi P., et al., 2001, ApJ 551, 624

3. Hasinger G., Altieri B., Arnaud M., et al., 2001, A&A 365, 45
4. Alexander D.M., Bauer F.E., Brandt W.N., et al., 2003, AJ 126, 539
5. Schmidt M., Hasinger G., Gunn J.E., et al., 1998, A&A 329, 495
6. Lehmann I., Hasinger G., Schmidt M., et al., 2001, A&A 371, 833
7. Barger A. J., Cowie L. L., Mushotzky R. F., Richards, E. A., 2001, AJ 121, 662
8. Barger A.J., Cowie L.L., Capak P., et al., 2003, AJ 126, 632
9. Szokoly G., Bergeron J., Hasinger G., et al., 2004 ApJS submitted (astro-ph/0312324)
10. Fiore F., Brusa M., Cocchia F., et al., 2003, A&A 409, 79
11. Hornschemeier A.E., Brandt W.N., Garmire G.P., et al., 2000, ApJ 541
12. Rosati P., Tozzi P., Giacconi R., et al., 2002, ApJ 566, 667
13. Norman, C., Ptak, A., Hornschemeier, A., et al., 2003, ApJ in press (astro-ph/0402140)
14. Zheng W., Mikles, V.J., Mainieri, V., et al., 2004, ApJ subm.
15. Comastri A., Setti G., Zamorani G., Hasinger G., 1995, A&A 296, 1
16. Fabian A.C., Barcons X., Almaini O., Iwasawa K., 1998, MNRAS 297, L11
17. Gilli R., Salvati M., Hasinger G., 2001, A&A 366, 407
18. Franceschini A., Braito V. & Fadda D., 2002, MNRAS 335, L51
19. Gandhi P. & Fabian A.C., 2003, MNRAS 339, 1095
20. Gilli R., 2003, Adv. Space Res. in press (astro-ph/0303115)
21. Norman C., Hasinger G., Giacconi R., et al. 2002, ApJ 571, 218
22. Stern D., Moran E.C., Coil A.L., et al., 2002, ApJ 568, 71
23. Ueda Y., Akiyama M., Ohta K., Miyaji T., 2003, ApJ 598, 886
24. Hasinger G. & the CDFS team, 2003, AIP Conf.Proc. 666, p. 227 (astro-ph/0302574)
25. Cowie L.L., Barger A.J., Bautz M.W., et al., 2003, ApJ 584, L57
26. Hasinger G., Miyaji T., Schmidt, M., 2004, in prep
27. Gilli, R., Cimatti, A., Daddi, E., et al., 2003, ApJ 592, 721
28. Barger, A. J., Cowie, L. L., Brandt, W.N. et al., 2002, AJ 124, 1839
29. Hasinger, G., astro-ph/0310804
30. Brunner, H., Hasinger G., Barcons, X., et al., 2004, in prep.
31. Hashimoto Y., Hasinger G., Arnaud, M. et al., 2002, A&A 381, 841
32. Hashimoto Y., Barcons, X., Böhringer H., et al., 2004, A&A in press (astro-ph/0401297)
33. Mainieri V., Bergeron J., Rosati P., et al., 2002, A&A 393, 425
34. Streblyanska, A, Bergeron, J., Brunner H., et al., 2003, astro-ph/0309089
35. Worsley, M., Fabian, A.C., Mateos, S., et al., 2004, MNRAS subm.
36. Wolf C., Dye, S., Kleinheinrich, M., et al., 2001, A&A 377, 442
37. Dickinson, M. et al., 2003 in "The Mass of Galaxies at low and high redshift", eds. R. Bender and A. Renzini (New York: Springer), 324
38. Schmidt, M., Schneider, D.P. & Gunn J.E., 1995, AJ 114, 36
39. Fan X., Strauss M., Richards G., et al., 2001, AJ 121, 31
40. Huchra J. & Burg R., 1992, ApJ 393, 90
41. Miyaji, T., Hasinger, G., Schmidt, M., 2000, A&A 353, 25
42. Gioia, I., Henry, J.P., Mullis, C.R., et al., 2003, ApLS 149, 29
43. Duschl, W.J., Strittmatter, P.A., in Active Galactic Nuclei: from Central Engine to Host Galaxy Abstract Book, meeting held in Meudon, France, July 23-27, 2002, Eds.: S. Collin, F. Combes and I. Shlosman. To be published in ASP (Astronomical Society of the Pacific), Conference Series, p. 76.
44. Madau, P., Pozzetti, L., Dickinson, M., 1998, ApJ 498, 106
45. Franceschini, A., Hasinger, G., Miyaji, T., Malquori, D., 1999, MNRAS 310, L5

Obscuration and Circumnuclear Medium in Nearby and Distant AGN

R. Maiolino

INAF - Osservatorio Astrofisico di Arcetri, Firenze, Italy

Abstract. Some recent results on the physical and statistical properties of nearby and distant AGN are presented. I first discuss the properties of "elusive" AGNs, i.e. obscured AGNs which do not show a Seyfert-like spectrum in the optical. Then I present preliminary results from a detailed study of the contribution of obscured AGNs and of their host galaxies to the infrared cosmic background. Finally I discuss an observational program aimed at investigating the properties of the most distant quasars, of their circumnuclear medium and the implications for their host galaxies.

1 Introduction

Obscured Active Galactic Nuclei (type 2 AGNs) have been investigated in great detail in the local universe. The recent deep surveys have also found several obscured AGN at high and intermediate redshift, and have shed new light on their evolution [12]. However, there are still some open issues on the physical and statistical properties of obscured AGNs. There is growing evidence for a population of obscured AGNs which do not show the classical AGN signatures in their optical spectra, which might require a revision of the unified model as well as reassessment of the density of AGNs in the local universe. At higher redshift it is now clear that a mixture of obscured and unobscured AGN produce most of the X-ray background, but it is much less clear what is their contribution to the IR background and the possible connection with the evolution of galaxies. At the highest redshift (z$\sim$6) probed so far by quasar surveys, it is not clear what are the properties of the circumnuclear medium and, specifically, if the gas has the requirements to produce obscuration, both in terms of metallicity and dust content. In this paper I shortly summarize some recent work aimed at tackling these issues.

2 Elusive AGNs

A fraction of active galactic nuclei do not show the classical Seyfert-type signatures in their optical spectra, i.e. they are optically "elusive".The closest example of this class of objects is NGC4945. This galaxy hosts a nuclear starburst and its optical spectrum is characterized by faint LINER-like emission lines associated with the starburst superwind. However, its hard X-ray spectrum has revealed the presence of a heavily obscured AGN [11]. Another clear case has been reported by [4], who detected a heavily obscured AGN in the starburst/HII system NGC3690. We specifically define "elusive AGN" as those nuclei which do not show Seyfert-like emission lines in their optical spectra, but where a relatively luminous AGN (i.e. in the Seyfert range) is detected at other wavelengths. Although this class of AGNs clearly exist, it is not clear how common they are, nor it is clear their nature (i.e. why they are optically elusive).

We have started a program aimed at assessing the fraction of elusive AGN in the local universe and to investigate their nature. We are mostly exploiting hard X-ray data from Chandra and XMM, but also near- and mid-IR spectroscopy, to detect obscured AGNs not identified by optical spectroscopy. Preliminary results were reported in [13] and summarized here. There are about 20 elusive AGN identified so far (though not all of them can be used for statistical purposes, see [13]). Once selection effects are taken into account we estimate that elusive AGNs may be as numerous as (or even outnumber) classical, optically identified Seyfert nuclei. The estimated fraction of elusive AGNs as a function of infrared luminosity is shown in Fig.1. Obviously the statistics are still poor and more data are required to secure this result. If confirmed, an important implication would be that the overall fraction of galaxies hosting an AGN in the local universe is significantly higher than estimated previously by optical surveys. This would nicely match the recent results from the hard X-ray surveys [20][12][9], which are finding that the evolution of Seyfert nuclei peaks at much lower redshifts than quasars and probably requiring a high density of Seyferts at z=0. Note that the higher redshift counterparts of elusive AGN may be the so-called XBONGs (X-ray Bright Optically Normal Galaxies) found in the hard X-ray surveys [6].

A most interesting result of the X-ray spectra of elusive AGNs is that they are heavily absorbed and, in particular, most of them are Compton thick, i.e. absorbed by column of gas $N_H > 10^{24}$ cm^{-2}. This suggests that their elusive nature is associated with heavy obscuration. In [13] we suggested that probably in elusive AGNs the nuclear radiation source is obscured in all directions thus preventing UV photons to escape and to produce a Narrow Line Region (NLR).

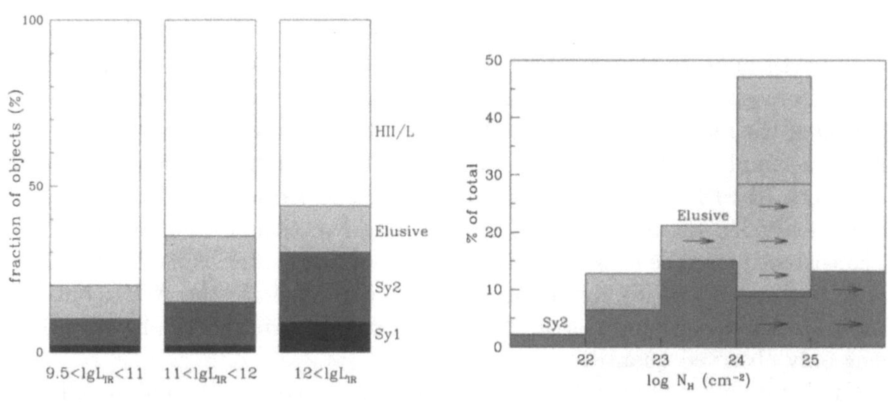

Fig. 1. *Left.* Fraction of elusive AGN and Seyfert nuclei as a function of IR luminosity. *Right.* Cumulative absorbing N_H distribution including elusive AGNs. From [13]

However, for some of the elusive AGNs other explanations are viable, such as heavy obscuration of the NLR and/or dilution by the circumnuclear starburst (which may apply to the more distant cases)[7]. An important implication of the heavy obscuration affecting elusive AGNs is that, if they are included in the local census of AGNs, then the overall distribution of absorbing N_H would be strongly skewed towards higher values with respect to what previously estimated for optically identified Seyfert 2, as shown in Fig.1.

Another interesting feature is that those elusive AGNs which were observed also in the 10–100 keV band result to be absorbed by columns in the range $10^{24} < N_H < 10^{25}$ cm^{-2}. In this case the observed X-ray spectrum is characterized by a prominent bump peaking at about 30–40 keV. If elusive AGNs are common also at higher redshift, then they could contribute significantly to the 30 keV bump of the X-ray background. Within this context, one should keep in mind that the X-ray background has been resolved at energies <10 keV, but the bulk of the X-ray background, which is produced at >10 keV, has not been resolved yet.

3 The Contribution of AGNs to the Infrared Background

Although it is clear that most of the X-ray background is produced by a mixture of obscured and unobscured AGN (the latter dominating at higher energies), it is less clear what is their contribution to the infrared cosmic background. This is a most important issue, since the infrared is the spectral region were most of the cosmic background is produced (after the CMB) and which is expected to trace a significant fraction of the global star formation history. Dust in the circumnuclear region of AGN absorbs most of the optical-UV light emitted by the nucleus and which is reprocessed into the infrared; therefore the contribution of AGN (and in particular *obscured* AGNs) to the IR background may be significant. Some previous studies have obtained contradictory results depending on the assumed Spectral Energy Distribution (SED) and evolution of the AGNs.

We have approached the problem by minimizing the number of assumptions and by using known and measured quantities to bridge the (AGN-produced) X-ray background to the IR background. In particular, we used the AGN hard X-ray luminosity functions and evolution recently derived by [20] using a compilation of the most recent X-ray surveys. We then derived the *nuclear* infrared SED of a sample of about 30 Seyferts by using high resolution near- and mid-IR data. Such IR SEDs were then normalized to the intrinsic hard X-ray emission. Then we divided the SEDs in bins of absorbing N_H; this novel approach allows to consistently use the same method adopted for the synthesis modelling of the X-ray background. With this information we can estimate the AGN contribution to the IR background starting from the X-ray background and by essentially using only observed quantities.

Further details will be provided in a forthcoming paper [19], here we only discuss some preliminary results. Fig.2 shows the estimated fractional contribution of AGNs to the cosmic background at various infrared wavelengths. The contribution is low at all wavelengths: the maximum contribution (but still $\lesssim$10%)

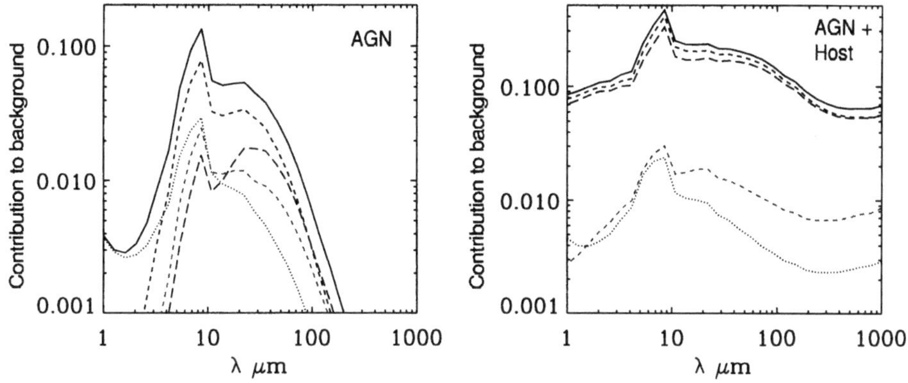

Fig. 2. *Left.* Fractional contribution of AGN to the IR background. Dotted and dashed lines show the relative contribution by AGN affected by different degrees of obscuration, while the solid line is the total contribution. *Right.* The contribution to the IR background when AGN host galaxies are also included.

is in the mid-IR, between 5μm and 40μm; while the contribution is negligible at longer wavelengths. However, AGNs are hosted in galaxies which may contribute significantly to the global infrared light emitted by these systems. If the host galaxies co-evolve with the AGN (and there are observational indications in favor of this scenario), then they could contribute significantly to the IR background. We have investigated this possibility by associating the SED of the host galaxies to the nuclear SEDs. We have then assumed that the host galaxies are linked to the AGN by the same cosmic evolution. The resulting contribution to the IR background is shown in Fig.2. AGNs and their host galaxies appear to contribute significantly to the IR background: about 20% in the mid-IR, exceeding 40% at 8μm, and ~10% in the far-IR and sub-mm. Therefore, the cosmic IR background may have the imprint of the co-evolution of AGN and galaxies. Our results are in good agreement with the contribution at 15μm obtained by [8][1] by cross-correlating ISO and Chandra/XMM sources. Additional implications, especially for what concerns to contribution to the IR/submm source number counts will be discussed in [19].

4 Absorption and Circumnuclear Medium in the Most Distant Quasars

The most distant quasars known are at a redshift (z~6) which is approaching the epoch of reionization. At early epochs it is not obvious that the circumnuclear gas has the requirements to obscure AGNs. Indeed, obscuration in the X-rays requires that the gas metallicity has already evolved enough to provide the opacity associated to various elements, and in particular iron (whose edge is shifted into the soft X-rays). Optical and UV absorption require that dust has already been produced in large quantities at such high redshifts.

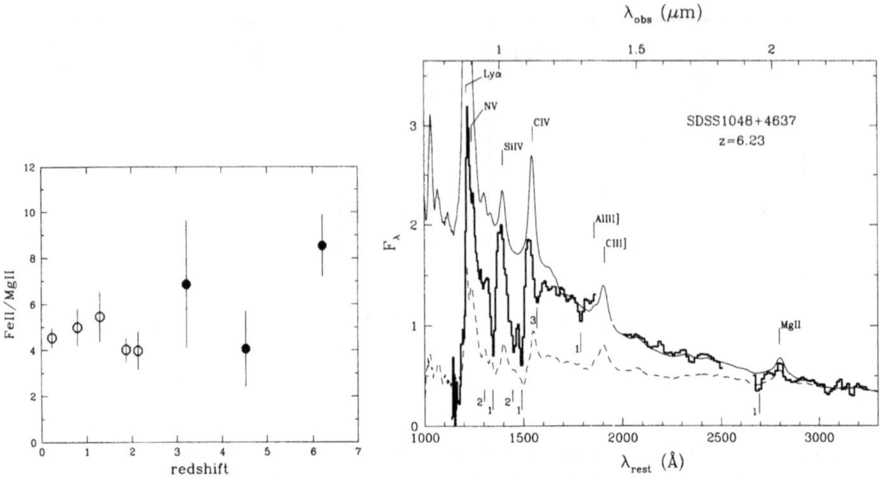

Fig. 3. *Left.* FeII/MgII observed in quasars as a function of redshift. Solid symbols at high redshift are from our survey, while open symbols at lower redshift are from [5] *Right.* The most distant LoBAL at z=6.23 found in our sample. Note the extremely deep, blueshifted CIV absorption and the associated absorptions of SIV, AlIII] and MgII. The thin solid line is the average spectrum of lower redshift non-BALs, while the dashed line is the average spectrum of lower redshift LoBAL from the SLOAN [17].

We have investigated the metallicity and absorption of the most distant quasars by means of the Near Infrared Camera Spectrometer (NICS [2]) at the Telescopio Nazionale Galileo (TNG), with its low resolution, high sensitivity IR spectroscopic mode, which allows to obtain the full near-IR spectrum from 0.8μm to 2.4μm in one shot. Thanks to the large spectral coverage and to the high sensitivity we could measure the intensity of the UV FeII-bump (redshifted into the near-IR) relative to the MgII doublet (2798Å) in a sample of 22 quasars at $3 < z < 6.4$. To a first approximation, the ratio FeII/MgII can be assumed as an indicator of the Fe abundance relative to the α elements. This is also an indicator of the age of the stellar population since α elements are predominantly produced by type II SN, while the production of Fe is delayed by type Ia SNe [16][18]. Fig.3 shows the ratio FeII/MgII in quasars as a function of redshift, indicating no decrease of the iron fraction up to z$\sim$6 (and actually there is a marginal indication for an increase of Fe). This result has two implications: 1) at the redshift of the most distant quasars known (z$\sim$6) the circumnuclear medium was already highly enriched (in particular plenty of Fe was available to provide obscuration in the X-rays); 2) the high ratio of Fe/α suggests that the hosts of the most distant quasars were formed at z>9, to allow the required time for the SN Ia to enrich the gas. Further details are given in [14].

For eight quasars at 4.9<z<6.4 our spectra include the resonant CIV line at 1549Å. Half of these quasars are characterized by deep, broad and blueshifted absorption of CIV, i.e. they belong to the class of Broad Absorption Line (BAL) quasars, which are associated with powerful outflows of dense gas. Fig.3 shows the spectrum of the most distant of these BAL quasars, characterized by a very deep absorption of CIV and also of other high and low ionization lines (SiIV, AlIII, MgII), actually identifying this as a Low ionization BAL (LoBAL). Although the sample is small, the large fraction of BAL quasars, the depth and ionization state of the absorption features suggest that these most distant quasars are surrounded by a much larger amount of dense gas than lower redshift quasars. As discussed in [15], such a result may indicate that the highest redshift quasars are characterized by extremely high accretion rates and associated with the early formation of quasars and of their host galaxies [10].

Another interesting finding is that all these distant BAL quasars are bluer than lower redshift BALs (which are generally reddened by dust). The dashed line of Fig.3 shows the average spectrum of LoBALs at lower redshift from the SLOAN [17], which is clearly much redder than our highest redshift LoBAL (although at $\lambda_{rest} < 1500$Å there is a bending of the spectrum, partly due to the prominent absorption lines, this issue is discussed in [15]). The lack of significant reddening in the most distant BALs cannot be ascribed to lack of dust, since the presence of large amounts of dust were inferred by the submm/mm detections of these objects [3]. Therefore, these results may indicate a different extinction, and possibly reflect a different evolution and formation mechanism of dust grains at z>5. This issue will be discussed in [15]

Acknowledgments. The results presented in this paper were obtained in collaboration with several people and more specifically: S. Bianchi, T. Böker, E. Colbert, A. Comastri, R. Gilli, F. Ghinassi, G.L. Granato, Y. Juarez, A. Krabbe, F. Mannucci, A. Marconi, G. Matt, R. Mujica, N. Nagar, E. Oliva, M. Pedani, M. Salvati and L. Silva.

References

1. Alexander, D. et al.: ApJ, **568**, L85 (2002)
2. Baffa, C. et al.: A&A, **378**, 722 (2001)
3. Bertoldi, F. et al.: A&A, **406**, L55 (2003)
4. Della Ceca, R. et al.: ApJ, **581**, L9 (2002)
5. Dietrich, M. et al.: ApJ, **596**, 817 (2003)
6. Comastri, A. et al.: in press (astro-ph/0203019)
7. Gonçalves, A., Véron-Cetty, M., & Véron, P.: A&AS, **135**, 437 (1999)
8. Fadda, D.: A&A, **383**, 838 (2002)
9. Fiore, F.: A&A, **409**, 79 (2003)
10. Granato, G. et al.: ApJ, in press (astro-ph/0309420)
11. Guainazzi, M. et al.: A&A, **356**, 463
12. Hasinger, G.: this volume
13. Maiolino, R. et al.: MNRAS, **344**, L59 (2003)
14. Maiolino, R. et al.: ApJ, **596**, L155 (2003)
15. Maiolino, R. et al.: A&A, submitted
16. Matteucci, F. & Recchi, S.: ApJ, 558, 351
17. Reichard, T. et al.: AJ, **125**, 1711 (2003)
18. Romano, D. et al.: MNRAS, **334**, 444 (2002)
19. Silva, L. et al.: in prep.
20. Ueda, Y. et al.: ApJ, in press (astro-ph/0308140)

Multiwavelength Observations of the Subaru/XMM-Newton Deep Field

K. Sekiguchi[1], M. Akiyama[1], H. Furusawa[1], C. Simpson[2], T. Takata[1], Y. Ueda[3], M.W. Watson[4], and the SXDS Team

[1] Subaru Telescope, NAOJ, Hilo, HI 96720, USA
[2] Department of Physics, University of Durham, Durham DH1 3LE, UK
[3] Institute of Space and Astronautical Science, Kanagawa 229-8510, Japan
[4] Department of Physics and Astronomy, University of Leicester LE1 7RH, UK

Abstract. The Subaru/XMM-Newton Deep Survey (SXDS) is a new deep Optical/X-ray survey, whose purpose is to provide an accurate census of the contents of the Universe without suffering from the biasing effects of large-scale structures. The SXDS is part of an ambitious project to obtain extensive multi-wavelength data across a $\sim 1.3\,\mathrm{deg}^2$ region of sky. Other wavelength observations ranging from the X-ray to the radio are being pursued and/or planned through various facilities. These multiwavelength data would allow us a comprehensive approach to investigate the mass assembly history of galaxies over a broad span of cosmic history.

1 Introduction

Understanding the contents and its distribution in the universe is a key issue of the study of cosmology. We have initiated a multiwavelength wide area and deep survey to provide an accurate census of objects in the Universe hence allow us to investigate the formation and evolution of structure. The available galaxy survey data are either sampling wide area but bright limiting magnitude or deep limiting magnitude but covering a small area. The wide area surveys like the 2dFGRS ([1]) and the SDSS ([2]) show us only the end result of galaxy formation. The ultra-deep pencil-beam surveys from space, like the HDFs, can provide the shape of the luminosity function of distant objects. However, the results drawn from the small volume is limited not only by the small number statistics, but also by the intrinsic spatial correlations between galaxies, which causes fluctuations of measurements much larger than they would be. A statistically sound, large volume, large numbers of high redshift sample of various different classes of objects are needed to determine the global properties and the formation and evolution of these objects.

Detailed simulations ([3]) indicate substantial Large Scale Structure (LSS), including voids, filaments, clusters and groups, occurs on scales a few x 10 Mpc up to ~ 100 Mpc (comoving). Thus, we must cover a large enough area ($> 1\,\mathrm{deg}^2$) to minimize the effects by the cosmic variance and at the same time we must observe sufficiently deep limiting magnitude so that we can accurately determine the global properties of different class of objects. Furthermore, the value of the survey dataset grows exponentially by adding the data at other wavelengths.

Therefore, we took a multiwavelength approach for the SXDS from the very start. In this paper we describe the status of the SXDS and its multiwavelength companion surveys. See [4] for the location, size and depth of the survey. Detailed information on the strategies, the observations and the data set of the SXDS will be described in elsewhere.

2 The SXDS

Our initial survey consists of optical imaging (B, R, i', z') with Suprime-Cam ([5]) on Subaru Telescope and X-ray imaging with the EPIC camera ([6]) on the ESA XMM-Newton satellite. These instruments provide the fastest survey speeds available at present in their wavelength ranges of operation. The survey field is centered at J0218−05. The field is accessible and suitable for observations at all wavelengths by making use of the wide range of currently available and of planned major observational facilities.

2.1 The X-Ray Imaging

The X-ray observations were taken as part of the XMM Survey Science Centre (SSC) guaranteed time. The X-ray survey consists of mosaic of 7 partially overlapping pointings of the EPIC camera. A total of 400 ksec were allocated for this program. The nominal exposure times of 100 ksec for the central field, and 50 ksec each for the six flanking fields. The image of the central field approaches the confusion limit for the XMM-Newton in the soft (0.5-2 keV) band ([7]), while together these data form one of the largest contiguous area over which deep X-ray observations have been performed. Fig. 1 gives a composite X-ray image of the SXDS field.

Source detection and parameterization of the X-ray data were completed. Preliminary analysis of the X-ray colours, spectra, morphology, and X-ray variability were performed. Over 1,000 X-ray sources were detected down to $\sim 10^{-15}$ $\mathrm{erg^{-1}\,s^{-1}\,cm^{-1}}$ for point sources in the 0.5-2 keV band. We obtained the X-ray colours for nearly 1,000 sources and the X-ray spectra for ~ 300 sources. There are ~ 25 cluster candidates selected on the basis of their extended X-ray emission.

2.2 The Optical Imaging

Most of the optical imaging observations were conducted as one of three Subaru Telescope's Observatory Key Projects. Five Suprime-Cam fields, as indicated in Fig. 2, were observed to cover a $\sim 1.3\,\mathrm{deg}^2$ region of sky. Our observations reached limiting magnitudes at $B = 28.1$, $R = 27.5$, $i' = 27.4$ & $z' = 26.8$ (AB magnitude, 3σ detection with 2" aperture). The observations were completed in mid-November 2003. Currently, complementary V-band observations are pursued.

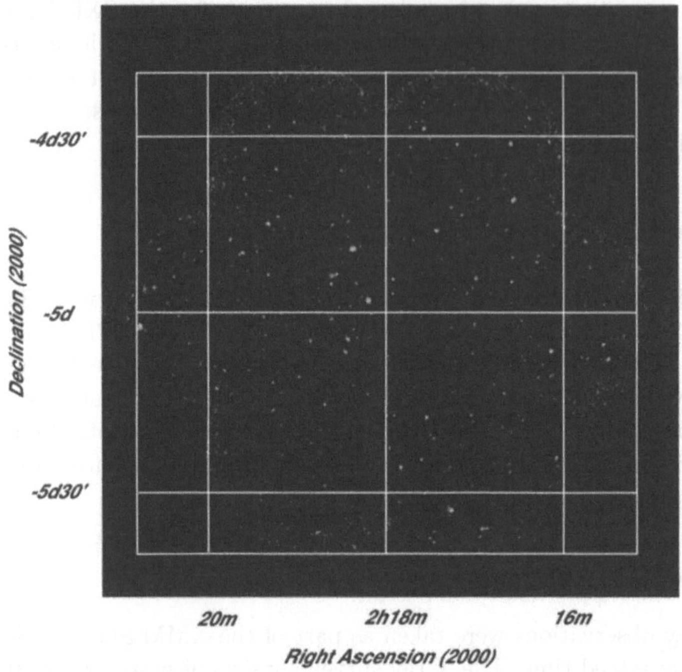

Fig. 1. 7 pointings of the XMM-Newton observations of the SXDS field. The X-ray observations cover a $\sim 1.3\,\mathrm{deg}^2$ field centered at R.A. = 02h18m, decl. = -05d (J2000).

Preliminary analysis of the optical 4-colour data from 2002 observations indicates > 1 million objects should be seen in the survey area. Fig. 3 gives the B- and i'-band number counts of the SXDS field from the 2002 dataset, compared with several previous studies.

3 Companion Surveys

Wide range of multiwavelength companion surveys to the SXDS are underway or planned by various groups using several of the major observational facilities around the world and in the space. These include:

- The VLA 1.4GHz, which covers the entire SXDS field to $S_{1.4} = 15\ \mu$Jy with B-array and 3 position (which covers around 1,000 arcmin^2) A-configuration at the central region to ~10 μJy. These observations were completed by the end of August 2003 and the data has been analyzed. Preliminary results have been presented at this meeting[8]

- The JCMT/SCUBA 850μm (as a part of SCUBA HAlf-Degree Extraglactic Survey (SHADES)) will survey a $0.25\,\mathrm{deg}^2$ regions at the centre of the SXDS

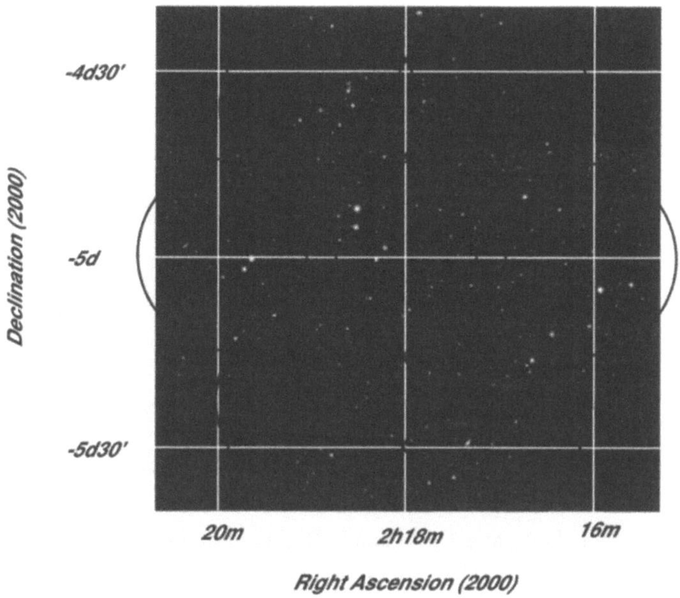

Fig. 2. The SXDS field. Five pointings of the Suprime-Cam images were combined to cover a $\sim 1.3\,\mathrm{deg}^2$ field centered at R.A. = 02h18m, decl. = -05d (J2000). Circles represent 7 pointings of the XMM-Newton observation (see Fig. 1)

field to depths of 2.5 mJy (r.m.s.). By the end of October 2003, SHADES covered 237 arcmin^2 in the SXDS field, from which ~113 sources with $S/N > 3$ were extracted.

- The SXDS field will be imaged in the U-band to limiting magnitudes of ~ 27 by the VLT/VIMOS as an ESO Public Imaging Survey.

- The UK Infrared Deep Sky Survey (UKIDSS) Ultra-Deep Survey (UDS; [9]) will use the UKIRT Wide-Field Camera (WFCAM; [10]) to survey an area $0.77\,\mathrm{deg}^2$ in the centre of the SXDS field to depths of $J = 25$, $H = 24$, $K = 23$.

- The *SIRTF* Wide-area InfraRed Extragalactic Survey (SWIRE; [11],[12]) will include a $3 \times 3\,\mathrm{deg}^2$ region containing the entire SXDS field with both IRAC ([13]) and MIPS ([14]).

- Observations are also planned with the Balloon-borne Large Aperture Submillimeter Telescope (BLAST; [15]) which will operate at 350, 450, and 750μm and it will cover $\sim 1\,\mathrm{deg}^2$.

- The near infrared 3 colour (JHK') observations using the Simultaneous InfraRed Imager for Unbiased Surveys (SIRIUS; [16]) were conducted by our

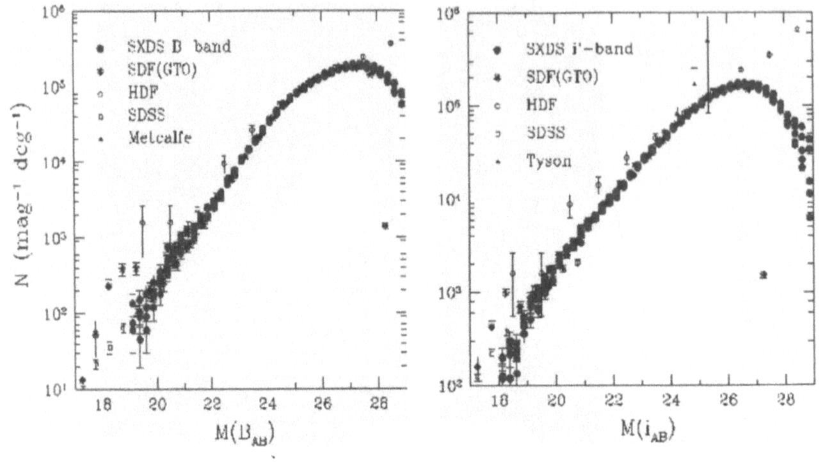

Fig. 3. The B-band (Left) and the i'-band (Right) number counts of the SXDS field. The data from several other previous studies were plotted for comparison.

team. It covered ~800 arcmin^2 to the magnitude limits of $K' = 19.5$.

Furthermore, we have started an extensive multiwavelength followup programme. The optical spectroscopy of the X-ray and radio sources and the Lyman Break galaxies with the Faint Object Camera And Spectrograph (FOCAS; [17]) on Subaru Telescope and the 2-degree field spectrograph (2dF; [18]) on the Anglo-Australian Telescope have been conducted. Also, some spectroscopic observations to calibrate the photometric redshifts have been done. The SIRTF IRAC imaging deeper than SWIRE survey is planned. We would certainly pursue the millimeter observations using ALMA, when it becomes available.

4 Summary

We have initiated a multiwavelength wide area and deep survey to provide an accurate census of objects in the Universe hence advance our understanding of the formation and evolution of structure. We have completed initial survey of the optical imaging by the Suprime-Cam on Subaru Telescope and the X-ray imaging with the EPIC camera on the ESA XMM-Newton satellite. This survey covers a $\sim 1.3\,\mathrm{deg}^2$ region of sky centered at R.A. = 02h18m, decl. = -05d (J2000), down to the limiting magnitudes at $B = 28.1$, $R = 27.5$, $i' = 27.4$ & $z' = 26.8$ (AB magnitude, 3σ detection with 2" aperture) and to $\sim 10^{-15}$ erg^{-1} s^{-1} cm^{-1} in the 0.5-2 keV band. Companion Surveys with VLA, JCMT/SCUBA, VLT/VIMOS, UKIRT/WFCAM, SIRTF/SWIRE, BLAST, and SIRIUS are being pursued. The combination of these observations will produce an unprecedented study of a large enough volume of the Universe to not be subject to cosmic variance. Our approach is to employ the photometric redshifts, using a wide range of wavelength coverage, to construct a large numbers (up to $\sim$ 1 million) of high redshift (z>1) samples which cannot be obtained by the spectroscopic surveys of the 8-10 m class telescopes.

References

1. P. Norberg, et al.: MNRAS 336, 907 (2002)
2. D. York: 'The Sloan Digital Sky Survey: Technical Summary', Astron. J. 120, 1579 (2000)
3. E. Gaztanaga, D. Hughes: 'Clustering in Deep (Submillimeter) Surveys'. In: *Deep millimeter surveys: implications for galaxy formation and evolution, the UMass/INAOE conference, University of Massachusetts, Amherst, MA, USA, 19-21 June 2000*, ed. by J. D. Lowenthal, D. Hughes (World Scientific Publishing, Singapore 2001) 131
4. K. Sekiguchi, et al.: 'Subaru/XMM-Newton Deep Survey (SXDS)'. In: *Multiwavelength Cosmology, Mykonos, Greece, 17-20 June 2003* (Kluwer Academic Publishers, Dordrecht), in press.
5. S. Miyazaki, et al.: PASJ 54, 833 (2002)
6. L. Strüder, et al.: A&A 365, L18 (2001)
7. M. G. Watson, et al.: A&A 365, L51 (2001)
8. C. Simpson, et al.: 'Radio observations of the Subaru/XMM-Newton Deep Survey field'. In: *ESO Astrophysics Symposia: Multiwavelength Mapping of Galaxy Formation and Evolution, Venice, Italy, 13-16 October 2003* ed. by R. Bender, A. Renzini(Springer-Verlag, Berlin Heidelberg), in press.
9. A. J. Adamson: 'Operational Models for the UKIRT WFCAM'. In *The New Era of Wide Field Astronomy*, ed. by R. G. Clowes, A. J. Adamson, G. E. Bromage (APS, San Francisco 2001), 232, 364
10. M. Casali: 'The UKIRT IR Wide-Field Camera (WFCAM)'. In *The New Era of Wide Field Astronomy*, ed. by R. G. Clowes, A. J. Adamson, G. E. Bromage (APS, San Francisco 2001), 232, 357
11. C. J. Lonsdale: BAAS 198, 25.02 (2001)
12. A. Franceschini, C. J. Lonsdale: 'The SWIRE SIRTF Legacy Program: Studying the Evolutionary Mass Function and Clustering of Galaxies'. In *The Mass of Galaxies at Low and High Redshift, ESO Workshop, Venice, Italy, 24-26 October 2001*, ed. by R. Bender, A. Renzini (Springer–Verlag, Berlin, 2003) 338.
13. G. G. Fazio, et al.: 'Infrared array camera (IRAC) for the Space Infrared Telescope Facility (SIRTF)'. In *Infrared Astronomical Instrumentation, Proc. SPIE 3354*, ed. by A. Fowler (SPIE, Bellingham, 1998) pp1024-1031
14. G. B. Heim, et al.:'Multiband imaging photometer for SIRTF'. In *Space Telescopes and Instruments V, Proc. SPIE 3356*, ed. by Pierre Y. Bely; James B. Breckinridge (SPIE, Bellingham, 1998) pp 985-1000
15. D. Scott, et al.: 'The Balloon-borne Large Aperture Sub-millimetre Telescope'. In *The Promise of the Herschel Space Observatory, ESA-SP 460*, ed. by G.L. Pilbratt, J. Cernicharo, A.M. Heras, T. Prusti, R. Harris. (ESA, 2001) 305
16. T. Nagayama, et al.: 'SIRIUS - Simultaneous three-color InfraRed Imager for Unbiased Surveys'. In *The New Era of Wide Field Astronomy*, ed. by R. G. Clowes, A. J. Adamson, G. E. Bromage (APS, San Francisco 2001), 232, 389
17. N. Kashikawa, et al.: PASJ 54, 819 (2002)
18. I. J. Lewis, et al.: MNRAS 333, 279 (2002)

Probing the High Redshift Universe with Extreme X-Ray / Optical Sources (EXOs)

A.M. Koekemoer[1], D.M. Alexander[2], F.E. Bauer[3], J. Bergeron[4], W.N. Brandt[3], E. Chatzichristou[5], S. Cristiani[6], S.M. Fall[1], N. Grogin[7], M. Livio[1], V. Mainieri[8], L. Moustakas[1], P. Rosati[8], E.J. Schreier[9], and C.M. Urry[5]

[1] STScI, 3700 San Martin Drive, Baltimore, MD 21218, USA
[2] Institute of Astronomy, Madingley Road, Cambridge, CB3 0HA, UK
[3] Penn State University, 525 Davey Lab, University Park, PA 16802 USA
[4] Institut d'Astrophysique de Paris, 98bis Bd Arago, F-75014 Paris, France
[5] Yale University, P.O. Box 208121, New Haven CT 06520-8121, USA
[6] INAF-Osservatorio Astronomico, via Tiepolo 11, I-34131 Trieste, Italy
[7] Department of Physics and Astronomy, JHU, Baltimore, MD 21218, USA
[8] ESO, Karl-Schwarzschild-Str. 2, Garching D-85748, Germany
[9] AUI, 1400-16th St, NW, Ste 730, Washington, DC 20036, USA

Abstract. Extreme X-ray / Optical ratio sources (EXOs) are a class of source that are robustly detected in our Chandra Deep Field X-ray data, while being completely undetected in our deep multi-band GOODS HST/ACS data, including the long-wavelength z-band data, yet are detected in the deep GOODS near-IR VLT imaging. These sources have values of Fx/Fopt that are factors of $10 - 100$ times higher than those generally found for other AGN. We thus infer two plausible scenarios: (1) if these sources lie at moderately high redshifts ($z \sim 3 - 5$) then their hosts need to be exceedingly underluminous and/or excessively reddened compared with other AGN hosts; (2) if they lie above z $6 - 7$, such that their Lyman-alpha emission is redshifted out of the z-band filter, then their K-band and X-ray properties can be accounted for in terms of relatively normal hosts and moderate-luminosity AGN. In this case, these objects can serve as an invaluable probe of black hole growth and accretion activity in the early universe.

1 Introduction

We describe a sample of sources with extreme X-ray / Optical flux ratios (or "*EXO*"s), that are robustly detected ($25 - 89$ counts) in the 2 Msec HDF-N and reprocessed 1 Msec CDF-S main catalogs (Brandt et al. 2001, Giacconi et al. 2002, Alexander et al. 2003). They are also detected in near-IR JHK imaging, but completely undetected in our deep GOODS *HST*/ACS survey, which to date is the most sensitive and detailed optical imaging of these fields (Giavalisco et al. 2004). Their X-ray emission may provide a powerful tool to investigate their physical nature, since these X-ray surveys are sufficiently sensitive to reveal AGN beyond the tentative reionization epoch ($z \sim 6 - 7$, Fan et al. 2001, 2003) where optical information is no longer available. They may thus provide valuable tracers of AGN evolution at early cosmological epochs. Throughout this work we adopt $H_0 = 70$ km s^{-1} Mpc^{-1}, $\Omega_{\mathrm{M}} = 0.3$, and $\Omega_\Lambda = 0.7$.

2 Observations and Sample Description

We began by matching the X-ray and ACS z_{850} catalogs; the matched sources are mostly moderate-luminosity AGN at $z \sim 0.5 - 4$, or star-forming galaxies at $z < 0.5 - 1$ (e.g., Hornschemeier et al. 2001; Koekemoer et al. 2002). The remaining sources were inspected in the z_{850} and the combined $B_{435}+V_{606}+i_{775}+z_{850}$ images. Most of these had faint optical counterparts, below the formal 10-σ z_{850} catalog threshold or undetected in z_{850} but detected in another band (perhaps high equivalent width Lyman-α emitters). Detection in z_{850} or bluer bands suggests these sources are at $z < 6$ (Barger et al. 2003; Cristiani et al. 2004).

Finally, there remained sources with no ACS counterparts within 2.4″ (> 10 times the positional uncertainty) in z_{850} and the combined $B_{435}+V_{606}+i_{775}+z_{850}$ images, despite robust detections in the X-ray main catalogs (25 – 89 counts). We focus on CDF-S where we have complete ACS and near-IR coverage. Five of these 7 CDF-S sources have fairly well behaved X-ray exposure maps; the other two are still detected in *JHK* thus are likely also real. The sources are also at a wide range of off-axis angles in the *Chandra* image. We computed z_{850} upper limits from the pixel r.m.s. over a 0.2″×0.2″ aperture (4×4 ACS-pixel box), to yield 3-σ upper limits of $z_{850} \sim 27.9 - 28.4$ AB magnitudes.

Table 1. Optically Undetected X-Ray Sources

R.A.[a] (J2000)	Dec.[a] (J2000)	Tot. counts (0.5–8 keV)	$F_{0.5-8\text{keV}}$ (erg/s/cm^2)	z_{850}	J	H	K
03 32 08.39	−27 40 47.0	89^{+15}_{-14}	4.3×10^{-15}	>27.9	25.2±1.3	23.3±0.2	24.8±1.1
03 32 08.89	−27 44 24.3	27^{+9}_{-8}	3.8×10^{-16}	>28.3	>25.9	24.8±0.5	23.8±0.6
03 32 13.92	−27 50 00.7	44^{+9}_{-8}	6.1×10^{-16}	>28.3	23.9±0.4	22.7±0.1	22.4±0.2
03 32 20.36	−27 42 28.5	42^{+11}_{-10}	8.7×10^{-16}	>28.3	>25.9	>25.1	>25.0
03 32 25.83	−27 51 20.3	25^{+7}_{-6}	3.3×10^{-16}	>28.4	>25.9	>25.1	23.3±0.4
03 32 33.14	−27 52 05.9	44^{+9}_{-7}	5.9×10^{-16}	>28.4	25.7±1.1	24.8±0.5	25.4±1.5
03 32 51.64	−27 52 12.8	45^{+11}_{-9}	1.3×10^{-15}	>28.4	>25.9	24.1±0.3	23.5±0.6

[a] Positions for the X-ray sources, from Alexander et al. (2003).

3 "EXO"s: Extreme X-Ray / Optical Ratio Sources

The 7 sources with no ACS counterparts are shown in Figure 1. We carried out photometry at the X-ray positions, and present the ACS upper limits and near-IR detections in Table 1. We next examine the relationship between their X-ray and optical fluxes (e.g., Maccacaro et al. 1988; Stocke et al. 1991). In Figure 2 we plot $F_{0.5-8\,\text{keV}}$ versus z_{850} for all the X-ray sources, including the ACS non-detections, also showing lines of constant F_X/F_{opt} (derived from z_{850}). At soft X-ray energies this ratio is complicated by absorption and thermal gas emission (e.g., Beuermann et al. 1999) but these no longer dominate above a

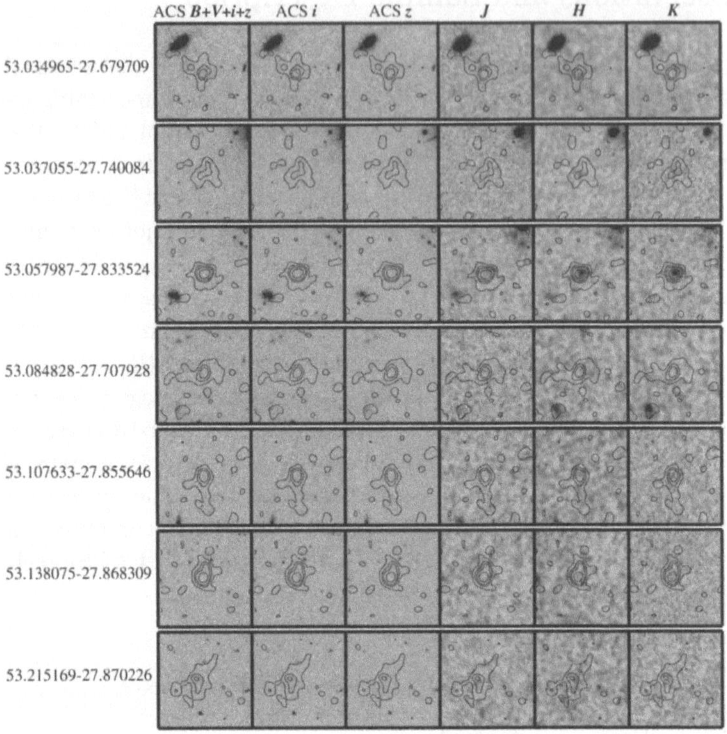

Fig. 1. The 7 CDF-S X-ray sources with no ACS counterparts, also showing the SOFI JHK detections. Each panel is 15″ on a side. Contours show 0.5 − 8 keV *Chandra* data, starting at 1,2,3-σ, and doubling thereafter.

few keV, which is redshifted into the soft band for $z > 1 - 2$. Sources with $F_X/F_{opt} > 0.1 - 10$ are generally identified with AGN; these deep X-ray data also reveal normal galaxies and starbursts (e.g., Hornschemeier et al. 2003) with F_X/F_{opt} extending a few orders of magnitude below AGN, but at $z_{850} > 24$ these are lost from the X-ray sample and the AGN dominate.

Figure 2 also reveals that our z_{850}-undetected objects occupy the high end of the F_X/F_{opt} plane, above most of the detected optically faint sources. This is interesting since it is not expected *a priori* that optically undetected objects should have strong X-ray flux. We also plot our measured ACS magnitudes for CDF-S sources reported as undetected in ground-based data by Yan et al. (2003), and show EROs from Alexander et al. (2002) and Stevens et al. (2003).

In Figure 3 we plot F_X/F_{opt} against $z_{850} - K$; the normal galaxies and starbursts with $F_X/F_{opt} < 0.1 - 1$ are blue, while the only objects with $z_{850} - K > 4$ are those with high F_X/F_{opt} (see also Brusa et al. 2002; Cagnoni et al. 2002). Not *all* the high F_X/F_{opt} objects are red; some are blue, with $z_{850} - K \sim 0 - 2$, and may be unobscured AGN. However, at high F_X/F_{opt} we also find the reddest objects, with most of our undetected z_{850} sources having $z_{850} - K > 4.2 - 6.2$.

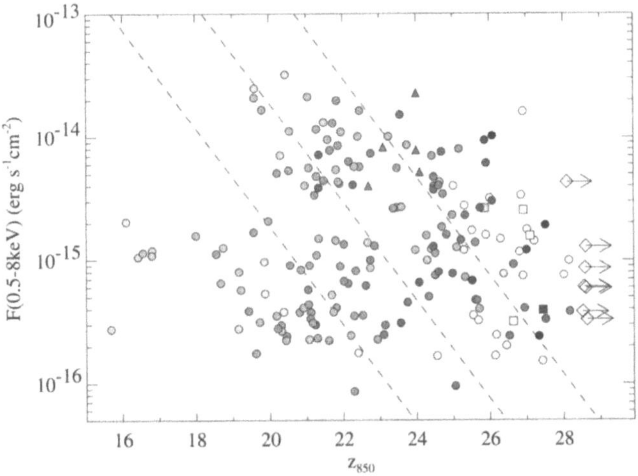

Fig. 2. Total X-ray flux ($0.5-8\,\mathrm{keV}$) against z_{850} magnitude for all the X-ray sources, including those unidentified in z_{850} (indicated by diamonds with arrows). Lines indicate $F_X/F_{\mathrm{opt}} = 0.1, 1, 10$. Symbol shading shows the $z_{850}-K$ colour of each source, ranging from light to dark for $z_{850} - K \sim 0 - 5$. White symbols (no shading) were undetected in K. Squares show our measured ACS magnitudes for sources reported as undetected in ground-based data by Yan et al. (2003), and triangles indicate X-ray selected EROs in the Lockman Hole (Stevens et al. 2003).

4 Possible Constituents of the "EXO" Population

We now examine various possibilities for these sources in order to explain: (1) deep optical non-detection; (2) robust X-ray detection; and (3) extremely red colors. We rule out main-sequence stars as these typically have $F_X/F_{\mathrm{opt}} \sim 10^{-5} - 0.1$ (e.g., Stocke et al. 1991). More exotic galactic sources include low-mass X-ray binaries, cataclysmic variables, or neutron stars, which generally have $L_X \sim 10^{30} - 10^{35}\,\mathrm{erg\,s^{-1}}$ (e.g., Verbunt & Johnston 1996 and references therein). At distances $\sim 10-100$ kpc they would have X-ray fluxes $< 10^{-14}\,\mathrm{erg\,s^{-1}\,cm^{-2}}$, consistent with what is observed. However, the number of sources in our small survey area would imply total counts $\sim 10^4 - 10^5$ times above what is known for our galaxy (e.g. Howell & Szkody 1990; Alexander et al. 2001). Thus we turn to an extragalactic origin for the *EXOs*.

Low-luminosity galaxies can host relatively luminous X-ray sources, for example the local dwarf NGC 4395 ($M_B \sim -16.5$) with a central X-ray source $L_{2-10\mathrm{keV}} \sim 3 \times 10^{38}\,\mathrm{erg\,s^{-1}}$ (Ho et al. 2001). However, such objects do not have the extremely red colors of our sources. Moreover, our z_{850} limits would imply a distance modulus > 45.3 mag, hence $L_X > 3 \times 10^{42} - 8 \times 10^{43}\,\mathrm{erg\,s^{-1}}$, at least 10^4 times above those for such dwarfs locally. It would be highly interesting if the *EXOs* were low-luminosity dwarfs with such extreme AGN, but since we know of no local analogs, we explore other alternatives.

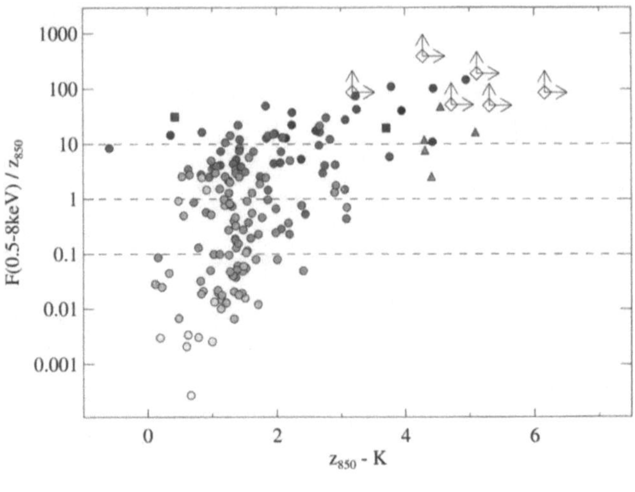

Fig. 3. Ratio of $F_{0.5-8\,\mathrm{keV}}$ to z_{850} for all the X-ray sources, plotted against $z_{850} - K$. Symbol shapes are as in Figure 2; however shading represents z_{850}, fainter sources being darker. Normal galaxies and starbursts ($F_X/F_{\mathrm{opt}} < 0.1 - 1$) are relatively blue, and that the only redder objects are those with higher F_X/F_{opt}.

4.1 Evolved or Dusty Galaxies

A possible alternative is that the *EXO*s may be at moderate redshifts and extremely evolved. At $z \sim 1, 2, 3, 4, 5, 6$, the oldest possible populations (5.7, 3.2, 2.1, 1.5, 1.2, 0.9 Gyr) would have observed colors $z_{850} - K \sim 1, 4, 4.5, 3.5, 3.2, 3$ respectively, for a single-burst model with 0.3 solar metallicity (Charlot & Bruzual 2003, in preparation). At least $\sim 1 - 2$ magnitudes of extinction would be required since we observe no UV excess. These properties are comparable to those derived for other X-ray ERO's (e.g., Alexander et al. 2002; Brusa et al. 2002; Stevens et al. 2003). However, it is interesting to note that the K-band magnitudes are ~ 100 times too low compared to what is needed to place these objects on the $M_{BH} - \sigma$ relation (see Woo & Urry 2002).

We next explore the amount of dust needed to achieve the high $z_{850} - K$ colors. If these sources were local ($z < 0.1$), then $z_{850} - K > 4.5$ would imply $A_V > 10 - 14$ (Rieke & Lubofsky 1985; Calzetti 1997), depending on the stellar populations. This would require $N_H > 2 - 5 \times 10^{22}$ cm^{-2} which is too high for the observed soft X-ray fluxes unless we invoke an unusual scenario where the AGN is unobscured while the rest of the galaxy is completely obscured.

At higher redshifts, the A_V limits decrease since z_{850} and K sample bluer rest-frame wavelengths. Between $z \sim 1 - 2.5$ we infer $A_V > 5 - 2$ respectively, comparable to similar objects discussed by Alexander et al. (2002); Smail et al. (2002); Stevens et al. (2003), and Yan et al. (2003). Again it is interesting that the *EXO* host luminosities at these redshifts, from their K magnitudes, would again place them at least 2 orders of magnitude away from the $M_{BH} - \sigma$ relation, suggesting atypical accretion rates (Woo & Urry 2002).

4.2 High-Redshift Lyman Break Galaxies

Finally we investigate the possibility that these sources are at high enough redshifts to move the Lyman break out of the z_{850} filter. Yan et al. (2003) investigated similar possibilities for their CDF-S sources undetected in ground-based R imaging but detected with ISAAC. We have in fact detected their sources in our ACS z_{850} data (Figure 2) and agree with them that those sources are unlikely to be at $z > 5$. However, their sources are the only ones with F_X/F_{opt} comparable to our *EXOs*. Thus, if the *EXOs* belong to the same population as the Yan et al. (2003) sources, then their distance modulus is about $1.5 - 2$ magnitudes higher. If the Yan et al. (2003) sources are luminous E/S0 galaxies with $z \sim 1-2.5$, this would translate to $z \sim 2.5 - 6$ for the *EXO* population. The other possibility considered by Yan et al. (2003) is that their sources may be higher redshift AGN up to $z \sim 3-5$. If *EXO*'s are the high redshift extension of this population, this would place them at $z \sim 6 - 10$. Then their host galaxy absolute magnitudes (from our K-band data) would be -20 to -21, comparable to $\sim L_*$, while their X-ray luminosities would be $\sim 2 - 6 \times 10^{44}$ erg s^{-1}. These are not unusually high luminosities for AGN, and the number of objects that we detect may thus imply fairly flat LF evolution up to and beyond $z \sim 6$.

References

1. Alexander, D., et al. 2001, AJ 122, 2156
2. Alexander, D., et al. 2002, AJ 123, 1149
3. Alexander, D., et al. 2003, astro-ph/0304392
4. Barger, A. J. et al. 2003, ApJ 584, L61
5. Bauer, F. E. et al. 2004, in preparation
6. Beuermann, K. et al. 1999, A&A 347, 7
7. Brandt, W. N., et al. 2001, AJ 122, 1
8. Cagnoni, I. et al. 2002, ApJ 579, 148
9. Cristiani, S. et al. 2004, ApJL, 600, 119
10. Fan, X. et al. 2001, AJ 121, 54
11. Fan, X. et al. 2003, AJ 125, 1649
12. Ferrarese, L. & Merritt, D. 2000, ApJ, 539, 9
13. Gebhardt, K., et al. 2000, ApJ, 539, 13
14. Giacconi, R., et al. 2002, ApJS 139, 369
15. Giavalisco, M. et al. 2004, ApJL, 600, 93
16. Ho, L. C. et al. 2001, ApJ 549, L51
17. Hornschemeier, A. E., et al. 2001, ApJ 554, 742
18. Howell, S. B. & Szkody, P. 1990, ApJ 356, 623
19. Koekemoer, A. M. et al. 2002, ApJ, 567, 657
20. Koekemoer, A. M. et al. 2004, ApJL 600, 123
21. Maccacaro, T., et al. 1988, ApJ 326, 680
22. Rieke, G. H. & Lebofsky, M. J. 1985, ApJ 288, 618
23. Schreier, E. J. et al. 2001, ApJ, 560, 127
24. Smail, I., et al. 2002, ApJ 581, 844
25. Steidel, C. C., et al. 1999, ApJ 519, 1
26. Stevens, J. A., et al. 2003, astro-ph/0302289
27. Stocke, J. T., et al. 1991, ApJS, 76, 813
28. Woo, J.-H. & Urry, C. M. 2002, ApJ 579, 530
29. Yan, H., et al. 2003, ApJ 585, 67

Clustering of Submillimetre-Selected Galaxies

Andrew W. Blain[1], Scott C. Chapman[1], Ian Smail[2], and Rob Ivison[3]

[1] California Institute of Technology, Pasadena, CA91125, USA
[2] Institute for Computational Cosmology, University of Durham, Durham, UK
[3] UK-ATC, Edinburgh, UK

Abstract. Using accurate positions from very deep radio observations to guide multi-object Keck spectroscopy, we have determined a substantially complete redshift distribution for very luminous, distant submillimetre(submm)-selected galaxies (SMGs). A sample of 75 redshifts for SMGs in 7 fields contains a surprisingly large number of 'associations': galaxies with redshifts within $1200\,\mathrm{km\,s^{-1}}$. This small sample provides tentative evidence of strong clustering at $z \simeq 2-3$ with a correlation length of $7.8 \pm 2.6 h_{100}^{-1}$ Mpc, using a simple pair-counting approach that is appropriate to the small, sparse SMG samples. The large correlation length could either indicate that SMGs trace the densest large-scale structures in the high-redshift Universe, and are evolutionarily distinct from optical-color-selected $z \simeq 3$ Lyman-break galaxies (LBGs) and QSOs, or that they are subject to a complex bias.

1 Submillimetre-Selected Galaxies (SMGs)

Since 1997 several hundred galaxies have been discovered using imaging arrays at mm and submm wavelengths (Blain et al. 2002; Smail et al. 2002; Bertoldi et al. 2004). The difficulties of identifying these galaxies at other wavelengths, and moreover determining their redshifts are significant, owing largely to the coarse positional accuracy ($\simeq 8$ arcsec) of the discovery images. By exploiting the high-redshift analogue of the far-infrared–radio correlation between hot dust and radio synchrotron emission both powered by young, hot stars, sub-arcsec radio positions can be found for a substantial fraction of SMGs, enabling efficient multi-object optical spectroscopy to search for redshifts, given their surface density on the sky is around $500\,\mathrm{deg^{-2}}$ (Chapman et al. 2003). The redshift distribution of a large sample of 75 of these galaxies in seven independent fields (Table 1) – about 45% of all SMGs and 70% of those with radio detections – is presented by Chapman et al. (2004 & this volume).

This reasonably complete redshift distribution confirms absolutely SMGs as an important high-redshift galaxy population, that dominates the luminosity density from galaxies at redshifts $z \simeq 2$. The distribution can be represented reasonably well by a Gaussian with $\bar{z} = 2.4$ and $\sigma = 0.65$. The remaining $\sim 35\%$ of SMGs with no redshifts are likely split between those at higher redshifts and any with cooler dust temperatures and lower luminosities, neither of which yield radio detections, and galaxies in the redshift range $1.2 < z < 1.8$, whose optical continua are too faint to detect rest-frame ultraviolet absorption lines, and which have no emission lines detectable using LRIS-B (Chapman et al. 2004).

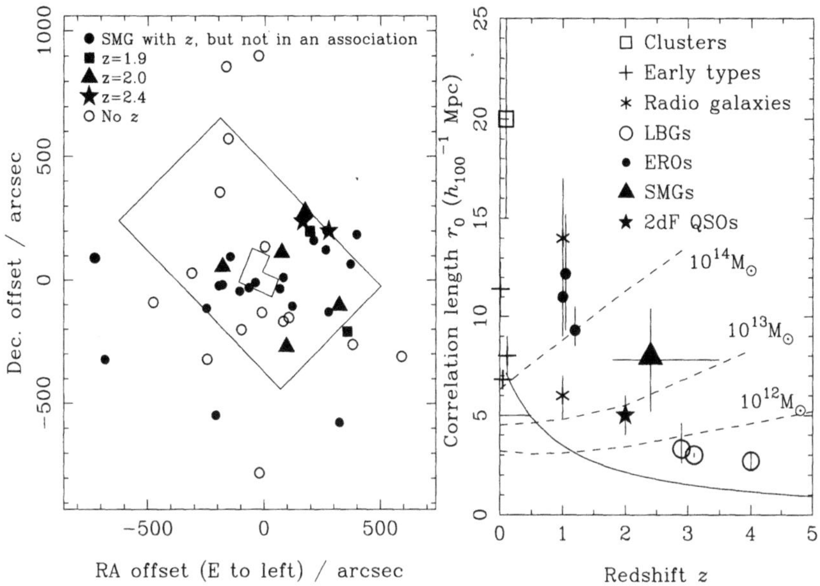

Fig. 1. Left: The positions of SMGs in the richest field in our redshift survey, surrounding the plotted boundaries of the HDF-N and GOODS fields. Galaxies in our catalog with and without redshifts, and those in the three associations present are shown by different point styles. **Right:** The correlation length of the SMGs inferred from our survey in contrast with other populations of low and high-redshift galaxies (see Overzier et al. 2003). The solid line shows a reasonable model for the evolution of an overdensity with decreasing redshift. The dashed lines show the expected correlation length of dark matter halos as a function of mass and redshift

2 The Importance of Galaxy Clustering

The detection of a clustering signal is important for understanding the distribution of different populations of galaxies as a function of dark matter density. In principle, in a dark-matter dominated Universe it provides an insight into the masses of the dark halos that host luminous galaxies (see Fig. 1 right panel).

Without redshifts, only limits to the projected clustering strength of SMGs in 2-dimensions have been obtained (Carilli et al. 2002; Scott et al. 2002; Borys et al. 2003; Webb et al. 2003). This is unsurprising given the narrow, deep geometry of SMG surveys: multiple structures along the line of sight overlap and dilute the clustering signal, as clearly shown in the left panel of Fig. 1.

The clustering strength of many types of high- and low-redshift galaxies have been probed by a wide variety of methods (see Fig. 1), as summarized by Overzier et al. 2003). At redshifts below those of SMGs, $z < 1.3$, large-scale structure of galaxies is being mapped in great detail by the DEEP2 survey (Coil et al. 2003). At higher redshifts, the clustering of many hundreds of LBGs can be measured (Adelberger et al. 1998). Note that published LBGs lie at higher redshifts than the SMGs: $z = 3 \pm 0.3$. This explains the lack of observed correlation between

LBGs and SMGs (Webb et al. 2003). The lower-redshift counterparts to LBGs (Steidel et al. 2004) in fields that overlap with the SMG surveys listed in Table 1 should provide an excellent opportunity to determine the relative 3-dimensional distribution of LBGs and SMGs. The clustering of QSOs and radio galaxies at the same redshifts as the SMGs has been probed by Croom et al. (2002) and Overzier et al. (2003). However, these are sparser, larger surveys in which the typical source separation is much greater than the correlation length.

3 Signs of Clustering in the SMGs

In five of the seven fields 'associations' of galaxies are found, with redshifts separated by less than 1200 km s^{-1} in radial velocity (Table 1). This corresponds to an approximate velocity dispersion expected for a rich cluster of galaxies, and larger than the offset velocity expected from galactic winds. Most of these associations are pairs, but there is a group of three in the highest overdensity in the survey of optically selected Lyman-break galaxies (LBGs) (Steidel et al. 2000), and a group of five in the HDF (Fig. 1). The associations are not close interacting pairs, but have separations up to scales of many Mpc; they may finally reside in the densest clusters, but at the observed epoch, they are much more widely spread.

The SMGs are unusual for a high-redshift population in being found in relatively small fields, less than 10 arcmin in extent, but spread very widely in redshift from at least 0.8 to 3.4. This corresponds to a uniquely long thin pencil geometry, which extends in radius by 2–3 Gpc, but across the sky only as far as about 5 Mpc, the scale of the correlation length r_0 of the galaxy distribution.[1]

That SMGs can definitely be found in close proximity to other classes of high-redshift galaxies is confirmed by the detection of optically-selected galaxies at redshifts matching those of known SMGs within about 100 kpc in the backgrounds of two lensing clusters of galaxies (at $z = 2.80$ in Abell 370; Santos et al. 2004; and $z = 2.55$ in Abell 2218 Kneib et al. 2004). Overdensities of bright SMGs near the targeted positions of some of the most extreme high-redshift radio galaxies at redshifts as high as $z = 3.8$ (Stevens et al. 2003) provide circumstantial evidence for some strong overdensities within the population. Redshifts for optical-selected galaxies obtained in parallel to our survey, and for $z \simeq 2-3$ color-selected optical galaxies (Steidel et al. 2004) in the SMG survey fields will provide more information on the cross correlation between the SMGs and the more numerous optically-selected galaxies.

4 The Correlation Length of SMG Clustering

To interpret our sparse data we make use of the limited information as best we can, comparing the number of pairs of galaxies measured in each field (Table 1)

[1] The two-point correlation function of galaxies that describes the excess probability of finding a pair of galaxies separated by r over an unclustered distribution is observed to be a decreasing power-law with distance $\xi(r) = (r/r_0)^{\simeq -1.8}$ (Coil et al. 2003)

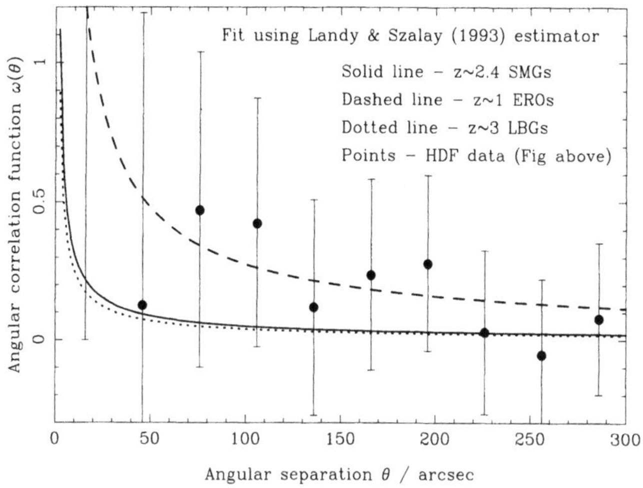

Fig. 2. The angular correlation functions (ACFs) observed for 45 SMGs in the field shown in Fig. 1. See also Borys et al. (2003), Carilli et al. (2000), Scott et al. (2002) and Webb et al. (2003). The correlation functions of EROs (Daddi et al. 2003), LBGs and SMGs from our correlation length are also shown. After projection SMGs and LBGs have similar ACFs, much weaker than for EROs. The errors on the ACF data are much too large to resolve the true model without much larger SMG samples. Redshifts are essential to discern the clustering of SMGs.

against the number expected assuming a redshift-independent correlation length, the known redshift distribution of the population and a space density normalised to match the observed the number of SMGs in each field. This requires a double integral over volume, with one radial variable being redshift and the other the distance corresponding to relative velocity in the Hubble flow, with limits set to the limiting velocity to define a pair: $1200\,\mathrm{km\,s^{-1}}$. The results are not very sensitive to reasonable changes in the choice of velocity range, but do depend on the geometry of the field: smaller fields yield a larger boost in the expected number of pairs, as the correlation function is sampled on smaller scales.

Only for the SA22 and HDF fields are the results significant, as the other fields include at most a single pair. The number of pairs expected in our fields with no clustering are listed in Table 1, along with the correlation length required to boost this number to match the observed result. The definition of the field size within the HDF region makes little difference to the final result (Table 1). The correlation length for the HDF field is thus $r_0 = 7.8 \pm 3\,h^{-1}\,\mathrm{Mpc}$. The correlation length for SA22 is greater, but note that SSA22 includes a known strong overdensity at the redshift of the association of SMGs (Steidel et al. 2000), and so we do not claim that including the r_0 value for SA22 would give a fair representation of the SMGs in a typical field. The Lockman field promises to yield a comparable number of redshifts in the next year.

Table 1. The radio-pinpointed SMG survey fields. The total number of spectroscopic redshifts (N_{gal}) and the numbers of pairs of galaxies N_{pair} within 1200 km s^{-1} in radial velocity at redshift z_{as} in each are listed. In the HDF (Fig. 1) brackets correspond to the values in an inner square that includes most of the galaxies. There is a 5-galaxy association in the HDF at $z = 2.0$ and a triplet[a] in SA22 at $z = 3.1$. The pair abundance predicted in the absence of clustering N'_{pair} and the correlation length r_0 required to match this to the observed number are also given.

Field	RA	dec	Size(" × ")	N_{gal}	N_{pair}	z_{as}	N'_{pair}	r_0/h(Mpc)
03hr	03 02	00 06	320 × 310	6	0	...	0.21	...
Lockman	10 52	57 20	360 × 410	5	0	...	0.21	...
HDF	12 36	66 10	1450 × 1750	28	9	1.9, 2.0	6.46	7.5
			(800 × 800)	(24)	(9)	2.4	(4.35)	(8.2)
SA13	13 12	42 40	370 × 500	9	1	2.6	0.67	3.7
14hr	14 18	52 29	320 × 210	8	1	2.1	0.41	5.2
N2	16 37	40 55	480 × 700	8	1	2.4	0.53	6.3
SA22	22 18	00 12	800 × 760	11	3	3.1[a]	1.0	11.5

[a] These SMGs lie in the richest spectroscopically confirmed high-redshift large-scale structure (Steidel et al. 2000), which contains 29 of the 76 LBGs in this field. If the geometry matched the SMG surveys, then only 15% of LBGs would lie in this structure, not the 3/11 found. The density of Westphal LBGs and HDF SMGs is 8×10^{-3} and 3.2×10^{-4} Mpc^{-3} respectively,

The robustness of our results can be checked using the same procedure in richly sampled LBG fields. The largest 850×850" Westphal field[2] contains 192 galaxies with redshifts, and 912 pairs using our definition. As the LBG redshift distribution is approximately a Gaussian with $\bar{z} = 3.0$ and $\sigma_z = 0.28$, 594 pairs would be expected with no clustering, and thus a correlation length of $6h^{-1}$ Mpc is found, consistent with Adelberger et al. (1998) in a ΛCDM cosmology. The correlation length of $z \simeq 2$ QSOs is less, 5 h^{-1} Mpc (Croom et al. 2002).

Redshifts are essential to reveal clustering properties of the SMGs (Fig. 1). The clustering length is similar to the projected extent of the fields, and so the relative spatial position of the members of each 'association' includes little information on the nature of the clustering present.

It is interesting that the masses of the dark-matter halos surrounding LBGs inferred from their clustering properties (Abelberger et al. 1998) exceed the central dynamical masses inferred from near-IR spectra (Erb et al. 2003). This could indicate that more complex astrophysics is at work in SMGs to break a simple link between clustering amplitude and mass. For example, the time-dependence of merging has been claimed to both boost (Scannapieco & Thacker 2003; from infall), and leave unaffected (Percival et al. 2003) the clustering properties of

[2] ftp://ftp.astro.caltech.edu/pub/ccs/lbgsurvey

galaxies. Intuitively, briefly-visible, very luminous galaxies appearing in overdense regions might be expected to dilute the clustering signal, but perhaps they are triggered preferentially in dense regions. Understanding these relationship between the most extreme visible objects and the underlying dark matter distribution is essential to account for galaxy evolution.

5 Summary

Redshifts for SMGs hint that they are clustered with a correlation length of order $8h^{-1}$ Mpc. Much larger samples and cross-correlations with more numerous populations are required to better quantify this value, but it offers the potential to signpost the densest regions of the high-redshift Universe, provide a comparison between their dynamical and halo masses and investigate the mechanisms that trigger their extreme luminosities.

Acknowledgements

AWB is supported by NSF grant AST-0205937. The Keck telescopes are a scientific partnership among Caltech, UC and NASA, and were made possible by generous financial support from the W.M. Keck Foundation.

References

1. Bertoldi F. et al.: ApJ submitted (2004)
2. Blain A. et al.: Phys. Rep. **369** 111 (2002)
3. Borys C. et al.: MNRAS **344** 385 (2003)
4. Carilli C. L. et al.: conference proceeding (2000; astro-ph/0009298)
5. Chapman S. C. et al.: Nature **422** 695 (2003)
6. Chapman S. C. et al.: ApJ submitted
7. Coil A. et al.: ApJ in press (2003; astro-ph/0305586)
8. Croom S. et al: MNRAS **335** 459 (2002)
9. Daddi E. et al.: ApJ **588** L83 (2003)
10. Erb D. et al.: ApJ **591** 101 (2003)
11. Kneib J.-P. et al.: MNRAS submitted
12. Landy S., Szalay A.: ApJ **412** 63 (1993)
13. Overzier R. A. et al.: A&A **405** 53 (2003)
14. Percival W. J. et al.: MNRAS **338** L12 (2002)
15. Santos M. et al.: ApJ in press (2004; astro-ph/0310478)
16. Scannapieco E., Thacker R.J.: ApJ **590** L69 (2003)
17. Scott S. E. et al.: MNRAS **331** 839 (2002)
18. Smail I. et al.: MNRAS **331** 495 (2002)
19. Steidel C.C. et al.: ApJ **532** 170 (2000)
20. Steidel C.C. et al.: ApJ **592** 728 (2003)
21. Steidel C.C. et al.: ApJ in press (2004)
22. Stevens J.A. et al.: Nature **425** 264 (2003)
23. Webb T. et al.: ApJ **582** 6 (2003)

Star Formation and AGN in the Early Universe: Quasars in the MAMBO Deep Field Survey

F. Bertoldi[1], C.L. Carilli[2], H. Voss[1], F. Owen[2], D. Lutz[4], H. Dannerbauer[4], K.M. Menten[1], and M.A. Holdaway[3]

[1] Max-Planck-Institut für Radioastronomie, 53121 Bonn, Germany
[2] NRAO, Soccoro, NM 87801, USA
[3] NRAO, Tucson, AZ 85721, USA
[4] Max-Planck-Institut für extraterrestrische Physik, 85741 Garching, Germany

Abstract. Blank field surveys with SCUBA and MAMBO revealed a population of (sub)mm background galaxies which are massive, $\sim 10^{11} M_\odot$ starbursts at median redshifts of ~ 2.5. Due to a radio-selection bias, for one third of the SCUBA/MAMBO sources a clear optical identification or redshift estimate has not yet been possible, but it is likely that these sources lie at redshifts > 3. We find evidence for a different population dominating the counts at $S_{1.2\mathrm{mm}} >$ 10 mJy. The three brightest sources yet found in a wide field MAMBO survey are of lower redshifts ($\sim$0.8 to 1.4), are X-ray bright, radio-loud, and time-variable. Their areal density is significantly higher than that expected from an extrapolation of the flat spectrum radio source population to 1.2 mm. Our preliminary analysis of the the brightest MAMBO source with a thermal submm spectrum (10 mJy at 1.2 mm) suggests it to be a radio-quiet, X-ray bright quasar at $z \sim 3$, thereby is possibly the most luminous starburst ever found.

1 The (Sub)mm Background Resolved

At least half of the cosmic extragalactic energy density (not counting the cosmic microwave background) emerges at far-infrared wavelengths [12,16]. During the past five years, blank field surveys and cluster lens assisted surveys [4], with SCUBA at 850 μm and MAMBO at 1200 μm have resolved a significant fraction of this background into marginally detectable ($\sim$ 3-5 mJy at 1.2mm, a factor 3 brighter at 850μm) sources that are likely to be ultraluminous infrared galaxies at redshifts $\sim 2-4$ [26,19,1,3,8,9]. This previously unknown population of dusty and therefore optically inconspicuous galaxies contributes significantly to the cosmic energy production (and star formation) in the early Universe, although this estimate much depends on the yet ill-constrained dust temperature (and initial mass function).

The high star formation rates ($\approx 1000\, M_\odot \mathrm{yr}^{-1}$) implied by the estimated far-IR luminosities ($L_{\mathrm{IR}} \sim 10^{12-13} L_\odot$), and the sizes and linewidths (300 – 800 kms^{-1}) of some (sub)mm sources recently detected in CO line emission [22,14] implies that these galaxies are very massive ($\sim 10^{11} M_\odot$) systems. The similarity of the co-moving space density of (sub)mm sources and local ellipticals suggests that they are massive spheroids caught during their formation.

The properties and mass function of these high redshift massive galaxies provide robust tests for models of galaxy formation. There is mounting evidence

[13,22,14] that current models [15,20,28,2] need modifications to reproduce the space densities of the massive high redshift (sub)mm galaxies and their quiescent phases, that must exist since their star formation rates and gas content suggest a low duty cycle of the bright phase. With starburst duty cycles of order 10^8 yr, the extrapolated average volume density of the massive galaxies traced by the MAMBO/SCUBA surveys by far exceeds that predicted by semi-analytic models of galaxy formation through hierarchical mergers [13,14]. With this discrepancy the (sub)mm sources hold important clues to understand the formation and evolution of galaxies. They provide a means to discriminate between an early formation of spheroids via monolithic collapse, and a late formation via hierarchical merging.

Since the (sub)mm galaxies are the most massive of the high redshift star forming galaxy populations, they are also potential beacons to highly biased (i.e., clustered) regions of galaxy formation in the early Universe [27].

2 Redshifts: Missing the High End

While submillimeter and millimeter blank field surveys have by now identified over one hundred sources, spectroscopic redshifts have been difficult to obtain, mainly because of the faintness of their optical/near-IR counterparts and the lack of accurate positions associating them uniquely to an optical source. Accurate source positions could be obtained for about half of the secure (sub)mm sources from deep VLA 20 cm radio maps, which show the most likely counterparts, although ambiguities remain due to the faintness of the radio counterparts (typically $30 - 100\mu$Jy, often detected with low significance) and the apparent clustering of galaxies around the (sub)mm sources.

If the (sub)mm galaxies have radio-to-far-infrared spectral energy distributions similar to those of local star forming galaxies, their observed radio-to-(sub)mm spectral indices provide a rough estimate of the populations' redshift distribution [5]. However, the thereby derived redshift distribution introduces a bias toward low redshifts: due to distance dimming, even our deepest (6.5 μJy rms for Abell 2125) 20 cm radio maps fail to detect counterparts at redshifts larger ~ 3, unless the radio emission is enhanced by an AGN. Therefore currently at least one-third of the MAMBO/SCUBA sources remain without positional or redshift identification. It is this population of (sub)mm galaxies that are most likely to lie at very high (> 3) redshift, and hence provide the most stringent challenges to hierarchical galaxy formation models, and better means must be found to identify them properly.

As an alternative photometric redshift measure, we observed a sample of 23 MAMBO galaxies (with radio counterparts) with SCUBA at 850 μm [11]. Within the uncertainty arising from photometric errors and the degeneracy of dust temperature and redshift, the mm/submm photometric redshift distribution is consistent with the radio-(sub)mm photometric redshift distribution of the larger MAMBO/SCUBA sample (i.e. most sources at redshift 1.5-3.5, median ~ 2.5). However, many of the sources show significantly flatter SEDs than expected from the radio-mm photometric redshift: half of the ten sources we observed in the Abell 2125 field yield mm/submm photometric redshifts of 4 to 10. The flat SEDs could partly be due to a bias towards higher redshift in MAMBO- rather than SCUBA-selected sources, which would also contribute to the observed differences in optical/near-IR counterpart brightness [8,9].

The radio-mm-submm photometric redshift distributions were recently confirmed by optical follow-up spectroscopy for some of the radio-detected submm sources, resulting in a (again, low redshift-biased) sample of now > 90 sources with optical redshifts [6]. Recently 14 of these sources have been observed with the IRAM Plateau de Bure interferometer, detecting CO emission lines from 7 of them [22,14], thereby confirming the radio-optical identifications. The observed CO linewidths and source sizes establish that these galaxies must be very massive systems. Interestingly, half of the targets either show no CO emission near the optical redshifts, and in one case, the CO emission is associated with a galaxy neighboring that for which the optical redshift was measured. The latter underlines mounting evidence that (sub)mm galaxies are part of a complex environment of multiple sources of different morphologies and colors (Fig.1).

Although much has been learned about the (sub)mm background, the most interesting part of the harvest remains poorly explored: the highest redshift (sub)mm sources. All our detailed knowledge of the (sub)mm galaxies derives from those which could be uniquely identified in radio maps. For those the optical counterparts (if present) could be studied, optical redshifts measured, and CO detected. About half of the population, however, remains unexplored.

3 The Brightest Millimeter Sources: Quasars

One aim of the MAMBO ("Max-Planck Millimeter Bolometer" array at the IRAM 30m telescope) survey was to cover a very large field in order to constrain the abundance and nature of the very brightest mm background sources. We imaged three fields with the MAMBO 37- and 117-element receivers: a 1600 arcmin^2 region surrounding the low-redshift Abell 2125 cluster (Fig. 1), a $\sim$400 arcmin^2 region in the Lockman Hole, and a $\sim$200 arcmin^2 region centered on the NTT Deep Field, summing up to about 2200 arcmin^2 imaged to a rms noise level ranging from 0.5 to 3 mJy per 11 arcsec beam [3,8,9,11]. More than 50 sources were detected at $> 4\sigma$, most of which show a faint ($< 100\mu$Jy) 1.4 GHz radio counterpart. Their radio-to-mm spectral energy distributions are similar to those of starburst galaxies, which are dominated in the radio by synchrotron emission, and in the far-infrared by a thermal dust emission spectrum.

Interestingly, we discovered an unexpected large number of bright objects (circled in Fig. 1), with 1.2 mm flux densities in the range 10-100 mJy. The three brightest of these show bright optical, radio, and X-ray counterparts and a high degree of variability in the mm flux density on timescales of months. Their flat radio-to-mm spectral energy distributions and their variability implies a non-thermal nature of their mm emission, arising from a radio-loud AGN. For two of the brighter sources we have optical redshifts, placing them at $z = 1.38$ [23]

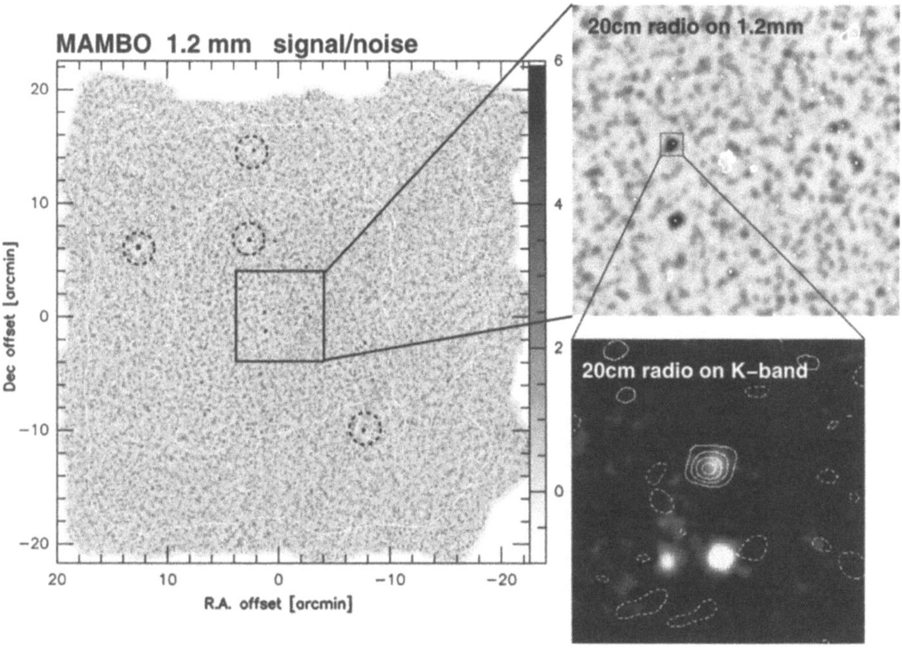

Fig. 1. *Left:* The Abell 2125 MAMBO deep field, displayed as a signal/noise map. White contours show where the rms noise level is below 0.9 mJy and 1.8 mJy. About 50 sources are identified with flux densities between 3 and 50 mJy. The three non-thermal sources (circled) stand out clearly, the thermal 10 mJy source, the northern of the circled sources, less so. *Upper right:* Radio 20cm contours on the MAMBO S/N map. At least half of the reliable MAMBO sources have likely radio counterparts. *Lower right:* Deep 15 arcsec K-band image (+radio contours) on one of the (4 mJy) MAMBO sources, revealing a typical multiple-source counterpart [3].

and 1.24. The third has a more uncertain redshift ~ 0.8. The X-ray luminosities are in the range $0.7 - 6 \times 10^{44}$ erg s^{-1}, classifying them as quasars.

The fainter object, $S_{1.2\mathrm{mm}} \approx 10$ mJy, shows no optical counterpart, and a 1.4 GHz radio flux density that is consistent (using the tight relation found in local star forming galaxies [7,5]) with star formation activity if the object is at $z \sim 3$. Assuming a thermal dust emission spectrum with a typical $T \sim 50$ K, this object would have an infrared luminosity of $\sim 3 \times 10^{13} L_\odot$, with an implied star formation rate of $\sim 5000\ M_\odot \mathrm{yr}^{-1}$, making it the most luminous starburst yet known. In particular, this object appears to be more luminous in the (sub)mm and X-rays than the unusual type-1 quasar SMMJ04135, recently discovered in SCUBA blank fields [21], or the type-2 QSO/submm source SMMJ02399 [13].

Its X-ray emission reveals the presences of a radio-quiet AGN, which is not unusual for (sub)mm galaxies. However, the implied X-ray luminosity of $\approx 6 \times 10^{44}$erg s^{-1} classifies this object as a quasar. The relation of the (sub)mm background population with the optical quasar or X-ray background population

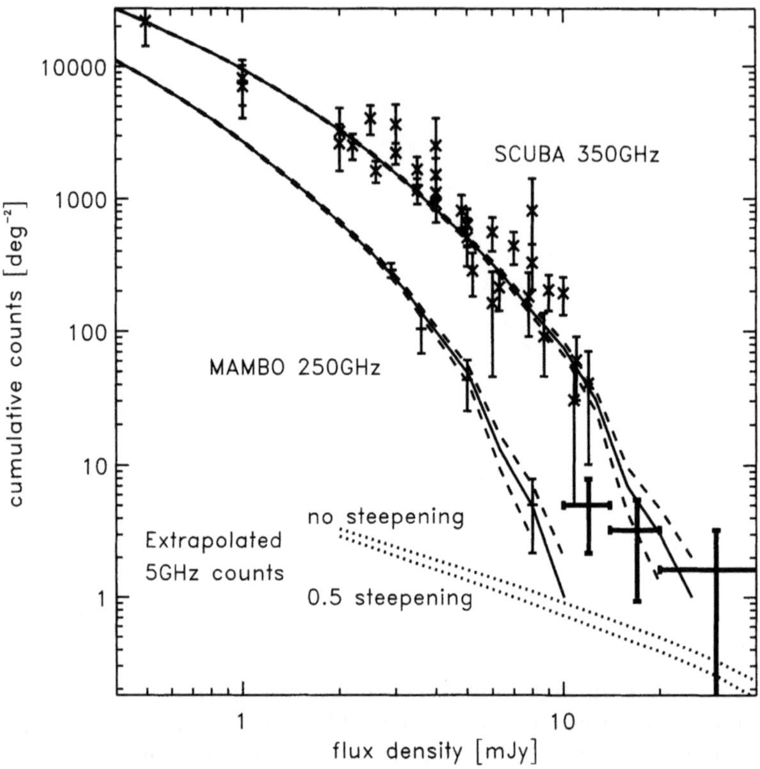

Fig. 2. The cumulative source counts of the MAMBO and SCUBA surveys. The rightmost three data points correspond to the non-thermal sources, which are *not* included in the counts below 11 mJy. The curves show our model fit to the thermal population's number counts and redshift distribution, using the Saunders [25] 60 μm luminosity function of local galaxies with an additional exponential cutoff at $\sim 10^{12}\ L_{\odot}$. The dotted lines show the expected numbers of flat spectrum radio sources extrapolated from 5 GHz counts, and spectral properties extrapolated from 90 GHz upward, with or without a steepening above 90 GHz [17].

remains unclear, and only a minority of all (sub)mm sources could be identified with X-ray or broad emission line objects. The four bright sources in the Abell 2125 MAMBO survey are among the first quasars to be discovered in mm or submm blank field surveys.

Figure 2 shows the cumulative source counts in the MAMBO and SCUBA surveys. The three brightest sources found in our survey do not match an extrapolation of our model fit to $>$ 10 mJy (Fig. 2). The mm background appears to change from a declining population of distant thermal starbursts to a shallower distribution of less distant, radio-loud quasars.

We compared the source counts at 1.2 mm with predictions based on models constrained by lower frequency observations of known flat spectrum radio source (FSS). The areal density of mm FSS implied by the three bright MAMBO sources exceeds by a factor of 6 what is expected from 5 GHz FSS source counts [7], for which the spectral properties were extrapolated upwards from 90GHz. This could mean that either the Abell 2125 field is unusual, or that the assumed spectral index distribution does not apply to sources much fainter than the $S_{5\mathrm{GHz}} > 100$ mJy FSS sample of Patnaik [24], for which the spectral index distribution seems to be independent of brightness. If the high areal density of bright mm sources is not a statistical fluke, this unexpected population may prove useful for ALMA calibrations [18].

References

1. A. Barger, L.L. Cowie, D.B. Sanders, et al.: Nature, **394**, 248 (1998)
2. C.M. Baugh, A.J. Benson, S. Cole, et al.: In *The Masses of Galaxies at Low and High Redshift*, eds. R.Bender & A. Renzini (Berlin: Springer), p.91 (2003)
3. F. Bertoldi, C.L. Carilli, K.M. Menten, F. Owen, et al.: A&A, **360**, 92 (2000)
4. A.W. Blain, I. Smail, R.J. Ivison, et al.: Physics Reports, **369**, 111-176 (2002)
5. C.L. Carilli, M. Yun: ApJ **513**, L13 (1999)
6. S.C. Chapman, A.W. Blain, R.J. Ivison, I.R. Smail: Nature, **422**, 695 (2003)
7. J.J. Condon: ARAA **30**, 575 (1992)
8. H. Dannerbauer, M.D. Lehnert, D. Lutz, L. Tacconi, F. Bertoldi, C. Carilli, R. Genzel, K.M. Menten: ApJ, **573**, 473 (2002)
9. H. Dannerbauer, M.D. Lehnert, D. Lutz, L. Tacconi, F. Bertoldi, C. Carilli, R. Genzel, K.M. Menten: ApJ, in press (2004), astro-ph/0402549
10. H. Dole, G. Lagache, J.L. Puget: ApJ, **585**, 617 (2003)
11. S. Eales, F. Bertoldi, R. Ivison, C.L. Carilli, L. Dunne, F. Owen: MNRAS **344**, 169 (2003)
12. D.J. Fixsen, C.L. Bennett, J.C. Mather: ApJ, **526**, 207 (1999)
13. R. Genzel, A.J. Baker, L.J. Tacconi, D. Lutz, et al.: ApJ, **584**, 633 (2003)
14. R. Genzel, A.J. Baker, R.J. Ivison, F. Bertoldi, et al.: this volume (2004)
15. B. Guiderdoni, E. Hivon, F.R. Bouchet, B. Maffei: MNRAS **295**, 877 (1998)
16. M.G. Hauser, R.G. Arendt, T. Kelsall, et al.: ApJ, **508**, 25 (1998)
17. M. Holdaway, F. Owen, M. Rupen: MMA Memo **123** (1994)
18. M. Holdaway, C.L. Carilli, F. Bertoldi: LAMA Memo **805** (2004)
19. D. Hughes, S. Serjeant, J. Dunlop, et al.: Nature, **394**, 241 (1998)
20. G. Kauffmann, J.M. Colberg, A. Diaferio, S.D.M. White: MNRAS, **303**, 188 (1999)
21. K.K. Knudsen, P.P. van der Werf, W. Jaffe: A&A, in press (2004)
22. R. Neri, R. Genzel, R. Ivison, F. Bertoldi, et al.: ApJ **597**, L113 (2003)
23. F.N. Owen, M.J. Ledlow, W.C. Keel, G.E. Morrison: AJ **118**, 633 (1999)
24. A. Patnaik et al.: MNRAS, **254**, 655 (1992)
25. W. Saunders, M. Rowan-Robinson, A. Lawrence, et al.: MNRAS **242**, 318 (1990)
26. I. Smail, R.J. Ivison, A.W. Blain: ApJ **490**, L5 (1997)
27. I. Smail, C.A. Scharf, R.J. Ivison, et al.: ApJ **599**, 86 (2003)
28. R.S. Somerville, J.R. Primack, S.M. Faber: MNRAS **320**, 504 (2001)

The CDF-S Viewed with SIMBA

Tommy Wiklind[1], Stefan Bergström[2], Gustaf Rydbeck[2], Felipe Barrientos[3], Duilia de Mello[4,5], Leopoldo Infante[3], Ken I. Kellerman[6], Colin Norman[7], and Piero Rosati[8]

[1] ESA Space Telescope Division, STScI , 3700 San Martin Dr., Baltimore, MD 21218, USA
[2] Department of Astronomy and Astrophysics, Onsala Space Observatory, SE-43992 Onsala, Sweden
[3] Departamento de Astronomia y Astrofisica, Pontificia Universidad Catolica de Chile, Casilla 306, Santiago 22, Chile
[4] Laboratory for Astronomy and Solar Physics, Code 681, Goddard Space Flight Center, Greenbelt, MD 20771, USA
[5] Department of Physics, Catholic University of America, 200 Hannan Hall, Washington, DC 20064, USA
[6] National Radio Astronomy Observatory, 520 Edgemont Road, Charlottesville, VA 22903, USA
[7] Department of Physics and Astronomy, Johns Hopkins University, 3400 North Charles Street, Baltimore, MD 21218, USA
[8] European Southern Observatory, Karl-Schwarzschild-Strasse 2, D-85748 Garching bei München, Germany

Abstract. Observations of the Chandra Deep Field South (CDF-S) with the bolometer array SIMBA on the SEST 15m telescope is presented. SIMBA operates at 1250μm and uses a fast-scanning observing technique suitable for large fields. An effective area of 110 sq arcmin was observed to a homogeneous noise level of 1.5 mJy.

1 Introduction

Observations at sub-millimeter wavelengths have revealed a population of dusty high redshift far-infrared luminous (FIR) galaxies [4] [12] [13] [19] [25], emitting most of their bolometric luminosity in the FIR. This population accounts for a large portion of the far-infrared background [7]. The main questions associated with this population has from the beginning been the nature of their high FIR luminosities, the nature of their power source, their redshift distribution and their dynamical masses. The question of their redshift has been addressed through the use of 'photometric' techniques, such as the radio-FIR flux ratio [8], and FIR/submm flux ratios [3] [19] [20] [30]. These techniques have shown that the redshift distribution of the submillimeter galaxies (SMGs) is dominated by objects at redshifts z$\geq$1-2. Recently however, direct optical spectroscopy has provided a first derivation of the redshift distribution, showing it to have a median redshift of z=2.4 [10].

Redshift information has allowed estimations of the masses of the SMGs, both through observations of CO emission [14] [15] [16] [21], and optical/NIR photometric measurements [27]. Both the gas and stellar masses are found to be $\sim 10^{11}$ $M_{\odot}$ and the co-moving density, corrected for an estimated duty cycle of the high FIR luminosity of $\sim 10^8$ years, is $\sim 10^{-4}$ Mpc^{-3} [16]. Recent studies show a correlation between X-ray sources detected with Chandra and SMGs [1] [2] [28], where ~40% of SMGs contain an AGN. The AGN activity, however, accounts for only 10-20% of the bolometric luminosity [2], leaving the rest to be powered by massive star formation. A seemingly robust signature of clustering has been detected among the SMGs [27] and in the cases where optical/NIR counterparts have been securely identified, they often show strongly interacting systems [7] [9] [11]. The co-moving space density, mass and environment together suggest that the SMG population constitutes the formation of massive elliptical galaxies in the central region of galactic proto-clusters. The inferred star formation rates, given the modest contribution from AGNs, further implies that the formation occurs on a relatively short time scale, followed by passive evolution.

The SCUBA and MAMBO bolometers operate at 850μm and 1250μm, respectively. The longer wavelength of MAMBO means that it is more sensitive to higher redshifts than SCUBA, assuming all dust properties remain constant for the sources. Unfortunately, there has been relatively little overlap between the SCUBA and MAMBO fields. One exception is the Lockman Hole [18] [24], where some sources detected at 1250μm are not seen at 850μm, and vice-versa [18]. It is presently unclear if this means that a different, higher redshift and/or cooler, population is seen at longer wavelengths or if these sources represent spurious detections.

Here we present the first results from a deep survey made with the SEST IMaging Bolometer Array (SIMBA) on the 15m SEST telescope at ESO La Silla. The survey covered a substantial part of the Chandra Deep Field South (CDF-S) and was done at 1250μm. A more detailed account will be presented in [5].

2 The SIMBA Bolometer

SIMBA was installed on the SEST in June 2001 as a collaborative project between the University of Bochum, the Max-Planck Institute for Radio Astronomy in Bonn, the Swedish National Facility for Radio Astronomy and ESO. SIMBA is a 37-channel bolometer array operating at wavelengths of 1.2 mm. The telescope beam size of the SEST is 24″ at the operating wavelength. The 37 channels are arranged in a hexagonal , non-overlapping pattern, with a total extent of slightly more than 4′.

As SEST did not have a nutating subreflector, images were produced in a fast-scanning mode, where the modulation is achieved through rapid, ~15 ms, sampling as the array is moved over the sky. The atmospheric fluctuations are assumed to affect the entire array elements in a similar manner over one sampling period, creating correlated noise over the array elements. This correlated signal is removed during the data reduction stage. A description of the fast-scanning observing mode can be found in [23][29]. The typical scanning speed was 60″ – 120″/sec. A map is created by scanning the telescope in azimuth, shifting 8″ in

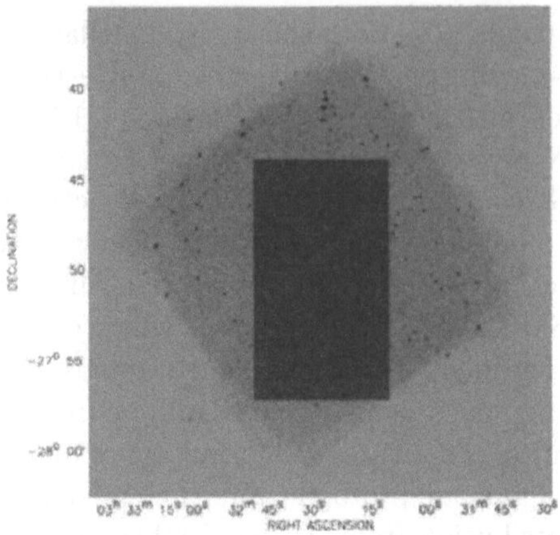

Fig. 1. The region observed with SIMBA (shaded) overlayed on the 1Msec Chandra hard X-ray image of the Chandra Deep Field South (CDF-S) [17]. The area covered by SIMBA also covers a substantial part of the GOODS field and the Ultra Deep Field (UDF).

elevation and repeating the scan. A minimum of 52 shifted scans are needed to let all array elements cover the same point on the sky. The fast-scanning method ensures a fast mapping capability, well suited for blank field surveys. For the CDF-S observations we used a scanning speed of $80''$/sec with a length of $800''$ and used $8'' \times 60$ shifts resulted in a map of size $800'' \times 500''$. The time to finish one map was 12 minutes and the typical noise rms per map was 30-40 mJy.

The observing mode affects the final map in two ways: (1) the edges of the mapped region will be covered by less array elements than the central parts, creating a drop-off in sensitivity towards the edges, and (2) since the scan direction is fixed in azimuth the astronomical object will rotate as the source is tracked. The field rotation is negligible for individual maps, but significant when adding maps obtained at different hour angles. The final result is an irregular area which represents the best and most homogeneous noise characteristics.

2.1 Data Reduction

The data reduction of the data obtained from SIMBA is done with the reduction package MOPSI[1]. After removing bad data, caused by the detector, and those obtained with zenith opacities >0.3 the data set contains 485 individual maps. In addition we obtained at least one map of Uranus per night for calibration purposes. Each map is reduced separately and then added together, weighted according to the noise rms. The data reduction consists of baseline fitting of each array element in each scan, removal of spikes caused by cosmic rays and internal vibrations (including the occasional earthquake), flat fielding and removal of the

[1] MOPSI has been developed by Robert Zylka, IRAM, Grenoble, France.

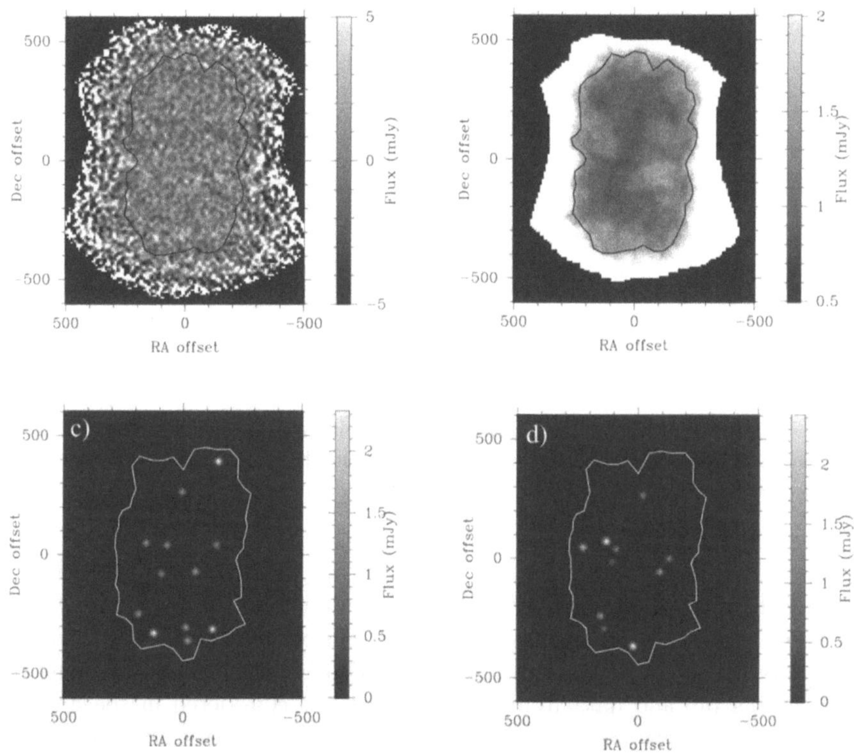

Fig. 2. **a)** The SIMBA image of the CDF-S. The irregular shape of the map is due to field rotation. Only the area inside the irregular polygon is considered in the analysis. **b)** The noise rms map. Inside the polygon the rms is 1.5 mJy. **c)** Extracted sources above 2.5σ. **d)** Extracted sources above 2.5σ from the inverted map (see text).

sky noise. Each map is individually calibrated using the corresponding maps of Uranus. The sky noise is removed by searching for correlation between the individual array elements. The correlated signal is removed and the process is repeated until the correlation signal is below a user defined threshold.

The fast-scanning technique results in an effective beam which has a response function that has both a negative and positive part. The true gaussian response function is restored through a deconvolution process. As the scan direction is fixed with respect to the telescope, the field rotates during the observations. This results in an effective area which is less than that covered by each individual map. A more detailed account of the data reduction procedure will be given in [5].

3 Results and Discussion

The SIMBA map is centered on a Type II QSO detected in hard X-rays by Chandra [22]. The observed area is shown in Fig. 1 where the SIMBA coverage is shown shaded on the Chandra hard X-ray image [17]. The effective area,

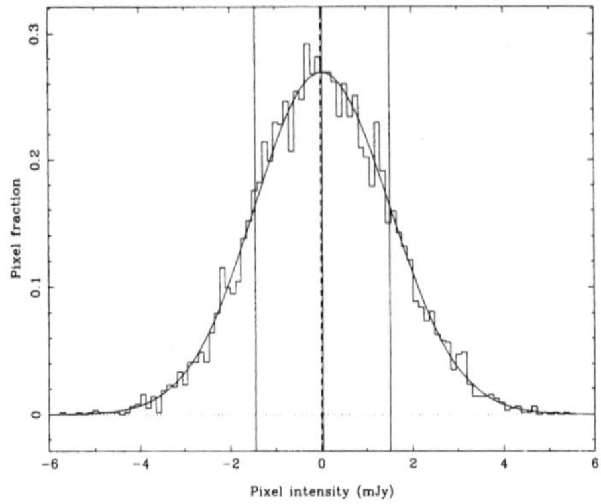

Fig. 3. Distribution of the pixel fluxes over the entire area outlined by the irregular polygon in Fig. 2a-d. The distribution shows that the noise characteristics is well described by a normal distribution.

i.e. the area with homogeneous and low noise rms, is 110 $arcmin^2$. In Fig. 2 it is outlined by the irregular polygon. Inside the polygon the noise rms is quite homogeneous at a level of 1.5 mJy (cf. Fig 2b). The observed flux, smoothed to an effective resolution of 34″ is shown in Fig. 2a. Outside the polygon the noise rms increases sharply and data from this region is not considered in the analysis. We used an iterative source extraction algorithm to identify pixels inside the polygon area above 2.5σ. These are shown in Fig. 2c. Since the signal-to-noise is relatively low, we also performed the source extraction on an inverted version of the image, Fig. 2d. Although the number of 'positive' detections is slightly higher than the number of 'negative' detections, the reality of the weaker sources is debatable. The stronger 'positive' sources, however, are more likely to be real, amounting 3 ± 1 sources. Due to the low significance of the peaks in the SIMBA map, the source count is essentially an exercise in statistics, where the number of 'up-peaks' is compared with the number of 'down-peaks'. In Fig. 3 we show a histogram of the pixel fluxes. The distribution adheres strongly to a normal distribution, showing that simple confidence limits can be put on the detections.

The SIMBA map of the CDF-S does not contain any strong submm sources. The maximum flux is 4.5 mJy/beam. Assuming a dust spectral energy distribution similar to low-z ultraluminous infrared galaxies, the corresponding 850μm flux would be ~11 mJy. The most likely detections, three in number, does not correspond to any obvious X-ray counterparts., including the Type II QSO at z=3.7 upon which the map is centered [22]. A comparison with radio continuum observations with the ACTA, optical and NIR data from GOODS as well as the Chandra data will be presented in [5].

4 Summary and Future Outlook

The SIMBA map of the CDF-S presented here represents the first map the region in the southern hemisphere covered by the CDF-S, GOODS and UDF. The number of sources detected with a reasonable confidence is 3 ± 1, corresponding to 98 ± 33 sources per square degree. This is consistent with results obtained from MAMBO at the same wavelength for other fields [18]. The use of SIMBA to achieve very low noise rms levels has shown that the fast-scanning technique is a viable method to obtain large submm maps.

There will not be any further SIMBA results as the SEST closed in October 2003. Instead we are looking forward to the 12m APEX antenna, to be operational during 2004. APEX is situated at the Chajnantor site at an elevation of 5000m, with one of the most transparent atmospheres known. It will be ideal for submm observations and the Bochum/Max-Planck groups are building LABOCA, a 300-element bolometer array for 850μm. A 37-element array operating at 350μm is also planned. LABOCA will allow much deeper and wider surveys than what was possible with SIMBA, and thereby add important data to the study of SMGs. It will be especially important for locating interesting targets for molecular line observations with the ALMA interferometer.

References

1. D.M. Alexander, F.E. Bauer, W.N. Brandt, et al.: AJ 125, 383 (2003)
2. D.M. Alexander, F.E. Bauer, S.C. Chapman, et al.: These proceedings (2004)
3. I. Aretxaga, D.H. Hughes, E.L. Chapin, et al.: MNRAS, 342, 759 (2003)
4. A.J. Barger, L.L. Cowie, D.B. Sanders, et al.: Nature, 394, 248 (1998)
5. S. Bergström, T. Wiklind, G. Rydbeck, et al.: A&A, in preparation (2004)
6. A.W. Blain: 1999, MNRAS, 309, 955
7. A.W. Blain, I. Smail, R.J. Ivison, et al.: Physics Reports, 369, 111 (2002)
8. C.L. Carilli, M.S. Yun: ApJ, 513, L13 (1999)
9. S.C. Chapman, R. Windhorst, S. Odewahn, et al.: ApJ, 599, 92 (2003a)
10. S.C. Chapman, A.W. Blain, R.J. Ivison, I.R. Smail: Nature, 422, 695 (2003b)
11. C.J. Conselice, S.C. Chapman, R.A. Windhorst: ApJ, 596, L5 (2003)
12. L.L. Cowie, A.J. Barger, J.-P. Kneib: AJ, 123, 2197 (2002)
13. S. Eales, S. Lilly, W. Gear, et al.: ApJ, 515, 518 (1999)
14. D.T. Frayer, R.J. Ivison, N.Z. Scoville, et al.: ApJ, 506, L7 (1998)
15. D.T. Frayer, R.J. Ivison, N.Z. Scoville, et al. ApJ, 514, L13 (1999)
16. R. Genzel, A.J. Baker, R.J. Ivison, et al.: these proceedings (2004)
17. R. Giacconi, A. Zirm, J. Wang, et al.: ApJS, 139, 369 (2002)
18. T.R Greve, R.J. Ivison, F. Bertoldi, et al.: these proceedings (2004)
19. D.H. Hughes, S. Serjeant, J. Dunlop, et al.: Nature, 394, 241 (1998)
20. D.H. Hughes, et al.: MNRAS, 335, 871 (2002)
21. R. Neri, R. Genzel, R.J. Ivison, et al.: astro-ph/0307310
22. C. Norman, G. Hasinger, R. Giacconi, et al.: ApJ, 571, 218 (2002)
23. L.A. Reichertz, B. Weferling, W. Esch, E. Kreysa: A&A, 379, 735 (2001)
24. S.E. Scott, et al.: MNRAS, 331, 817 (2002)
25. I. Smail, R.J. Ivison, A.W. Blain: ApJ, 490, L5 (1997)
26. I. Smail, R.J. Ivison, A.W. Blain, et al.: MNRAS, 331, 495 (2002)
27. I. Smail, S.C. Chapman, A.W. Blain, R.J., Ivison: in *Maps of the Cosmos*, ASP Conference Series, Vol., eds. M. Colless, L. Stiavely-Smith (2004)
28. T.J. Waskett, S.A. Eales, W.K. Gear, et al.: MNRAS, 341, 1217 (2003)
29. B. Weferling, L.A. Reichertz, J. Schmid-Burgk, E. Kreysa: A&A, 383, 1088 (2002)
30. T. Wiklind: ApJ, 588, 736 (2003)

Submillimeter Galaxies as Tracers of Mass Assembly at Large M

R. Genzel[1,2], A.J. Baker[1], R.J. Ivison[3], F. Bertoldi[4], A.W. Blain[5], S.C. Chapman[5], P. Cox[6], R.I. Davies[1], F. Eisenhauer[1], D.T. Frayer[7], T. Greve[8], M.D. Lehnert[1], D. Lutz[1], N. Nesvadba[1], R. Neri[9], A. Omont[10], S. Seitz[11], I. Smail[12], L.J. Tacconi[1], M. Tecza[1], N.A. Thatte[13], and R. Bender[1,11]

[1] Max-Planck-Institut für extraterrestrische Physik (MPE), Garching, Germany
[2] Department of Physics, University of California, Berkeley, CA, USA
[3] Astronomy Technology Centre, Royal Observatory, Edinburgh, UK
[4] Max-Planck-Institut für Radioastronomie (MPIfR), Bonn, Germany
[5] California Institute of Technology, Pasadena, USA
[6] Institut d'Astrophysique Spatiale, Université de Paris Sud, Orsay, France
[7] Spitzer Science Center, California Institute of Technology, Pasadena, USA
[8] Institute for Astronomy, University of Edinburgh, Edinburgh, UK
[9] Institut de Radio Astronomie Millimétrique (IRAM), Grenoble, France
[10] CNRS & Université de Paris, Paris, France
[11] Munich University Observatory (USM), München, Germany
[12] Institute for Computational Cosmology, University of Durham, Durham, UK
[13] University of Oxford Astrophysics, Oxford, UK

1 Introduction

Deep imaging in the rest-frame UV has constrained both the evolution of the cosmic star formation rate density [1] and its time integral, the growth of the cosmic stellar mass density [2]. Short-wavelength studies give an incomplete picture, however, since an important population of high-redshift galaxies is heavily dust-obscured. The strength of the extragalactic mid- and far-IR/submillimeter background indicates that about half of the cosmic energy density comes from dusty luminous and ultra-luminous infrared galaxies (LIRGs/ULIRGs: $L_{\rm IR} \sim 10^{11.5}$ to $10^{13.5}\,L_\odot$) at $z \geq 1$ [3,4]. Because the brightest of these submillimeter galaxies (SMGs; see [5] and references therein) tend to lack strong X-ray emission [6], their large IR luminosities probably correspond to high star formation rates [7]. As the strikingly different appearances of the Hubble Deep Field at 0.83 μm [8] and 850 μm [9] exemplify, SMGs are rarer and forming stars much more intensely than typical optically selected systems [8,9]. Here we discuss new observations that shed light on the importance of SMGs in the history of galaxy mass assembly (all numbers assuming a flat $\Omega_\Lambda = 0.7$ cosmology with $H_0 = 70\,{\rm km\,s^{-1}\,Mpc^{-1}}$).

2 Observations: PdBI Millimeter Interferometry

Since SMGs in most cases have relatively poorly known positions and frequently have only weak counterparts in the rest-frame UV and optical, redshifts and

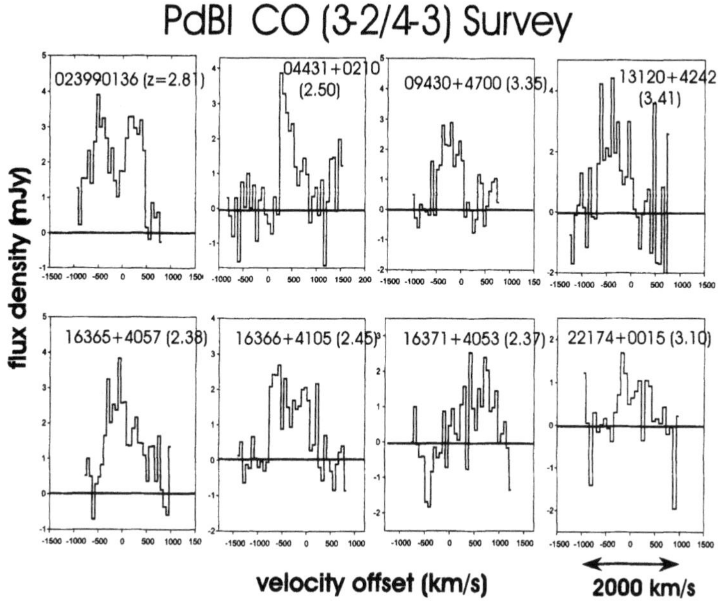

Fig. 1. Integrated CO spectra of the seven detected PdBI survey sources [11,12] plus SMMJ 02399-0136 [13]. Source names and redshifts are listed in the insets; flux density and velocity scales are the same for all sources.

spectroscopic parameters were initially confirmed with CO interferometry for only two of the few hundred detected sources [5]. Recently, a subset of the authors have initiated a new program to obtain spectroscopic redshifts with the Keck telescope for a number of sources detected with SCUBA at 850 μm, in most cases aided by more precise positions derived from deep 1.4 GHz VLA maps [10]. Here we report the first results of the millimeter follow-up of these SMGs, undertaken with the IRAM Plateau de Bure Interferometer (PdBI). In the B, C, and D configurations of the PdBI, we searched for the CO(3–2) and CO(4–3) rotational transitions (redshifted to the 3 mm atmospheric window) in a sample of so far twelve SMGs with $S_{850} \geq 5$ mJy and reliable UV/optical spectroscopic redshifts. Integration times per source ranged from 13 to 30 hours, for a total integration time of about 1Ms in the 2002/2003 observing season. We have detected significant CO emission in 7 out of the 12 sources, thus roughly quadrupling the number of CO-confirmed SMG redshifts [11,12]. Figure 1 shows the integrated CO(3–2/4–3) spectra for the seven new sources and for the $z =$ 2.81 system SMM J02399-0136 [13]. Figure 2 shows the integrated CO maps superposed on rest-frame UV (I-band) and optical (K-band) images of three of the sources [11].

Our PdBI observations confirm the median $z \sim 2.4$ determined from rest-frame UV/optical spectroscopy [10], although in several cases, the UV redshift is inferred from a source that is physically distinct from the SMG (e.g.,

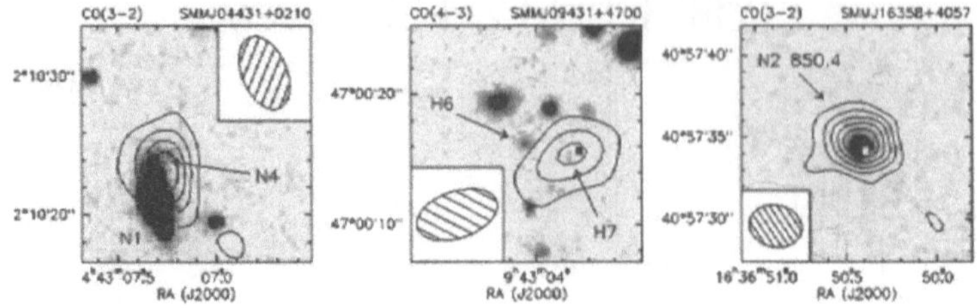

Fig. 2. Integrated CO maps of three SMGs studied with the PdBI [11], superposed on greyscale images of rest-frame UV (center panel) and optical (left and right panels) emission. FWHM synthesized beams are shown as hatched ellipses. In the left panel, the asterisk marks the position of the ERO N4 (uncertainty $\pm 0.5''$ – see text), while the edge-on spiral galaxy N1 ($2''$ SE of the CO source) lies in the foreground cluster at $z = 0.18$. In the center and right panels, filled squares mark millimeter continuum positions. Arrows show the positions of radio sources H6 and H7; the stronger of these (H6) is a narrow-line Seyfert galaxy with Ly α indicating $z = 3.349$. CO(4–3) emission from N2 850.4 remains largely unresolved.

SMM J09431+4700 in Fig. 2). SMGs likely sample very dense environments that may harbor additional, UV-bright star-forming galaxies.

SMGs are rich in molecular gas. Correcting for foreground lensing wherever appropriate (and known), and using the $M_{\rm gas}/L_{\rm CO}$ conversion factor appropriate for $z \sim 0.1$ ULIRGs, the median gas mass of the eight SMGs in Fig. 1 (plus the $z = 2.56$ galaxy SMM J14011+0252 [14,15]) is $2.2 \times 10^{10}\,M_\odot$. This value is similar to the median molecular gas mass found in high-z QSOs and about three times greater than in local ULIRGs. The inferred median far-IR luminosity of the sample (again corrected for lensing) is about is $1.3 \times 10^{13}\,L_\odot$, corresponding to a star formation rate of about $1100\,M_\odot\,{\rm yr}^{-1}$.

Our most important finding is that the brightest SMGs ($S_{850} \geq 5\,{\rm mJy}$), contributing about 25% of the submillimeter background, have large dynamical masses. The median velocity dispersion of our sample is $255(\pm 38)\,{\rm km\,s}^{-1}$. The ratio of the squares of the median velocity dispersions then implies that our SMGs have dynamical masses at least $13(\pm 4)$ times greater than the $z \sim 3$ Lyman break galaxy (LBG) population [16], and at least $5(\pm 1.5)$ times greater than optically selected populations at $z \sim 2$ ([17] – see also this volume). There are indications that SMGs are larger than LBGs in which case these estimates are lower limits [13,11,18]. Assuming a rotating disk geometry at an inclination of $40°$, the median velocity dispersion corresponds to a dynamical mass of $1.1 \times 10^{11}\,M_\odot$ within $R = 4\,{\rm kpc}$. Based on its spatially resolved rotation curve, SMM J02399-0136 has a total dynamical mass of $3 \times 10^{11}\,M_\odot$ within $R = 8\,{\rm kpc}$, half of which is probably stellar [13]. Near-IR photometry indicates that the median stellar mass for the $z \sim 3$ LBG population is $1.1 \times 10^{10}\,M_\odot$[19], consistent with the dynamical mass ratio estimated above. Dynamical and stellar masses of the SMGs in our sample are thus comparable to those of local m^* galaxies.

3 Observations: SPIFFI NIR Integral Field Spectroscopy

We have also conducted a detailed rest-frame optical case study of the bright source SMM J14011+0252 (hereafter J14011), a $z = 2.565$ SMG lying behind the $z = 0.25$ cluster A1835 [20]. For this study we used SPIFFI, the SPectrometer for Infrared Faint Field Imaging [21], at the ESO VLT. In three runs between February and April 2003, we observed J14011 in the J, H, and K bands, with a FWHM spatial resolution of about $0.6''$ and spectral resolutions ranging from 125 to $200\,\mathrm{km\,s^{-1}}$. SPIFFI obtains simultaneous spectra for each of 32×32 contiguous $0.25''$ pixels. The total on-source integration times were 60 minutes in J, 95 minutes in H, and 340 minutes in K [22].

Within the bright eastern J1 component of J14011, the SPIFFI data reveal substantial spatial variation in Hα equivalent width: nebular line emission is extended over about $1''$ (lensing corrected) and is strongest away from the central continuum peak. The integrated spectrum of J1 (Fig. 3) appears to exhibit a continuum break between the observed J and H bands. The feature's most likely origin is the Balmer break of a ≥ 100 Myr old stellar population at $z \simeq 2.565$. The J1 line fluxes allow us to measure the R_{23} estimator for its oxygen abundance. After correcting for the $E(B-V) = 0.18$ implied by the Balmer decrement, we derive an oxygen abundance of 8.96 ± 0.10, or 1.9 times solar. The observed high [NII]/Hα ratio implies that J14011 cannot fall on the lower-metallicity branch of the [O/H] vs. R_{23} relation. In contrast to the high abundances claimed for some high-z QSOs, this metallicity refers to the entire galaxy on ~ 10 kpc scales.

For a closed-box model with solar yield, this metallicity corresponds to a gas/baryonic mass fraction of $\exp(-Z/Z_\odot) \simeq 0.16$. From its intrinsic (corrected for a lensing magnification of 5) gas mass of $\simeq 1.3 \times 10^{10}\,M_\odot$, we therefore infer a total baryonic mass $\simeq 8 \times 10^{10}\,M_\odot$ and stellar mass $\simeq 7 \times 10^{10}\,M_\odot$. This last value is empirically supported by the location of J14011 in the local mass-metallicity relation. Adopting a recent version of that relation [23], we would expect a stellar mass of $\sim 6 \times 10^{10}\,M_\odot$. From the closed-box model with

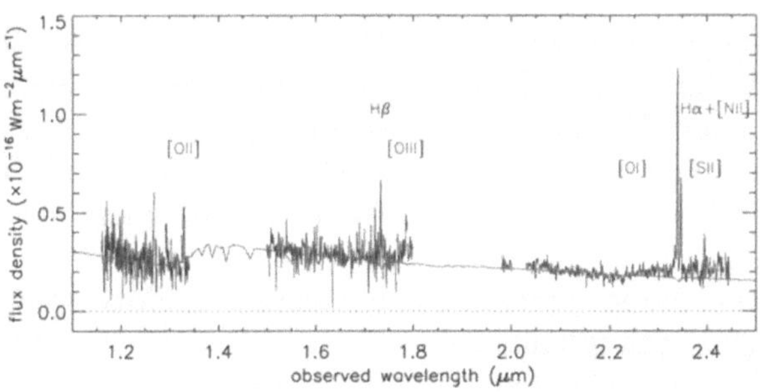

Fig. 3. J, H, and K spectra of J14011 extracted from a $2'' \times 1.5''$ aperture [22]. Superposed is the model STARS spectrum for a 200 Myr old continuous star formation episode with solar metallicity, a $1-100\,M_\odot$ Salpeter IMF, and extinction $A_V = 0.7$.

continuous star formation at the current rate ($380\,M_\odot\,\mathrm{yr}^{-1}$) we infer the overall duration of the star forming activity in J14011 to be $t_{\mathrm{SF}} \simeq 220\,\mathrm{Myr}$. This age is entirely consistent with the strength of the Balmer break seen in Fig. 3. As a further consistency check, we can estimate a dynamical mass for J14011 from the $\sim 2.2''$ and $\sim 180\,\mathrm{km\,s^{-1}}$ separations between J1 and [second component] J2, assuming these orbit around a common center. For a predominantly north-south lensing shear, the result is $\sim 3.4 \times 10^{10} \sin^{-2} i\, M_\odot$, consistent with the moderately low inclination also implied by the Hα morphology.

4 SMGs and Mass Assembly

Our IRAM and VLT results on metallicity and gas/dynamical mass build a compelling case that the most luminous SMGs ($S_{850} \sim 5$ to $10\,\mathrm{mJy}$) have dynamical and – within the factor 2 uncertainties of the present estimates – also stellar masses of $\sim 10^{11}\,M_\odot$. This stellar component has plausibly formed in the current starburst of duration $\sim 100\,\mathrm{Myr}$. How does the inferred comoving volume density of SMGs compare to theoretical models of galaxy formation?

From the observed surface density of SMGs with $S_{850} > 5\,\mathrm{mJy}$ [24], we infer that radio-detected SMGs with baryonic masses $\geq 10^{11}\,M_\odot$ in the range $1.8 \leq z \leq 3.5$ have a comoving number density of $1.6^{+1.0}_{-0.6} \times 10^{-5}\,\mathrm{Mpc}^{-3}$ [22]. This density needs to be corrected upward to account for sources that are in the same mass bin but no longer luminous enough to have been detected at $850\,\mu\mathrm{m}$. Relative to the $1.8\,\mathrm{Gyr}$ elapsed over the range $1.8 \leq z \leq 3.5$, the $\sim 220\,\mathrm{Myr}$ age and $\sim 30\,\mathrm{Myr}$ gas exhaustion timescale for the starburst in J14011 imply a lifetime correction factor of about 7. Figure 4 plots this prediction, together with the comoving number densities of galaxies with stellar masses above the $10^{11}\,M_\odot$ threshold at $z \sim 0$ [25] and $z \sim 1$ [26]. Also plotted are the predictions of two semi-analytic models [27,28]. Since all theoretical and observational values assume the same cosmology and a consistent (either Miller-Scalo or $1 - 100\,M_\odot$ Salpeter) IMF, we are secure in concluding that the models underpredict the number densities of massive galaxies at $z \sim 2.5$ [22]. Couching Fig. 4 in terms of *mass* emphasizes that the observed surface densities of SMGs cannot be explained using a very flat IMF alone. We also note that the models contain enough dark halos of total mass $\geq 10^{11}\,(\Omega_{\mathrm{M}}/\Omega_{\mathrm{b}})\,M_\odot$ to account for the observed comoving number densities of massive SMGs, provided all of their baryons can be rapidly and efficiently assembled into galaxies. SMGs thus impose a powerful "baryonic mass assembly" test at the upper end of the galaxy mass function [13].

Our correction of the observed number density of SMGs to account for their less dust-luminous descendants raises the question of what types of galaxies SMGs can plausibly evolve into. Although LBGs are clearly excluded due to their smaller masses, an intriguing alternative is the newly identified population whose red $J - K$ colors can stem from a strong Balmer break at $z \sim 2.5$ [29]. These galaxies appear to be strongly clustered [30] and – assuming a uniform distribution in the redshift range $2 \leq z \leq 3.5$ [31] – have a comoving number

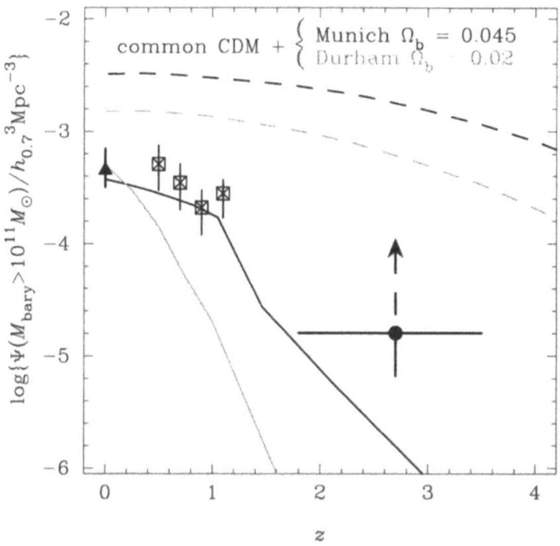

Fig. 4. Comoving number densities of galaxies with baryonic masses $\geq 10^{11}\,M_\odot$ as a function of redshift [22]. Triangle and open squares show densities of massive stellar systems at $z = 0$ [25] and $z \sim 1$ [26]; circle shows density for massive SMGs at $z \sim 2.7$, with a factor of 7 correction for burst lifetime. Solid curves show the predictions of semi-analytic modelling by the "Munich" and "Durham" groups [27,28]; dashed curves show the corresponding number densities of halos with *available* baryonic masses $\geq 10^{11}\,M_\odot$. The two models use the same halo simulations but assume different Ω_b.

density $\sim 1.8 \times 10^{-4}\,\mathrm{Mpc}^{-3}$. Relatively few sources with red $J - K$ colors are also luminous (sub)millimeter emitters, indicating that the two populations have little direct overlap. With three examples in the HDF-S having (for a Miller-Scalo IMF) stellar masses $(0.6 - 1.4) \times 10^{11}\,M_\odot$ [32], it would seem plausible that the more luminous objects with red $J - K$ colors at this epoch could have passed through a SMG phase at higher redshift.

References

1. P. Madau, H. C. Ferguson, M. E. Dickinson, et al.: MNRAS, **283**, 1388 (1996)
2. M. Dickinson, C. Papovich, H. C. Ferguson, T. Budavári: ApJ, **587**, 25 (2003)
3. J.-L. Puget, A. Abergel, J. P. Bernard, F. Boulanger, et al.: A&A, **308**, L5 (1996)
4. Y. C. Pei, M. S. Fall, M. G. Hauser: ApJ, **522**, 604 (1999)
5. A. W. Blain, I. Smail, R. J. Ivison, et al.: Physics Reports, **369**, 111 (2002)
6. O. Almaini, S. E. Scott, J. S. Dunlop, et al.: MNRAS, **338**, 303 (2003)
7. A. J. Barger, L. L. Cowie, E. A. Richards: AJ, **119**, 2092 (2000)
8. R. E. Williams, B. Blacker, M. Dickinson, et al.: AJ, **112**, 1335 (1996)
9. D. H. Hughes, S. Serjeant, J. Dunlop, et al.: Nature, **394**, 241 (1998)
10. S. C. Chapman, A. W. Blain, R. J. Ivison, I. Smail: Nature, **422**, 695 (2003)
11. R. Neri, R. Genzel, R. J. Ivison, F. Bertoldi, et al.: ApJ, **597**, L113 (2003)
12. T. Greve, et al.: in preparation (2004)

13. R. Genzel, A. J. Baker, L. J. Tacconi, D. Lutz, P. Cox, et al.: ApJ, **584**, 633 (2003)
14. D. T. Frayer, R. J. Ivison, N. Z. Scoville, A. S. Evans, et al.: ApJ, **514**, L13 (1999)
15. D. Downes, P. M. Solomon: ApJ, **582**, 37 (2003)
16. M. Pettini, A. E. Shapley, C. C. Steidel, J.-G. Cuby, et al.: ApJ, **554**, 981 (2001)
17. D. K. Erb, A. E. Shapley, C. C. Steidel, M. Pettini, et al.: ApJ, **591**, 101 (2003)
18. S. C. Chapman, et al.: in preparation (2004)
19. A. E. Shapley, C. C. Steidel, K. L. Adelberger, et al.: ApJ, **562**, 95 (2001)
20. R. J. Ivison, I. Smail, D. T. Frayer, J.-P. Kneib, A. W. Blain: ApJ, **561**, L45 (2001)
21. F. Eisenhauer, M. Tecza, N. A. Thatte, et al.: The ESO Messenger, **113**, 17 (2003)
22. M. Tecza, A. J. Baker, R. I. Davies, R. Genzel, et al.: ApJL, submitted (2004)
23. C. A. Tremonti, T. M. Heckman, G. Kauffmann, et al.: ApJ, submitted (2004)
24. I. Smail, R. J. Ivison, A. W. Blain, J.-P. Kneib: MNRAS, **331**, 495 (2002)
25. S. Cole, P. Norberg, C. M. Baugh, C. S. Frenk, et al.: MNRAS, **326**, 255, (2001)
26. N. Drory, R. Bender, J. Snigula, G. Feulner, U. Hopp, C. Maraston, et al.: 'The Mass Function of Field Galaxies at $0.4 < z < 1.2$ as Derived From the MUNICS K-Selected Sample', in *The Masses of Galaxies at Low and High Redshift*, ed. by R. Bender, A. Renzini (Berlin, Springer Verlag, 2003), pp. 140–145
27. G. Kauffmann, J. M. Colberg, et al.: MNRAS, **303**, 188 (1999)
28. C. M. Baugh, A. J. Benson, S. Cole, C. S. Frenk, C. Lacey: 'The Evolution of Galaxy Mass in Hierarchical Models', in *The Masses of Galaxies at Low and High Redshift*, ed. by R. Bender, A. Renzini (Berlin, Springer Verlag, 2003), pp. 91–96
29. M. Franx, I. Labbé, G. Rudnick, P. G. van Dokkum, et al.: ApJ, **587**, L79 (2003)
30. E. Daddi, H. J. A. Röttgering, I. Labbé, G. Rudnick, et al.: ApJ, **588**, 50 (2003)
31. P. G. van Dokkum, N. M. Förster Schreiber, M. Franx, et al.: ApJ, **587**, L83 (2003)
32. P. Saracco, M. Longhetti, et al.: A&A, submitted (2004, astro-ph/0310131)

A Spectroscopic Survey of the Submillimeter Galaxy Population: 85 Redshifts Using Keck/LRIS-B

Scott C. Chapman[1], Andrew Blain[1], Rob Ivison[2], and Ian Smail[3]

[1] Caltech, Pasadena, CA, 91125, USA
[2] Royal Observatory, Edinburgh EH9 3HJ, UK
[3] University of Durham, Durham DH1 3LE, UK

Abstract. We present spectroscopic identifications for a sample of 85 millimeter and submillimeter galaxies detected with the SCUBA/JCMT and MAMBO/IRAM30m facilities, all identified with deep 1.4GHz VLA radio observations. The galaxies lie at redshifts $z = 0.5$ to 3.5, We describe their properties, especially the presence of active galactic nuclei (AGN) in the sample, and discuss the connection of the mm/submm galaxies and the formation of spheroidal components of galaxies, which requires knowledge of the timescales of their very luminous activity. For a subset of the galaxies, we show their disturbed and diverse *Hubble Space Telescope (HST)* optical morphologies.

1 Introduction

Millimeter/Submillimeter galaxies – hereafter SMGs (Smail, Ivison & Blain 1997; Barger et al. 1998; Hughes et al. 1998; Eales et al. 1999; Bertoldi et al. 2000) are an important population (Blain et al. 1999, 2002). They are both numerous, with a surface density about 10% that of optically-selected Lyman-break galaxies (LBGs; Steidel et al., 2003), and luminous (possibly because of very high SFRs), with typical luminosities about 10 times greater than LBGs, assuming plausible spectral energy distributions (SEDs). The existing SMGs produce most of the extragalactic mm/submm background radiation intensity, and a significant fraction of the background at far-infrared (FIR) wavelengths (Blain et al. 2002; Smail et al. 2002). They represent a population which is generally difficult to detect at optical wavelengths, with at least the fainter half clearly being missed in current optical cosmological surveys. Unfortunately, the SMG population is notoriously difficult to study! Until now, we have been able to gather almost nothing about their redshifts and morphologies as fundamental observable properties. As a consequence, deriving properties such as total mass and dust temperature, and finding both the fraction of power contributed by AGN and starbursts, and their connection to optically-selected star-forming LBGs, have been the topic of largely idle speculation over the five years since their discovery.

The principal hurdle has always been identification at other wavelengths. SCUBA/MAMBO surveys have large beam sizes (10"–15"), a situation that will remain true for upcoming single-antenna, wide-field instruments: Bolocam, SHARC-II, a bolometer camera on the APEX telescope and SCUBA2. Candidate counterparts to SMGs cannot therefore be identified unambiguously without interferometry to pinpoint their positions. However, mm/submm interferometry is currently an arduous process, with tens of hours of integration required to detect a single object at the OVRO MMA or IRAM PdB interferometers.

As a consequence, the 20-cm radio emission from SMGs has become an important, but *second best* surrogate with which to probe both the energy generation processes and morphology of submm galaxies (Ivison et al. 1998; Smail et al. 2000; Barger, Cowie & Richards 2000; Chapman et al. 2001, 2002a, 2003a; Ivison et al. 2002) The problem remains that in order to use a radio wavelength as a surrogate for submm/FIR emission, we would like a clear physical principle connecting the emission at the different wavelengths. We don't yet, but we do have a strong empirical connection: the far-IR–radio correlation (e.g., Helou et al. 1985), which has an RMS dispersion of only 0.2 dex over a large range in luminosity at low redshifts. Radio identification of SMGs has allowed their optical properties to be explored in detail (Chapman et al. 2003a; Ivison et al. 2002). A large range of optical properties is observed for the radio-selected SCUBA galaxies, with 65% fainter than $I > 23.5$. They have red optical-IR colors with $I - K = 3$ to 6 and $< I - K >= 4.3$, but they are not all extremely red objects (EROs). Deep radio observations allow SMGs to be pre-selected with an efficiency of about 40%.

2 Redshifts for SMGs

We have been able to secure redshifts for 85 radio-identified SMGs through deep Keck/LRIS observations (Chapman et al. 2003b, 2004; Ivison et al. in preparation), and the sample is growing rapidly.[1] The spectrum of the brightest example is shown in Fig. 1a. The redshifts allow, for the first time, accurate calculation of luminosities and dust temperatures for the SCUBA galaxy population. We emphasize that obtaining redshifts is not easy, relying heavily on the superb blue sensitivity of the new LRIS-B multiobject spectrograph (Steidel et al. 2004), and often have no detectable continuum emission with which to extract the spectrum. In addition the galaxies are hard to identify, being faint, messy, composite objects in optical images. Because of their small radio/optical offsets (~ 0.5") it is often difficult to assess the best position at which to align the slit. Often we designated several slit positions on different masks for each target. Our spectroscopic completeness is 74% over the magnitude range of the sample from $I = 22$ to 27.

While the issue of correctly identifying the submm galaxy is concerning, 8 out of 12 CO detections with IRAM-PdB have already been made, realizing an unequivocal confirmation of the redshifts (Neri et al. 2003, Greve et al. in preparation). We also note that beyond the radio positional identification, we

[1] Data presented herein were obtained at the W.M. Keck Observatory, which is operated as a scientific partnership between Caltech, the University of California and NASA. The Observatory was made possible by the generous financial support of the W.M. Keck Foundation.

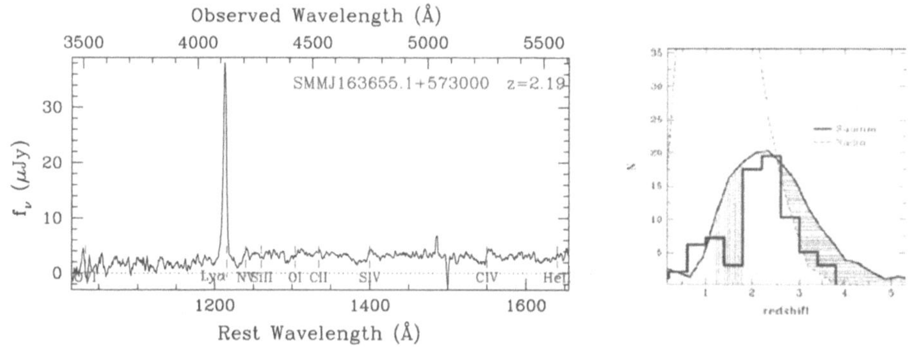

Fig. 1. **(a)** The rest-UV spectrum of the radio–SCUBA galaxy N2.4, showing P-Cygni absorption line profiles, and a host of stellar and interstellar features. The spectrum is best fit by a ~5 Myr-old starburst, with an inferred star-formation rate of $1000\,M_\odot\,yr^{-1}$. **(b)** The observed histogram of the redshift distribution for our radio-identified SCUBA galaxies. Curves derived for a model of the radio/SMG populations are overplotted, suggesting that the redshifts of the sources missed in the radio identification process lie at moderate redshifts between the radio and submm model tracks. The redshift distribution of radio selected QSOs (Shaver et al. 1998) are shown with a dashed line, and reveal a very similar evolution to the SMGs.

are finding the correct *type* of object through optical spectroscopy. The identifications are at high-z, have apparent SFRs of several $100\,M_\odot\,yr^{-1}$, often show AGN features, and are very rare objects in the LBG distribution. We ask, what else can they be?

We further note that we believe we have spectroscopically identified a sample of blank-field SCUBA galaxy counterpart candidates without radio detections, effectively through trial and error by targeting faint optical sources lying within the SCUBA beam using otherwise redundant spectrograph slits. Observations of these identifications were tried based on our cumulative experience of with the properties of the radio-identified sample: often faint, distorted blue/red composite sources near the SCUBA beam centroid, are found to have Type-II AGN spectra and inferred star formation rates of order $100\,M_\odot\,yr^{-1}$.

In Fig. 1b, we show the observed redshift distribution and a model (Chapman et al. 2003c) for the radio and submm distributions derived from an evolving far-IR luminosity function. The model is very useful for understanding selection effects, and in particular the bias against the highest-redshift galaxies due to the requirement of a radio selection. This bias has a strong dependence on dust temperature (T_d). The sources missed by the radio are expected to fill the region lying between the model radio and submm distributions, overlapping significantly in redshift with the radio-identified sample if the dust temperatures are in the cold to warm regime ($T_d < 35\,K$).

3 Spheroids in Formation?

We have presented a plausible redshift distribution for submm galaxies, the most important ingredient for addressing whether they could be spheroids in formation. Encouragingly, they lie at the correct redshifts to be proto-ellipticals or the forming bulges of spiral galaxies. (all the stellar population constraints point to most stars being formed at z=2–3). In particular, we have demonstrated that most submm galaxies do not lie at $z > 4$, and thus most are not very high redshift Population-III sources.

However, to assess what type of formation mechanism is at work we must assess the AGN/starburst contribution to the dust heating, the timescales of their luminous phase, and compare their *HST* morphologies. The AGN versus starburst issue is always difficult to address. We have firstly from our restframe UV spectra, the possibility of diagnosing the presence of an AGN through high ionization lines. Indeed, we find signs of CIV, and other lines which cannot easily be excited in a starburst, in approximately 40% of our sources. Our near-IR spectra of SMGs (e.g., Smail et al. 2003) can reveal OIII/Hβ ratios typical of Seyfert-2 galaxies as well as strong NeV emission. We have evidence in the X-ray from the Chandra Deep Field North that ~90% of the radio-identified SCUBA galaxies have X-ray detections (Alexander et al. 2003, 2004), but only ~35% of these are too bright to be generated by star formation alone (scaling from Nandra et al. 2002 for LBGs). Finally, ~20% of our sources appear to be unusually bright at radio wavelengths, departing significantly from the far-IR–radio distribution for low-redshift starburst galaxies. Together, this suggests that while AGN are prevalent in the sample (as expected from the likely coeval evolution of the BH and bulge required to generate the tight correlation between their masses; Magorrian et al. 1998), but not necessarily dominant in the bolometric energy released by SCUBA galaxies.

The duration of the very luminous activity of the SCUBA galaxies are important: are their timescales 10 Myr or 1 Gyr? If the former, then several bursts would be required to form a massive elliptical galaxy. In the spectra with the highest signal-to-noise ratios, stellar and interstellar features have been observed. These can be fitted using starburst synthesis models, to yield a timescale for the starburst activity visible at ultraviolet (UV) wavelengths, and a relation between the star-formation rates at observed optical and submm wavelengths. For our brightest object, N2.4 (Fig. 1a; Smail et al. 2003) the best fit is for a ~5 Myr-long instantaneous burst of star formation. The implied star formation rate is about $1000\,M_\odot\,yr^{-1}$, which is similar to that inferred directly from its submm and radio flux densities. While not all sources may contain such a short burst of star-formation dominated energy output, this example is likely to define the paradigm for the nature of a significant fraction of SCUBA galaxies.

The morphologies of a subset of SCUBA galaxies have recently been studied with *HST*-STIS by Chapman et al. (2003d). Fig. 2 shows a montage of several representative examples, typically revealing multi-component, distorted galaxy systems that are reminiscent of mergers in progress. There are no examples of isolated, compact sources, and we conclude generally that the morphologies

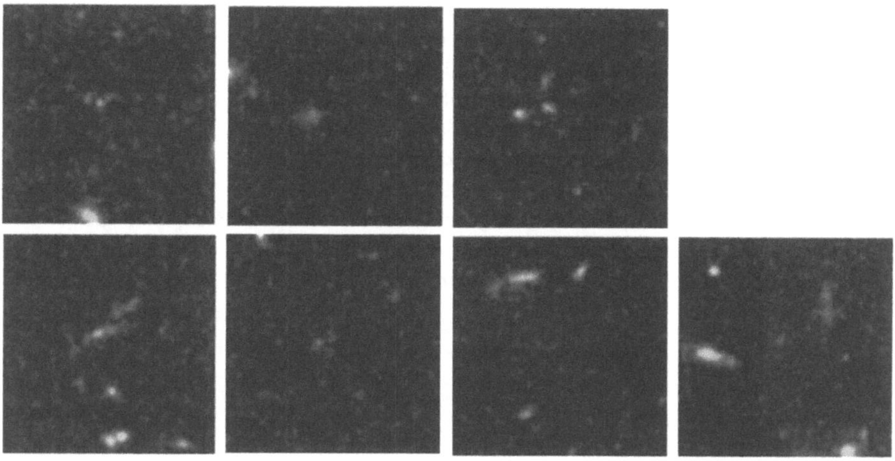

Fig. 2. The morphologies of 8 SCUBA galaxies as studied with *HST*-STIS (Chapman et al. 2003d; 4" images centered on the radio position). All are multi-component, distorted galaxy systems reminiscent of mergers in progress.

of SCUBA galaxies are generally consistent with hierarchical galaxy formation scenarios in which the most intense activity occurs when gas-rich, high-redshift galaxies collide and merge. Although they could be the sites of spheroid formation, most spheroids do not form in a monolithic collapse of primordial gas clouds at extreme redshifts.

References

1. Alexander, D., et al., ApJ, 2003, in press (astro-ph/0211267)
2. Alexander, D., Smail, I., Chapman, S., Blain, A., Ivison, R., ApJ, 2004, submitted
3. Barger, A., et al., Nature, 1998, 394, 293
4. Barger, A., Cowie, L., Richards, E., 2000, AJ 119, 2092
5. Bertoldi, F., et al., 2000, A&A, 476, 124
6. Blain, A.W., Smail, I., Ivison, Kneib, J.-P., 1999, MNRAS, 302, 632
7. Blain, A.W., Smail, I., Ivison, Kneib, J.-P., Frayer D.T., 2002, Physics Reports, 369, 111 (astro-ph/0202228)
8. Chapman, S., Richards, E., Lewis, G., Wilson, G., Barger, A., 2001, ApJ, 548, L147
9. Chapman, S.C., Lewis, G.F., Scott, D., Borys, C., Richards, E., 2002a, ApJ, 570, 557
10. Chapman S.C., Barger A., Cowie L., et al., 2003a, ApJ, in press (astro-ph/0211075)
11. Chapman, S., Blain, A., Ivison, R., Smail, I., 2003b, Nature, 422, 695
12. Chapman, S., Helou, G., Lewis, G., Dale, D., ApJ, in press
13. Chapman, S., Windhorst, R., Odewahn, S., Haojing, Y., Conselice, C., ApJ, 2003d, in press
14. Chapman, S., Blain, A., Ivison, R., Smail, I., 2004, ApJ, submitted
15. Eales, S., et al., 1999, ApJ 515, 518
16. Helou, P., et al., 1985, ApJ 440, 35

17. Hughes, D., et al., Nature, 1998, 394, 241
18. Ivison, R., et al., 1998, MNRAS, 298, 583
19. Ivison, R., et al., 2002, MNRAS, 337, 1
20. Magorrian J. et al., 1998, AJ, 115, 2285
21. Nandra, P., et al. 2002, ApJ, in press (astro-ph/0205215)
22. Neri, R., et al. 2003, A&A, in preparation
23. Smail, I., Ivison, R.J., Blain, A.W., 1997, ApJ 490, L5
24. Smail, I., Ivison, R.J., Blain, A.W., Kneib, J.-P., 2002, MNRAS, 331, 495
25. Smail, I., et al., 2003, MNRAS, submitted
26. Steidel, C., et al., 2003, ApJS, submitted

Extremely Red Galaxies: Dust Attenuation and Classification

Daniele Pierini[1], Claudia Maraston[1], Ralf Bender[12], and Adolf N. Witt[3]

[1] MPE, Giessenbachstr., D-85748 Garching, Germany
[2] Universitäts-Sternwarte München, Scheinerstr. 1, D-81679 München, Germany
[3] The University of Toledo, Toledo OH 43606, USA

Abstract. The "extremely red galaxies" (ERGs) constitute a controversial population in deep near-IR surveys. Their extremely-red observed colours ($I_c - K \geq 4$ or $R_c - K \geq 5.3$) are mostly attributed to either passively evolving, old (i.e. ≥ 1 Gyr) stellar populations (associated with ellipticals) or to reddening by internal dust of the light produced by young stellar populations (as in galaxies forming stars continuously). Recently Pozzetti & Mannucci have shown that two-colour plots may separate these two classes of galaxies when they lie at $1 \leq z \leq 2$. Here we show that the early, dusty post-starburst phases of an intermediate-mass galaxy, that is formed in a 0.1 Gyr–long burst with a star formation rate of $10^3\,M_\odot\,yr^{-1}$ at $1 \leq z \leq 2$ or so, meet the colour–selection criteria for the ERGs. Dusty post-starburst galaxies with intermediate-age (i.e. 0.1–1 Gyr) stellar populations may be misclassified, since they have optical/near-IR colours similar to those of either ellipticals or dusty starbursts, whether they lie at $1.3 < z \leq 2$ or at $z \sim 1$, respectively. The amount and distribution of dust, and the star formation history make the total mid-IR–mm emission of the model dusty post-starburst and starburst galaxies range from 10^{10} to $5.6 \times 10^{12}\,L_\odot$, consistent with the submm observations of dusty ERGs.

1 Introduction

The "extremely red galaxies" (ERGs) form the bulk of the objects with $I_c - K \geq 4$ or $R_c - K \geq 5.3$ selected from deep near-IR surveys [3] [2] [15]. They lack an uncontroversial classification. Half or more of the entire ERG sample in the literature is held to contain high-z ellipticals, as identified by passively evolving, old (≥ 1 Gyr) stellar populations in a dust–free environment; the rest mostly owes its extremely red colours to attenuation by internal dust of light produced during an on-going process of star formation activity at $1 \leq z \leq 2$ [11] [8] [1].

Old, massive ellipticals (Es) at $z > 1$ are a problem for the standard hierarchical merging scenario of galaxy formation and evolution [20], where the most massive systems are assembled through a merging tree of less massive ones, even though the bulk of their stars may already be very old at the time of the last major merging [7]. Conversely, the association of ERGs with dusty starburst galaxies (DSGs) at intermediate redshifts does not necessarily favour the previous scenario, if galaxies are assembled within times comparable to their star formation timescales and the latter are shorter the more massive the galaxies [5] [17].

The ERG dichotomy is enforced by a two-colour classification criterion [13], based on the rather neat difference in the distribution of model Es and DSGs at

$1 \leq z \leq 2$ in the $I_c - K$ (or $R_c - K$) vs. $J - K$ colour–colour diagram. In fact passively evolving, old stellar populations at these redshifts have their strong 0.4 μm-break redshifted shortward of 1.2 μm (J band), while coeval dusty, star-forming galaxies show smoother spectral energy distributions (SEDs) and, thus, redder $J - K$ colours. This classification criterion is statistically effective [8] [15]; however, recent observations show that ERGs represent a heterogeneous family of objects [15], where even the dusty ones span a huge range in total rest-frame mid-IR–mm luminosity [10]. Hence it is important to consider new potential ERG candidates and to investigate how the radiative transfer through a dusty interstellar medium (ISM) determines the optical/near-IR colours and the total mid-IR–mm emission of the dusty ERGs. We address this here (see also [12]).

2 Models of ERG Candidates

We consider as ERG candidates galaxies with passively evolving, old stellar populations (i.e. Es), dusty starbursts and dusty post-starburst galaxies (DPSGs). For all models, we fix metallicity (solar) and initial mass function and consider three representative redshifts (1.07, 1.39 and 1.85). We link z and age of the models via the cosmological model ($\Omega_m = 0.3$, $\Omega_\lambda = 0.7$, $H_0 = 70\,\mathrm{km\,s^{-1}\,Mpc^{-1}}$).

2.1 Passively Evolving, Old Elliptical Galaxies

Es are simulated through simple stellar population (SSP) models [9]. We assume a maximal formation redshift $z_f = 10$; thus an elliptical galaxy has a maximum age of about 5, 4 and 3 Gyr at $z = 1.07$, 1.39 and 1.85, respectively. For model Es we adopt a minimum stellar mass of $7.5 \times 10^{10}\,M_\odot$, which is about the residual (i.e. after 25 per cent mass–loss owing to stellar evolution) stellar mass of a 1 Gyr–old SSP [9] as well as of a 1 Gyr–old model DSG/DPSG (cf. Sect. 2.2), both originating from a $10^{11}\,M_\odot$ cloud of primordial gas.

We make the new hypothesis of Es at $1 \leq z \leq 2$ being possibly as dusty as nearby ones. The attenuation by internal dust is described by Monte Carlo simulations of the radiative transfer [22]. The geometry of a dusty elliptical assumes a spherical stellar distribution that declines from the core as r^{-3} and the presence of homogeneous Milky Way–type dust only in the inner portions of the model, from the center to two thirds of the galaxy radius, with a distribution that declines as r^{-1}. The value of the radial extinction optical depth from the center to the edge of the dust environment at V band, assuming a constant density, homogeneous distribution, (τ_V) is set to 0.5 [22].

2.2 Dusty Starburst/Post-Starburst Galaxies

Intrinsic SEDs of young (≤ 0.1 Gyr)/intermediate-age (0.1–1 Gyr) starburst and intermediate-age post-starburst galaxies at $1 \leq z \leq 2$ are computed with the evolutionary synthesis code PÉGASE 2.0 [4], because this code includes nebular emission. Gas is assumed to be transformed into stars at a constant rate over a period $t_{\rm burst}$ of 0.1 Gyr (i.e. *short burst*) or 1 Gyr (i.e. *long burst*). For a total mass of transformed gas $M_{\rm gas} = 10^{11}\,{\rm M}_\odot$, a star formation rate (SFR) equal to $M_{\rm gas}/t_{\rm burst}$ corresponds to $10^2\,{\rm M}_\odot\,{\rm yr}^{-1}$ (long burst) or $10^3\,{\rm M}_\odot\,{\rm yr}^{-1}$ (short burst). Note that SFRs $\sim 10^3\,{\rm M}_\odot\,{\rm yr}^{-1}$ have been inferred from observations of high-z SCUBA submm sources [14]. These star formation histories and SFRs are complementary to those considered by [13] for their DSGs.

The difference between a post-starburst galaxy and a passively evolving elliptical galaxy younger than 1 Gyr may sound a matter of semantics, since a DSG at $1 \leq z \leq 2$ or so may evolve into an elliptical at $z = 0$ (e.g. [18]) and, eventually, already at these redshifts. As far as stellar properties are concerned, the SEDs of a post-starburst stellar population under investigation and of an SSP with the same age and metallicity are different even after 1 Gyr from the onset of star formation. However, it is the presence, distribution and structure of the dusty ISM which provides fundamental physical differences.

We apply the Monte Carlo simulations of the radiative transfer of [21] for the SHELL geometry to the model DSGs/DPSGs. This geometry assumes a dust/stars distribution where the stars extend to 0.3 of the system radius and the dust extends from 0.3 to 1 of the system radius. The dust in the shell has a Small Magellanic Cloud–like extinction curve. It is distributed in a two-phase clumpy medium, comprised of molecular clouds and diffuse gas, where the filling factor of the clumps is set to 0.15 and the low-density to high-density ratio is set to 0.01. Models of dusty starbursts based on similar considerations well reproduce the UV/optical colours of nearby starburst galaxies [6] and the optical colour–redshift distribution of Lyman-Break Galaxies at $2 < z < 4$ [19]. We also consider the extreme hypothesis that the local distribution of the dust is homogeneous. Finally, we consider values of $\tau_{\rm V}$ equal to 0.25 through 50 (8) for the two-phase clumpy (homogeneous) dust distribution. We recall that $\tau_{\rm V}$ gives the total amount of dust in the shell.

3 Results

Hereafter we briefly present the main results of [12].

3.1 When a Model Is Selected as an ERG

Existing ground-based near-IR surveys [3] [2] [15] reach a typical limit of 20.5 K-mag. K < 20–21 is also the limit of validity of the method of [13], introduced in order to avoid the statistical contamination by $z > 2$ Es in the optical/near-IR colour–colour region populated by DSGs.

We require our models to match the observed photometry of ERGs, i.e. K < 20.5 and e.g. $4 \leq {\rm I_c} - {\rm K} < 8$. We find that passively evolving, old Es at $1 \leq z \leq 2$ are always ERG candidates, whether dusty or not, if their stellar mass is larger than $\sim 10^{11}\,{\rm M}_\odot$. Conversely, a coeval DSG/DPSG may be an ERG candidate as a function of its star formation history and/or amount of dust and structure of its dusty ISM (see Figs. 1 and 2, and [12] for a quantitative analysis). This is largely due to the existence of the age/opacity degeneracy [16].

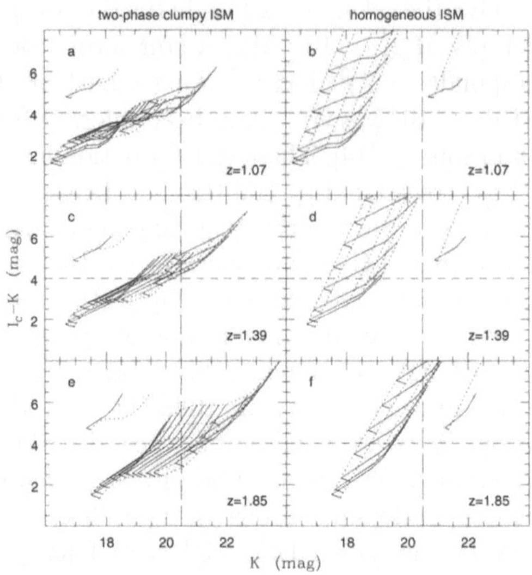

Fig. 1. The distribution of model DSGs with a short burst and DPSGs at $z = 1.07$ (**a**, **b**), 1.39 (**c**, **d**) or 1.85 (**e**, **f**) in the $I_c - K$ vs. K colour–magnitude plane for the two-phase clumpy (**a**, **c**, **e**) and homogeneous (**b**, **d**, **f**) dust distributions. In each panel we reproduce the limiting magnitude (*long-dashed line*) and the $I_c - K$ colour selection criterion (*short-dashed line*) for ERGs, as well as models with constant τ_V (*solid lines*) or age (*dotted lines*). In the upper left (right) corners of panels a,c,e (b,d,f) we reproduce the typical trend of models when age (*solid line*) or τ_V (*dotted line*) increases

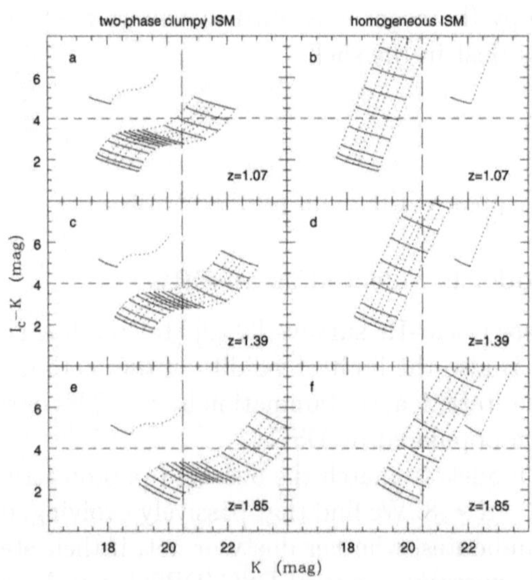

Fig. 2. The same as in Fig. 1 but for model DSGs with a long burst

3.2 Ambiguity in the $I_c - K$ vs. $J - K$ Colour–Colour Diagram

Model Es obey the classification criterion of [13], even when dusty; however, the attenuation by internal dust may explain extremely red optical/near-IR colours without the need of invoking extremely high ages, metallicities or redshifts.

Model DSGs behave as in [13], whatever the duration of their burst. This holds whether the nebular emission is considered or not.

However, DPSGs with ages between 0.2 and 1 Gyr, at $1.3 < z \leq 2$, may mix with galaxies with passively evolving, old stellar populations at $1 \leq z \leq 2$ in the $I_c - K$ vs. $J - K$ colour–colour diagram (see Fig. 3). Conversely, intermediate-age DPSGs at $z \sim 1$ may mix with DSGs younger than 1 Gyr at $1 \leq z \leq 2$ in the same colour–colour plot. This holds for a large range in amount of dust and whatever the structure of the dusty ISM. These results could explain in part the complex nature of ERGs, that emerges when their classification is based on multi-wavelength photometry [15] and not on two-colour plots [13].

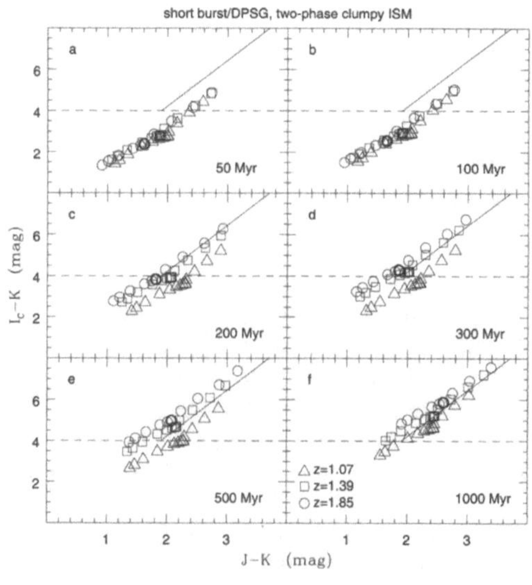

Fig. 3. $I_c - K$ vs. $J - K$ for model DSGs with a short burst and DPSGs, in case of a two-phase, clumpy, dusty ISM. In each panel the selection function (*short-dashed line*) and the classification criterion (*solid line*) for ERGs [13], and models of selected age with $z = 1.07$ (*triangles*), 1.39 (*squares*) or 1.85 (*circles*) are reproduced; τ_V increases from 0.25 to 50 from the lower left to the upper right

3.3 On the Photometric Definition of ERGs

We find that model Es (dusty or not) at $1 \leq z \leq 2$ are always selected as ERGs (if massive enough), whether selection is via the $I_c - K$ or $R_c - K$ colour. Conversely, selection as ERGs of observable (i.e. with $K < 20.5$), coeval DSGs/DPSGs is a complex function of the properties of their dusty ISM. Hence, selection via the $R_c - K$ colour may produce a sample of ERGs biased towards a larger ratio of Es at $1 \leq z \leq 2$ to DSGs/DPSGs at $1.3 < z \leq 2$; vice versa, selection via the $I_c - K$ colour may bias the ERG sample towards a larger ratio of Es at $1 \leq z \leq 2$ to DSGs/DPSGs at $z \sim 1$. This result confirms and complements that of [23].

3.4 The Mid-IR–mm Luminosity of Model DSGs/DPSGs

We find that the re-emission by dust in the rest-frame mid-IR–mm ($L_{\rm dust}$) has a dynamic range of more than 500 (from 10^{10} to $5.6 \times 10^{12}\,L_\odot$) for DPSGs through DSGs. In particular, $L_{\rm dust}$ ranges between 2 and $5.6 \times 10^{12}\,L_\odot$ for DSGs with a short burst, between 2 and $7.6 \times 10^{11}\,L_\odot$ for DSGs with a long burst, and between 10^{10} and $5.7 \times 10^{11}\,L_\odot$ for DPSGs. This high dynamic range is consistent with the large range in total rest-frame mid-IR–mm luminosity estimated for ERGs observed in the submm [10]. It may be explained as follows. The contribution of UV, optical or near-IR photons to dust heating depends on the properties of the stellar populations and on the absorption efficiency of the dust at those wavelengths. This efficiency depends on the type and amount of dust, the structure of the dusty ISM and the dust/stars configuration. These aspects may differ a lot among the observed dusty ERGs.

References

1. A. Cimatti, E. Daddi et al.: A&A **381**, L68 (2002a)
2. A. Cimatti, M. Mignoli et al.: A&A **392**, 395 (2002b)
3. N. Drory, G. Feulner et al.: MNRAS **325**, 550 (2001)
4. M. Fioc, B. Rocca-Volmerange: A&A **326**, 950 (1997)
5. G. Gavazzi, D. Pierini, A. Boselli: A&A **312**, 397 (1996)
6. K.D. Gordon, D. Calzetti, A.N. Witt: ApJ **487**, 625 (1997)
7. G. Kauffmann, S. Charlot, S.D.M. White: MNRAS **283**, 117 (1996)
8. F. Mannucci, L. Pozzetti et al.: MNRAS **329**, L57 (2002)
9. C. Maraston: MNRAS **300**, 872 (1998)
10. N.R. Mohan, A. Cimatti et al.: A&A **383**, 440 (2002)
11. G. Moriondo, A. Cimatti, E. Daddi: A&A **364**, 26 (2000)
12. D. Pierini, C. Maraston, R. Bender, A.N. Witt: astro-ph/0309223 (2003)
13. L. Pozzetti, F. Mannucci: MNRAS **317**, L17 (2000)
14. I. Smail, R.J. Ivison, A.W. Blain: ApJ **490**, L5 (1997)
15. I. Smail, F.N. Owen et al.: ApJ **581**, 844 (2002)
16. T. Takagi, N. Arimoto, V. Vansevicius: ApJ **523**, 107 (1999)
17. D. Thomas, C. Maraston, R. Bender: Ap&SS **281**, 371 (2002)
18. T. Totani, Y. Yoshii et al.: ApJ **558**, L87 (2001)
19. U. Vijh, K.D. Gordon, A.N. Witt: ApJ **587**, 533 (2003)
20. S.D.M. White, M.J. Rees: MNRAS **183**, 341 (1978)
21. A.N. Witt, K.D. Gordon: ApJ **528**, 799 (2000)
22. A.N. Witt, H.A. Thronson, J.M. Capuano: ApJ **393**, 611 (1992)
23. L. Yan, D. Thompson: ApJ **586**, 765 (2003)

Modeling the Multi-Wavelength Universe: The Assembly of Massive Galaxies

Rachel S. Somerville

Space Telescope Science Institute

Abstract. Pan-chromatic galaxy surveys are providing tightening constraints on the global mass assembly history, and high-resolution imaging of large fields is telling us when and where stars were formed. How well are state-of-the-art hierarchical galaxy formation models currently doing at reproducing these observations? I present results here that suggest that hierarchical models are doing quite well at reproducing the *global* star formation and stellar mass assembly history obtained from galaxies selected in the optical and Near IR. However, the same models fail to reproduce two very important populations at high redshift: quiescent red spheroids and vigorously star-forming, dust-enshrouded starbursts. This mismatch carries important lessons about how star formation is triggered and regulated in early galaxies, and may force us to consider new ideas about the formation of massive spheroids.

1 Tracing Galaxy Assembly

Probing the high-mass end of the galaxy mass function at any epoch provides a particularly strong test of theories of galaxy formation. A zoo of apparently massive objects at high redshift has been discovered in recent years, in a broad range of wavelengths and utilizing a variety of selection techniques. For example, at this meeting, we heard about massive galaxies at $z \sim 1$ in the COMBO-17/GEMS survey (Bell, Rix), VIMOS (Le Fevre), and DEEP (Koo, Newman), near IR selected galaxies and Extremely Red Objects (EROs) at $z \sim 1$–2 (Fontana, Daddi, Drory, Cimatti), color-selected galaxies in the spectroscopic redshift 'desert' at $z \sim 1.5$-2.5 (Steidel, Chen), and sub-mm and mm selected galaxies at $z \sim 2.5$ (Chapman, Bertoldi, Genzel). The challenge posed to us by the multi-wavelength universe is to unify this zoo of objects into a coherent picture of galaxy evolution.

Another exciting observational development is the imaging of relatively large fields with the Advanced Camera for Surveys (ACS) on the Hubble Space Telescope (HST), combined with multi-wavelength ground-based observations. These new surveys, such as GOODS (Great Observatories Origins Deep Survey), GEMS (Galaxy Evolution from Morphology and SEDs), and COSMOS, will for the first time allow us to study the connection between galaxy morphology (through structural parameters such as bulge-to-disk ratio) and stellar populations (through broad band colors and spectroscopy) for a statistically meaningful sample, over a large range of cosmic time. We can then go beyond a global census of star formation or the build-up of stellar mass, to understand *where and how* stars were formed, and perhaps which processes stimulated or regulated star formation. In this paper, I shall attempt to briefly summarize the status of our theoretical understanding of some of these observational results and what we may learn about galaxy formation from them.

2 Do Massive Galaxies at High Redshift Pose a 'Crisis' for CDM?

It has become a familiar story: in the Cold Dark Matter (CDM) model of structure formation, small mass objects form first, and larger mass objects form hierarchically through mergers and accretion. Therefore, measurements of the number density of massive structures at high redshift pose *potentially* strong constraints on this class of models. The extent to which this potential is realized, however, depends on how massive and how common are the objects at any given epoch. Moreover, such arguments also require making a connection between an observed population and the dark matter halos in which they are expected to reside. For most, if not all, of the populations mentioned above, this is far from straightforward. It is a common perception that the detection of massive galaxies at high redshift is a serious problem for CDM. It is interesting to take a step back for a moment from the complexities of modeling gas physics, star formation, feedback, radiative transfer, stellar populations, etc., and to simply ask whether the 'concordance' ΛCDM model produces enough massive *dark matter halos* to plausibly host the objects that have been detected.

This simple exercise is carried out in Fig. 1, which shows the predicted cumulative comoving number density of dark matter halos above a given mass, from 10^{11}–$10^{15}\,\mathrm{M}_\odot$ as labeled, as a function of redshift[1]. These were obtained from the Sheth-Tormen modified Press-Schechter model [28]. Also shown are the estimated comoving number densities of various observed populations, as described in the figure caption. It has been claimed that the sub-mm galaxies, EROs and K20 $z \sim 2$ objects have stellar masses of a few times $10^{11}\mathrm{M}_\odot$ [6,15,23,10]. Making the plausible assumptions that about fifteen percent of the total mass is in the form of baryons, and a few tenths of the baryons are in stars, one can easily accommodate the observed numbers of objects within suitably massive dark matter halos of $\sim 10^{13}\mathrm{M}_\odot$. The Lyman break galaxies, which are more numerous but probably have stellar masses about an order of magnitude smaller (a few times $10^{10}\mathrm{M}_\odot$; [27,24]) are also easily accommodated within halos of a few times $10^{11}\mathrm{M}_\odot$. Even the $z \sim 6$ quasars detected by SDSS [13], which must harbor black holes of $\sim 10^{9}\mathrm{M}_\odot$ if they are Eddington limited, can be easily accommodated within $10^{13}\mathrm{M}_\odot$ halos (see [5] for more detailed modeling of these objects). So far, none of the observed populations poses a *fundamental* problem for concordance ΛCDM. Had the observed number density of $10^{11}\mathrm{M}_\odot$ galaxies at $z \sim 2$ been an order of magnitude higher than the current estimates, theorists would have good cause to squirm. As it is, these results suggest that there were fewer massive galaxies in the past, in qualitative agreement with hierarchical

[1] All results presented in this paper assume the 'concordance' ΛCDM cosmology: $\Omega_m = 0.3$, $\Omega_\Lambda = 0.7$, $H_0 = 70$ km/s/Mpc, σ_8=0.9.

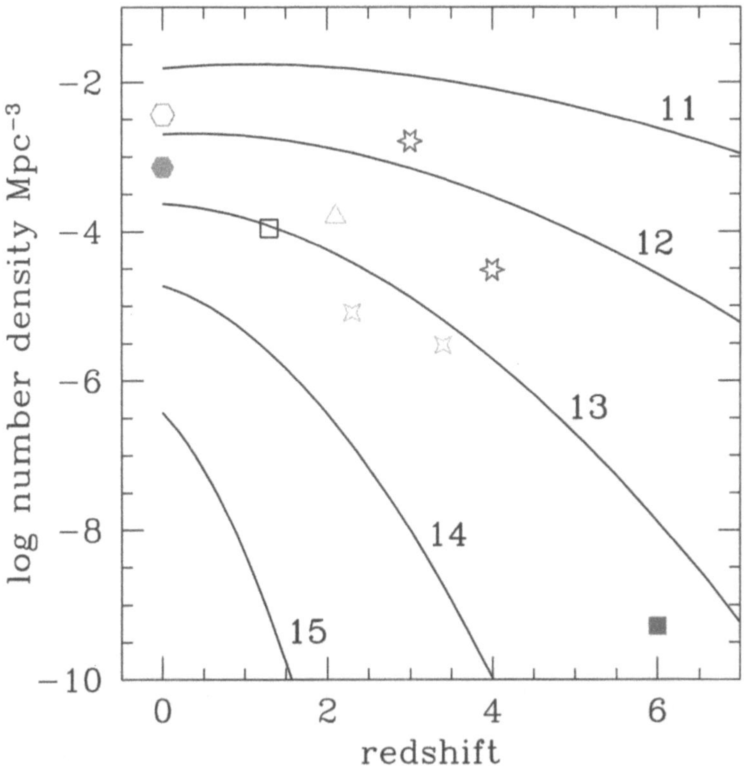

Fig. 1. Curves show the cumulative number density of dark matter halos more massive than 10^{11}, 10^{12}, 10^{13}, 10^{14}, and 10^{15} $M_{\odot}$(from top to bottom, as labeled), as a function of redshift. Points show the estimated comoving number densities of several observed populations, as follows. Hexagons: galaxies with stellar masses greater than $10^{11}M_{\odot}$ (lower point) and $2.5 \times 10^{10}M_{\odot}$ (higher point), obtained by integrating the stellar mass function of [2] from SDSS+2MASS; open square: Extremely Red Objects [22]; open triangle: K20 galaxies [10]; crosses: sub-mm galaxies [6]; six pointed stars: Lyman break galaxies [32], filled square: quasars [13]. For all populations, ΛCDM predicts that there are enough dark matter halos massive enough to plausibly host the observed objects (see text).

structure formation. The challenge now lies in understanding how, where and why the stars and dust that make these objects visible to us are formed. This requires getting our hands a bit more dirty.

3 The Global Mass Assembly History

There are several distinct methods of tracing the global history of star formation and the build-up of stellar mass in the Universe observationally. One is to measure star formation rates by directly counting actively star forming galaxies

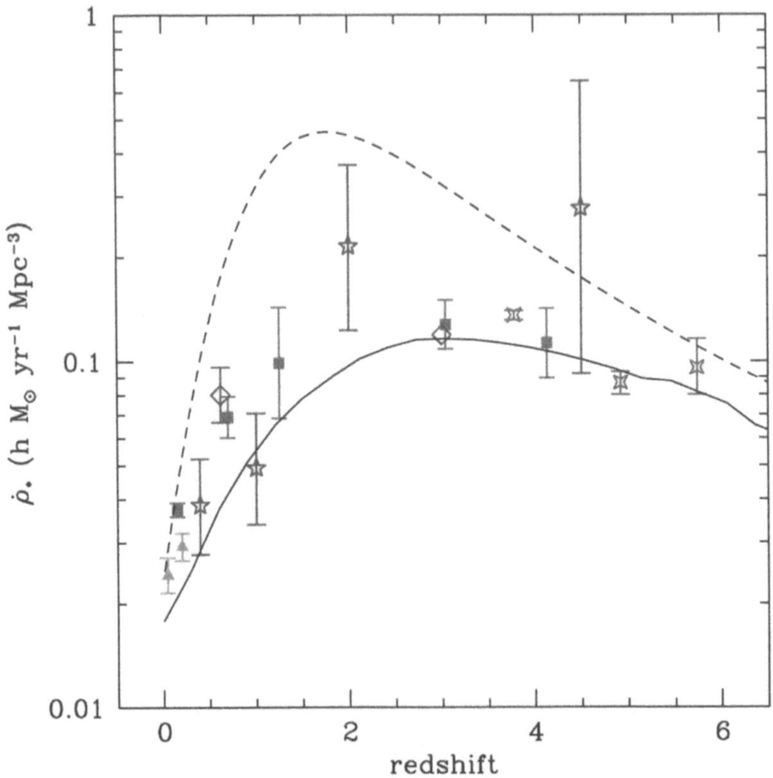

Fig. 2. The global star formation history. The solid curve shows the prediction from a semi-analytic model (similar to the 'accelerated quiescent' model of [30]). The dashed line shows the model from [3], which was designed to reproduce the counts of bright sub-mm galaxies and the Far-IR extragalactic background. Triangles are observational estimates based on Hα, squares are based on rest UV, and open diamonds are based on mid-IR or sub-mm observations (see [30] for references). The crosses are new measurements, based on rest UV, from [16]. Star symbols show measurements based on sub-mm observations from [1]. All optical/UV measurements are corrected for dust extinction and converted to a ΛCDM cosmology where necessary, as described in [30].

via various tracers (e.g., [20,21]). Another is to detect established galaxies at different redshifts, estimate their stellar masses from their colors or spectra, and sum the masses up to obtain a global average [12,26,17]. Yet a third approach is to measure the ages of stellar populations and use galactic 'archeology' to deduce something about star formation at earlier times (e.g. [25]). It is useful to compare the results obtained from all three methods, as the attendant errors and selection effects are different.

Fig. 2 shows a compilation of measurements of the star formation history of the Universe from $z \sim 6$ to the present, including the recent results from GOODS [16]. The solid curve shows the integrated star formation rate density

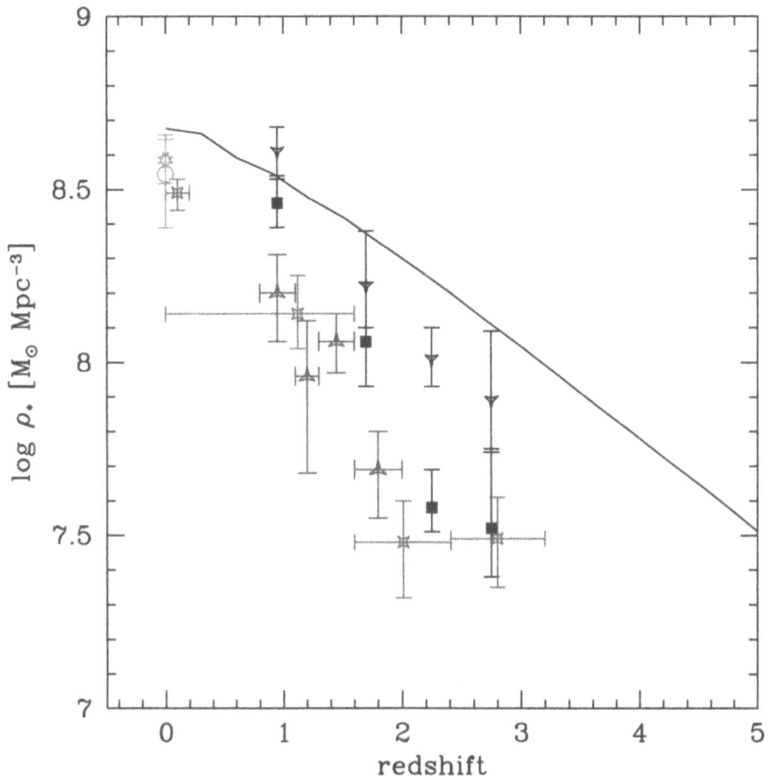

Fig. 3. Global stellar mass assembly history. The solid line shows the total comoving density of stellar mass as a function of redshift, from the same semi-analytic model shown in Fig. 2. Symbols show observational estimates of the stellar mass density at various redshifts from summing the mass in established galaxies. Open circle and many-pointed star: local estimates from [8] and [2]. Filled squares: [12]; downward-pointing triangles: 'maximum mass' estimates from [12]; crosses: [26]; upward-pointing arrows: [17]. The Dickinson et al. and $z = 0$ points are meant to represent *total* masses, while the Rudnick et al. and Glazebrook et al. points are integrated above a fixed rest-frame luminosity or stellar mass, respectively.

predicted by a fairly standard semi-analytic model (similar to the 'accelerated quiescent' model of [30]). It was shown in [30] that this model produced good agreement with existing (dust-corrected) observational estimates of the star formation history to $z \sim 4$; it is encouraging that the same model agrees almost uncannily well with the new measurements out to $z \sim 6$ from [16]. The dashed curve is the updated model of [3], which I shall return to later. Let us now turn to Fig. 3, which shows the total *stellar mass density* as a function of redshift. The solid line is the mass density of all stars from the same semi-analytic model shown in Fig. 2. Symbols show observational estimates, as described in the figure caption.

Obviously, in an ideal world, and neglecting recycling from stellar mass loss, this curve should be the integral of the curve shown in Fig. 2. For the semi-analytic model, this is indeed the case. Given that the semi-analytic model is quite a good fit to the observational results shown in Fig. 2, we would obtain something quite similar if we simply integrated a curve fit to the observed star formation history. However, although if anything the SFR produced by the semi-analytic model is a bit on the low side at low redshift (which dominates the integral), the total mass produced by the present day is on the high side. At $z \gtrsim 1$, the observational estimates of the total mass in established systems are more than 50 percent lower than the model, or than the integrated star formation curve. This could be due to several factors. One possibility is that the IMF changes with time (for example, is more top-heavy at early times). Another is that the total star formation rate has been overestimated, perhaps because of dust corrections, or completeness corrections. A third is that galaxies are missing from the stellar mass census ([19] show that this is likely to be the case at high redshift) or that the stellar masses are systematically underestimated.

This shows that a ΛCDM-based galaxy formation model has no difficulty producing enough early stars or star formation *globally*. Similar results have been found based on hydrodynamic N-body simulations by e.g. [31]. Similarly, the model predicts that roughly half of the present day stellar mass was in place by $z \sim 2$, and seventy percent by $z \sim 1$, implying that there should be enough old stars ($\gtrsim 8 - 10$ Gyr) at $z = 0$ to satisfy the 'fossil evidence' in present day galaxies.

4 Trouble: Old Red Galaxies and Big Dusty Starbursts

The new challenge is to go beyond global quantities like those discussed above, and to ask what kind of galaxies host those stars and where the star formation is occurring. For example, we know that most of the stellar mass in the local universe is contained in old, massive, spheroids, while most of the star formation is occurring in smaller-mass, disk-dominated objects. The new generation of multi-wavelength surveys is allowing us to determine whether this was true at earlier times. The results will pose much more stringent constraints on models.

Current indications are that the same hierarchical models that do well at reproducing global quantities, as discussed above, do much less well at producing the correct proportions of different types of galaxies as a function of redshift. In particular, while the models seem to produce *enough stellar mass globally*, at redshifts of about unity and above they do not produce as many massive/luminous *red* galaxies as are observed. For example, it has been noted in several different studies (e.g. [14,7,29]) that semi-analytic models do not produce enough 'extremely red objects' (EROs), objects with observed-frame optical-infrared colors redder than a passively evolving elliptical (e.g. $(R - K)_{\rm VEGA} > 5$–6), which tend to be at redshifts $0.8 \lesssim z \lesssim 1.5$ (see e.g. [22]). Semi-analytic models underproduce these objects by factors of two to ten, depending on the precise color and magnitude cut, and the specific star formation recipes used. Also, although semi-analytic models can be tuned to give passable agreement with the rest frame $U - V$ color-magnitude relation at $z = 0$, the model fails to produce enough luminous/massive, red, bulge-dominated galaxies at $z \gtrsim 0.5$–1, based on measurements from COMBO-17 and the DEEP redshift survey.

A possibly related problem is that, when absorption and emission by dust are modeled in detail, semi-analytic models also fail to produce enough luminous sub-mm galaxies [18,11], now believed to be predominantly at redshifts $z \sim 2.5$ [6]. This problem is demonstrated in Fig. 2: the dashed curve shows the star formation density needed to account for the sub-mm sources and the Far-IR extragalactic background [3,4]. The required star formation at $z \sim 1.5$–2.5 greatly exceeds that predicted by the semi-analytic models or hydro simulations. One way to solve this problem is to assume that the IMF was very top-heavy at early times, or in starbursts (which account for a larger fraction of star formation at high redshift), as we heard from Cedric Lacey at this meeting. It is possible, though, that these two problems are reflections of the same underlying difficulty: perhaps not enough massive objects are being created (in huge dust-enshrouded bursts) at $z \gtrsim 2$, which is why not enough old, red, quiescent massive objects are in place at $z \sim 1$–2. However, the integrated stellar mass density at low redshift poses a strong constraint — it would clearly be exceeded if the Blain et al. curve were integrated. This remains a puzzle.

5 Summary

To summarize, hierarchical models of galaxy formation (whether based on semi-analytic or N-body+hydro techniques) are currently able to satisfy observational constraints on the global history of star formation and stellar mass assembly from *optically selected* galaxies, from $z \sim 6$ to the present. However, they fail badly at reproducing two very different kinds of massive objects at high redshift: quiescent red spheroids, and vigorously star forming, heavily dust-obscured ULIRG-like objects detected in the mid-IR and sub-mm. It is possible, though not certain, that these problems are connected. One can think of many physical mechanisms, neglected in the current modeling, that could *quench* star formation and produce more red objects, but these mechanisms alone could only make galaxies dimmer and less massive. What seems to be needed is a mechanism that *boosts* star formation but then rapidly *quenches* it — in massive systems preferentially. This may sound like an old story (monolithic dissipative collapse). What is needed is a *physical mechanism* that produces such a process within the framework of the otherwise largely successful hierarchical structure formation model. There is a growing belief that AGN-driven feedback will probably play a major role in solving these problems, but many details remain to be worked out. I conclude that the multi-wavelength universe is a messy but exciting place to live.

References

1. A. J. Barger, L. L. Cowie, E. A. Richards: AJ, **119**, 2092 (2000)
2. E. F. Bell, D. H. McIntosh, N. Katz, M. D. Weinberg: ApJS, **149**, 289 (2003)
3. A. W. Blain, I. Smail, R. J. Ivison, J.-P. Kneib: MNRAS, **302**, 632 (1999)
4. A. W. Blain, I. Smail, R. J. Ivison, J.-P. Kneib, D. T. Frayer, PhR, **369**, 111 (2002)
5. J.M. Bromley, R.S. Somerville, A. C. Fabian: MNRAS, in press (astro-ph/0311008)
6. S.C. Chapman, A.W. Blain, R.J. Ivison, I.R. Smail: Nature, **422**, 695 (2003)
7. A. Cimatti et al.: A&A, **391**, L1 (2002)
8. S. Cole et al.: MNRAS, **326**, 255 (2001)
9. E. Daddi, A. Cimatti, A. Renzini: A&A, **362**, L45 (2000)
10. E. Daddi et al.: ApJL, GOODS special edition, in press (2004)
11. J.E.G. Devriendt, B. Guiderdoni, A&A, **363**, 851 (2000)
12. M. Dickinson, C. Papovich, H.C. Ferguson, T. Budavari: ApJ, **587**, 25 (2003)
13. X. Fan et al.: AJ, **122**, 2833 (2001)
14. A. E. Firth et al.: MNRAS, **332**, 617 (2002)
15. R. Genzel, this meeting
16. M. Giavalisco et al.: ApJL, GOODS special edition, in press (2004)
17. K. Glazebrook et al. (astro-ph/0401037)
18. B. Guiderdoni, Hivon, Eric; Bouchet, Francois R.; Maffei, Bruno: MNRAS, **295**, 877 (1998)
19. R. Idzi, R. S. Somerville, C. Papovich, H. C. Ferguson, M. Giavalisco, C. Kretchmer, J. Lotz: ApJL, GOODS special edition, in press (2004)
20. Lilly, S. J.; Le Fevre, O.; Hammer, F.; Crampton, David: ApJL, **460**, 1 (1996)
21. P. Madau, H.C. Ferguson, M. E. Dickinson, M. Giavalisco, C. C. Steidel, A. Fruchter: MNRAS, **283**, 1388 (1996)
22. L. A. Moustakas et al.: ApJL, GOODS special edition, in press (2004)
23. L. A. Moustakas et al., poster
24. C. Papovich, M. Dickinson, H. C. Ferguson: ApJ, **559**, 620 (2001)
25. A. Renzini, in 'The Young Universe', D'Odorico, Fontana, Giallongo, eds, ASP Conf. Ser., Vol. Astron. Soc. Pac., San Francisco
26. G. Rudnick et al.: ApJ, **599**, 847 (2003)
27. A. E. Shapley, C. C. Steidel, K. L. Adelberger, M. Dickinson, M. Giavalisco, M. Pettini: ApJ, **562**, 95 (2001)
28. R. K. Sheth, G. Tormen: MNRAS, **308**, 119 (1999)
29. R.S. Somerville et al.: ApJL, GOODS special edition, in press
30. R.S. Somerville, J.R. Primack, S.M. Faber: MNRAS, **320**, 504 (2001)
31. V. Springel, L. Hernquist: MNRAS, **339**, 312 (2003)
32. C.C. Steidel, K. L. Adelberger, M. Giavalisco, M. Dickinson, M. Pettini: ApJ, **519**, 1 (1999)

The Size Evolution of High-Redshift Galaxies

Henry C. Ferguson, Swara Ravindranath, and the GOODS Team

Space Telescope Science Institute, 3700 San Martin Drive, Baltimore, MD 21212, USA

Abstract. It is appropriate in a conference on the growth of galaxies to examine what we know about the evolution of galaxy sizes. Quantifying galaxy size evolution is challenging because most high-redshift galaxies are only barely resolved from the ground. The Hubble Space Telescope (HST) Advanced Camera for Surveys (ACS) has greatly improved our ability to measure the spatial structure of distant galaxies. We briefly review previous HST measurements of size evolution and present early results from analysis of the first three-fifths of the Great Observatories Origins Deep Survey (GOODS) ACS data. The observed size evolution appears to follow the basic trends expected from hierarchical models.

1 Introduction

Observations from HST over the past several years have provided us with clear evidence for the evolution in galaxy morphology and size. The challenge now, with larger and deeper samples, is to quantify that evolution to an accuracy useful for testing theoretical models. We focus here on galaxy sizes, which are perhaps the simplest morphological quantity to measure and interpret.

The basic framework of galaxy formation within the hierarchical Cold-Dark-Matter (CDM) cosmology was set out by White & Rees[25], and has been refined by numerous N-body and semi-analytical studies [24,10,4,20]. Most studies follow the basic framework of Fall & Efstathiou [8], in which dark halos acquire their angular momenta via tidal torques, the angular momentum per unit mass of the baryons and the dark matter are initially the same, and angular momentum is conserved as the baryons collapse and cool to form a disk. With these simplifying assumptions the baryons typically collapse by factors of ~ 10 and the resulting disks have rotation curves, surface-density profiles, and scale radii similar to those observed locally. Further analytical studies have calculated the distribution function of disk-galaxy sizes and the disk-galaxy size-redshift relation [5,13]. Meanwhile cosmological N-body + hydrodynamical simulations have formed disks that appear similar to spiral galaxies, but that tend to be too small to match present-day galaxies ([14]; but see [7]). It is not yet known whether the source of this discrepancy lies in baryonic, dark-matter physics, or numerical issues in the simulations.

If the general Fall & Efstathiou view of disk formation is correct, there are two rather robust expectations that are worth exploring through the observations of high-redshift galaxies. (1) The sizes of galactic disks forming at a redshift z should be a fixed fraction of the size of the dark-matter halo, and therefore scale with redshift as $R_s \propto H^{-1}(z)$ at fixed circular velocity, or $R_s \propto H^{-2/3}(z)$ at fixed mass. (2) If governed primarily by the angular momenta of their halos, the sizes of disk galaxies should should show a log-normal distribution.

The theory described above is specific to disk galaxies. However, it is not unlikely that, at least at early times, the scaling with redshift is similar for early-type galaxies (i.e. that the characteristic radii of the luminous parts of these non-rotationally supported galaxies are directly proportional to the virial radii of the dark-matter halos). For both types of galaxies there are likely to be a variety of effects, only partially included in theoretical models, that may influence the size distribution or scaling with redshift. Merging, adiabatic contraction, the sensitivity of star-formation thresholds to the the metagalactic background and chemical evolution, are all likely to be relevant. Furthermore, one should differentiate between galaxy radii at the "formation epoch", with the expected simple redshift scaling discussed above and galaxy radii at some later epoch. This distinction may become quite important for understanding trends at redshifts $z < 1$. In any case, it is certainly worth measuring the trends, even if there is not a rock-solid set of predictions. We adopt the cosmology defined by $H_0 = 70$ km s^{-1} Mpc^{-1}, $\Omega_M = 0.3$, and $\Omega_\Lambda = 0.7$ throughout this paper.

2 Previous Observations

Redshift $z \sim 1.3$ provides a natural dividing line for studying size evolution. At lower redshifts it is possible to work at rest-frame optical wavelengths and it is interesting and sensible to restrict the comparisons to specific Hubble types. At higher redshifts, the dearth of high-resolution near-IR data forces us to work in the rest-frame ultraviolet and the morphological connection to present-day Hubble types becomes more tenuous. The evolution of disk galaxies to $z \sim 1$ has been explored previously via the magnitude–size ($M_B - r_e$) relation [17,11,16,18,2] and the Tully-Fisher ($M_B - V_c$) relation [23]. Based on *HST* imaging, most studies found evidence for a significant increase ($\sim$ 1–1.3 magnitude) in the rest–frame B-band surface brightness of disk galaxies to $z = 1$, while Simard et al. [18] found no evidence for surface brightness evolution once the selection effects of the survey were taken into account. The luminosity–size evolution of disks remains a controversial issue and the interpretation of any observed evolution with redshift depends crucially on accounting for the selection biases of the survey [18,6,2]. Lilly et al. [11] reported that the abundance and size distribution remains constant out to $z \sim 1$, for the large disks with scalelengths greater than ~ 5 kpc, for which their sample is fairly complete. The space densities at different look-back times provides a key observable to help determine how and when large galaxies like the Milky Way were formed.

At higher redshifts early results from the Hubble Deep Field (HDF) indicated that Lyman-break galaxies were in general smaller than local early-type galaxies [12]. Bouwens et al. [1] have done pioneering work quantifying the selection biases, using the technique of "cloning" dropouts from lower redshifts. They find a trend from redshift $z \sim 3$ to $z \sim 5$ that is consistent with $r(z) \propto (1+z)^{-1}$

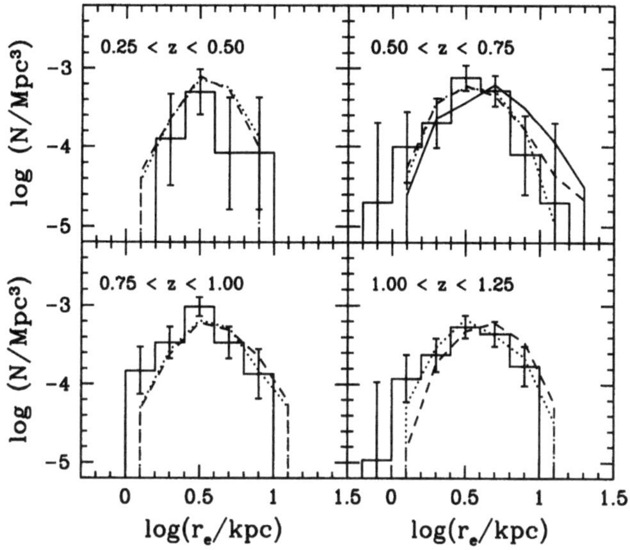

Fig. 1. Disk-galaxy size evolution to $z = 1.25$. The histogram shows the observed size distribution along with the Poisson error bars for the disk galaxies which are largely **free of selection biases**. The curves show the effect of selection (*dotted line*) and measurement (*dashed line*) on an input "non-evolving" SD (*solid line*) based on simulations. The input log–normal distribution (prior to selection and measurement) with a peak at $r_e = 6$ kpc and width $\sigma(\ln r_e) \sim 0.5$ provides a good match to the size distribution observed at $0.50 < z < 0.75$ after the selection criteria is applied, and is held fixed for the other redshifts; the normalization has been adjusted to match the number of galaxies in this redshift bin. The selection causes the peak of the log-normal distribution to shift to smaller sizes ($r_e = 4$ kpc). Measurement biases are significant only in the highest redshift bin.

with large uncertainties. However the study was limited to the small samples with high-resolution imaging from the HDF-N and HDF-S WFPC2 campaigns. The wealth of data now available from ACS should greatly improve constraints in this area.

3 GOODS Results at $z < 1.3$

Ravindranath et al. [15] present an analysis of the first three epochs of the GOODS Chandra Deep Field South (CDF-S) observations. The sample consists of 1508 objects identified as disk galaxies on the basis of two-dimensional fits to their surface-brightness profiles (i.e. galaxies with a Sersic index $n < 2$.) Sizes of the galaxies are measured in the rest-frame B band, and the sample extends to an absolute magnitude $M_B = -19.5$. The completeness of the sample is a function of surface-brightness (or equivalently size at fixed magnitude). The analysis is restricted to the portion of the parameter space where the survey is at least 80%

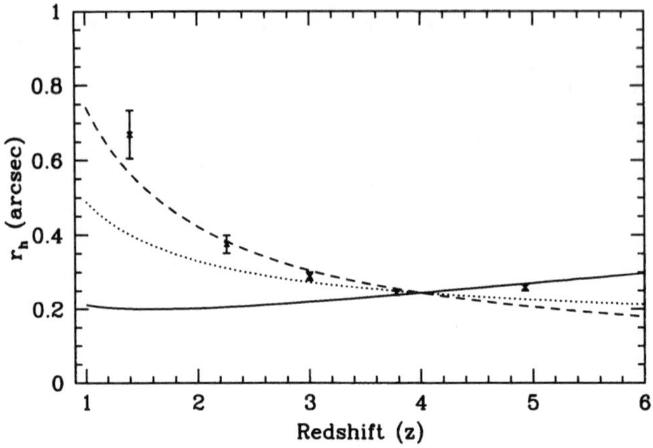

Fig. 2. Size vs. redshift relation at $z > 1$. Mean SExtractor half-light radii are plotted with errorbars indicating the standard error of the mean (i.e. the sample standard deviation divided by the square-root of the sample size). The solid curve shows the expected trend if physical (proper) sizes do not evolve. The dashed curve shows the trend if sizes evolve as $H^{-1}(z)$ and the dotted curve shows $H^{-2/3}(z)$. The curves are all normalized to the mean size at $z \approx 4$ (approximately $r_h = 1.7$ kpc).

complete in all redshift bins. In practice this means that the results are relevant for the evolution of the *high-surface-brightness tail* of the disk-galaxy surface-brightness distribution, and are not necessarily applicable to more typical disk galaxies near the peak of the distribution (e.g. Freeman disks).

Figure 1 shows the observed disk-galaxy size distribution in four redshift bins. The peak of the distribution shifts very little over the redshift range $0.25 < z < 1.25$. The number-densities also show only mild changes as well. The best-fit evolution in number-density is $n(z) \propto (1+z)^{0.08 \pm 0.23}$. The observed trends are milder than those derived by Bouwens & Silk [2] from earlier data, but are consistent with some of the models they summarize in their paper. In particular, the hierarchical model with $\Omega_M, \Omega_\Lambda = 0.3, 0.7$ and the Chiappini et al. [3] infall model both show trends similar to the observed trends.

4 GOODS Results at Higher Redshift

Ferguson et al. [9] use both Lyman-break selection (at $z > 2.7$) and and photometric redshifts (at lower z) to examine the size-redshift relation at more extreme lookback times, once again using the preliminary GOODS data set. Samples are confined to non-evolving rest-frame luminosities between $0.7L^*$ and $5L*$, where L^* is the characteristic luminosity of a $z = 3$ Lyman break galaxy from [22]. Five samples in different bins of redshift ($z \approx 1.4, 2.3, 3.0, 4.0$, and 5.0) are constructed. After rejecting sources classified as stellar (SExtractor `CLASS_STAR`> 0.9) or with unreliable photometry the sample sizes (for the lowest to highest redshift bins, respectively) are 21, 73, 140, 386, and 153 galaxies.

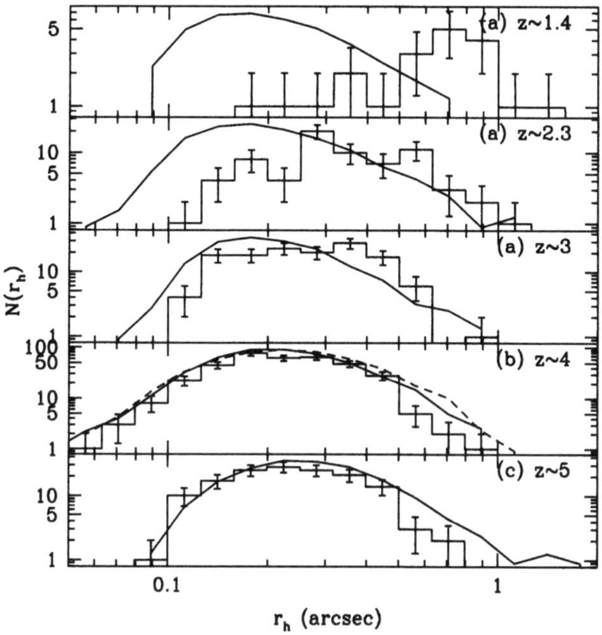

Fig. 3. Observed size distributions of the high-redshift samples. Data for the different samples are shown with Poisson error bars. The curves show a "no-evolution" model of the size distribution, tuned to match approximately the observed distribution at $z \approx 4$ (see [9] for details). Model galaxies have been inserted into the images in a Monte-Carlo fashion and detected and measured using SExtractor in the same way as the real galaxies. Therefore the model curves faithfully represent the effects of both sample incompleteness and measurement biases. The input distribution (prior to selection and measurement biases) for the $z \approx 4$ bin is shown as the dashed curve.

(Sample sizes and S/N will increase significantly with use of the full GOODS data set.) The galaxy sizes are measured in the band closest to rest-frame 1500Å.

The measured size-redshift relation for these samples is shown in Fig. 2. The observed points are uncorrected for biases or incompleteness, and the errorbars ignore possible effects of clustering [19] or crowding. For comparison, in Fig. 2 we show three size-redshift relations. The solid line is the angular-size redshift relation for a standard measuring-rod in the WMAP cosmology [21]. In this cosmology, angular diameters increase by 23% from $z = 3$ to $z = 5$ (for objects with fixed proper diameters). The two lines that decrease with redshift are the expected scalings if size at fixed luminosity tracks size at fixed virial velocity $r \propto H^{-1}(z)$ or fixed mass $r \propto H^{-2/3}(z)$. The mean half-light radius of 0.24″ at $z \approx 4$ corresponds to 1.7 kpc.

The size distributions of galaxies in the different redshift bins are shown in Fig. 3. The curves show a "no-evolution" model tuned to match the observed size distribution at $z = 4$. This model accounts for selection and measurement biases (including cosmological surface-brightness dimming) as described in [9]. The galaxies in the two low-redshift bins are clearly systematically larger than the no-evolution model would predict. However, given the sample sizes and measurement biases, the trend between redshifts $z = 3$ and $z = 5$ is not easily distinguished from the no-evolution model. Nevertheless the overall trend appears remarkably close the generic expectations of hierarchical models of galaxy formation.

References

1. Bouwens, R., Broadhurst, T., & Illingworth, G. 2003, ApJ, 593, 640
2. Bouwens, R., & Silk, J. 2002, ApJ, 568, 522
3. Chiappini, C., Matteucci, F., & Gratton, R. 1997, ApJ, 477, 765
4. Cole, S., Aragon-Salamanca, A., Frenk, C. S., Navarro, J. F., & Zepf, S. E. 1994, MNRAS, 271, 781
5. Dalcanton, J. J., Spergel, D. N., & Summers, F. J. 1997, ApJ, 482, 659
6. de Jong, R. S., & Lacey, C. 2000, ApJ, 545, 781
7. Eke, V., Efstathiou, G., & Wright, L. 2000, MNRAS, 315, L18
8. Fall, S. M., & Efstathiou, G. 1980, MNRAS, 193, 189
9. Ferguson, H. C., et al. 2004, ApJ, 600, L107
10. Kauffmann, G., White, S. D. M., & Guiderdoni, B. 1993, MNRAS, 264, 201
11. Lilly, S., et al. 1998, ApJ, 500, 75
12. Lowenthal, J. D., et al. 1997, ApJ, 481, 673
13. Mo, H. J., Mao, S., & White, S. D. M. 1998, MNRAS, 295, 319
14. Navarro, J. F., & Steinmetz, M. 1997, ApJ, 478, 13
15. Ravindranath, S. et al. 2004, ApJ, 604, L9
16. Roche, N., Ratnatunga, K., Griffiths, R. E., Im, M., & Naim, A. 1998, MNRAS, 293, 157
17. Schade, D., Lilly, S. J., Le Fevre, O., Hammer, F., & Crampton, D. 1996, ApJ, 464, 79
18. Simard, L., et al. 1999, ApJ, 519, 563
19. Somerville, R. S., et al. 2004, ApJ, 600, L171
20. Somerville, R. S., & Primack, J. R. 1999, MNRAS, 310, 1087
21. Spergel, D., et al. 2003, ApJS, 148, 175
22. Steidel, C. C., Adelberger, K. L., Giavalisco, M., Dickinson, M., & Pettini, M. 1999, ApJ, 519, 1
23. Vogt, N. P., Forbes, D. A., Phillips, A. C., Gronwall, C., Faber, S. M., Illingworth, G. D., & Koo, D. C. 1996, ApJ, 465, 15
24. White, S. D. M., & Frenk, C. S. 1991, ApJ, 379, 52
25. White, S. D. M., & Rees, M. J. 1978, MNRAS, 183, 341

High-Redshift QSOs in the GOODS

S. Cristiani[1], D.M. Alexander[2,10], F. Bauer[2], W.N. Brandt[2], E.T. Chatzichristou[3], F. Fontanot[4], A. Grazian[5], A. Koekemoer[6], R.A. Lucas[6], J. Mao[11], P. Monaco[4], M. Nonino[1], P. Padovani[6,9], D. Stern[7], P. Tozzi[1], E. Treister[3], C.M. Urry[3], and E. Vanzella[8]

[1] INAF-Osservatorio Astronomico, Via Tiepolo 11, I-34131 Trieste, Italy
[2] Department of Astronomy and Astrophysics, Pennsylvania State University, 525 Davey Lab, University Park, PA 16802
[3] Department of Astronomy, Yale University, PO Box 208101, New Haven, CT06520
[4] Dipartimento di Astronomia dell'Università, Via Tiepolo 11, I-34131 Trieste, Italy
[5] INAF-Osservatorio Astronomico di Roma, via Frascati 33, I-00040 Monteporzio, Italy
[6] Space Telescope Science Institute, 3700 San Martin Dr., Baltimore, MD 21218
[7] Jet Propulsion Laboratory, California Institute of Technology, Mail Stop 169-506, Pasadena, CA 91109
[8] European Southern Observatory, K.Schwarzschild-Straße 2, D-85748 Garching, Germany
[9] ESA Space Telescope Division
[10] Institute of Astronomy, Madingley Road, Cambridge, CB3 0HA, UK
[11] SISSA, via Beirut 4, I-34014 Trieste, Italy

1 Introduction

QSOs are intrinsically luminous and therefore can be seen rather easily at large distances; but they are rare, and finding them requires surveys over large areas. As a consequence, at present, the number density of QSOs at high redshift is not well known. Recently, the Sloan Digital Sky Survey (SDSS) has produced a breakthrough, discovering QSOs up to $z = 6.43$ [8] and building a sample of six QSOs with $z > 5.7$. The SDSS, however, has provided information only about very luminous QSOs ($M_{1450} \lesssim -26.5$), leaving unconstrained the faint end of the high-z QSO Luminosity Function (LF), which is particularly important to understand the interplay between the formation of galaxies and super-massive black holes (SMBH) and to measure the QSO contribution to the UV ionizing background [16]. New deep multi-wavelength surveys like the Great Observatories Origins Deep Survey (GOODS), described by M.Dickinson, M.Giavalisco and H.Ferguson in this conference, provide significant constraints on the space density of less luminous QSOs at high redshift. Here we present a search for high-z QSOs, identified in the two GOODS fields on the basis of deep imaging in the optical (with *HST*) and X-ray (with *Chandra*), and discuss the allowed space density of QSOs in the early universe, updating the results presented in [6].

2 The Database

The optical data used in this search have been obtained with the ACS onboard HST, as described by M.Giavalisco. Mosaics have been created from the first three epochs of observations, out of a total of five, in the bands $F435W(B_{435})$, $F606W(V_{606})$, $F775W(i_{775})$, $F850LP(z_{850})$. The catalogs used to select high-z QSOs have been generated using the SExtractor software, performing the detection in the z_{850} band and then using the isophotes defined during this process as apertures for photometry in the other bands. This is a common practice avoiding biases due to aperture mismatch coming from independent detections.

The X-ray observations of the HDF-N and CDF-S consist of 2 Ms and 1 Ms exposures, respectively, providing the deepest views of the Universe in the 0.5–8.0 keV band. The X-ray completeness limits over 90% of the area of the GOODS fields are similar, with flux limits (S/N= 5) of 1.7×10^{-16} erg cm^{-2} s^{-1} (0.5–2.0 keV) and 1.2×10^{-15} erg cm^{-2} s^{-1} (2–8 keV) in the HDF-N field, and $\approx 2.2\times10^{-16}$ erg cm^{-2} s^{-1} (0.5–2.0 keV) and $\approx 1.5\times10^{-15}$ erg cm^{-2} s^{-1} (2–8 keV) in the CDF-S field. The sensitivity at the aim point is about 2 and 4 times better for the CDF-S and HDF-N, respectively. As an example, assuming an X-ray spectral slope of $\Gamma = 2.0$, a source detected with a flux of 1.0×10^{-16} erg cm^{-2} s^{-1} would have both observed and rest-frame luminosities of 8×10^{42} erg s^{-1}, and 3×10^{43} erg s^{-1} at $z = 3$, and $z = 5$, respectively (assuming no Galactic absorption). [1] produced point-source catalogs for the HDF-N and CDF-S and [9] for the CDF-S with more than 1000 detected sources in total.

3 The Selection of the QSO Candidates

We have carried out the selection of the QSO candidates in the magnitude interval $22.45 < z_{850} < 25.25$. Four optical criteria have been tailored on the basis of typical QSO SEDs in order to select QSOs at progressively higher redshift in the interval $3.5 \lesssim z \lesssim 5.2$. The criteria have been applied independently and produced in total 645 candidates in the CDF-S and 557 in the HDF-N. They select a broad range of high-z AGN, not limited to broad-lined (type-1) QSOs, and are less stringent than those typically used to identify high-z galaxies. We have checked the criteria against QSOs and galaxies known in the literature within the magnitude and redshift ranges of interest confirming the high completeness of the adopted criteria.

Below $z \simeq 3.5$ the typical QSO colors in the ACS bands move close to the locus of stars and low-redshift galaxies. Beyond $z \simeq 5.2$ the $i - z$ color starts increasing and infrared bands would be needed to identify QSOs efficiently with an "i-dropout" technique.

4 Match with Chandra Sources

The optical candidates have been matched with X-ray sources detected by Chandra within an error radius corresponding to the 3 σ X-ray positional uncertainty.

With this tolerance the expected number of false matches is five and indeed two misidentifications, i.e. cases in which a brighter optical source lies closer to the X-ray position, have been rejected (both in the CDF-S).

The sample has been reduced in this way to 11 objects in the CDF-S and 6 in the HDF-N. Type-1 QSOs with $M_{1450} < -21$, given the measured dispersion in their optical-to-X-ray flux ratio [19], are detectable in our X-ray observation up to $z \gtrsim 5.2$. Conversely, any $z > 3.5$ source in the GOODS region detected in the X-rays must harbor an AGN ($L_x(0.5-2\ \text{keV}) \gtrsim 10^{43}$ erg s^{-1}).

5 Redshifts of the QSO Candidates

Twelve objects out of the 17 selected have spectroscopic confirmations. Nine are QSOs with redshifts between 2.6 and 5.2. Three are reported to be galaxies, and the relatively large offsets between the X-ray and optical positions suggest that they could be misidentifications. Both quasars of Type I and II are detected.

Photometric redshifts of the 17 QSO candidates have been estimated by comparing with a χ^2 technique (see [2] for details) the observed ACS colors to those expected on the basis of *a)* the typical QSO SEDs; *b)* a library of template SEDs of galaxies (the "extended Coleman" of [2]). For the nine QSOs with spectroscopic confirmation the photometric redshifts are in good agreement with the observed ones. In general the estimates *a)* and *b)* are similar, since the color selection criteria both for galaxies and QSOs rely on a strong flux decrement in the blue part of the spectrum - due to the IGM and possibly an intrinsic Lyman limit absorption - superimposed on an otherwise blue continuum. If we limit ourselves to the redshift range $z > 4$ - where the selection criteria and the photometric redshifts are expected to be most complete and reliable - in addition to one spectroscopically confirmed QSO (HDF-N 123647.9+620941, $z = 5.186$) we estimate that between 1-2 more QSOs are present in the GOODS, depending on whether galaxy or QSO SEDs are adopted for the photometric redshifts. This brings the total to no more than three QSOs with redshift larger than 4.

6 Comparison with the Models

Let us compare now the QSO counts observed in the z_{850} band with two phenomenological and two more physically motivated models. The double power-law fit of the 2QZ QSO LF [5] has been extrapolated for $z > 2.7$ (the peak of QSO activity) in a way to produce a power-law decrease of the number of bright QSOs by a factor 3.5 per unit redshift interval, consistent with the $3.0^{+1.3}_{-0.9}$ factor found by [7] for the bright part of the LF. The extrapolation is carried out either as a Pure Luminosity Evolution (PLE) or a Pure Density Evolution (PDE). The PLE model predicts about 17 QSOs with $z_{850} < 25.25$ at redshift $z > 4$ (27 at $z > 3.5$) in the 320 arcmin^2 of the two GOODS fields, and is inconsistent with the observations at a more than 3σ level. The PDE estimate is 2.9 QSOs at $z > 4$ (6.7 at $z > 3.5$).

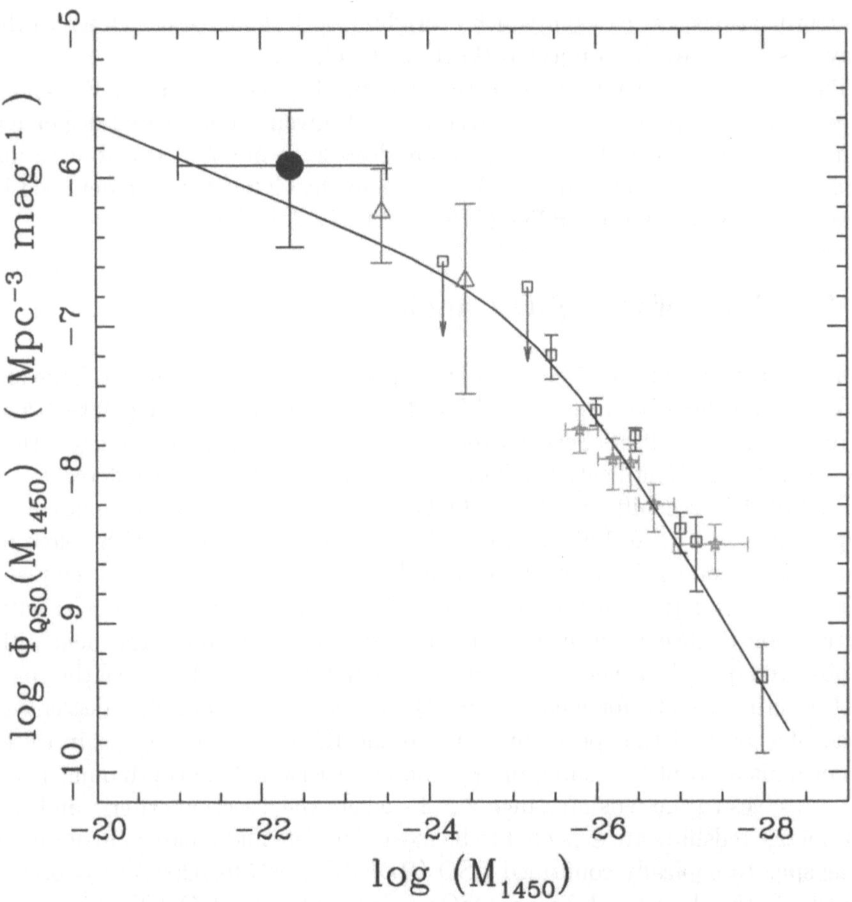

Fig. 1. Differential QSO luminosity function at $z \sim 4-5$. Squares: data from [13] and upper limits from [14] at $z \sim 4.4$. Stars: SDSS [7] computed at $3.5 < z < 5$ assuming a power-law decline and rescaled to $z = 4.3$. Triangles: COMBO-17 LF [20], for $4.2 < z < 4.8$. Circle: GOODS LF at $4 < z < 5.2$. The continuous line shows the prediction of the PDE model described in the text. All LFs have been rescaled to a cosmology with $h, \Omega_{\rm tot}, \Omega_m, \Omega_\Lambda = 0.7, 1.0, 0.3, 0.7$.

It is possible to connect QSOs with dark matter halos (DMH) formed in hierarchical cosmologies with a minimal set of assumptions (MIN model, e.g. [11]): *a)* QSOs are hosted in newly formed halos with *b)* a constant SMBH/DMH mass ratio ϵ and *c)* accretion at the Eddington rate. The bolometric LF of QSOs is then expressed by:

$$\Phi(L|z)dL = n_{\rm PS}(M_H|z) \int_{t(z)-t_{\rm duty}}^{t(z)} P(t_f|M_H, t(z))dt_f \epsilon^{-1} \frac{dL}{L_{\rm Edd}} \quad (1)$$

where the abundance of DMHs of mass M_H is computed using the Press & Schechter [18] recipe, the distribution $P(t_f|M_H, t(z))dt_f$ of the formation times t_f follows [15], $t_{\rm duty}$ is the QSO duty cycle and $L_{Edd} = 10^{4.53}\ L_\odot$ is the Eddington luminosity of a 1 $M_\odot$ SMBH. This model is known to overproduce the number of high-z QSOs [12] and in the present case predicts 151 QSOs at $z > 4$ (189 at $z > 3.5$).

To cure the problems of the MIN model feedback effects have been invoked, for example assuming that the QSOs shine a $t_{\rm delay}$ time after DMH formation (DEL model; [17], [10]. The QSO LF is then computed at a redshift z' corresponding to $t(z) - t_{\rm delay}$. The predictions of the DEL model are very close to the PDE, with 3.2 QSOs expected at $z > 4$ (4.8 at $z > 3.5$), in agreement with the observations. While all the models presented here are consistent with the recent QSO LF measurement of the COMBO-17 survey [20] at $z \simeq 2-3$, only the PDE and DEL models fit the COMBO-17 LF in the range $4.2 < z < 4.8$ and $M_{1450} < -26$.

7 Conclusions

At $z > 4$ the space density of moderate luminosity ($M_{1450} \simeq -23$) QSOs is significantly lower than the prediction of simple recipes matched to the SDSS data, such as a PLE evolution of the LF or a constant universal efficiency in the formation of SMBH in DMH. A flattening of the observed high-z LF is required below the typical luminosity regime ($M_{1450} \lesssim -26.5$) probed by the SDSS. An independent indication that this flattening must occur comes from the statistics of bright lensed QSOs observed in the SDSS [21] that would be much larger if the LF remains steep in the faint end. A similar result has been obtained at $5 \lesssim z \lesssim 6.5$, by [4]. The QSO contribution to the UV background is insufficient to ionize the IGM at these redshifts. This is an indication that at these early epochs the formation or the feeding of SMBH is strongly suppressed in relatively low-mass DMH, as a consequence of feedback from star formation [10] and/or photoionization heating of the gas by the UV background [12], accomplishing a kind of inverse hierarchical scenario.

References

1. Alexander, D.M., et al. 2003, AJ 126, 539
2. Arnouts, S., Cristiani, S., Moscardini, L., Matarrese, S., Lucchin, F., Fontana, A., Giallongo, E., 1999, MNRAS 310, 540
3. Barger, A.J., Cowie, L.L., Richards, E.A., 2000, AJ 119, 2092
4. Barger, A.J., Cowie, L.L., Capak, P., Alexander, D.M., Bauer, F.E., Brandt, W.N., Garmire, G.P., Hornschemeier, A.E., 2003, ApJ 584, L61
5. Boyle, B.J.,Shanks, T., Croom, S.M., Smith, R.J., Miller, L., Loaring, N., & Heymans, C., 2000, MNRAS 317, 1014
6. Cristiani, S., et al. 2004, ApJ 600, L119
7. Fan, X., Strauss, M.A., Schneider, D., et al., 2001, AJ 121,54

8. Fan, X., Strauss, M.A., Schneider, D.P., Becker, R.H., et al., 2003, AJ 125, 1649
9. Giacconi, R., et al. 2002, ApJS 139, 369, G02
10. Granato, G.L., Silva, L., Monaco, P., Panuzzo, P., Salucci, P., De Zotti, G., Danese, L., 2001, MNRAS 324, 757
11. Haiman, Z., & Hui, L., 2001, ApJ 547, 27
12. Haiman, Z., Madau, P., & Loeb, A., 1999, ApJ 514, 535
13. Kennefick, J.D., Djorgovski, S.G., & de Carvalho, 1995, AJ 110, 2553
14. Kennefick, J.D., Djorgovski, S.G., & Meylan, G., 1996, AJ 11, 1816
15. Lacey, C., & Cole, S., 1993, MNRAS 262, 627
16. Madau, P., Haardt, F., Rees, M.J., 1999, ApJ 514, 648
17. Monaco, P., Salucci, P., Danese, L., 2000, MNRAS 311, 279
18. Press, W.H., & Schechter, P., 1974, ApJ 187, 425
19. Vignali, C., et al. 2003, AJ 125, 2876
20. Wolf, C., Wisotzki, L., Borch, A., Dye, S., Kleinheinrich, M., Meisenheimer, K., 2003, A&A 408, 499
21. Wyithe, J.S.B., Loeb, A., 2002, ApJ 577, 57

Constraining the Evolutionary Mass Function and Star-Formation Activity in Galaxies from the Spizer Wide-Area Infrared Extragalactic Survey (SWIRE)

Alberto Franceschini[1], Carol Lonsdale[2], Stefano Berta[1], Giulia Rodighiero[1], and the SWIRE Co-Investigator Team

[1] Astronomy Department, Padova University, Vicolo Osservatorio 2, I-35122 Padova, Italy
[2] IPAC, Pasadena, USA

Abstract. The Wide-area Infrared Extragalactic (SWIRE) survey is a *Legacy* Program of the *Spizer Space Telescope (SST)* exploiting 851 hours of observatory time to conduct a set of large-area (currently foreseen $\sim$ 49 square degrees split into 6 high Galactic latitude fields) imaging surveys, achieving 5σ sensitivities of 4.3/6.5/33/36 μJy at 3.6/4.5/5.8/8.0 μm with IRAC and 0.19/15/40 mJy at 24/70/160 μm with MIPS. SWIRE will allow studying evolved galaxies (from IRAC data) and actively star-forming galaxies and quasars (from MIPS) over co-moving scales up to several hundreds Mpc and down to substantial cosmological depths ($z \simeq 2.5$ for luminous sources). Thanks to the large total survey area, SWIRE will provide a complement to smaller and deeper observations in the Spitzer Guaranteed Time and the Legacy Program, by allowing direct comparison of local and high-z galaxy samples and the investigation of the effects of environment on galaxy evolution.

1 Introduction

There have been reported indications that galaxy evolution might depend on the environment, in particular that it might have been faster in cosmic time within high density condensations (galaxy clusters) than in the field (e.g. Treu et al. 2000, also Stanford et al. 1998, Franceschini et al. 1998). Indeed, old red galaxies appear to cluster much more strongly than the younger blue galaxy population. Furthermore, some discrepant published results on the high-z galaxy redshift distributions and cosmological evolution based on deep surveys in different sky regions are interpreted as due to cosmic variance affecting the typically small survey areas. So any conclusions based on deep investigations of small fields have to be considered as tentative until they are proven in representative cosmic volumes.

It is also clear that any self-consistent picture of galaxy formation has not only to account for the evolution with cosmic time of the volume emissivity and luminosity functions of young stars in galaxies, as inferred from directly detected stellar light, but it needs also to match the evolutionary galaxy mass function, which is essentially unaffected by the dust extinction and therefore a particularly robust physical parameter.

A powerful alternative to the time-expensive spectroscopic investigations for the study of the evolutionary mass function exploits broad-band observations of the spectral intensity of galaxies in the restframe near-IR and its weak dependence on the age of the contributing stars (compared with that of the optical SED). The photometric estimate of the stellar mass in high-redshift galaxies is a particularly attractive approach to recover the history of star formation, in consideration of the relatively simple and fairly well understood underlying physics, as opposed to the extremely more complex and less known physical processes ruling star formation and emission by young stars within dusty media.

The recently launched Spitzer IR Space Telescope is allowing to image with unprecedented sensitivity in the 3−10 μm spectral domain. This is being achieved with the Infrared Array Camera (IRAC), one of the three Spitzer science instruments, providing simultaneous 5.12×5.12 arcmin images at 3.6, 4.5, 5.8, and 8 μm, using two InSb and two SiAs 256x256 detector arrays. The availability of data in this region of the e.m. spectrum will allow accurate photometric estimates of stellar masses in galaxies to redshifts $z>2$ (Section 3).

At the same time, deep sky probes with the *SST*'s Multiband Imaging Photometer (MIPS) at 24, 70, 160 μm will detect dust re-radiation from primeval and actively star-forming galaxies and will measure with high accuracy the rate of star-formation in high-z galaxies (Section 4).

The recent December 2003 NASA Press Release (www.spitzer.caltech.edu/Media/releases/ssc2003-06) has provided in-orbit demonstration of the capability of the observatory to fully accomplish the tasks of deep cosmological survey originally planned. It turns out in particular that the imaging sensitivities in the critical wavebands at 3.6, 4.5 and 24 μm are significantly better than expected.

The Spizer Wide-Area Infrared Extragalactic Survey (SWIRE), part of the *SST* Legacy Program, will survey the emissions of evolved stellar populations and dust re-radiation from star-forming galaxies with enough cosmic volume and galaxy statistics to cover the whole range from rich galaxy clusters to the cosmic voids.

2 The SWIRE Legacy Program

SWIRE, the largest among the six Spitzer Legacy programs, is a wide-area imaging survey designed to trace the evolution of dusty, star-forming galaxies, evolved stellar populations, and active galactic nuclei (AGNs) as a function of environment, from redshifts $z \sim 3$ to the current epoch.

SWIRE will provide an unprecedented view of the evolution of galaxies, large-scale structure, and AGNs. The key scientific goals of SWIRE are (1) to determine the evolution of actively star forming and passively evolving galaxies in order to understand the history of galaxy formation in the context of cosmic structure formation; (2) to determine the evolution of the spatial distribution and clustering of evolved galaxies, starbursts, and AGNs in the key redshift range $0.5<z<3$ over which much of cosmic evolution has occurred; and (3) to determine the evolutionary relationship between normal galaxies and AGNs and the contribution of AGN accretion energy versus stellar nucleosynthesis to the cosmic backgrounds.

SWIRE will survey six high-latitude sky areas, totaling 49 deg^2 (see Table 1) in all seven Spitzer bands of IRAC and MIPS. The total area coverage was selected such that the survey would probe several hundred volume cells of scale ~100 Mpc, enough to sample many different environments within the cosmic web. The number of lines of sight was a trade-off between cosmic variance considerations (more fields) and maximizing the physical size scale probed in each field (fewer, larger fields), moderated by the availability of large sky areas with acceptable cirrus noise levels. The resulting six fields project between ~130 and 250 100 h^{-1} comoving Mpc at z=2, with about 50 radial-distance cells of 100 Mpc comoving scale along each line of sight to that epoch.

Table 1. SWIRE Survey Areas

Field name	Field centers (J2000)	β [deg.]	I(100μm) [MJy/sr]	E(B-V)	Area [sq.deg.]
ELAIS-S1	00h38m30s -44d00m00s	−43	0.42	0.008	8
XMM-LSS	02h21m00s -05d00m00s	−18	1.30	0.027	9
CHANDRA-S	03h32m00s -28d16m00s	−48	0.46	0.001	7.2
Lockman Hole	10h45m00s +58d00m00s	+44	0.38	0.006	12.3
ELAIS-N1	16h11m00s +55d00m00s	+74	0.44	0.007	8
ELASI-N2	16h36m48s +41d01m45s	+62	0.42	0.007	4.5

These SWIRE fields have been selected to minimize Galactic cirrus emission over large scales and to include sky regions with relevant complementary data from round-based and space observatories. SWIRE includes ~9 deg^2 of the unique large-area XMM Large Scale Structure hard X-ray imaging survey and is partly covered by the UKIDSS deep J and K survey. An extensive optical/near-IR imaging program is underway from the ground.

The expected SWIRE photometric sensitivities are reported in Table 2. The large area of SWIRE is important to establish statistically significant population samples over enough volume cells that we can resolve the star formation history as a function of epoch and environment, i.e., in the context of structure formation. The large total volume is also optimized for finding rare objects and to allow comparison of galaxy luminosity functions at high- and low-redshifts, and local, which is needed for robust determinations of the evolutionary histories.

Extensive modeling suggests that the SWIRE Extragalactic Catalog may contain in excess of 2 million IR-selected galaxies, including ~100000 luminous infrared galaxies (LIRGs; LFIR >10^{11} $L_{\odot}$) detected by MIPS (and significantly more detected by IRAC), ~7000 of these with z>2; 1 million IRAC-detected early-type galaxies ($\sim 2\,10^5$ with z> 1 and ~10000 with z> 2); and~20000 classical AGNs detected with MIPS, plus significantly more dust-obscured quasi-stellar objects/AGNs among the LIRGs. Images and source catalogues, together with ancillary ground-based data, will be released starting from July 2004. More details on the SWIRE project may be found in Lonsdale et al. (2003).

Table 2. SWIRE Sensitivities and Confusion Limits

Band (μm)	3.6	4.5	5.6	8	24	70	160
Sensitivity [mJy], 5σ	0.003	0.006	0.040	0.050	0.19	15	40
Confusion[a] [mJy], 5σ	0.002	0.002	0.003	0.003	0.090	4.0	36.

[a] Based on models by Franceschini et al. (2001), Rodighiero et al. (2003), Shupe et al. (2004).

3 Constraining the Build-Up of Stellar Masses in High-z Galaxies with IRAC

SWIRE will detect the emission of evolved stars in galaxies, through the IRAC bands, reaching z ~2.5. The stellar mass of these systems can be estimated by SED fitting using a library of stellar population synthesis models (e.g. Papovich et al. 2001, Dickinson et al. 2003). The galaxy near-IR light is dominated by the emission of old low-mass stars, which provide the bulk of the assembled stellar mass.

This method has been prototyped for SWIRE using the models of Poggianti et al. (2001), Franceschini et al. (2003) and Berta et al. (2004). The latter found quite large uncertainties in the estimated stellar masses for ISO-detected HDF starbursts (at z$\sim 0.5 - 1.5$) and Ly-break galaxies when using only optical/NIR data. These uncertainties are especially due to a degeneracy between red old and young extinguished stellar populations. An illustration of the uncertainty in such determinations of the baryonic mass for an ISO-discovered galaxy at z=1.27 is given in Fig. 1 (Berta et al. 2004). The figure also shows how sampling the emission longwards of the atmospheric limit at $\lambda = 2.5$ μm is needed for a proper mass determination, particularly at such high or higher redshift. The optical-NIR observed broad band SED of the ISO starburst S40 — based on the available data (filled dots) up to 2.2 μm — is equally well fitted by the three different models shown. On the other hand IRAC data (open squares) will allow to distinguish among the three solutions by detecting the restframe $\sim 1 - 2$ μm emission of the system, dominated by old stars. In this case the synthetic SED data have been obtained by simulating a SWIRE observation based on the optical-NIR best-fit model.

For galactic systems at higher redshifts, the photometric method for estimating the stellar mass content suffers for an increasingly poorer sampling of

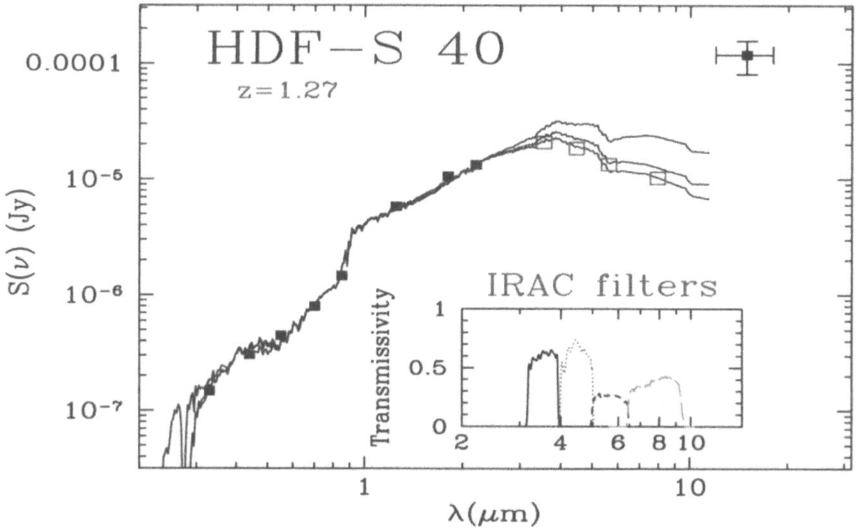

Fig. 1. Simulations of Spitzer/IRAC observations of the HDFS/ISO starburst S40 (Franceschini et al. 2003) to constrain the estimate of stellar mass (Berta et al., 2004). Three different spectral fits to the observed data on S40 (black filled dots) are compared with simulated IRAC fluxes (open squares): the latter are clearly needed to disentangle among the three. Small boxes show the transmission curves of the four IRAC filters.

Table 3. Photometric estimate of stellar mass and its uncertainty for a Lyman-break galaxy sample, based on our spectro-photometric synthesis code. For each source we report the redshift, best-fit χ^2, stellar mass estimate and 2σ uncertainties (excluding or including simulated Spitzer/IRAC data). Optical+near-IR photometry from Papovich et al. (2001).

Obj #	z	χ^2_{opt} b.f.	Mass (bf) $[10^{11}M_\odot]$	ΔM $[10^{11}M_\odot]$	$\Delta M_{Spitzer}$ $[10^{11}M_\odot]$
Ly1	2.005	5.13	0.21	0.16−2.47	0.27−0.55
Ly2	2.233	4.24	0.13	0.04−0.24	0.09−0.19
Ly3	2.267	1.47	0.04	0.03−0.84	0.04−0.09
Ly4	2.591	10.99	0.09	0.03−0.28	0.03−0.09
Ly5	2.800	5.87	0.02	0.02−0.29	0.02−0.06
Ly6	2.980	9.02	0.18	0.05−0.35	0.06−0.20
Ly7	2.991	7.54	0.04	0.02−0.13	0.03−0.06
Ly8	3.160	2.29	0.56	0.27−1.35	0.38−0.81

the near-IR restframe emission, when only ground-based data up to the K band are available. The longer wavelength IRAC bands enable robust mass estimation over a much larger redshift interval. Using SED data from the optical to 5μm, we will assess the build-up rate of stellar mass in massive stellar systems since z∼2.5. Our simulations (see Fig. 1 and Table 3) show that such an extension of the wavelength dynamic range will reduce the 2σ mass uncertainties to factors of 2−3 for various classes of galaxies up to z∼3 (Berta et al. 2004). This promises to be good enough for statistically reliable determinations of the galaxy evolutionary mass functions. The dependence of the results on the matter density field and the nearby galaxy environment will be of primary importance to constraining galaxy formation models.

4 Sampling Old Galactic Stars and Ongoing Star Formation in SWIRE Survey Galaxies

The simultaneous deep coverage of the large SWIRE survey area with MIPS will add the capability of measuring the star-formation rate for a large number of high-z starbursts, particularly using the 24 μm channel. Although the 24 μm flux is not directly a measure of the bolometric far-IR emission, various analyses have empirically shown that the mid-IR flux is tightly correlated with the galaxy bolometric luminosity and radio flux in both local IRAS galaxies (Elbaz et al. 2002) and low-z ISO samples (Gruppioni et al. 2003).

From their analysis of an ISO-selected sample of dusty starbursts at redshift z∼1 in the Hubble Deep Field South, Franceschini et al. (2003) have found a very good match between the rates of star-formation estimated from the radio centimetric flux and those based on the 15 μm emission. From this, and considering the photometric sensitivity and spatial resolution achievable with Spitzer at somewhat longer wavelengths (24 μm), we conclude that SWIRE is going to provide us with a rather unique capability of sampling simultaneously old galactic stars and the ongoing star-formation in galaxies over a very representative cosmic volume and spanning a substantial fraction of the Hubble time.

References

1. Dickinson, M:, Papovich, C., Ferguson, H.C., Budavari, T.: ApJ **587**, 25 (2003)
2. Elbaz, D.; Cesarsky, C.; Chanial, P.; Aussel, H.; Franceschini, A.; Fadda, D. & Chary, R.: A&A **384**, 848 (2002)
3. Franceschini, A., et al.: ApJ **506**, 600 (1998)
4. Franceschini A., Aussel H., Cesarsky C., Elbaz D., Fadda D.: A&A **378**, 1 (2001)
5. Franceschini A., et al.: A&A **403**, 501 (2003)
6. Gruppioni, C., et al.: MNRAS **341**, L1 (2003)
7. Lonsdale, C., et al. PASP **115**, 897 (2003);
   see also http://www.ipac.caltech.edu/SWIRE/swire_v2.html (2001)
8. Papovich, C., Dickinson, M., & Ferguson, H. C., ApJ **559**, 620 (2001)
9. Rodighiero, G., Lari, C., Franceschini, A., Gregnanin, A., Fadda, D.: MNRAS **343**, 1155 (2003)
10. Shupe, D., Surace, J., Castro, S., et al., ApJ submitted (2004)
11. Stanford, S.A., Eisenhardt, P.R., Dickinson, M.: ApJ **492**, 461 (1998)

The European Large Area ISO Survey: A Pathfinder for SIRTF

Seb Oliver[1] and the ELAIS Consortium

Astronomy Centre, Dept. of Physics & Astronomy, University of Sussex, Brighton, BN1 9QH, UK

Abstract. The European Large Area ISO Survey, ELAIS, was the largest Open Time survey on the Infrared Space Observatory, ISO. It was designed to explore obscured galaxies and hence quantify the recent star-formation history of the Universe. We have recently completed our final reanalysis of the data, compiled band-merged catalogue products with associations across many wavelengths and released the data to the global astronomical community (http://astro.imperial.ac.uk/elais/). This paper summarizes some of the key results.

1 The ELAIS Project

The European Large Area ISO Survey was the largest non-serendipitous project on ESA's Infrared Space Observatory (ISO, [1]) using ISOCAM [2] and ISOPHOT [3]. The survey covered around 12 square degrees, with ISO data at 6.7, 15, 90 and 175 μm. The aims of the project and a detailed description of the observations can be found in [4]. A Preliminary Analysis was undertaken to provide early scientific results and to provide follow-up source lists at 6.7 and 15 μm [5], 90 μm [6] – the 175μm data was processed by the FIRBACK consortium [7]. To extract the maximum information from the ISO data we have now completed a very careful Final Analysis of the 15 μm data [8], [9], [10], and at 90 μm [11]. We have also cross-correlated the infrared catalogues and associated them with complementary optical, NIR and radio catalogues (including [12], [13], [14],[15], [16]). The final ISO and Radio selected compilation comprises nearly 4000 sources with limiting flux densities at 6.7, 15, 90 and 175 μm of $\sim$ 1.0, 0.7, 70, 223, 70 mJy [17], these "band-merged" catalogues have been made public (http://astro.imperial.ac.uk/elais/).

2 Motivation

The motivation for studying infrared galaxy evolution is also discussed in [18]. The original motivation for studying galaxy evolution in the infrared comes from our understanding of star formation within our own galaxy. In these regions the intrinsically bright optical and UV emission from young massive stars is strongly affected by dust. The effects of obscuration are less significant in the thermal infrared (1-3 μm) and in the mid and far infrared we actually see the reprocessed emission from the dust. Thus to properly understand star-formation and hence galaxy formation we must study both optical and infrared wavelengths (appropriately for the theme of this workshop). Indeed actively star forming galaxies such as M82 emit significantly more power at far infrared wavelengths than in the optical and UV. The discovery that as much or more power in the extra-galactic background emerges at far-infrared/sub-mm wavelengths as in optical/UV bands [19] emphasizes the cosmological importance of these far-infrared emitting populations. The unexpectedly numerous galaxies discovered by SCUBA (e.g. [20] and [21], [22]) are expected to be star forming galaxies at high redshifts where the rest-frame far-infrared galaxies has been shifted to sub-mm wavelengths.

The ELAIS survey was designed to study the comparatively recent evolution of obscured star formation providing a bridge between local IRAS samples and SCUBA. It also aimed to understand the AGN phenomenon by exploring emission from the dusty tori.

3 Number Counts

The first science from any new survey inevitably comes from the number counts. We analysed the preliminary counts at 6.7, 15 μm [5], 90μm [6] and the FIRBACK team presented the 175 μm counts [7]. The Final Analysis counts at 15 μm in S1 and 90 μm have been discussed by [26] and [11] respectively, an analysis of all the 15 μm data will be given by [27]. Whichever wavelength you choose the result is the same: the counts are inconsistent with any no evolution predictions, i.e. we detect significant evolution at all wavelengths. This is illustrated at 15 and 90 μm in 1. At 15 μm the departure from no evolution models is particularly abrupt at around 1 mJy. This result is enhanced by the deeper ISO-CAM surveys (e.g. [23], [24], [25], [18]) but is clear with ELAIS data alone which could not be affected by small sample biases.

4 Optical Identifications

To investigate what sort of galaxies are responsible for the excessive numbers of galaxies seen at faint fluxes requires data at other wavelengths. There have been extensive observing imaging survey campaigns or comparisons with archival data at radio [15], [16], [34], optical [12], [13], [29], NIR [14], and X-ray wavelengths [30], [31], [32], [33]. There is no room here to discuss the radio, NIR or xray associations and the reader is referred to the above publications. 80% have optical identifications to $R \sim 23$ [12] and 92% to $R \sim 24$ [13]. The optical photometry suggest a significant fraction of the sources have a high infrared to optical ratio and that this fraction increases as we move to fainter infrared fluxes ([13]). Redshifts have been obtained 500 galaxies of our infrared samples (47% of the 15μm galaxies with identifications), primarily on AAT, ESO 3.6m, WHT, and other smaller telescopes. The optical spectra can be used to classify sources as AGN (Type-1 and Type-2); star burst galaxies and normal galaxies ([12], [35]). We have dissected the number counts in S1 using these classifications (Fig.

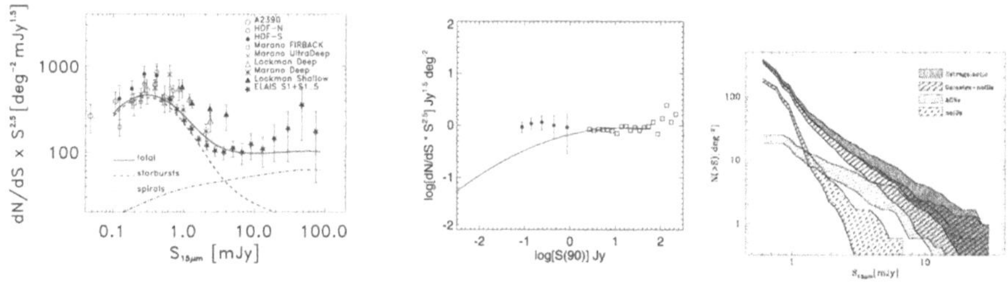

Fig. 1. Number counts from ELAIS and other ISO surveys. Left: Differential–Euclidean normalised 15 μm [26] with an evolving star-burst model [28], spirals are non-evolving and illustrate the generic shape of a no-evolution model. Centre: 90μm [11] compared with a no-evolution model. Both show significant departures from no evolution models. Right: Integral 15μm number counts from the ELAIS S1 field broken down by optical classes. Notice that the star-burst/AGN fraction increases at fainter fluxes and, naturally, the unidentified fraction increases sharply at the faintest fluxes. Taken from [12]

1, [12]). It is clear that the unidentified fraction increases as we go to fainter fluxes (where the excess over no evolution predictions becomes more extreme). Assuming that these unidentified sources are galaxies (as seen in deeper ISO surveys) rather than AGN then the galaxy fraction increases at fainter ISO fluxes. Overall we have an identified AGN fraction of around 25%.

5 Luminosity Functions

We have calculated 15 μm luminosity functions for Type-1 AGN [36] and galaxies [28]. Both show very significant evolution (see Fig. 2). As yet we have too few AGN to determine the luminosity function in different redshift bins, but the IRAS sample of [37] gives a low-z counterpart. The luminosity function of galaxies shows clear evolution within our current spectroscopic samples. The Type-1 AGN evolution is consistent with pure luminosity evolution at a rate of $(1+z)^Q$ with $Q \sim 3$ while for star-forming galaxies a combination of luminosity ($Q \sim 3.5$) and density evolution $(1+z)^{3.8}$ provides the best fit. We have also calculated the 90 μm luminosity function [38] and from the Final Analysis and with more spectroscopic redshifts in [39]. This function is also significantly different from the redshift zero IRAS counterpart.

6 Hyper-Luminous Galaxies

Objects as luminous as those seen in the SCUBA surveys are very rare. The wide areal coverage and large volume (5×10^6 Mpc^{-3}) of ELAIS makes it ideal for discovering rare objects such as these. So far we have discovered 10 galaxies whose infrared luminosities probably exceed $L > 10^{13.22} L_\odot$ [17] (60% of these

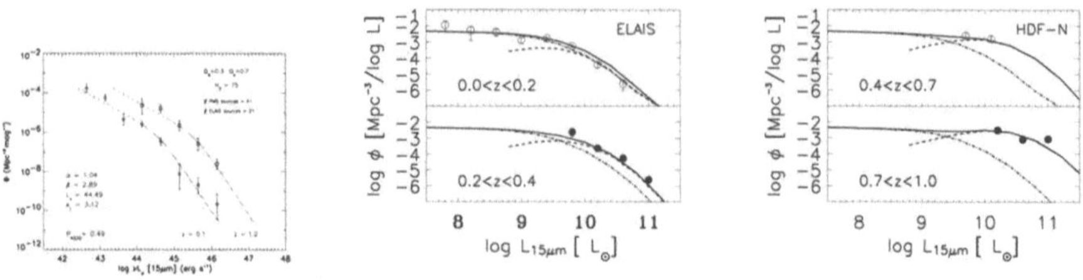

Fig. 2. Luminosity functions at 15 μm from ELAIS and other ISO surveys. Left [36]: including only those objects classified optically as Type-1 AGN. Centre: [28]: including only those objects classified optically as galaxies, split into $0 < z < 0.2$ and $0.2 < z < 0.$ intervals with a model with non-evolving spirals and strongly evolving starbursts illustrated. Right: as centre but for HDF North sample and $0.4 < z < 0.7$ and $0.7 < z < 01.0$. All plots show significant evolution

are based on photometric redshifts and await spectroscopic confirmation). It is very likely that this number increases as the follow-up proceeds but this is already almost as many as have ever been discovered from the IRAS database. The bolometric emission from first of these, ELAISP90 J1640101410502 [40], has been modelled with an almost equal contribution from star formation and accretion disk processes. Such objects may provide us with accessible "local" counterparts to the elusive SCUBA galaxies or, if they are unrelated, will provide interesting additional challenges to galaxy evolution models.

7 Large-Scale Structure

Very recently we have been able to analyse the clustering within ELAIS. In the southern field, S1, we have estimated the angular correlation function (Fig. 3, [41]). Using the median redshift [12] and inverting Limber's Equation we can estimate the infrared galaxy-galaxy autocorrelation function. Assuming it to be a power law we have estimated the slope γ and amplitude (encapsulated in the correlation length r_0). These results are given in Table 1 where they are contrasted with other surveys. The slopes are all consistent with each other. It is well known that local infrared samples are less strongly clustered than optical samples. Our results conform to this picture i.e. the clustering strength is consistent with more local infrared samples and weaker than optically selected samples. This result appears to contradict that presented earlier in this volume [18] who found that in deep infrared surveys the infrared galaxies were more strongly clustered that the optical galaxies. This may suggest that within the ISO surveys we have already detected the evolution of clustering. This would be qualitatively consistent with the predictions of hierarchical models where star-formation occurs first in the strongly clustered peaks of the density field and takes longer to initiate in the lower density, more weakly clustered "field".

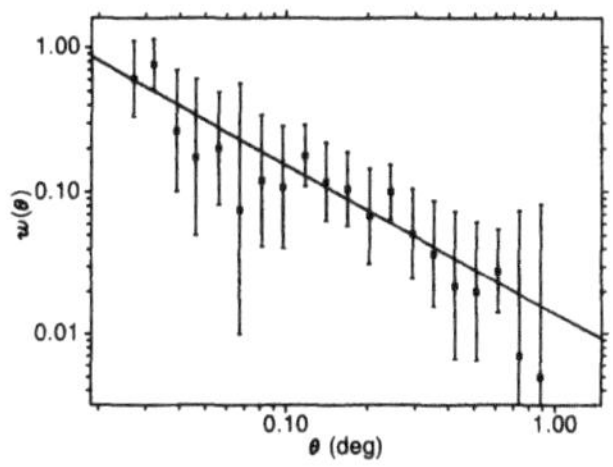

Fig. 3. The Angular correlation function measured in our Southern field S1. Taken from [41]

Survey	z_m	r_0	γ	**Ref.**
APM	0.05	5.7	1.67	[42]
SDSS	0.18	5.7±0.2	1.75±0.03	[43]
2dFGRS	0.08	4.92±0.27	1.71±0.06	[44]
IRAS	0.02	3.79±0.14	1.57±0.03	[45]
PSCz	0.02	3.7	1.69	[46]
ELAIS	0.2	$4.3^{+1.0}_{-0.6}$	2.04±0.18	[41]

Table 1. Summary of correlation lengths values obtained from different surveys, taken from [41].

8 Conclusions

The ELAIS survey has been completed and a catalogue with multi-wavelength made public. We have demonstrated that infrared emitting galaxies evolve strongly. The indications are that the star forming galaxies make up the bulk of the excess over no evolution predictions. Luminosity functions of star forming galaxies and AGN both evolve strongly. So far we have discovered 10 hyper luminous galaxies, comparable with the total number discovered by IRAS. We detect clustering with an amplitude consistent with lower redshift infrared samples and lower than optical samples, this contrasts with the higher clustering of infrared galaxies in a deeper sample suggesting that an evolution of clustering may have been detected in the ISO surveys.

The strong evolution in star forming galaxies and AGN provides further motivation for the extra-galactic surveys of SIRTF. The ELAIS samples will provide invaluable templates for understanding the deeper samples detected by the GTO and legacy surveys (e.g. [47], [48]). The preliminary results on clustering provide a particularly keen incentive to explore the clustering of infrared galaxies as SWIRE will be doing [47]. Three of the ELAIS fields are core component of the SWIRE fields [47]. This is because the ELAIS fields had been carefully chosen to avoid contamination from dust emission from our own Galaxy [4] and also because the existing data in the ELAIS fields, particularly the 15 μm band (inaccessible to SIRTF) will be extremely valuable. Thus ELAIS is both a scientific and practical path-finder for SIRTF in general and SWIRE in particular.

References

1. M. Kessler, et al.: A&A, **315**, L27 (1996)
2. C. Cesarskey, et al.: A&A, **315**, L32 (1996)
3. D. Lemke, et al.: A&A, **315**, L64 (1998)
4. S. Oliver, et al.: MNRAS **316**, 749 (2000)
5. S. Serjeant, et al.: MNRAS **316**, 768 (2000)
6. A. Efstathiou, et al.: MNRAS **316**, 1169 (2000)
7. H. Dole, et al.: A&A, **372**, 364 (2001)
8. C. Lari, et al.: MNRAS **325**,1173 (2001)
9. M. Vaccari, et al.: in prep (2004),
10. M. Vaccari, et al.: this volume (2004)
11. P. Héraudeau, et al.: MNRAS submitted (2003)
12. F. La Franca, et al.: AJ, submitted (2003)
13. E. Gonzales-Solares, et al.:in prep (2004)
14. P. Vaisanen et al.,: MNRAS **337**, 1043 (2002)
15. P. Ciliegi et al.,: MNRAS **302**, 222 (1999)
16. C. Gruppioni et al.,: MNRAS **335**, 297 (2001)
17. M. Rowan-Robinson, et al.: MNRAS submitted astro-ph/0308283 (2003)
18. D. Elbaz, et al.: this volume (2004)
19. J-L. Puget, et al.: A&A **308L**, 5 (1996)
20. D. Hughes, et al.: Nature, **394**, 241 (1998)
21. S. Chapman: this volume (2004)
22. A. Blain: this volume (2004)
23. S. Oliver, et al.: MNRAS, **289**, 471 (1997)
24. D. Elbaz, et al.: A&A **351**, L37 (1999)
25. S. Oliver, et al.: MNRAS, **332**, 356, (2002)
26. C. Gruppioni, et al.: MNRAS **335**, 831 (2002)
27. C. Lari, et al.: in prep. (2004)
28. F. Pozzi, et al.: ApJ, submitted (2003)
29. R. G. McMahon, et al.: New Astronomy Reviews, **45**, Issue 1-2, 97 (2001)
30. D. Alexander, et al.: AJ **554**, 18 (2001)
31. S. Basilakos, et al.: MNRAS **331**, 417 (2002)
32. C. Willott, et al.: MNRAS **339**, 297 (2003)
33. J. Manners, et al.: MNRAS **343**, 293 (2003)
34. C. Gruppioni, et al.: MNRAS
35. F. Pozzi, et al.: MNRAS, **343**, 1348 (2003)
36. I. Matute, et al.: MNRAS, **332**, L11 (2002)
37. B. Rush, M.A. Malkan, L. Spinoglio L., ApJS, **89**, 1 (1993)
38. S. Serjeant, et al.: MNRAS, **322**, 262 (2001)
39. S. Serjeant, et al.: MNRAS, submitted (2003)
40. T. Morel, et al.: MNRAS **327**, 1187 (2001)
41. E. Gonzales-Solares, et al.: MNRAS submitted (2003)
42. Maddox S. J., Efstathiou G., Sutherland W. J., Loveday J., MNRAS, **242**, 43P (1990)
43. Zehavi I., Blanton M. R., Frieman J. A., et al., ApJ, **571**, 172 (2002)
44. Norberg P., Baugh C. M., Hawkins E., et al., MNRAS, **328**, 64 (2001)
45. Saunders W., Rowan-Robinson M., Lawrence A., MNRAS, **258**, 134 (1992)
46. Jing, Y. P., Börner, G., & Suto, Y. , ApJ, **564**, 15 (2002)
47. C. Lonsdale, et al.: PASP (2003)
48. M. Dickenson, et al.: PASP (2003)
49. I. Perez-Fournon, et al.: in prep. (2004).

Properties of Spiral and Elliptical Galaxy Progenitors at $z > 1$

Christopher J. Conselice

California Institute of Technology, Pasadena, CA, USA

Abstract. We present the results of a Hubble Space Telescope and ground based optical and near infrared study to identify progenitors of spirals and ellipticals at $z > 1$. We identify these systems through photometric and spectroscopic redshifts, deep K-band imaging, stellar mass measurements, and high resolution imaging. The major modes of galaxy formation, including: major mergers, minor mergers, and accretion of intergalactic gas, and their relative contribution towards building up the stellar masses of galaxies, can now be directly measured using these data.

1 Introduction

The generally accepted modern hierarchical galaxy formation picture consists of galaxies forming in dark matter halos that later merge to form larger halos and more massive galaxies. The end result of this evolution is the morphological and stellar population mix in the nearby universe. This picture however remains largely untested. Understanding how the modern galaxy population was put into place requires understanding when and how stars (and hence galaxies) formed. Based on several decades of observations and modeling of stellar evolution we know that the stars in nearby galaxies contain a wide diversity of ages and metallicities. To first order these differences correlate with galaxy type and environment. Generally, early-types or elliptical galaxies are dominated by old stars and are found in dense environments, while later type galaxies have a mix of young and old stellar populations and are found in lower density areas.

Directly measuring the galaxy mass assembly and star formation history has now been accomplished out to $z \sim 3 - 6$. However, these measurements do not tell us *how* galaxies formed. One way to address this question is to include high resolution imaging, such as from deep Hubble Space Telescopes images. Imaging surveys with Hubble show that galaxies evolve into normal systems from peculiars between $z \sim 1 - 2$ (~ 10 Gyrs ago) (Figure 1; e.g., van den Bergh et al. 2001; Conselice et al. 2003). At redshifts higher than $z \sim 1.5$ most galaxies are distorted and asymmetric (Abraham et al. 1996; Conselice et al. 2003). Deep NICMOS observations of the Hubble Deep Field North, which samples the rest frame B-band morphologies of galaxies at $z > 1.2$, demonstrates that these galaxies are intrinsically distorted in the rest-frame optical, and that we are not witnessing morphological k-correction effects (Papovich et al. 2003).

These high redshift galaxies are also undergoing large amounts of star formation (e.g., Madau et al. 1998), creating the normal bright galaxies we see today.

What causes the structural peculiarities in these galaxies, and presumably also the induced star formation? If we can answer this it will reveal the formation modes of galaxies, and allow us to quantify the relative contributions of different formation processes.

2 Mass Assembly as a Function of Morphology

The first step towards understanding the formation mechanisms of galaxies is to have a robust determination of how the structures of galaxies change with redshift, and how these changes relate to the star formation and mass assembly history. Figure 1 shows co-moving number, luminosity, and stellar mass densities of ellipticals, spirals, and peculiar galaxies as a function of redshift. One interesting trend from this figure, besides the dominance of spirals and ellipticals at $z < 1.5$ and peculiars at $z > 1.5$, is the that there exists an equilibrium point at $z \sim 1.5$ where the relative fraction of luminosity, mass and number densities for normal galaxies (disks/ellipticals) and peculiars are nearly equal. In all regards this is the redshift in galaxy evolution where modes of forming galaxies are rapidly transitioning. This trend is also seen when studying the NICMOS observations of the Hubble Deep Field North and in the Hubble Deep Field South.

There is a growth in both the stellar mass density and number density of spirals and ellipticals at $z < 1$, with a corresponding decrease in the number of peculiars (see also Brinchmann & Ellis 2000). Star formation is still occurring in disks and ellipticals at $z < 1$ as the stellar mass densities for these types grows by a factor of two or more (Figure 1). The luminosity density for both ellipticals and spirals also peaks at $z \sim 1$, and mass to light ratios for these systems increase with lower redshift at $z < 1$. At $z > 2$ peculiar galaxies consistent with major merging (Conselice et al. 2003) are dominating the luminosity and stellar mass density, suggesting that this is the dominate star formation process at early times.

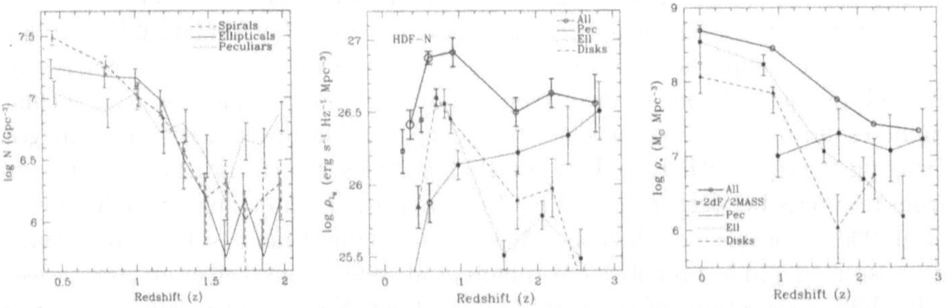

Fig. 1. The relative co-moving number, rest-frame B-band luminosity, and stellar mass density of galaxies as a function of redshift from deep NICMOS images of the Hubble Deep Field North (Conselice et al. 2004, in prep). Points at redshifts $z < 0.5$ are taken from Brinchmann & Ellis (2000), Fukugita et al. (1998) and the 2dF/2MASS survey.

3 Tracing Galaxy Formation

There are a few major global methods by which galaxies can form, or said in another way, how star formation is triggered. One is through major mergers where galaxies are built up by merging with systems of similar mass. The other is through minor mergers where the mass of a galaxy is built up by the accretion of less massive galaxies. The third method is through accretion of intergalactic gas which forms stars around galaxies in disk like structures (Abadi et al. 2003). A forth is through tidal interactions with nearby galaxies. This is perhaps an over simplification of the star formation triggering process, but it is a valuable starting point for understanding galaxy formation.

The quantitative structural properties of galaxies reveal the formation mechanisms responsible for producing galaxy and star formation (e.g., Conselice 2003). We can use the structures of galaxies, including their sizes and stellar masses, to argue which systems at high redshifts are likely progenitors of either disks or spheroidal components. Figure 2 shows the brightest galaxies at $1 < z < 2$ selected in the near infrared in the GOODS south field (Giavalisco et al. 2004).

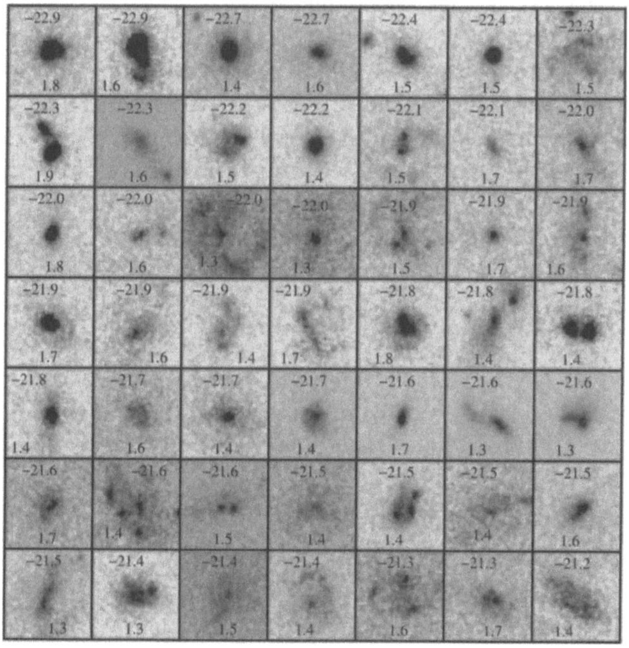

Fig. 2. Examples of galaxies in ACS GOODS images whose photometric redshifts place them at $1 < z < 2$. These are ordered from brightest to faintest down to $M_B = -21$. The upper number is the M_B of each galaxy and the lower number is its redshift. There is a large diversity of properties, from systems that appear very peculiar to those that look similar to normal galaxies. Scale of these images is $\sim 2''$ on each side, which corresponds to about 17 kpc at these redshifts.

These systems must be the progenitors of disks and ellipticals and many show structures suggestive of this.

3.1 Major Mergers – Spheroid Formation

Major mergers occur when two galaxies of similar mass merge together to form a more massive system. Galaxies which are undergoing major mergers can be identified in the rest-frame optical through their large asymmetries in comparison to other parameters, such as color or clumpiness (Conselice et al. 2000; Conselice 2003). We can use the asymmetry index to determine how common major mergers are at various redshifts and to measure properties of these mergers. The results of this are shown in Figure 3 which plots the inferred merger fraction as a function of redshift for galaxies at two luminosity limits. It appears that most major mergers at high redshift occur in luminous systems. This is also the case for the most massive galaxies at the same redshifts (Conselice et al. 2003). About 50% of the most luminous and massive galaxies at high redshift are undergoing major mergers. The merger rate for the most luminous and massive systems however rapidly declines at lower redshifts. A smaller fraction of fainter and less luminous systems are undergoing mergers, suggesting that some other process is responsible for their formation.

Do these mergers add enough material to produce a massive galaxy at low redshift from the most massive and brightest systems found at $z > 2$? This is uncertain, as it depends not only on the merger rate, but also on the induced star formation produced through the merger. Integrating the mass obtained through major mergers (Figure 3) with reasonable and empirical assumptions about star

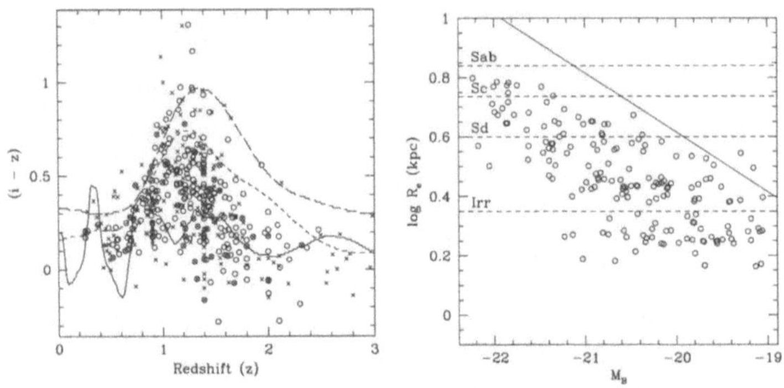

Fig. 3. Left side - The distribution of $(i - z)$ colors for LDOs (circles) as a function of redshift with two Coleman, Wu and Weedman spectral energy distributions and a Kinney et al. starburst model plotted (see Conselice et al. 2004). These are from bluest to reddest - starburst (solid line), Scd (dashed), Sbc (long dashed). Right side - Absolute magnitude effective radius relationship for LDOs. The solid line is the canonical Freeman disk relationship at $z \sim 0$. The dashed horizontal lines show the effective radii of different nearby galaxy types.

formation histories induced by mergers, it appears that massive ellipticals can be formed through multiple merger/starbursts. What is clear however is that a large fraction of stars cannot have formed through major mergers, given that the fraction of galaxies involved in major mergers declines rapidly at $z < 2$ and stellar mass is still assembled at these redshifts (e.g., Dickinson et al. 2003). Figure 3 also shows that the measured merger fraction is lower at low-z than what is predicted in CDM models, such that there are too many major mergers produced in these models at low redshift. That is, massive galaxy formation occurs earlier than predicted in semi-analytic CDM models.

3.2 Minor Mergers

Minor mergers are likely a major method for building up the masses of galaxies, as predicted in simulations (Murali et al. 2002). The observed role of minor mergers is however not know with certainty, simply because it is difficult to identify them, as they do not produce major structural effects on galaxies. One clue that minor mergers might be occurring in large numbers at $z < 1$ is that about 50% of the total stellar mass in the universe appears to form at $z < 1$; yet major mergers are not producing this increase (Conselice et al. 2003). The properties of minor mergers can be studied out to $z \sim 1$ most effectively using pair studies of galaxies either in the optical (Patton et al. 2002) or in the near infrared (Bundy et al. 2004). The optical pair fraction suggests that minor merger rates are high enough to produce a suitable increases in stellar mass. The merger rate is however lower when studying pairs in the near infrared (Bundy et al. 2004). The reason for this is that pre-accreted satellite galaxies are star forming with low mass to light ratios, resulting in a lower near infrared pair fraction. The amount of stellar mass added to galaxies through minor mergers at $z < 1$ is roughly equal to the gain in stellar mass at these redshifts, although this is likely a coincidence (Bundy et al. 2004). However, if these merging/interacting satellites are creating new stars with a total stellar mass equal to that of the original satellite, minor mergers would produce enough mass through tidally induced star formation to account for the observed growth in stellar mass. Detailed determinations of the star formation and stellar mass build-up induced in pairs of galaxies at $z < 1$ are necessary to quantify the importance of this effect.

3.3 Disk Formation

Another major type of galaxy formation is through smooth accretion of intergalactic gas. Morphologically systems undergoing accretion will appear as disks, with a relatively higher amount of star formation than what is found in modern spirals. Some forming disks have possibly been found in deep HST imaging in the Hubble Deep Field South (Labbé et al. 2003) and the GOODS South field (Conselice et al. 2004). These are typically found between redshifts $1 < z < 2$ and are symmetric galaxies containing bright outer regions with low central light concentrations. These bright regions create unconcentrated structures, which is how these systems are identified. Figure 2 shows examples of these luminous diffuse objects (LDOs) found in the GOODS South field (Conselice et al. 2004).

These systems have star formation rates consistent with starbursts (Figure 4) with typical uncorrected for extinction UV derived star formation rates of $\sim 4\ M_0$ yr^{-1}. The integrated star formation rate in these galaxies accounts for 35-40% of the total star formation rate between redshifts $1.5 < z < 2.5$. These systems also have sizes and scaling relationships consistent with disk galaxies in formation (Figure 4). Some systems also contain structures such as spiral arms and bars (Labbé et al. 2003; Conselice et al. 2004). The star formation rate peak found between $1 < z < 2$ appears to relate to the formation of these disk-like systems through accretion, while at lower/higher redshifts, minor/major mergers are the dominate processes.

References

1. Abraham, R., et al. 1996, ApJS, 107, 1
2. Abadi, M.G., Navarro, J.F., Steinmetz, M., Eke, V.R. 2003, ApJ, 591, 499
3. Brinchmann, J., Ellis, R.S. 2000, ApJ, 536, 77L
4. Bundy, K. et al. 2004, ApJ, 601, 123L
5. Conselice, C.J., Bershady, M.A., Jangren, A. 2000, ApJ, 529, 886
6. Conselice, C.J. 2003, ApJS, 147, 1
7. Conselice, C.J., Bershady, M.A., Dickinson, M., Papovich, C. 2003, AJ, 126, 1183
8. Conselice, C.J., et al. 2004, ApJ, 600, 139L
9. Dickinson, M., Papovich, C., Ferguson, H.C., Budavari, T. 2003, ApJ, 587, 25
10. Fukugita, M., Hogan, C.J., Peebles, P.J.E. 1998, ApJ, 503, 518
11. Giavalisco, M. et al. 2004, 600, 93L
12. Labbé et al. 2003, ApJ, 591, 95L
13. Madau, P., Pozzetti, L., Dickinson, M. 1998, ApJ, 498, 106
14. Murali, C., Katz, N., Hernquist, L., Weinberg, D.H., Dave, R. 2002, ApJ, 571, 1
15. Papovich, C., et al. 2003, ApJ, 598, 827
16. Patton, D.R., et al. 2002, ApJ, 565, 208
17. van den Bergh, S., Cohen, J.G., Crabbe, C. 2001, AJ, 122, 611

Star Forming Galaxies in the 'Redshift Desert'

C. Steidel[1], A. Shapley[2], M. Pettini[3], K. Adelberger[4], D. Erb[1], N. Reddy[1], and M. Hunt[1]

[1] California Institute of Technology, MS 105-24, Pasadena, CA 91125, USA
[2] University of California, Berkeley
[3] Institute of Astronomy, Cambridge, UK
[4] Carnegie Observatories

Abstract. We describe results of optical and near-IR observations of a large spectroscopic sample of star-forming galaxies photometrically-selected to lie in the redshift range $1.4 \lesssim z \lesssim 2.5$, often called the "redshift desert" because of historical difficulty in obtaining spectroscopic redshifts in this range. We show that the former "redshift desert" is now very much open to observation.

1 Background

The "redshift desert" results from an accident of nature in which the windows of low atmospheric opacity and low terrestrial background are barren of familiar, strong spectroscopic features that make redshift identification easy using ground-based spectroscopy. At $z \sim 1.4$ the last of the strong nebular lines, [OII] $\lambda 3727$, passes well into the range where most optical spectrographs perform less well because of decreasing CCD quantum efficiency and rapidly increasing sky brightness. The result has been that there is a dearth of direct spectroscopic information on galaxies at $z \gtrsim 1.4$ until $z \gtrsim 2.5$, at which point (for star forming galaxies, at least) techniques like Lyman break selection coupled with normal optical spectroscopy have been quite successful. The difficulties with spectroscopy in the desert translate directly into larger uncertainties in photometric redshifts, since both methods depend on strong spectral features at observationally accessible wavelengths. Of course, nature has conspired to make this redshift range between $z \sim 1.5 - 2.5$ perhaps one of the most crucial in understanding the development of massive galaxies and of their central black holes, as we have heard from a number of talks at this meeting. There is thus a very strong impetus to gain access to galaxies in this range of redshifts, in spite of the difficulties that may be encountered.

Following the familiar rest-optical nebular lines into the near-IR is certainly possible in principle, but to date there has been little effort to do this in a wholesale manner because there have not been many instruments capable of multiplexed spectroscopy in the near-IR; this of course will change significantly over the next $\sim 3 - 5$ years as cryogenic multi-object near-IR spectrographs come on line. We have adopted an alternative strategy: to push faint object spectroscopy into the blue/UV portion of the observed spectrum, where the same spectral features used for the identification of galaxies at $z \sim 3$ [1] remain accessible down to $z \sim 1.4$, thereby closing the gap in redshift space that has been called the "desert".

Studying galaxies at $z \sim 2$ is quite rewarding for both scientific and practical reasons: This redshift range $1.5 \lesssim z \lesssim 2.5$ evidently contains the peak of the QSO epoch, and may well contain the formation era for most of the stars in today's massive galaxies, judging by the redshift distribution of bright sub-mm sources [2]. But it is much more than a crucial epoch for galaxies and AGN– at these redshifts, it is possible to simultaneously study the diffuse intergalactic medium (IGM) *and* star forming galaxies in the same volumes, allowing for a direct examination of the magnitude of the effects of supernova feedback on the properties of galaxies and the IGM. Because both background QSOs and spectroscopically accessible galaxies have much higher surface densities than at higher redshifts where such experiments have already been done [3], $z \sim 2$ may be the optimal redshift for joint IGM/galaxy studies. In terms of attaining physical understanding of the galaxies one finds, $z \sim 2$ offers significant advantages as well: first, the surface density of galaxies bright enough for detailed spectroscopic studies using 8m-class telescopes at both optical and near-IR wavelengths is high; secondly, as we describe below, one has access to diagnostic spectroscopy in both the rest-frame far-UV *and* the rest-frame optical, allowing independent means of measuring physical properties such as chemical abundances, stellar initial mass function, and mass.

2 Survey of the "Redshift Desert"

We began our survey of $z \sim 2$ galaxies in the fall of 2000 just after the commissioning of the LRIS-B instrument (see [4]) on the Keck I telescope. Our approach to identifying which galaxies to target in order to efficiently survey galaxies at $z \sim 2$ was largely empirical, in the sense that we chose the region in color-space that would be occupied by galaxies having the same rest-frame SEDs as $z \sim 3$ LBGs in the sample of objects that have been observed from $0.3 - 2.2\mu m$ (see [5] for a full description and motivation of the color selection). Generally speaking, this means that the galaxies we are selecting for spectroscopy have the same range of UV color as the better-studied LBGs at $z \sim 3$. Because we began our survey with the primary aim of studying the connection between the galaxies and the IGM, we have focused most of our efforts on the redshift range $2 \lesssim z \lesssim 2.5$ (we refer to this particular color selection as "BX" objects) where the Lyman α forest is easily studied from the ground, and in fields having more than 1 suitable $z \sim 2.5$ background QSO for studying the IGM component. However, we have recently obtained additional spectroscopy in both the GOODS-N and Groth/Westphal fields, where in addition to the $z \sim 2 - 2.5$ "BX" objects we have targeted objects expected to lie in the range $1.5 \lesssim z \lesssim 2$ (and which we refer to as "BM" objects). A more in-depth overview of the survey and its initial results is given in [4].

Figure 1 shows the results to date for the spectroscopy of the two new photometrically selected samples, with $\langle z \rangle = 2.20 \pm 0.32$ and $\langle z \rangle = 1.70 \pm 0.34$ for the

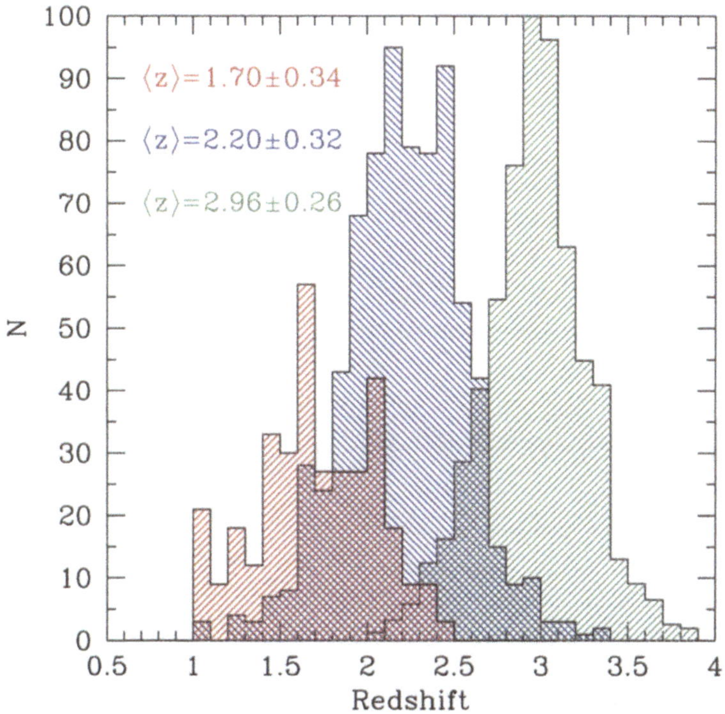

Fig. 1. Redshift histograms for rest-UV selected samples of star forming galaxies. The blue histogram is the 750 "BX" galaxies with spectroscopic redshifts at the time of this writing; the red histogram results from the "BM" selection targeting $z \simeq 1.5 - 2.0$ (118 redshifts, scaled by 3 for plotting). The green histogram is the $z \sim 3$ LBG sample of [1] (940 redshifts, scaled by 0.7 for plotting). The total number of new redshifts in the range $1.4 \lesssim z \lesssim 2.5$ is 737 at this time.

"BX" and "BM" samples, respectively. Figure 2 shows example LRIS-B spectra obtained since the science-grade detector system was installed in the instrument in June 2002.

The location of the $z = 1.5 - 2.5$ galaxies in color space (we use the same 3-band system used to isolate $z \sim 3$ galaxies) is easily estimated [5], and in fact galaxies in this range of redshifts are very common. After correcting for $\sim$ 8% contamination by stars and low-redshift galaxies in the photometric samples, the total surface density of BX+BM galaxies is ~ 9 arcmin^{-2} to $\mathcal{R} = 25.5$ (our adopted spectroscopic limit), or about 25% of the number counts to the same apparent magnitude limit. With only 90-minute total integration times, the LRIS-B spectroscopy achieves $\sim$ 70% success in identifying redshifts, primarily based on numerous interstellar absorption lines, as illustrated in Figure 2. Because LRIS-B is so efficient in the UV-visual range, and the sky is extremely dark, it is possible to collect high quality continuum spectroscopy in very reasonable integration times, so that the detection of strong emission lines

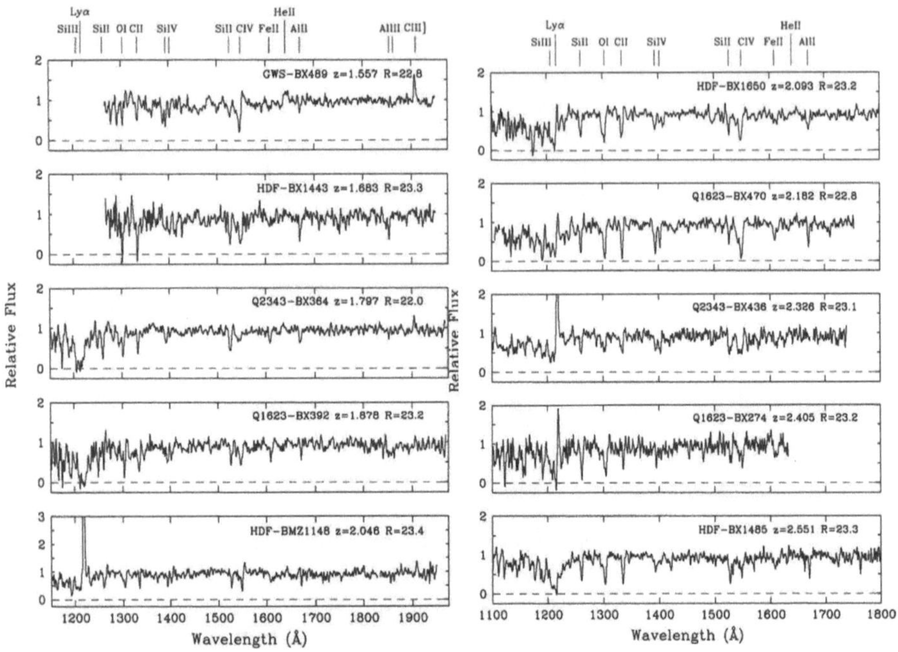

Fig. 2. Example LRIS-B spectra (shifted into the rest frame, but generally covering the wavelength range 3200 – 6000Å) of galaxies in the BX/BM sample covering the full range of redshifts and spectral properties seen in the sample. All spectra were obtained with 90-minute total integration times, in survey mode. The spectral resolution is $\sim 1.5 - 2.0$ Å in the rest frame.

is completely unnecessary for achieving high spectroscopic completeness. At the time of this writing, we have obtained a total of more than 900 spectroscopic redshifts in the range $1.4 \lesssim z \lesssim 2.6$, with the distributions as shown in Figure 1. We have had similar success using LRIS-B to obtain spectra of objects which failed to yield redshifts in the DEEP2 redshift survey (see Jeff Newman's contribution to these proceedings), where because of the spectroscopic configuration used, it is difficult to measure redshifts at $z > 1.4$. In general, LRIS-B is ideal for measuring redshifts for just the objects that generally fail to yield redshifts for spectroscopic setups focused on the red part of the optical spectrum.

3 Near-IR Observations of Optically Selected Desert Galaxies

3.1 Imaging

It is interesting to investigate the rest-frame optical properties of the rest-UV-selected "desert" galaxy population, particularly as a means of comparing them

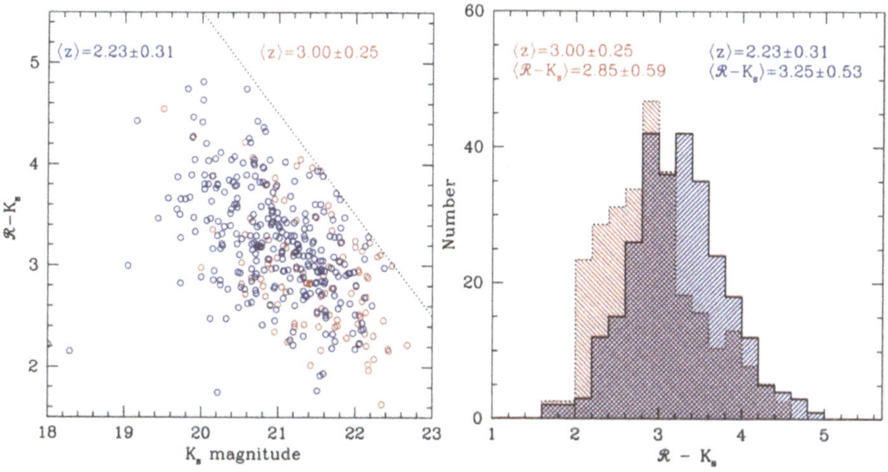

Fig. 3. Left: Color/magnitude diagram showing the 283 $z \sim 2$ galaxies with K-band measurements in our spectroscopic sample (blue points). The red points are taken from the $z \sim 3$ sample of [6]. The dotted line corresponds to our spectroscopic limit (for both surveys) of $\mathcal{R} = 25.5$. **Right:** Histograms of optical/IR color for the two samples, normalized to the same numbers. The $z \sim 2$ sample has many more galaxies of with ($\mathcal{R} - K_s \gtrsim 3.2$), which tend to be objects with significant Balmer breaks indicating extended star formation histories.

differentially to the similarly-selected $z \sim 3$ sample, and to various other surveys that use near-IR selection (e.g., FIRES, K20, GDDS–see contributions from Labbé, Cimatti, and Chen in these proceedings, respectively). Beginning in June 2003, with the commissioning of the Wide Field Infrared Camera (WIRC) on the Palomar 5m telescope, we have been obtaining very deep near-IR images of the regions of our survey fields with the densest spectroscopic coverage. In the first 3 pointings of the camera, which has a field of view of 8.7×8.7 arcmin, we have obtained K_s band measurements for a total of 283 'BX' galaxies that already have spectroscopic redshifts. Although it requires very long (~ 12 hour) integrations on a 5m telescope, we have been able to reach limiting magnitudes of $K \sim 22.3$, allowing us to detect $\sim 85\%$ of the galaxies with spectra that fall within each pointing. Figure 3 shows a color-magnitude diagram of the 283 detected galaxies in the current sample. About 8% of these galaxies have $K_s < 20$ (the limit for the K20 survey) and $\sim 30\%$ has $K_s < 20.6$ (the limit for the Gemini Deep Deep Survey).

A cursory comparison of the distribution of $\mathcal{R} - K_s$ colors of the samples of star forming galaxies at $z \sim 3$ and $z \sim 2$ indicates that the $z \sim 2$ galaxies are significantly redder. Since objects with identical star formation histories observed at $z \sim 3$ and $z \sim 2$ are predicted to have essentially identical $\mathcal{R} - K_s$ color, the result suggests a real difference in the rest-optical properties of objects selected using the same rest-UV criteria in the two redshift intervals. The most likely cause of this is an increase in the average stellar mass of star forming galaxies at

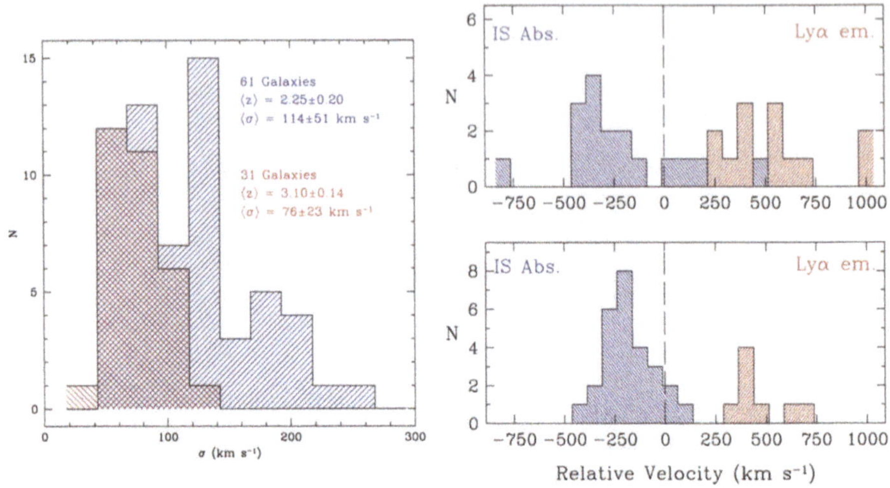

Fig. 4. Left: A comparison of the distribution of one-dimensional velocity dispersions (as measured using Hα or [OIII] lines) for a sample of LBGs at $z \sim 3$ [8] (red) and $z \sim 2$ (blue). More than 50% of the $z \sim 2$ galaxies have σ larger than the largest values observed at $z \sim 3$. **Right:** a comparison of the galactic outflow kinematics of $z \sim 2$ (bottom) and $z \sim 3$ (top, [8]) galaxies.

$z \sim 2$ compared to their counterparts at $z \sim 3$. These results will be quantified in future work.

3.2 Spectroscopy

Galaxies at $z \sim 2$, and especially those at redshifts $z = 2 - 2.5$, are extremely well-suited for follow-up spectroscopy in the near-IR because the various nebular lines fall at favorable redshifts with respect to the atmospheric J, H, and K band windows. Using primarily the NIRSPEC instrument on the Keck II telescope, we have focused on observing the Hα line primarily in the K-band window (initial results are presented in [7] and in the poster paper presented by Dawn Erb at this meeting). The scientific aims of these observations are manifold, and include 1) accurate measurement of the galaxy systemic redshift, crucial for analysis of the galaxy/IGM interface and for evaluating the velocities of supernova-driven outflows in the galaxies 2) measurement of the galaxy kinematics (line widths, rotation curves, etc.) 3) measuring chemical abundances in the galaxy's HII regions using various nebular line diagnostics and 4) additional estimates of star formation rates in the galaxies, possibly less affected by extinction than estimates based on UV continuum measurements.

Here we focus on the kinematics and chemical abundances. Our first observations ([7]) showed that the Hα measurements could be made with a high success rate for the UV-selected $z \sim 2$ galaxies, and that in a reasonably high fraction of the cases observed, spatially resolved velocity shear was observed,

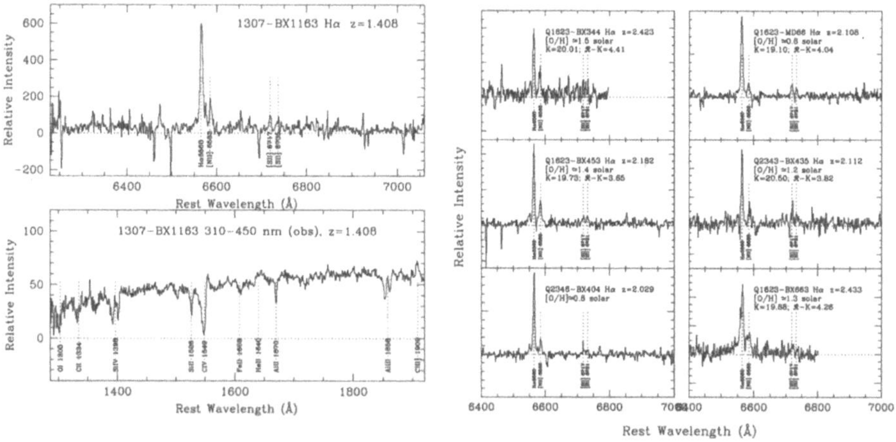

Fig. 5. Left: LRIS-B (lower) and NIRSPEC (upper) spectra of a galaxy at $z = 1.41$ illustrating the type of data from which detailed astrophysical properties of "desert" galaxies can be extracted. **Right:** Example NIRSPEC spectra of near-IR bright galaxies drawn from our UV-selected sample at $z \sim 2$. Using the calibration of [10] these have oxygen abundances that are consistent with being solar or greater.

such as might be expected in the case of rotating disks. The observability of such velocity shear in high redshift galaxies is very seeing dependent [9] but the overall line widths (σ) are also consistently larger than for the $z \sim 3$ galaxies in the sample of [8] despite the fact that they were selected on very similar UV properties. Comparisons of the $z \sim 2$ and $z \sim 3$ samples are shown in Figure 4– note that at $z \sim 2$, the line widths are much larger, while the galactic outflow kinematics, characterized by the velocities of the interstellar absorption lines and Lyman α emission with respect to the Hα-defined redshift, are quite similar to the $z \sim 3$ counterparts. The most straightforward interpretation of both the 1-d line widths and the K_s band properties (§3.1) is that the stellar masses of UV-selected galaxies have increased considerably between the epochs corresponding to $z \sim 3$ and $z \sim 2$.

We have begun a program to obtain data suitable for measuring chemical abundances in the $z \sim 2$ galaxies, using both nebular indicators such as the [NII]/Hα ratio ([10]) and also the rest far-UV features recorded in the optical LRIS-B spectra (see, e.g., Fig. 5a). Such observations allow significant insight into the physical nature of the galaxies: for example, from the spectra in Fig. 5a, we can infer [4] that both the nebular lines and the far-UV measurements indicate $\simeq$solar abundances, that the high mass end of the IMF is consistent with Salpeter, that the velocity dispersion of the HII regions of the galaxy is $\sigma = 126$ km s^{-1}, and that interstellar absorption lines due to outflowing gas are blue-shifted by a bulk velocity of $\simeq 300$ km s^{-1}and have a velocity width of 650 km s^{-1}. In figure 5b, we show Hα/[NII] spectra of objects selected to be either bright in the K_s band or red in $\mathcal{R} - K_s$; for these objects, which comprise a significant fraction of the UV-selected sample, there are indications that the galaxies are typically solar metallicity or greater, even at $z \sim 2$; the full results of such analysis will be presented in [11].

4 "Desert" Results in the GOODS-N Field

While most of our effort to date on "redshift desert" galaxies has been concentrated in specially chosen fields with multiple bright background QSOs for the IGM component of the survey (not discussed here), we have also worked in fields with extensive existing or planned multi-wavelength observations. One of these fields is GOODS-N (formerly known as HDF-N). This is one of two fields in which we have targeted significant numbers of galaxies in both the $z \simeq 1.5 - 2$ and $z \simeq 2 - 2.5$ redshift range, as well as at $z \sim 3$ (see Figure 6a). The existence of large spectroscopic samples of $z > 1.4$ galaxies together with existing SCUBA, Chandra, and VLA images, as well as the very high quality GOODS/ACS images and planned SIRTF/IRAC and SIRTF/MIPS observations, will allow for a large amount of progress and a deeper understanding of the energetics, masses, and stellar populations of star forming galaxies at high redshifts. Here we make only a few comments.

There has been quite a bit of discussion about the overlap, or lack thereof, between SCUBA galaxies detected at 850 μm and the UV-selected samples of LBGs at $z \gtrsim 3$ (and now the large samples in the "redshift desert"). In fact, it turns out that there is indeed substantial overlap between the BX/BM samples and the SCUBA/radio objects of [2], and the HDF region is no exception. There are at least 4 objects in common between our BX/BM sample in GOODS-N (fig. 6a) and the radio-detected "SMGs" in the Chapman et al sample: 3 of these have LRIS-B redshifts from our own survey, at $z = 2.098$, $z = 1.989$, and $z = 1.865$. A glance at fig. 6a shows that each of these is a member of one of the redshift desert "spikes" found from the LRIS-B spectroscopic results for BX/BM galaxies, qualitatively supporting the claim (see Blain's contributions to these proceedings) that the SMGs are strongly clustered objects. It is possible that the reason for the lack of overlap between SCUBA sources and the $z \sim 3$ LBGs may be due to the fact that the redshift distribution for SMGs peaks in the $z = 2 - 2.5$ range, very much like optically selected QSOs; $z > 3$ SMGs are relatively rare. The increase in overlap with SCUBA sources is yet another piece of evidence for significant evolution in the sample of UV-selected galaxies between $z \sim 3$ and $z \sim 2$.

Very recently we have examined the average X-ray and radio emission from the spectroscopic $z \sim 2$ galaxy sample in the GOODS-N field via a stacking analysis [12], providing a 3-way check on the average inferred star formation rate (and UV attenuation factor). The stack of the X-ray image at the positions of 160 $z \sim 2$ galaxies (after excluding all directly detected X-ray sources, among which are all 3 of the identified sub-mm/UV galaxies) in the Chandra 2 Ms image [13] is shown in fig. 6b. The stacked BX/BM galaxies are detected at

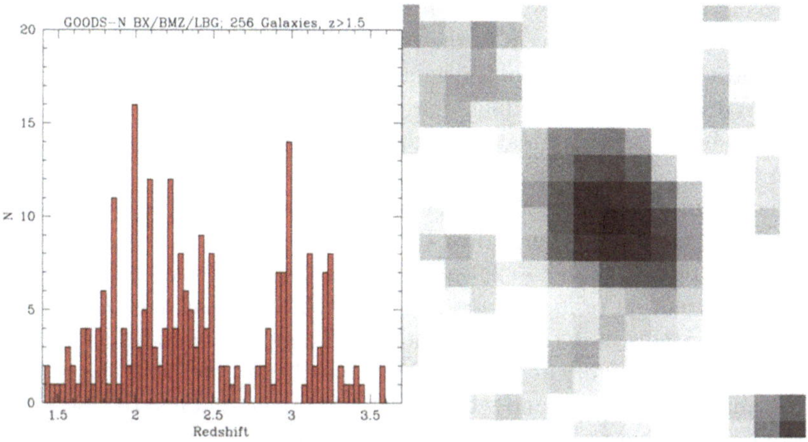

Fig. 6. Left: Redshift histogram of 256 $z > 1.5$ GOODS-N galaxies selected using various UV color selection techniques (those at $z < 2.6$ are based upon the BX and BM criteria). **Right:** X-ray stack of 160 $z \sim 2$ UV-selected galaxies with spectroscopic redshifts. The average X-ray flux implies an average star formation rate of $\simeq 50\,\mathrm{M}_\odot$ yr^{-1}, in excellent agreement with both the average SFR estimated from the radio stack and that estimated from the UV continuum slope.

the $\sim 10\sigma$ level. A similar exercise has been completed using the radio image of [14], and the estimated average star formation rate has been computed from the ground-based optical photometry assuming the [15] relation between far-UV color and far-IR luminosity. The result is that all 3 means of estimating SFR for the same sample suggest an *average* SFR of $\simeq 50\,\mathrm{M}_\odot\ \mathrm{yr}^{-1}$ ($h = 0.7$, $\Omega_m = 0.3$, $\Omega_\Lambda = 0.7$). The good agreement between UV, X-ray, and radio estimates of star formation rates suggests that, on average, the locally calibrated relationships between measurements at these wavelengths and the bolometric luminosity (i.e., SFR) still apply for starburst type systems at high redshift. The results also imply that the average $z \sim 2$ UV selected galaxy in our spectroscopic sample has a far-IR luminosity of $\langle L_{\rm bol} \rangle \simeq 2 \times 10^{11}\ \mathrm{L}_\odot$, which would qualify as a LIRG by the usual definition. The average implied UV extinction factor is $\simeq 5$, very much in line with the value that has been suggested for $z \sim 3$ galaxies [16][17]. With such typical luminosities, the spectroscopic $z \sim 2$ galaxies should be easily detectable by SIRTF/MIPS in the 24μm band on an individual basis (rather than by stacking a large number of objects).

5 Summary

We have highlighted the fact that, with suitable optical photometric selection and follow-up spectroscopy with a highly efficient UV/blue optimized spectrograph, the "redshift desert" ceases to exist. In fact, in many ways star forming galaxies at $z \sim 2$ offer more opportunities for measuring key physical quantities using current generation telescopes than at any redshift beyond the local universe, because of the simultaneous access to the rest-frame far-UV and rest-frame optical spectra, and the observability of the galaxies' effects on the diffuse intergalactic medium. With the addition of high quality space-based data, particularly from SIRTF, we should learn a great deal about galaxies in this important cosmic epoch in the very near future.

We have shown that there has been significant evolution in the kinematics and the optical/IR colors of (identically) UV-selected galaxies between the epochs corresponding to $z \sim 3$ and $z \sim 2$. Both of these observations are consistent with a significant evolution (a factor of ~ 2) in the typical stellar mass of vigorously star forming systems over this redshift range. There are significant numbers of galaxies *still forming stars* at $z \sim 2$ which have already enriched themselves to solar metallicity or greater, and which have stellar masses of $\sim 10^{11}$ $M_\odot$.

More quantitative results in the areas discussed above, and analyses of the clustering properties, luminosity functions, and the relationship to and interaction with the diffuse IGM are all in preparation. We would like to thank the David and Lucile Packard Foundation and the US National Science Foundation for support for this work.

References

1. C. Steidel, K. Adelberger, A. Shapley, M. Pettini, M. Dickinson, M. Giavalisco: ApJ **592**, 728 (2003)
2. S. Chapman, A. Blain, R. Ivison, I. Smail: Nature **422**, 695 (2003)
3. K. Adelberger, C. Steidel, A. Shapley, and M. Pettini: ApJ **584**, 45 (2003)
4. C. Steidel, A. Shapley, M. Pettini, K. Adelberger, D. Erb, N. Reddy, M. Hunt: ApJ in press (2004)
5. K. Adelberger, C. Steidel, A. Shapley, M. Hunt, D. Erb, N. Reddy, and M. Pettini: ApJ, in press. (2004)
6. A. Shapley, C. Steidel, K. Adelberger, M. Dickinson, and M. Giavalisco: ApJ **562**, 95 (2001)
7. D. Erb, A. Shapley, C. Steidel, M. Pettini, K. Adelberger, et al.: ApJ **591**, 101 (2003)
8. M. Pettini, Shapley, A., Steidel, C. et al: ApJ **554**, 981 (2001)
9. D. Erb, C. Steidel, A. Shapley, M. Pettini, and K. Adelberger: ApJ, submitted (2004)
10. G. Denicolo, R. Terlevich, and E. Terlevich: MNRAS **330**, 69
11. A. Shapley, et al., in preparation.
12. N. Reddy and C. Steidel: ApJL, in press (2004)
13. D. Alexander, F. Bauer, W. Brandt, et al: AJ **126**, 539 (2003)
14. E. Richards: ApJ **533**, 611 (2000)
15. G. Meurer, T. Heckman, and D. Calzetti: ApJ **521**, 64 (1999)
16. C. Steidel, K. Adelberger, M. Giavalisco, et al: ApJ **519**, 1 (1999)
17. K. Nandra, R. Mushotzky, K. Arnaud, et al: ApJ **576**, 625 (2002)

The Deepest Near-Infrared View of the Universe

Ivo Labbé[1], Marijn Franx[1], Gregory Rudnick[2], Natascha Förster Schreiber[1], Emanuele Daddi[3], Peter van Dokkum[4], Konrad Kuijken[1], Alan Moorwood[3], Hans-Walter Rix[5], Huub Röttgering[1], Lottie van Starkenburg[1], Ignacio Trujillo[5], Arjen van der Wel[1], and Paul van der Werf[1]

[1] Leiden Observatory, P.O. Box 9513, NL-2300 RA, Leiden, The Netherlands
[2] MPA, Karl-Schwarzschild-Str. 1, 85741 Garching, Germany
[3] ESO, Karl-Schwarzschild-Str. 2, D-85748, Garching, Germany
[4] Yale, P.O. Box 208101, New Haven, CT 06520-8101, USA
[5] MPI für Astronomie, Königstuhl 17, D-69117, Heidelberg, Germany

Abstract. We present very deep near-infrared imaging in the J_s, H and K_s-bands of the HDF-S and the field around the z=0.83 cluster MS1054–03 with ISAAC on the VLT. For the HDF-S this resulted in the deepest ground-based infrared observations to date and the deepest K_s-band in any field. We constructed K_s-selected multicolor catalogs in both fields, selecting high-redshift galaxies on their rest-frame optical light. We discovered a new, substantial population of optically faint galaxies with very red near-infrared colors ($J_s - K_s > 2.3$) which are underrepresented in standard U-dropout samples. Recently, these galaxies were spectroscopically confirmed at redshifts $z > 2$ and we estimate that they may contribute as much as 50% to the total stellar mass density at $z \sim 3$. Furthermore, the near-infrared data allow direct measurement of the evolution of the global rest-frame optical color, luminosity density, and stellar mass density from $z = 0$ to $z = 3$. We find that the average rest-frame optical color was monotonically bluer in the past and the universe had a ~10 times lower stellar mass density in bright galaxies at $z \sim 3$. In general, our results demonstrate the necessity of deep near-infrared imaging for a more complete picture of the early universe.

1 Introduction

In the past decade, our ability to routinely identify and systematically study distant galaxies has dramatically advanced our knowledge of the high-redshift universe. In particular, the efficient U-dropout technique [24,25] has enabled us to select distant galaxies using simple photometric criteria. Now more than 1000 of these Lyman break galaxies (LBGs) are spectroscopically confirmed at $z > 2$, and have been subject to targeted studies on spatial clustering [13], internal kinematics [19,20], dust properties [1], and stellar composition [23,18]. Although LBGs are among the best studied classes of distant galaxies to date, many of their properties are still unknown. More importantly, it is unclear whether the U-dropout technique alone will give us a fair census of the galaxy population at $z \sim 3$, as it requires galaxies to be bright in the far-ultraviolet, blue and relatively compact. We know that dust obscured systems and most present-day galaxies do not satisfy these criteria.

In this context we initiated the Faint InfraRed Extragalactic Survey (FIRES) [9], a large public program consisting of very deep near-infrared (NIR) imaging carried out at the *Very Large Telescope* (VLT). We observed fields with existing deep optical WFPC2 imaging from the *Hubble Space Telescope* (HST): the WPFC2-field of Hubble Deep Field South (HDF-S) and the field around the distant cluster MS1054–03. By selecting sources in the NIR K_s−band, we access the rest-frame optical light of objects at $1 < z < 4$ where long-lived stars may dominate the emission. Compared to the rest-frame far-UV, rest-frame optical light is less sensitive to the effects of dust extinction and on-going star formation, providing a better tracer of stellar mass. Now, we are for the first time obtaining a better census of the galaxies that dominate the stellar mass densities at different epochs, and hence are a step closer to tracing the assembly of massive galaxies directly.

We present here selected results of the FIRES survey. Throughout this paper, we assume a flat Λ-dominated cosmology ($\Omega_M = 0.3, \Lambda = 0.7, H_0 = 70$ km s^{-1} Mpc^{-1}). All magnitudes are expressed in the Johnson [15] photometric system.

2 Observations and Photometry

The NIR observations were obtained using the Infrared Spectrograph and Array camera (ISAAC) [17] on the VLT/ANTU at Paranal. On the HDF-S, the total integration times in the J_s (1.24μm), H (1.65μm) and K_s (2.17μm) passbands are 34, 32, and 36 hours respectively, and we reach limiting depths of 25.9, 24.8, and 24.4 (3σ, total magnitude for compact sources). Complemented with HST/WFPC2 observations in the passbands $F300W, F450W, F606, F814W$ [6] we have imaging in 7 bands over a survey area of 4.7 arcmin^2. In the field of MS1054 a similar amount of observation time in each NIR band was divided over a 4 times larger area resulting in ~ 1 magnitude shallower data. Here, we have imaging in 8 bands (VLT/FORS UBV; HST/WFPC2 $F606W, F814W$; VLT/ISAAC J_sHK_s) over an area of 23.6 arcmin^2. The median seeing in the ISAAC images of both fields is slightly better than $0.5''$.

We constructed K_s−band selected samples of galaxies using SExtractor software [2]. We accounted for resolution differences in the HST and VLT datasets and we measured colors in all passbands using specially tailored isophotal apertures defined from the detection image. In addition, we obtained a measurement of the total flux using an aperture based on the SExtractor *AUTO* aperture. The observations, reduction and analysis techniques are described in detail in [16], [21] and [11].

Finally, we identified stars and determined the photometric redshifts and rest-frame luminosities of all extragalactic sources (see [21,22]).

3 Optically Red Galaxies at $z > 2$

The most striking result of the deep NIR data is the discovery of galaxies that are barely detectable in the deep optical HST images, but quite bright in the

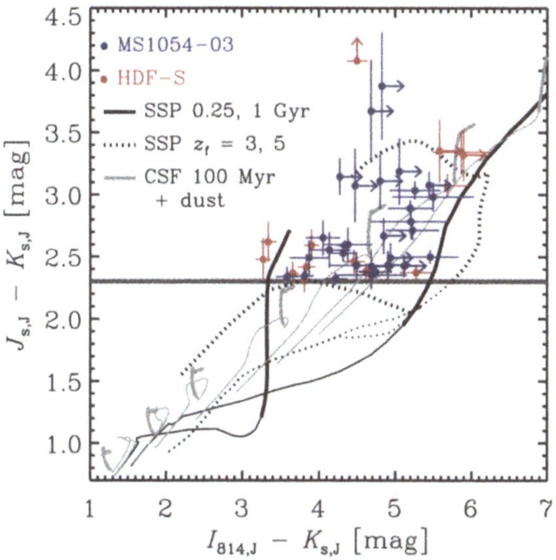

Fig. 1. Color-color diagram of the $J_s - K_s > 2.3$ red galaxies in the FIRES survey: the field of the HDF-S (*light gray circles*) and MS1054–03 (*dark gray circles*) . The range in $I_{814} - K_s$ and $J_s - K_s$ colors is comparable. Overplotted are tracks of several redshifted stellar population models (Bruzual & Charlot): simple stellar populations (SSPs) of age 0.25, 1 Gyr (*solid*), SSPs with formation redshifts $z_f = 3, 5$ (*dotted*), and constant star formation models with varying dust reddening (*solid gray*). The thick part of the curves indicates $z > 2$. The colors of the galaxies are consistent with evolved stellar populations, or strong reddening by dust, or a combination of both.

ISAAC K_s-band. These galaxies can be effectively selected by a simple photometric criterion $J_s - K_s > 2.3$ that is designed to isolate galaxies with prominent Balmer or 4000Å breaks at $z > 2$ (see Fig 1), but possibly also selects highly obscured distant galaxies. In both fields red $J_s - K_s > 2.3$ sources are found in large numbers [10,26] with surface densities ~ 1 arcmin^{-2} to $K_s = 21$ and ~ 3 arcmin^{-2} to $K_s = 22.5$. The first spectroscopic observations confirm the $z > 2$ redshifts of the galaxies [26], but progress is slow owing to their faintness in the observer's optical. The mean photometric redshift is $< z > \approx 2.5$ (rms 0.7) and their space density is about $30 - 50\%$ of that of LBGs [10] with large uncertainties. There is a variety of morphologies: most objects are very compact and unresolved in the NIR, some are large with exponential profiles and effective radii $r_e = 0.7 - 1.0''$ (5 – 8 kpc). To constrain the properties of the stellar populations we fit models to the observed broadband spectral energy distribution (SEDs). We used the HYPERZ fitting code [3] with Bruzual & Charlot [5] stellar population models and applied the Calzetti [7] reddening law. For simplicity we limit the discussion to constant star formation models with solar metallicity, characterized by three parameters: age (the time since the onset of star formation), star formation rate, and extinction. A full discussion will appear in [12]. The simple models provide a reasonable description to the majority of red galaxies, although we note that a major uncertainty is the parameterization of the star formation history (see, e.g., [23,18]). The best fitting SEDs indicate that the galaxies are relatively old (0.7 – 2 Gyr), actively star forming (50 – 200 $M_{\odot}yr^{-1}$ after extinction correction), substantially obscured by dust ($A_V = 1$ – 2 mag), and massive (1 – 2 x 10^{11} $M_{\odot}$). This suggest that these galaxies are unlike any other galaxies studied so far, with star formation rates comparable to LBGs (but older and more obscured) and stellar masses comparable to local elliptical galaxies. In the HDF-S, we have estimated that approximately 50% of the stellar mass at $z \sim 3$ is in luminous $J_s - K_s > 2.3$ galaxies [22]. While large uncertainties remain, the first analysis of the NIR spectra of a handful of red galaxies confirms the surprising results of the SED fits [27]. The available evidence suggests that they could be the most massive galaxies at high redshift and the progenitors of today's early-type galaxies.

4 The Evolution of the Cosmic Color, Luminosity, and Mass Density from z=0 to z=3

A primary goal of measuring the evolution of the globally averaged color and luminosity is to understand how the stellar content of the universe was assembled and changes over time. Using our accurate photometric redshifts – and spectroscopic redshifts where available – and rest-frame luminosities for all galaxies with $K_s^{tot} < 23$ and $z < 3.2$ (see [21,22]), we measured the rest-frame optical luminosity density j_{λ}^{rest} in the U, B, and $V-$bands, and the globally averaged $(U-B)_{rest}$ and $(B-V)_{rest}$ colors. Simple exponentially declining star formation histories (SFHs) with constant metallicity reproduce the evolution of the colors well if modest amounts of reddening (E(B-V) < 0.35) are applied. These simple models are used to estimate the mean stellar mass-to-light ratios M/L_V^* of the integrated cosmic stellar population. Combining these estimates with our j_{λ}^{rest} measurements yields a robust determination of the evolving stellar mass density. Nevertheless, cosmic variations and possible variations in metallicity, assumed IMF, and in particular dust reddening are large sources of uncertainty.

We find that the $(U-B)_{rest}$ and $(B-V)_{rest}$ colors of bright ($L_V > 1.4$ x 10^{10} $L_{\odot}$) galaxies bluen smoothly and monotonically towards higher redshift. Remarkably, the colors at all redshift are similar to those of nearby galaxies, implying that we can likely use "normal" models to infer their stellar masses. The evolution in the $V-$band luminosity density from $z = 0$ to $z \sim 3$ is mild (a factor of ~ 2), but progressively stronger in the B and $U-$bands. Our results indicate that at $z \sim 3$ the universe only contained 10% of the current stellar mass in bright galaxies, and that by $z \sim 1$ the stellar mass density had increased to approximately half of its local value (see Fig. 2). These values agree well with other determinations [8,4]. Interestingly, it implies that any population of galaxies which forms most of their stellar mass at $z > 2$ can at most contribute 10% to the present-day stellar mass density. Likewise, the epoch of $1 < z < 2$ was an important period in the stellar mass build-up of the universe and a large

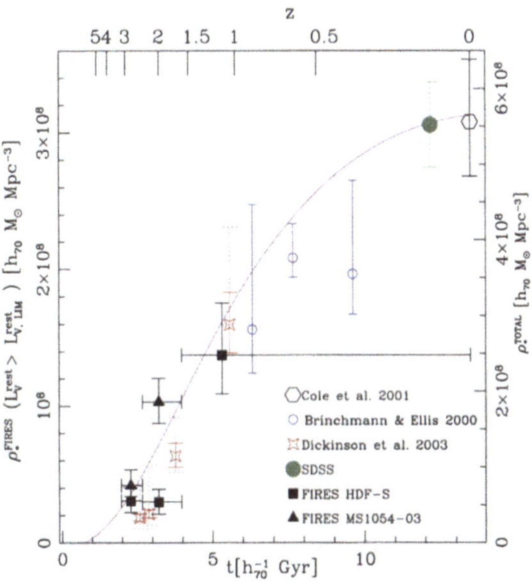

Fig. 2. The build-up of the stellar mass density as a function of redshift (Rudnick et al. 2003). The stellar mass density measurements for galaxies in the HDF-S and MS1054 are indicated by solid squares and solid triangles respectively. The mass density in the HDF-S at $z \sim 3$ is twice that in the HDF-N (Dickinson et al. 2003, *hollow stars*) and at $z \sim 2$ there is a factor 3 difference between the HDF-S and MS1054. More measurements are needed to estimate the effect of cosmic variance.

fraction of stellar mass today should reside in objects with mass-weighted ages corresponding to this formation period. The contribution to the stellar mass density at low redshift of early type galaxies, which likely formed most of their stars at redshifts $z > 1$, is estimated to be $\sim 50\%$ [14], consistent with our results of rapid mass growth during this epoch.

5 Conclusions

The first results of the FIRES survey demonstrate the necessity of extending optical observations to NIR wavelengths for a more complete census of the early universe. Our deepest K_s−band proves invaluable for it probes well into the rest-frame optical at $1 < z < 4$, where long-lived stars may dominate the light of galaxies. We find the following:

- A new substantial population of red ($J_s - K_s > 2.3$) galaxies at $z > 2$. They are underrepresented in UV-selected samples, such as U-dropout galaxies, because most of them are barely detectable in even the deepest optical images. The SEDs of many red galaxies show a prominent jump in the NIR, which may be caused by the Balmer/4000Å break. Modeling of the broadband SEDs indicates that the majority are old (0.7 – 2 Gyr), star forming at high rates (50 – 200 $M_{\odot}yr^{-1}$), substantially obscured ($A_V = 1 - 2$ mag), and massive (1 – 2 x 10^{11} $M_{\odot}$). They likely contribute significantly to the total stellar mass budget at redshifts $z \sim 3$.

- The cosmic color, luminosity density, and mass density evolve smoothly towards higher redshift. The globally averaged rest-frame colors get bluer, the V_{rest}-band luminosity density increases, and the mass density decreases. At $z \sim 3$ the universe only contained 10% of the current stellar mass density in bright galaxies, growing to roughly half of its local value by $z \sim 1$.

We are pursuing our follow-up programs to obtain spacebased NIR imaging with NICMOS to study the detailed morphologies and Spitzer Space Telescope imaging to better constrain the masses of high-redshift galaxies. We are also taking optical and NIR spectroscopy to better estimate the space density and stellar population properties of $J_s - K_s > 2.3$ selected high-redshift galaxies.

We wish to thank the ESO staff for their dedicated efforts and kind assistance in obtaining these data and making it available to us. This work was supported by a grant from the Netherlands Organization for Scientific Research.

References

1. Adelberger, K. L. & Steidel, C. C. 2000, ApJ, 544, 218
2. Bertin, E. & Arnouts, S. 1996, A&AS, 117, 393
3. Bolzonella, M., Miralles, J.-M., & Pelló, R. 2000, A&A, 363, 476
4. Brinchmann, J. & Ellis, R. S. 2000, ApJL, 536, L77
5. Bruzual, G. & Charlot, S. 2003, MNRAS, 344, 1000
6. Casertano, S. et al. 2000, AJ, 120, 2747
7. Calzetti, D. et al. 2000, ApJ, 533, 682
8. Dickinson, M., Papovich, C., Ferguson, H. C., & Budavári, T. 2003, ApJ, 587, 25
9. Franx, M. et al. 2000, The Messenger, 99, 20
10. Franx, M. et al. 2003, ApjL, 587, L79
11. Förster Schreiber, N.M. et al. 2004a, in preparation
12. Förster Schreiber, N.M. et al. 2004b, in preparation
13. Giavalisco, M. & Dickinson, M. 2001, ApJ, 550, 177
14. Hogg, D. W. et al. 2002, AJ, 124, 646
15. Johnson, H. L. 1966, ARA&A, 4, 193
16. Labbé, I. et al. 2003, AJ, 125, 1107
17. Moorwood, A. F. 1997, proc. SPIE, 2871, 1146
18. Papovich, C., Dickinson, M., & Ferguson, H. C. 2001, ApJ, 559, 620
19. Pettini, M. et al. 1998, ApJ, 508, 539
20. Pettini, M. et al. 2001, ApJ, 554, 981
21. Rudnick, G. et al. 2001, AJ, 122, 2205
22. Rudnick, G. et al. 2003, ApJ, 599, 847
23. Shapley, A. E. et al. 2001, ApJ, 562, 95
24. Steidel, C. C. et al. 1996, AJ, 112, 352
25. Steidel, C. C. et al. 1996, ApJ, 462, L17
26. van Dokkum, P. G. et al. 2003, ApJL, 587, L83
27. van Dokkum, P. G. et al. 2004, in preparation

Galaxy Evolution in Mass-Selected Samples

Adriano Fontana

INAF - Osservatorio Astronomico di Roma
Via Frascati 33, 00040 Monte Porzio, Rome, Italy

Abstract. I summarize the recent results from the HDFS and K20 surveys that allow to trace the evolution of massive galaxies up to $z \simeq 3$. I show that the evolution of both the specific star–formation rate and of the stellar mass density suggest that, at $z > 1$, we observe the epoch where the evolution of massive objects becomes appreciable. At $z \simeq 1.8$, we indeed find that about 40% of the local density has been locked in massive galaxies. This is associated with a change in the physical properties of massive galaxies, with star–forming objects that contribute to a substantial fraction of the observed mass density. I quantitatively show that these results do not pose challenges to current $\Lambda - CDM$ models, but represent valuable constraints to understand the complex physical mechanisms involved in the assembly of massive galaxies.

1 Introduction

Thanks to the recent - and hopefully definitive - consolidation of the "concordance" cosmological scenario, that has removed the ambiguity about the cosmological parameters, observations at high redshift can be used to better constraint the complex baryonic mechanisms involved in this process. Within this framework, K–band surveys have long been recognized as ideal tools to study the process of mass assembly at high redshift. Albeit the relation between near–IR luminosity and stellar mass is not univocal, indeed, mass–selected galaxy samples can be extracted from these samples and used to constrain the basic features of theoretical models. In this contribution, I review the recent results from the HDFS and K20 surveys, both based on deep K-selected samples, forwarding the reader to other papers for a full discussion ([6], [7], Cimatti this volume).

2 Massive Galaxies at $z > 2$

We will first discuss the physical properties of the sample of galaxies at $z \geq 2$, as obtained from a sample of 75 objects at $K \leq 25$ in the HDFS, using star–formation rates and stellar masses obtained from best–fit to the spectral distribution. In particular, the measured specific star–formation rate $\dot{M}/M_*$ for the K–selected $z > 2$ sample is shown in Fig.1. At lower redshifts, the evolution of $\dot{M}/M_*$ has been studied by [1], of which we plot in fig.1 their average relations at $0.2 < z < 0.5$ and at $0.75 < z < 1$. We find that the trend of increasing $\dot{M}/M_*$ with z detected at $z \leq 1$ by [1] still continues at $z \simeq 3$. At the median value of the $M_* > 10^{10} M_\odot$ complete sample the average specific star–formation rate

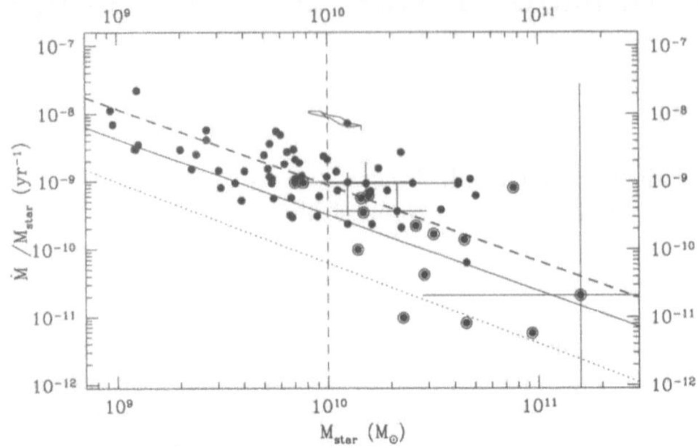

Fig. 1. Rest frame properties of $z > 2$ galaxies from spectral fitting of the HDFS sample. The vertical dashed line shows the rough estimated limit for mass completeness in the $z > 2$ sample. Specific star–formation rate as a function of stellar mass in the $z > 2$ sample of the HDFS, compared with the average relations at $0.2 < z < 0.5$ (thin dotted line) and at $0.75 < z < 1$ (solid line) from[1]. The thick dashed line correspond to our fit to the whole HDFS sample. Encircled points represent the objects with $J - K \geq 1.34$ discussed by [8]. Errorbars are shown for a few objects: the large errorbars on the most massive objects reflect the ambiguity in its nature, between passively–fading and dusty–star forming objects.

$< \dot{M}/M_* > \simeq 4 \times 10^{-10}$ yr^{-1} (note that the adoption of a Calzetti extinction curve would shift this it to a much higher value). The estimated average Scalo b parameter $SFR \times AGE/MASS$, is about 0.8 in this star–forming sample, suggesting a long duration of the starburst activity.

Only seven "red" objects are detected at both large $V - I \geq 1$ and $J - K \geq 1.5$, a combination that may be due to either a strongly absorbed star–forming galaxy or to a passively fading stellar population, an ambiguity that cannot be safely removed without spectroscopy, although we have found that the spectral fitting to the complete $UBVIJHK$ distribution slightly prefers the "passive" spectral model. These objects have $\dot{M}/M_* < 10^{-10}$ yr^{-1} i.e. more than 10 times lower that the typical $z \geq 2$ population if we adopt the "passively fading" fitting models. Given the ambiguity of their spectral classification, they can be used to obtain an upper limit to the fraction of "passively fading" galaxies: in the HDF–S, these objects contribute at most to about 25% of the total galaxy *number* density and up to $\simeq$ 40% of the total *stellar mass* density of the mass–complete subsample of $z \simeq 2$ galaxies.

3 The Cosmological Evolution of the Stellar Mass Density

While the properties of galaxies at the highest redshifts can be currently investigated only with small-area surveys like the HDFs, wider area surveys are

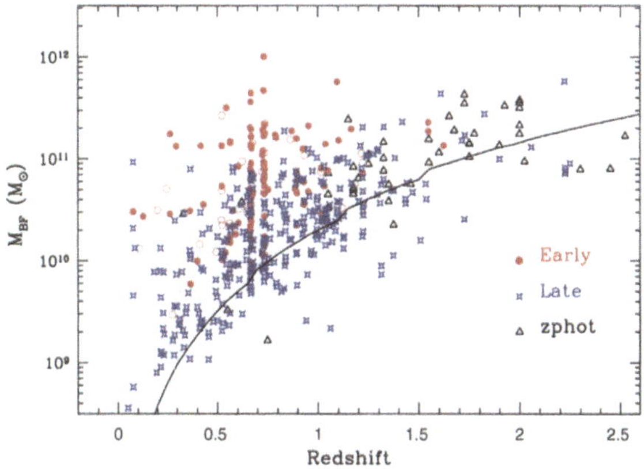

Fig. 2. Galaxy stellar masses in the K20 sample as a function of redshift. The thin solid line is an estimate of the completeness threshold in the sample, corresponding to a maximally massive, dust–free model with $K_s = 20$.

necessary to investigate the properties of the more massive (and rarer) galaxies over the whole Hubble time. The K20 survey (see Cimatti, this volume) is ideally suited, both for the large spectroscopic coverage as well as for the availability of deep multicolor coverage for each galaxy.

The derived masses for the K20 sample are shown in Fig.2 as a function of redshift: it is shown that a significant number of massive galaxies, in the range $10^{11}M_\odot < M_* < 10^{12}M_\odot$, is detected at least up to $z \simeq 2$. These stellar masses are significantly larger than those detected in the HDFs ([5], [6]), that are typically at most $M_* \simeq 10^{11}M_\odot$, as is to be expected given the much smaller area covered by the HDFs.

More interestingly, the HDFN data show the upper envelope of the mass distribution to decrease at high redshift, as naively expected in CDM models. Such a trend is much less clear in the similar analysis of the HDFS ([6]) as well as in the K20 sample. Unfortunately, strong physical conclusions are difficult to draw from the shape of this upper envelope, that is intrinsically plagued by the inevitably small statistics of the more massive galaxies, emphasized by the small cosmic volume sampled by these surveys. For such a reason, quantitative conclusions can be drawn only when appropriate statistics are computed, either in the comparison with the models (see [6]), or in the data, such as the stellar mass distributions or mass densities that we shall present below.

The global evolution of the stellar mass density, as obtained from the K20 and other surveys, is depicted in Fig.3. We have corrected for incompleteness effects, extrapolating the observed data assuming the best–fit Schechter functions (see [7] for the details). The wide errorbars in the HDFS and HDFN data reflect the uncertainty in the mass estimates of very high z galaxy, that will be settled with the forthcoming SIRTF mission. The K20 survey data are well placed within a global trend that witness a fast increase of the stellar mass density from $20^{+20}_{-10}\%$

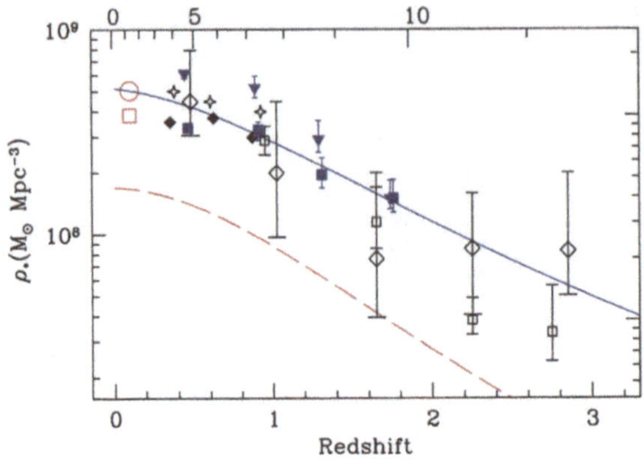

Fig. 3. Evolution of the cosmological mass density as a function of redshift as observed from the K20 and other surveys. Filled points represent the total values from the K20 survey: squares are the "Best Fit" values and triangles the "Maximal Age" estimates. The large open circle is the local value ([4], while the large open square is the same local value rescaled to our technique to estimate stellar masses. Open hexagons are from [2], open triangles from [1], open squares from [5], diamonds from [6]. Error bars in the latter two surveys take into account for the systematic uncertainties among the methods adopted, as in the case of our MA and BF approaches. The two lines are the mass densities expected by integrating the star formation histories of Madau et al 1998 for two different extinction coefficients (dashed: $E(B-V)=0$; solid: $E(B-V)=0.15$)

of the local value at $2<z<3$ to about unity at $z\leq 1$, i.e. in a relatively short amount of cosmic time.

Somewhat surprisingly, there appears to be a good agreement between the observed evolution of $\rho_*(z)$ and the integrated star–formation history due *only* to UV bright galaxies, after allowing for a reasonable amount of dust, as shown by the solid line in Fig.3. Given the uncertainties in the estimates of both the stellar masses and the UV–corrected star–formation rates, it is well possible that the detailed agreement between the two quantities is to some extent fortuitous, leaving space for other significant contributors to the global star–formation rate, as could be the case of dust–enshrouded sources. However, the overall agreement, and in particular the match with the local value of the stellar mass density, suggest that the star-formation rate inferred from UV–selected surveys provides a remarkable fraction of the observed stellar mass density.

4 Massive Galaxies and Λ–CDM Scenarios

What is the scenario for the evolution of massive galaxies that emerge from our analysis? Up to $z\simeq 1$, little evolution has occurred in these objects. Their number density is close to the local one (80% in our data) and to the prediction of passively evolving PLE models. At $z>1$, we enter in the epoch where the

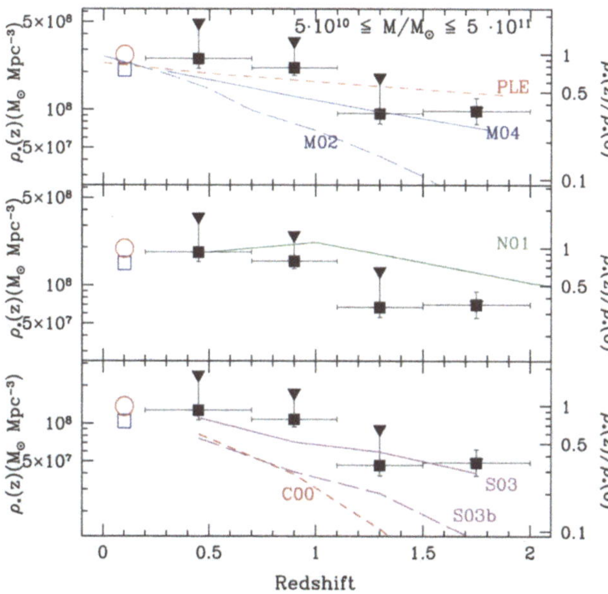

Fig. 4. Cosmological Stellar Mass Density for objects of $M_* \simeq 10^{11} M_\odot$ in the K20 survey, compared with the theoretical models discussed in the text, divided according to the relevant IMF: Salpeter (*upper panel*), Gould, Bahcall, & Flynn (*central panel*), Kennicutt (*lower panel*). The models plotted are: Menci et al 2002 (M02, thin dashed line) and Menci et al 2004 (M04, thin solid line) and PLE (short–dashed line) in the upper panel; Nagamine et al 2001(N01, solid line) in the central panel, and Cole et al 2000 (C00, short-dashed); Somerville et al 2003 (S03 and S03b, thick solid and dashed lines). See text for more details. The relative value with respect to the local value of Cole et al 2001 is shown in the right y–axis.

evolution of massive objects becomes much faster. At $z \simeq 1.8$, we find that about 35-40% of the local stellar mass density has been locked in massive galaxies. This is associated with a change in the physical properties of massive galaxies, with $z \simeq 1.5$ star–forming objects contributing to a more substantial fraction of the observed mass density. At even higher z, the number of passively evolving objects is even smaller, as emerge from the ultra deep IR observations of the HDFs (Fig.1 and [6], [5]), and star-forming galaxies are accreting stars at a rate that likely makes them substantial contributors to the present day mass density (Fig.1).

Overall, these features are in agreement with the basic elements of the Λ–CDM scenarios. In an accelerated Universe, galaxies have time to complete their assembly at $z \simeq 1$ without requiring too much merging at low redshifts, and star formation naturally occurs more efficiently at earlier times and in more massive objects. Despite these "built in" features, the predictions of the models for galaxy formation have often been presented - or perceived - as unable to provide a fair number of massive of high z galaxies.

To perform a quantitative check, we have focused on the the evolution of the stellar mass density computed only for galaxies in the range $5\times10^{10}M_{\odot} \leq M_* \leq 5\times10^{11}M_{\odot}$ (i.e. of about the typical Schechter mass in the local mass function of Cole et al. 2001), where our sample is nearly complete at all redshifts. We have compared the expectations of updated Λ–CDM hierarchical models in this mass range, including in our comparison the prediction of several renditions that differ both in the treatment of the physical processes as well as in the numerical technique adopted, with our data, scaled to the relevant IMF (Fig.4).

As expected, the gradual decrease of the density of massive objects with cosmic time is clearly built in these models. However, the models show a remarkable spread in the predicted mass density, reaching differences of more than 2 orders of magnitudes at the highest redshifts: the reason is tied to the different physical processes contained - or ignored - in the models and to the way they are described (see [7] for a full discussion).

This is clearly shown by the comparison between semianalytic models, where the ingredients are easier to compare: the most striking evidence is how sensitive the models are to the inclusion of interaction–driven starburst at high z. The inclusion of more refined description of galaxy merging ([9]) and most of all the treatment of starbursts associated with bound merging ([13]) and fly–by interactions ([10]) is progressively improving the quality of the theoretical predictions. Interestingly, fully hydrodynamical simulation ([11]), that in principle include these effects as well as the QSO feedback, even overproduce the high redshift stellar mass density.

Overall, it appears that the large number of massive galaxies that we detect at high z does not violate any fundamental Λ–CDM constraint, since it is in the range of the values predicted by the most recent models. At the same time, the galaxy stellar mass densities and mass functions appear to be critical observational quantities, that are necessary to provide constraints about the physical processes that are caught in action during the phase of galaxy formation.

References

1. Brinchmann,J. & Ellis R.S. 2000, ApJ 536L, 77B
2. Cohen, J., 2002, ApJ, 567, 672
3. Cole, S., Lacey, C.G., Baugh, C.M., Frenk, C.S., et al., 2000, MNRAS, 319, 168
4. Cole, S., Norberg, P., Baugh, C. M., et al., 2001, MNRAS, 326, 255
5. Dickinson, M., Papovich, C., Ferguson, H.C., Budavari, T., 2003, ApJ 587, 25
6. Fontana, A., Donnarumma, I., Vanzella, E., et al., 2003, ApJ, 594L, 9
7. Fontana, A., Pozzetti, L., Donnarumma, Cimatti, A., et al., 2004, A&A subm.
8. Franx, M., Labbe', I., Rudnick, G., et al., 2003, ApJL, 587, 79
9. Menci N., Cavaliere A., Fontana A., Giallongo E., Poli F.2002, ApJ, 575, 18
10. Menci, N., Cavaliere, A., Fontana, A., et al. 2004, in press
11. Nagamine, K. Fugucita, M. Cen, R., Ostriker, J.P. 2001, MNRAS, 327L, 10
12. Poli, F., Giallongo, E., Fontana, A., et al., 2003, ApJL 593, 1
13. Somerville, R. S., Moustakas, L. A., Baharam, M., et al., 2003, astro-ph/0309067

A Multiwavelength Survey of Luminous Compact Blue Galaxies from z=3 to z=0

Rafael Guzmán

Department of Astronomy, University of Florida, Gainesville, FL 32611-2055

Abstract. Luminous Compact Blue Galaxies (LCBGs) are small starburst systems that dominate the number density of galaxies at intermediate redshifts. LCBGs have evolved more than any other galaxy class in the last 8 Gyrs and are a major contributor to the observed enhancement of the UV luminosity density of the universe at z ¡ 1. Despite the key role LCBGs may play in galaxy evolution, their statistical properties are still largely unknown. We are currently conducting a multiwavelength study of LCBGs from z = 3 to the present, ranging from FUV to FIR. This study is unique in combining the capabilities of space observatories with ground-based large telescopes and large public databases of galaxy surveys. The main goals of this study are: (i) to characterize the global properties of LCBGs at 0 ¡ z ¡ 3 using statistically representative samples defined using the same selection criteria and over the same rest-frame wavelength range; and (ii) to investigate the role of this important population in galaxy formation and evolution.

1 Introduction

In the last few years, there has been substantial progress in our understanding of the numerous faint blue galaxies seen in deep images of the sky. The breakthrough came with the advent of HST and the new generation of 10-m class telescopes, which provided a detailed study of their morphology, internal kinematic, and mass. In particular, Keck spectra show that their emission-line velocity widths are only ≤ 60 km s^{-1} [16] [9], while their HST images reveal complex, distorted and knotted structures with typical half-light radii $r_e \leq 0.5''$, or $R_e \leq 3$Kpc [11]. The combined Keck and HST data indicate that these distant galaxies, although being intrinsically luminous ($\sim 0.2 - 5L^\star$), have mass-to-light ratios that are only $\sim 0.1 - 1.0 M_\odot/L_\odot$, i.e., *about 10 times smaller than a typical $L^\star$ galaxy today* [20] [10]. The low mass-to-light ratios are consistent with these galaxies being small but vigorously star-forming systems. Jangren et al. [14] find that most of these starbursts can be isolated quantitatively on the basis of their luminosity, color, and surface brightness as a class distinct from normal Hubble-types found in similarly bright samples in the nearby universe. We refer to such class as "Luminous Compact Blue Galaxies" (LCBGs).

The nature of LCBGs and their relation to today's galaxy population still remain largely unknown. The most comprehensive study of LCBGs at intermediate redshift to date –only 45 objects– [20] [10] concluded that the LCBG class is populated by a mixture of starbursts. About ~60% of galaxies in their sample are classified as "HII-like" since they are similar to today's population of luminous, young, star-forming HII galaxies. The remaining ~40% are classified as "SB disk-like" since they form a more heterogeneous class of evolved starbursts similar to local starburst disk nuclei and giant irregular galaxies. This classification is consistent with previously published results for other LCBG samples. For instance, some authors [15] [16] [9] [11] have established the association between HII galaxies and LCBGs for their sample of compact narrow emission-line galaxies, while other studies [18] [13] have concluded that their LCBG samples can be best identified with bright irregulars, late-type spirals or disk galaxies with a young bulge.

Recently, various observational studies have highlighted the key role that LCBGs play in galaxy evolution over cosmological time-scales. For instance:

(i) *LCBGs contribute the most to the evolution of blue L* galaxies in the last ~8 Gyrs:* the CFRS collaboration [17] first demonstrated that there is no significant evolution in the luminosity function of large galaxies (with typical sizes >8 kpc) all the way to $z \sim 1$. However, the number density of small but luminous stellar systems at $z > 0.5$ increase by a factor of ~10 compared to that observed at lower redshifts. In a follow-up study [18], they showed that this rapidly evolving population has global properties (i.e., blue colors, compact morphology, L* luminosities, high SFR, small sizes, and small velocity widths) identical to those of LCBGs.

(ii) *LCBGs are a major contributor to the observed increase in the star formation rate density of the universe at $z < 1$:* we [10] have shown that LCBGs are responsible for ~45% of the observed enhancement in the global star formation density of the universe at $z \sim 1$. Such galaxies contribute a negligible amount to the local rest-frame blue luminosity density [19] but equal the contribution of regular spirals at $z \sim 1$ [1].

2 Description of the LCBG Survey

We are currently carrying out an extensive multiwavelength study of LCBGs from $z \sim 3$ to the present epoch. Our study will include observations ranging from far-UV to far-IR at high redshifts, and from far-UV to cm wavelengths at $z \sim 0$. The main goals are:

(1) to characterize the global properties of LCBGs as a class using statistically representative samples defined using *the same selection criteria* and over *the same rest-frame wavelength range* from $z \sim 3$ to $z = 0$. In particular we will:

• investigate their morphology, surface brightness, half-light radius, and color profile at various wavelengths. Systematic differences among FUV, optical and near-IR morphologies and structural parameters will allow us to study the spatial distribution and relative contribution of the young starburst and the underlying older population, as well as the extent and effect of dust extinction.

• measure their emission spectral features in the rest-frame wavelength range 1000-8000Å, including Lyα, [OII]3727, Hβ, [OIII]4959,5007, Hα, [NII], [SII], as

well as the stellar absorption from SiII, OI, CII, SiIV, CIV in the UV, and Mg2, Fe, and CaII in the optical. The spectral line analysis will provide direct information on the ages and metallicities of their composite stellar populations.
• estimate their internal kinematics using emission-line velocity widths and rotational velocities traced by the ionized gas, as well as velocity dispersions measured from stellar absorption lines.
• determine their kinematical masses (from the virial theorem applied to the kinematic and size measurements) and stellar masses (from rest-frame near-IR fluxes). Galaxy mass has proven to be a key parameter in studies of galaxy evolution over cosmological timescales since it is independent on the amount of luminosity evolution. We can thus compare directly distant young galaxies at look-back times with today's stellar systems of similar mass and identify unambiguously their local counterparts.
• parameterize their star formation rates using the UV continuum, [OII]3727 and Hα luminosities, and FIR emission. This will provide not only an accurate measure of their contribution to the global star formation rate density of the universe, but also a very needed calibration among the various star formation tracers currently used at different redshifts.
(2) to investigate the role of LCBGs in galaxy formation and evolution. Only after the global properties of the LCBG population as a class have been clearly defined it is possible to tackle the issue of evolution. A key aspect of this survey in this regard is the homogeneity in the selection criteria imposed to the LCBG samples at different redshifts. Evolution will be investigated in two different ways:
• "internally": comparing statistically the global properties of LCBGs at different redshifts in order to quantify their evolution in number density and star formation rate density, as well as any systematic changes in galaxy properties with redshift.
• "externally": comparing statistically the global properties of LCBGs at intermediate redshifts with those of representative samples of nearby galaxies *with similar masses*. This comparative study, together with evolutionary population synthesis models, will allow us to shed light on their evolutionary sequence and understand how distant LCBGs relate to today's normal galaxy population.

3 The Sample Selection

The most novel aspect of our selection criterion is that it ensures the selected region of the luminosity-surface brightness-color 3-D parameter space that defines the LCBG galaxy class can be observed over a very wide range in redshift, considering the magnitude and surface brightness limits of the photometric data, k-correction and cosmological dimming. Thus we first selected the LCBG population at high-redshifts from deep HST F814W images, and then applied the same criteria to define the LCBG samples at lower redshifts in the same rest-frame wavelengths and with the same magnitude and surface brightness isophotal cutoff. The current datasets allow us to study LCBGs with $M_B < -18.5$ mag,

$SBe_B < 21$ mag arcsec^{-2}, and rest-frame $(B-V)_0 < 0.6$ mag all the way from $z \sim 1$ to $z \sim 0$. However, only the brightest LCBGs thus selected can be studied at $z > 1$.

4 The Reference Sample of Local LCBGs

We have identified a sample of ~100 LCBGs at $z < 0.03$ from the SDSS using the same selection criterion used to select LCBG at redshift $z \sim 1$ summarized above. This sample was selected after investigating the properties of over 10^6 nearby galaxies in the luminosity-surface brightness-color parameter space using the SDSS database. LCBGs are indeed very rare in the nearby universe. However, this sample, although small, is of the greatest importance since it will provide an excellent reference for comparison with our studies of the LCBG population at higher redshifts, and to ensure the correct interpretation of the lower S/N, less detailed observations of distant LCBGs.

We are currently carrying out a detailed photometric and spectroscopic follow-up at UV, optical, infrared, mm, and cm wavelengths using a wide array of public databases and telescopes, including GALEX, KPNO, IRAM, and Arecibo, among other facilities. This comprehensive study includes number density [2], age and metallicity of the gas and the stellar populations [4], kinematics (gas and stars) and masses (stellar, virial, gas) [5] [4] [6]. Preliminary results reveal a very heterogeneous population of vigorously star-forming galaxies exhibiting a wide range of morphologies (from S0-type to spiral disks and highly irregular, merging-like stellar systems), masses (from 10^9 to 10^{11} $M_\odot$), and gas depletion timescales (from 10^6 to 10^9 years). This combined data set aims to be a cornerstone for future surveys of similar galaxies at high redshifts.

5 The Intermediate-Redshift Sample of LCBGs

The intermediate redshift sample consists of ~1000 LCBGs at $0.3 < z < 1.5$ selected from several HST fields with deep F814W and F606W images [7] (e.g., Groth, GOODS-N, CFRS03+00) that will also be observed with GALEX and SIRTF. A detailed spectroscopic study of the properties of this galaxy sample from rest-frame FUV to the optical is currently underway using GMOS at GEMINI, DEIMOS at Keck, and GIRAFFE at VLT. The number density of intermediate-z LCBGs is about two orders of magnitude higher than that of the reference sample of LCBGs in the nearby universe [20] [7]. HST images reveal a wide variety of morphological types from spheroidal-like to disk-like objects, with an abundance of irregulars and peculiar/interacting stellar systems [15] [20] [11] [7]. Preliminary spectroscopic results confirm the heterogeneous nature of this galaxy class showing emission-line dominated spectra characteristic of nearby starburst galaxies such as HII-galaxies or starburst disk nuclei [10] [13]. Intermediate-z LCBGs also have dynamical masses of $\sim 10^{10}$ as derived from emission line velocity widths [9] [10]. These estimates are supported by absorption line velocity dispersion measurements [8], which confirm previous claims that LCBG at intermediate-z have mass-to-light ratios that are one order of magnitude smaller than the majority of galaxies with similar luminosities in the nearby universe [11] [12].

6 The High-Redshift Sample of LCBGs

LCBGs at $z > 2$ will be identified using the well-known photometric redshift technique using deep multiband ground-based photometry in the same fields selected for the intermediate-z LCBG studies. We note that *the vast majority* of galaxies at high-z selected from deep optical images lie in the same region of the luminosity-surface brightness-color 3-D parameter space that defines the LCBG galaxy class a lower redshifts. This is simply the result of the combined effects of the k-correction and the cosmological dimming. The deep UBRIJK imaging required to measure photometric redshifts is currently being obtained using a wide array of 4m-class telescopes in La Palma, and Calar Alto. The limiting magnitudes of the photometric survey, defined as the AB magnitude with S/N=5 within 1 arcsec aperture, are 26 mag in U, B, R, and I, and 22 mag in J, and K. This depth ensures the detection of the population of today's L^* galaxies at $z > 2$, considering a conservative luminosity evolution of ~2 magnitudes in rest-frame optical wavelengths since $z \sim 2$ till the present.

We are about 2/3 near completion of the photometric survey, including the most time-consuming k-band observations [3]. The complete catalog of ~1000 high-z LCBGs is expected to be finished by the end of 2004. The comprehensive spectroscopic follow-up will be carried out using mainly OSIRIS and EMIR at the GTC, a 10-m telescope currently under construction in La Palma by a consortium of institutions in Spain, Mexico, and the University of Florida. The OSIRIS spectra will provide observations of the rest-frame FUV, while EMIR will provide, for the first time, a complete description of the rest-frame optical (~3000-7000Å) properties of ~1000 LCBGs at $z > 2$, including Hα-based SFR, Oxygen abundances, Balmer decrement-based extinctions, and emission line velocity widths. This project is part of the scientific case behind EMIR, the first wide-field near-IR multiobject spectrograph for a 10-m class telescope. We expect to have at least 30 nights of guaranteed time with EMIR at the GTC to carry out this survey.

7 Summary and Acknowledgments

Our LCBG survey makes an optimum use of the unique array of space observatories and ground-based large telescopes available to our collaboration, including GALEX, HST, SIRTF, Keck, VLT and, in the future, GTC. Our study also benefits from the direct access to large databases of other major galaxy surveys such as SDSS and DEEP. This wealth of data will allow us, for the first time, to investigate in detail the physical properties from rest-frame far-UV to the far-IR of statistically representative samples of LCBGs since $z \sim 3$ to the present, and assess their key role in galaxy evolution over the last ~10 Gyrs of the history of the universe.

This multi-wavelength, multi-redshift LCBG survey is being carried out in collaboration with C. Garland, D.J. Pisano, J. Williams, F. Castander, D. Koo, M. Bershady, F. Hammer, S. Yi, N. Gruel, A. Graham, M. Balcells, M. Prieto, F. Garzón, J. Gallego, N. Cardiel, and R. Pelló. This work is funded by NASA grant LTSA NAG5-11635.

References

1. Brinchmann, J. et al. 1998, ApJ, 499, 112
2. Castander, F.C. & Guzmán, R., 2004, in prep.
3. Cristobal, D., Balcells, M., Prieto, M., Guzmán, R., Gallego, J., Cardiel, N., Serrano, A., & Pelló R., 2003, ApJ, 595, 71
4. Gallego, J.P., Guzmán, R., Castander, F.C., Garland, C., & Pisano, D.J., 2004, AJ, in prep.
5. Garland, C., Pisano, D.J., Williams, J., Guzmán, R., & Castander, F.C. 2004, AJ, submitted.
6. Garland, Williams, J., C., Pisano, D.J., Guzmán, R., & Castander, F.C. 2004, AJ, in prep.
7. Graham, A., Guzmán, R., Wu, K., Gruel, N., Trujillo, I., Aguirre, J.A.L, Koo D.C., & Bershady M.A., 2004, in prep.
8. Gruel, N., Guzmán, R., Koo D.C.,vBershady M.A., & Hammer, F., 2004, in prep.
9. Guzmán, R., Koo, D. C., Faber, S. M., Illingworth, G. D., Takamiya, M., Kron, R., & Bershady, M. A. 1996, ApJ, 460, L5
10. Guzmán R., Gallego J., Koo D. C., Phillips A. C., Lowenthal J. D., Faber S. M., Illingworth G. D. & Vogt N. P. 1997, ApJ, 489, 559
11. Guzmán R., Koo D. C., Jangren, A., Bershady M., Faber S. M. & Illingworth G. D., 1998, ApJ, 495, L13
12. Guzmán R., Ostlin G., Kunth D., Koo D.C., Bershady M.A., & Pahre M., 2003, ApJ, 586 L45
13. Hammer, F., Gruel, N., Thuan, T. X., Flores, H., & Infante, L., 2001, ApJ, 550, 570
14. Jangren A., Bershady M. A., Conselice C., Koo D. C. & Guzmán R., 2004, AJ, submitted
15. Koo, D. C., Bershady, M. A., Wirth, G. D., Stanford, S. A., & Majewski, S. R. 1994, ApJL, 427, L9
16. Koo, D. C. Guzmán, R., Faber, S. M., Illingworth, G. D., Bershady, M. A., Kron, R., & Takamiya, M. 1995, ApJ, 440, L49
17. Lilly, S. J., et al. 1998, ApJ, 500 L75
18. Mallén-Ornelas, G., Lilly, S. J., Crampton D., & Schade D. 1999, ApJ, 518, 83
19. Marzke, R. O., Geller, M. J., Huchra, J. P., Harold, G., & Corwin, J. 1994, AJ, 108, 437
20. Phillips A. C., Guzmán R., Gallego J., Koo D. C., Lowenthal J. D., Vogt N. P., Faber S. M. & Illingworth G. D. 1997, ApJ, 489, 543

The Galaxy Evolution Explorer – Early Data

Christopher Martin[1] and the GALEX Team

California Institute of Technology, Division of Physics, Mathematics, & Astronomy, Pasadena, California

Abstract. We report on early results from the Galaxy Evolution Explorer (GALEX), a NASA Explorer Mission launched on April 28, with a nominal mission start of June 19. GALEX is performing the first space UV sky-survey, including imaging and grism surveys in two bands (1350-1750Å and 1750-2800Å). The surveys include an all-sky imaging survey (limit AB∼20-21), a medium imaging survey of 1000 sq. deg (limit AB∼23.5), a deep imaging survey of 100 square degrees (limit AB∼25), and a nearby galaxy survey. Spectroscopic grism surveys (R=100-300) are underway with various depths and sky coverage. Many targets overlap existing or planned surveys. We will use the measured UV properties of local galaxies, along with corollary observations, to calibrate the UV-global star formation rate relationship in galaxies. We will apply this calibration to distant galaxies discovered in the deep imaging and spectroscopic surveys to map the history of star formation in the universe over the redshift range $0<z<2$. The GALEX mission will include a Guest Investigator program for primary observations and supporting data analysis. This will support a wide variety of investigations made possible by the first UV sky survey.

1 Science Overview

The Galaxy Evolution Explorer (GALEX), a NASA Small Explorer mission, is performing the first all-sky, deep imaging and spectroscopic surveys in the space ultraviolet. The prime goal of GALEX is to study star formation in galaxies and its evolution with time. Early results include detection of extended star formation in numerous disk galaxies, formation of tidal dwarf galaxies, luminous UV analogs to Lyman Break Galaxies in the local universe, and widespread star formation in red galaxies. GALEX primary mission surveys and dedicated observations for the Guest Investigator Program, beginning October 2004, will also support a broad array of other investigations. For example, and has already doubled the number of QSOs observable by STIS to probe the reionization of He II.

GALEX was launched on April 28, 2003. Its on-orbit performance is excellent, meeting or exceeding base-line requirements and identical to ground performance. To date, GALEX has completed 25% of its primary mission. An early data release was made available December 9, 2003, hosted by MAST (http://galex.stsci.edu/). The GALEX Guest Investigator (GI) program has issued a call for proposals, due April 16, 2004, for dedicated observations and archival research. The first major data release is scheduled for October, 2004, which will support archival research.

GALEX surveys the sky simultaneously in two bands, far ultraviolet (FUV, 1350-1800Å), and near ultraviolet (NUV, 1800-2700Å) (see [5] for a detailed description). It uses a dichroic beam splitter, large-format detectors, and a novel optical design to provide a circular field of view of 1.2 degrees. Observations are performed during 25-minute orbital nights outside the SAA. In imaging mode, GALEX obtains a resolution of ~5 arcsec FWHM, with 5 sigma photon-limited sensitivity for point sources of m_{AB} ~20.5 in 100 seconds. In grism mode, spectral resolution of R 100-250 is obtained over the full bandpass for every object in the field of view. The grism is rotated to a distinct position angle each eclipse to mitigate spectrum overlap. Time-tagged single photon data are stored on-board and telemetered to the ground during 3-4 daily ground station contacts. All pointings are staring and dithered in a 1 arcminute spiral pattern to minimize bright star fatigue and small scale efficiency fluctuations. Night sky backgrounds are low: the FUV[NUV] is dominated by diffuse galactic light[zodiacal light], with typical levels 27.5[26.5] mag arcsec^{-2} corresponding to 3[30] photon/PSF in one eclipse. Surveys become background limited at m ~23.5. Astrometric performance is about 1 arcsec (rms), and will be further improved. The GALEX data analysis pipeline uses aspect data and star fields to determine pointing vs. time and generate images, spectra and source catalogs. A spectroscopic module uses image source catalog inputs to extract spectra of these sources from the multiple slitless grism observations.

The GALEX primary mission addresses two overarching questions: How does the UV trace global star formation? and What is the star formation rate (SFR) in galaxies, and how does it evolve over $0 < z < 2$?. What are the physical drivers of star formation in galaxies over $0 < z < 2$?. To address these questions, the GALEX primary mission interleaves multiple imaging and spectroscopic surveys, sampling a wide range of sensitivities and angular coverage, and overlapping numerous surveys at other wavelengths.

All-sky Imaging Survey (AIS). The goal of the AIS is to survey the entire sky subject to a sensitivity of m_{AB}=20.5, comparable to the POSS II (m_{AB}=21) and SDSS spectroscopic (m_{AB}=17.6) limits. Several hundred to 1000 objects are detected in each 1 deg^2 field. The AIS is performed in roughly ten 100-second pointed exposures per eclipse (10 deg^2 per eclipse). AIS targets are constrained by the current bright star limit for the photon-counting detectors (5 kcps) which makes 50% of the sky inaccessible. As we gain confidence in the safety and calibration stability of the detectors, we plan to raise the count rate limit to recover these areas.

Medium Imaging Survey (MIS). In the baseline mission plan, the MIS will cover 1000 deg^2, with extensive overlap of the Sloan Digital Sky Survey North Galactic and Equatorial survey areas, along with the 2dF galaxy survey, ROSAT, and FIRST. MIS exposures are a single eclipse, typically 1500 seconds, obtaining limits of m_{AB}=23.5, net several thousand objects, and are well-matched to SDSS photometric limits.

Deep Imaging Survey (DIS). The DIS consists of 20 orbit (30 ksec, m_{AB} ~25) exposures, over 80 deg^2, located in regions where major multiwavelength efforts are already underway. DIS regions have high galactic latitude, low extinction, low diffuse galactic light, low zodiacal light, contiguous pointings of 10 deg^2 to obtain large cosmic volumes, and minimal bright stars. DIS regions overlap DEEP, VVDS, NOAO/DWS, CDFS, XMM/LSS, COSMOS, Keck/LBG, CFHTL, Spitzer/First Look Survey, SWIRE; Spitzer/GTO; and PHOENIX.

Nearby Galaxies Survey (NGS). The NGS targets well-resolved nearby galaxies for 1-2 eclipses. Surface brightness limits are m_{AB} 27.5 $arcsec^{-2}$, or a star formation rate of $10^{-3} M_{\odot} yr^{-1} kpc^{-2}$. The 200 targets include most galaxies from the Spitzer IR Nearby Galaxy Survey (SINGS), and many other ground-based continuum and emission-line, radio continuum, H I and CO surveys.

Spectroscopic Surveys. The suite of spectroscopic surveys includes (1) the Wide-field Spectroscopic Survey (WSS), which covers the full 80 deg^2 DIS footprint with comparable exposure time (30 ksec), and reaches m_{AB} 20 for S/N 10 spectra; (2) the Medium Spectroscopic Survey (MSS), which covers the high priority central field in each DIS survey region (total 8 deg^2) to m_{AB}=21.5-23, using 300 ksec exposures; and (3) Deep Spectroscopic Survey (DSS) covering 2 deg^2 with 1000 eclipses, to a depth of m_{AB}=23-24.

2 Early Results

The nearby galaxy survey has obtained images of ~100 galaxies. In Figure 1 we show the GALEX image of M31.

GALEX has discovered star formation occurring in vast extended disks around several spiral galaxies (M83, NGC628, Mrk 348, and others) [13]. Study of these disks will lead to new insights into galaxy disk building and the nature of star formation in low-density regimes, a clue to SF triggers in more general contexts. GALEX has also discovered strong UV emission associated with five major galaxy interactions [8]. UV emission is detected far beyond the extent of the visible light (10-100 kpc) and coincides with H-rich tails, particularly at the location of strong density enhancements. Remarkably, GALEX may be detecting the onset of star formation and the formation of new dwarf galaxies in these gas-rich regions.

A number of our initial investigations have concentrated on studying the UV properties of galaxies in the local universe. Color-color diagrams combining GALEX and SDSS bands provide excellent separation and initial object classification into stars, galaxies, and QSOs, as shown in Figure 2. Fundamental plots juxtaposing colors, luminosity, and surface brightness show a large range of properties in UV-selected galaxies, which vary from red to blue ($-2 < FUV - r < 6$), high-to-low surface brightness (with a 1000:1 dynamic range), and a luminosity range of $10^6 < L/L_{\odot} < 10^{11}$. Correlations are immediately apparent between the UV surface brightness, UV luminosity, and UV-to-optical color, as we show in Figure 3.

Using matches to the 2dF redshift survey over 58 deg^2, and corrections obtained by comparing to the SDSS DR1 overlap, we have made a preliminary determination of the local UV luminosity function for an FUV selected sam-

Fig. 1. The Nearby Galaxies Survey mosaic image of M31. NUV is given in orange, FUV in blue.

ple of galaxies [15]. The luminosity function based on this initial sample of 500 galaxies is well fit by a Schechter function over the absolute magnitude range -14>m_{AB}>-21 with a flatter slope and lower characteristic luminosity than previous measurements [10]. This provides the first GALEX measurement of the local universe anchor point for the star formation history plot. One intriguing and unexpected observation is that red UV selected galaxies show stronger evolution than blue [14].

GALEX has discovered a rare population of low redshift starbursts with properties remarkably similar to high-redshift Lyman Break Galaxies (LBGs). These UVLGs resemble Lyman Break Galaxies in terms of size (2-5 kpc), UV luminosity ($L > 2 \times 10^{10} L_{\odot}$), metallicity, dust content, and color. SDSS spectra of these galaxies show a remarkable diversity, from extraordinary emission line spectra, to post-starburst, to red spectral energy distributions with weak to

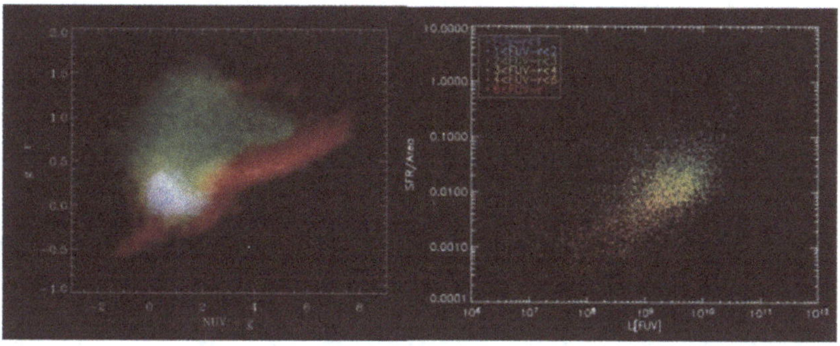

Fig. 2. LEFT: GALEX/SDSS color-color diagram showing galaxies [green] ranging over wide NUV-g color range with flat g-r response for red (NUV-g¿3) galaxies. GALEX/SDSS colors separate galaxies from stars [red] and QSOs [blue]. RIGHT: Plot of SFR per unit area vs. FUV Luminosity, with point colors given by FUV-r color. Significant trends are detected in surface brightness, color, and L[FUV]. For example, the highest SFR is found in the most UV-luminous galaxies, which are also the bluest.

absent emission lines. The latter could be objects with rapidly declining SFR seen at moderate redshift which will become present day E's and S0's.

UV extinction by dust remains the principle correction in converting UV luminosity directly to SFR. A major GALEX goal is to determine observational methods of correcting for dust and deriving the intrinsic UV luminosity and SFR. With IRAS data, we have begun to characterize the properties of FUV and Far-IR selected galaxy samples in the local universe [3]. As expected, flux-limited UV and Far-IR selected samples yield different projections of the bivariate UV/Far-IR luminosity function and distinct UV/Far-IR ratios. However, virtually all star-forming galaxies detected in local Far-IR-selected samples are also detected by GALEX. This is because galaxies with high Far IR/UV ratios are almost always very luminous, and therefore still moderately luminous even in the UV. Starburst galaxies in the local universe display a well-known relationship between UV slope and Far-IR-to-UV luminosity ratio (IRX). Early results from GALEX [11] indicate that 1) normal and quiescent star forming galaxies are likely to fall below the canonical IRX-β relation for starbursts, perhaps due to starburst age; and 2) the difference between starbursts and ULIRGs within IRX-β space is not as distinct as previously suggested, which is consistent with our observation that galaxies span a continuum in the bivariate UV/Far-IR luminosity function. Using the large and homogeneous GALEX samples, we are working to determine whether a generalized IRX-β law holds for all star-forming galaxies.

A major goal of GALEX is to determine the history of star formation over $0 < z < 2$. Step 1 is to determine the evolution of the UV luminosity function over this range. We combine several approaches. Candidate DIS fields overlap ongoing deep optical redshift surveys. We have obtained a preliminary luminosity function for $z < 1$ using 800 galaxies in the XMM/VIMOS field [1]. There is strong evidence for evolution of the UVLF and the cosmic SFR, but the evolution may be weaker than many previous determinations had suggested. We have also demonstrated, using overlap with HST images, that the GALEX angular resolution is sufficient to make an accurate census of the z∼1 universe [9].

UV extinction by dust remains the principle correction in converting UV luminosity directly to SFR. A major GALEX goal is to determine observational methods of correcting for dust and deriving the intrinsic UV luminosity and SFR. With IRAS data, we have begun to characterize the properties of FUV and Far-IR selected galaxy samples [3] in the local universe. As expected, flux-limited UV and Far-IR selected samples yield different projections of the bivariate UV/Far-IR luminosity function and distinct UV/Far-IR ratios. However, virtually all star-forming galaxies detected in local Far-IR selected samples are also detected by GALEX, since heavily extincted galaxies are almost always very luminous. Starburst galaxies in the local universe display a well-known relationship between UV slope (β) and Far-IR-to-UV luminosity ratio (IRX) [7]. Early results from GALEX [11] indicate that 1) normal and quiescent star forming galaxies are likely to fall below the canonical IRX-β relation for starbursts [2], perhaps due to starburst age [4]; and 2) the difference between starbursts and ULIRGS within IRX-β space is not as distinct as previously suggested, which is consistent with our observations that galaxies span a continuum in the bivariate UV/Far-IR luminosity function [6]. Using the large and homogeneous GALEX samples, we are working to determine whether a generalized IRX-β law holds for all star-forming galaxies [11].

GALEX (Galaxy Evolution Explorer) is a NASA Small Explorer, launched in April 2003. We gratefully acknowledge NASA's support for construction, operation, and science analysis for the GALEX mission. We also gratefully acknowledge the outstanding work done by the GALEX Implementation Team.

References

1. Arnouts, S. & The GALEX and VIMOS Teams, 2004, GALEX Ap.J.Letters, in prep.
2. Bell, E. F., 2002. Ap. J., 581, 1013.
3. Buat, V. & The GALEX Team, 2004, GALEX Ap.J.Letters, in prep.
4. Kong, X., Charlot, S., Brinchmann, J., & Fall, M. 2004, astro-ph/0312474.
5. Martin, C., et *al.*2003. Proc. SPIE 4854, 336.
6. Martin, C., & The GALEX Team, 2004, GALEX Ap.J.Letters, in prep.
7. Meurer, G. R., Heckman, T. M., & Calizetti, D., 1999. Ap. J., 521, 64.
8. Neff, S. & The GALEX Team, 2004, GALEX Ap.J.Letters, in prep.
9. Schiminovich, D. & The GALEX Team, 2004, GALEX Ap.J.Letters, in prep.
10. Sullivan, M. et *al.*, 2000. MNRAS, 312, 442.
11. Seibert, M. & The GALEX Team, 2004, GALEX Ap.J.Letters, in prep.
12. Steidel, C. C., Adelberger, K. L., Shapley, A. E., Pettini, M., Dickinson, M., & Giavalisco, M. 2003, Ap.J., 592, 728
13. Thilker, D. & The GALEX Team, 2004, GALEX Ap.J.Letters, in prep.
14. Treyer, M., Wyder, T. & The GALEX Team, 2004, GALEX Ap.J.Letters, in prep.
15. Wyder, T., Treyer, M., & The GALEX Team, 2004, GALEX Ap.J.Letters, in prep.

Science with the *Galaxy Evolution Explorer*: Starbursts and Stellar Populations

R. Michael Rich

Department of Physics and Astronomy, UCLA, Los Angeles CA 90095-1562

Abstract. The *Galaxy Evolution Explorer* was launched in April 2003 and has returned wide field UV imagery in a far UV (1450A) and near UV (2800A) channel. We report on new results in the study of starbursts and stellar populations. Galex detects light from recently formed massive stars, and also from hot evolved older stars. Combining SDSS and Galex photometry gives a large multiwavelength dataset. Fitting the SEDs to population models in a statistical sense gives constraints on the starburst history of galaxies. Galex is remarkably sensitive to low levels of star formation, some of which is detected in the red sequence of ellipticals and bulges, even to $10^{-3} M_{\odot} yr^{-1}$. The UV rising flux is clearly seen in nearby bulges and elliptical, and is also detected to $z \sim 0.1$ in A2670. Constructing a sample of ellipticals selected from SDSS to lack emission lines to appear in all respects "old, red, and dead" we find no correlation between the UV rising flux and the Lick Mg_2 index, or with any other measurable parameter such as velocity dispersion; this finding contradicts some earlier studies.

1 Introduction

The *Galaxy Evolution Explorer* was launched in April 2003 and began its first science observations in July. This review addresses our modeling of the starburst history of the local Universe and our study of the ultraviolet rising flux in old stellar populations. GALEX is described in the accompanying article [1]. Images in the far-UV (FUV; λ_{eff} = 1516 Å, $\Delta\lambda$=268Å) and near UV (NUV; λ_{eff}=2267Å, $\Delta\lambda$=268Å) bandpasses are obtained simultaneously over a 1.2° diameter circular field. The detector is a photon counting MAMA detector and consequently there is no "red leak": only UV photons can trigger the photoelectrons. By selecting a grism instead of the default dichroic, low resolution spectroscopy in both bandpasses can be obtained for every object in the field.

Although UV light suffers from extinction, observations in the UV are powerful probes of star formation history because the far UV flux of a starburst declines relatively slowly compared to $H\alpha$ emission. Only the most massive O stars produce copious numbers of the > 13.6eV photones needed to ionize hydrogen, while the UV continuum is produced by a wide range of stellar masses. While $H\alpha$ flux falls by an order of magnitude within few$\times 10^6$ yr, it takes 5×10^8 yr for the UV continuum to fall that much; GALEX may be most powerful in its study of these 'aged' starbursts. Given the importance of mergers and gas infall in driving the evolution of galaxies, the ultraviolet view of the Universe gives a powerful insight into galaxy evolution.

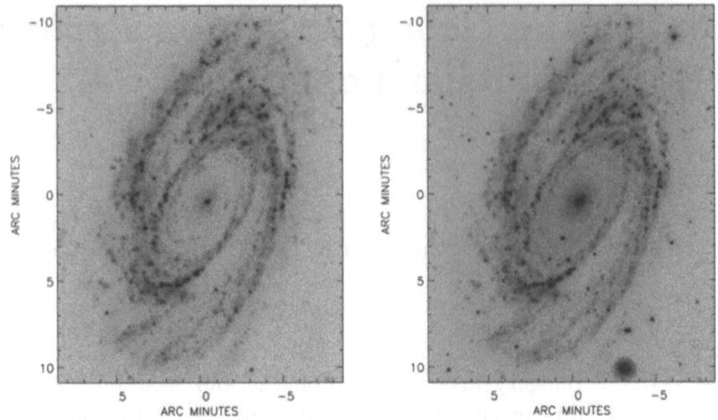

Fig. 1. M81 as imaged with *GALEX* showing the power of wide field UV imaging to delineate both the youngest and oldest stellar populations. The left panel illustrates the FUV (1200-1800A) image, and the right panel shows the NUV (1800-2800A) image– $12' \times 22'$, North up and East left. The star forming regions of the disk are detected clearly in both bands. The nuclear bulge is also detected, but the light of the bulge arises from evolved stars in > 10 Gyr old stars; the nuclear AGN may contribute a portion of the far UV light in the starlike nucleus. Halos around bright stars are artifacts of the detectors. The full 1.2° GALEX field of view is large enough to image both M81 and M82. Figure courtesy M. Seibert

Fig. 1 shows how GALEX views the Sb spiral galaxy M81. It is important to recall that while the light comprising the spiral arms is due to star formation, the nucleus and bulge shine by the light of > 10 Gyr old evolved stars (the UV rising flux). Note that the problem of the UV rising flux (UVX) in old stellar populations remains unsolved [2].

2 The Starburst History of the Local Universe

The Sloan Digital Sky Survey has proven crucial to using GALEX to interpret the properties of local starburst galaxies. SDSS gives optical counterparts and spectra. Our Medium Imaging Survey (MIS) data is an excellent complement with SDSS, reaching m(AB)$\sim$ 23.5 in one 1500 sec shadow "orbit". One of our major programs addresses the starburst history of the local Universe ($0 < z < 0.25$). We consider galaxies from the SDSS Main Galaxy Sample for which we also have GALEX MIS photometry and aim to constrain the star formation histories, stellar masses, and extinction in these galaxies. Following methods developed by Kauffmann et al. [4],[5] we generate a library of $> 10^4$ Monte Carlo realizations of different star formation histories based on the Bruzual-Charlot [6] models, including starbursts of varying strength and decay time spanning spanning $-1 < [Fe/H] < +0.3$. We fit the seven flux points (FUV, NUV, *ugriz*) and generate median likelihood estimates of burst mass fractions, the time since

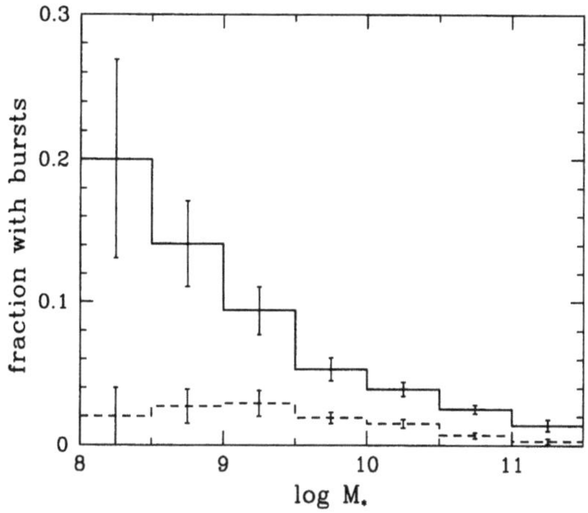

Fig. 2. Fraction of galaxies of a given stellar mass hosting a burst of star formation younger than a given age: *Solid line:* < 1 Gyr; *dashed line:* < 100 Myr [3]. These results from fitting 7 Galex+SDSS flux points with Bruzual-Charlot stellar population models (to get mass and age of the most recent starburst). The sample includes galaxies which have SDSS spectroscopic redshifts and Galex detections in the Medium Imaging survey (6500 galaxies). The highest redshift galaxy included in this plot is $z = 0.25$. Notice that bursts tend to be found in less massive galaxies. The trends in this plot confirm the findings by Kauffmann and collaborators [4],[5] that active star formation is largely confined to galaxies less massive than $3 \times 10^{10} M_{\odot}$. Courtesy S. Salim.

the most recent starburst, dust attenuation strengths, stellar masses and stellar mass-to-light ratios for our present sample of $\sim$ 6500 galaxies. This sample will ultimately grow dramatically, as the area of GALEX MIS coverage increases and as we include galaxies with photometric redshifts. Note that although it is in principle possible to estimate a photometric redshift along with these parameters, we believe that it is better to address the physical parameters independent of the redshift. The added UV flux points from GALEX result appear to improve estimates of attenuation and star formation rate, compared to doing the exercise using only the SDSS data. Figure 2 shows that the least massive galaxies in our sample are most likely to have experienced a recent (or ongoing) burst of star formation. The behavior in Figure 2 follow the general trends established by Kauffmann et al. [4],[5] that active star formation is generally found in galaxies less massive than $10^{10}M_{\odot}$.

2.1 Blue Light in Old Galaxies

GALEX detects as little as $10^{-3} M_{\odot}$ of star formation when there is relatively low extinction [3]. Galaxies nominally in the "red sequence" show UV light sufficiently strong that it cannot be ascribed to the UV rising flux. It is known that red galaxies show emission characteristic of star formation; from their emission lines on the BPT [7] diagram, these galaxies appear to be star forming. At present we are not sure whether ellipticals with star formation-related UV light also lack emission lines (i.e. a new class of star forming red galaxies). However our study [3] of morphological ellipticals finds that excess UV is always accompanied by emission lines.

Figure 3 shows that GALEX sees large dispersion in the NUV vs. r color-magnitude relationship [8]. There appear to be a subset of red galaxies with active current star formation comprising roughly 1% of the mass. In the sample of 41 luminous ($M_r < -22$) red galaxies considered, some 20% have UV flux strong enough to place them in this category. Only a subset of these would have experienced bursts strong enough so that Balmer lines could be detected > 1 Gyr on, resulting in E+A galaxies. Given the recent proposal that red sequence galaxies grow a factor of two in mass from $z = 1$ to the present [9], the discovery of star formation in the red sequence argues that we may be observing such growth "in the act". We aim to determine its prevalence and dependence on environment and lookback time. What are these galaxies and what is causing their star formation? E+A galaxies in the field [10] have companions, which suggests a merger-induced origin for these objects. While selected from a sample [11] of galaxies with elliptical-like morphology, low levels of star formation are found from the emission lines remaining after subtraction of the stellar continuum (see Figure 3).

3 Old Stellar Populations and the UV Rising Flux

The discovery of the UV rising flux (UVX) in elliptical galaxies in 1969 [12] was one of the first important space-based discoveries in stellar populations. Evolved horizontal branch (HB) stars are strongly suspected as the source; since the bluest HB stars are seen in only the oldest stellar systems one expects to see the UVX fade dramatically with only modest lookback time. Galex imaging of the rich M31 globular cluster system is our initial search for a clue, seeking metal rich globular clusters with a far-UV component, analogs of NGC 6388 and NGC 6441 [13]. A GALEX mosaic covering 15 sq. deg. on M32 detects 100 globular clusters. In their UV luminosities and in plots of UV-V color vs. [Fe/H], the clusters of M31 resemble closely those of the Milky Way, however among the ~ 100 brightest clusters we see no old metal rich clusters with far UV flux [14].

Do old, red, dead ellipticals have UVX and what is the dependence of UVX on galaxy properties? Our investigation [15] develops a sample of 1800 galaxies which overlap our GALEX imaging, but have concentrated elliptical-like morphology and $z < 0.25$. Of these only half have FUV flux. We require $S/N > 10$ for the SDSS spectrum; when the stellar continuum is subtracted from these galaxies using the approach of J. Brinchmann, low level residual emission lines reveal either AGN or star forming activity. Our present sample of 120 old, red, quiescent ellipticals with no emission lines and red spectra shows *no correlation between UVX and any Lick line strength index, velocity dispersion, or luminosity, in contradiction to the trends previously established for nearby galaxies [16].*

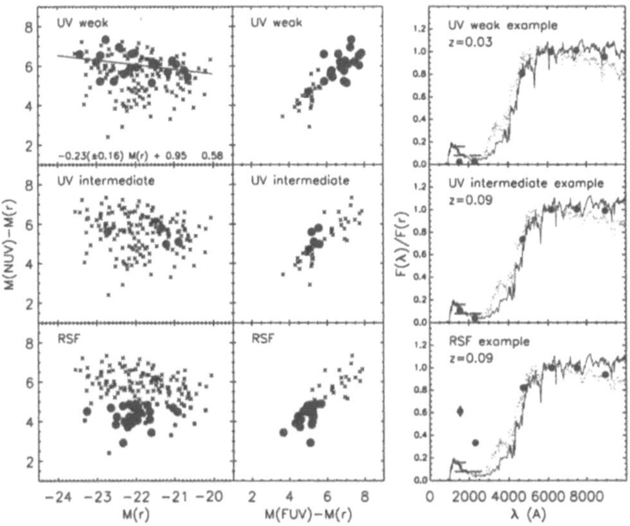

Fig. 3. GALEX photometry clearly reveals star forming galaxies in the red sequence [8]. The left panel shows the scatter in the GALEX NUV vs SDSS r color-magnitude relationship. The right panel shows three example SEDs with a red galaxy spectrum superposed. Quiescent red populations will either fall in the quiescent or UV rising flux category, which have UV light from evolved old stars. The 'RSFS' or recent star formation signature galaxy has more UV flux than can plausibly be accounted for by evolved stars; such galaxies also show the most dramatic excursion from the color-magnitude relationship. Courtesy S. Yi

The large range in FUV-g color, from 2 to > 8 is striking. The UVX would appear to have a huge range with little obvious controlling parameter in sight (certainly not metallicity or galaxy mass). With GALEX observations of the Virgo cluster just complete, we are undertaking a similar study which will focus on the nearest E, S0, and Sb bulge populations.

The origin of the UVX remains unclear. Direct UV imaging of M32 [17] finds no evolved bright HB stars or PAGB stars (attractive candidates due to their lifetimes). The lack of these stars is problematic for virtually every theoretical study [18],[21] to date. GALEX imaging of A2670 compared with that of the Fornax cluster finds a hint of evolution in the UVX [19] but the large dispersion in UVX at any given wavelength complicates greatly any secure detection of evolution (Figure 4).

GALEX gives us an unprecedented ultraviolet vision of the Universe: A huge 1.2° diameter field of view, solar blind imaging, extraordinary overall sensitivity, and special sensitivity to low surface brightness emission. In contrast to the Hα based surveys for star formation, GALEX is sensitive to bursts with timescales as old as 10^8 yr, capable of probing star formation over the widest range of hosts

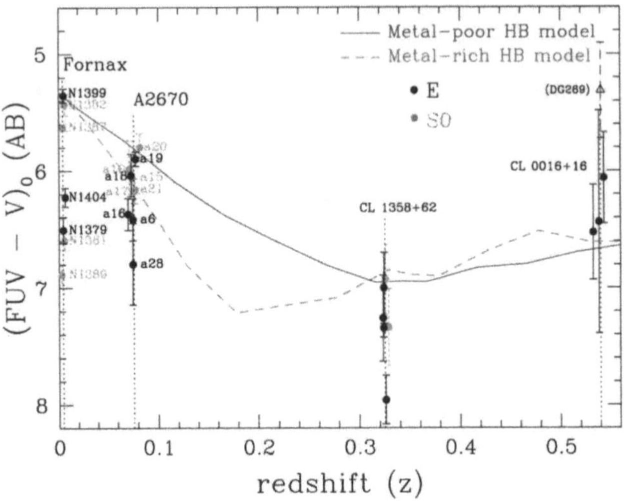

Fig. 4. UV rising flux $(1500 - V)$ as a function of redshift [19]. The $z \approx 0$ points are GALEX photometry of the Fornax cluster, while A2670 is derived from our 10 orbit integration. The cluster at $z = 0.3$ is measured from archival HST imaging [20]. *Models: [21]* for red HB models, the UVX is from metal rich HB blue HB stars in elliptical galaxies with median age younger or of order that of the Galactic halo [18]; in metal poor HB models, the UV flux arises from metal poor stars with age $\approx 30\% >$ than the median Galactic halo age. Courtesy Y-W.Lee.

and environments. Finally, GALEX offers the best possibility of studying the unsolved UVX emission in the Local Universe.

Acknowledgements

The *Galaxy Evolution Explorer* (GALEX) is a NASA Small Explorer, launched in April 2003. We gratefully acknowledge NASA's support for construction, operation, and science analysis for the GALEX mission. The author thanks GALEX collaborators and co-I's M. Seibert, S. Salim, S. Yi, Y. Lee, and T. Small for providing figures and data prior to publication.

References

1. C. Martin, this volume.
2. R.W.O'Connell: Ann.Rev.Astron.Astrophys. **37**, 603 (1999)
3. Salim, S., Rich, R.M., Charlot, S. et al. ApJL in press (2004)
4. G.Kauffmann, T.M.Heckman, S.D.M.White et al. MNRAS **341**, 33 (2003)
5. G.Kauffmann, T.M.Heckman, S.D.M.White et al. MNRAS **341**, 54 (2003)
6. G.Bruzual, S. Charlot, MNRAS, **344**, 1000 (2003)
7. J.A. Baldwin, M.M. Phillips, R. Terlevich: PASP **93**, 5 (1981)

8. S. Yi, J.-J. Yoon et al. ApJL in press (2004)
9. E.F. Bell, D.H. McIntosh, M. Barden et al. ApJ **600**, L11 (2004)
10. K-V.H.Tran, M. Franx, G.D. Illingworth et al. astro-ph/0403484
11. M. Bernardi, R.K. Sheth, J. Arnis, S. Burles et al. AJ **125**, 1817 (2003)
12. A.D.Code: PASP **81** 475
13. R.M.Rich,C.Sosin,S.G.Djorgovski,G.Piotto, I.King et al. ApJ **484**, L25 (1997)
14. S.-C.Rey, R.M. Rich et al. ApJ Letters in press (2004)
15. R.M.Rich, S.Salim, S. Charlot, J. Brinchmann et al. ApJ Letters in press (2004)
16. D. Burstein, F. Bertola, L.M. Buson, S.M.Faber, T.R. Lauer ApJ **328**, 440 (1988)
17. T.M.Brown astro-ph//0308509
18. L.Greggio, A. Renzini ApJ **364**, 35 (1990)
19. Y-W.Lee, R.M.Rich, Y-J.Sohn, S.-C.Rey et al. ApJ Letters in press (2004)
20. Brown, T.M., Ferguson,H.C.,Smith,E. et al. ApJ **592**, L17 (2003)
21. Yi,S., Lee,Y.-W., et al. ApJ **513**, 128 (1999)

Mid-Ultraviolet Spectral Diagnostics of Galaxy Evolution

Sara R. Heap[1] and Thierry Lanz[1,2]

[1] NASA's Goddard Space Flight Center, Greenbelt MD, USA
[2] University of Maryland, College Park MD, USA

Abstract. We describe how optical spectra of $z = 1 - 2$ galaxies can help test the hypothesis of hierarchical galaxy formation.

1 Introduction

The Redshift Desert of galaxies ($z \cong 1 - 2.5$) is rapidly being converted into the Redshift Riviera thanks to new, deep galaxy surveys at Keck, Gemini, VLT, and other large, ground-based telescopes. As we have seen (Steidel, this conference; Crampton, this conference), this redshift region is rich in star-forming galaxies, which can be used to track the progress of galaxy assembly. The Redshift Riviera is also rich in somewhat older galaxies, whose restframe mid-UV spectrum gives clues to the formation and evolution of galaxies. It is these intermediate-age galaxies that we wish to discuss today. As shown in Figure 1, intermediate-age galaxies can be at any age up to 3 Gyr old at $z \sim 2.5$, and up to 6 Gyr old at $z \sim 1$, depending on the redshift of formation. By our definition, intermediate-age galaxies are older than about 0.5 Gyr.

Figure 2 shows the isochrone of a typical intermediate-age stellar population. As happens for all populations younger than 6 Gyr, the hottest stars are those at the Main-Sequence Turn-off (MSTO). (Stellar populations older than 6 Gyr have horizontal-branch stars, which could be hotter than the MSTO stars.) Since the wavelength of peak flux decreases with increasing temperature, there will be a spectral region where MSTO stars – and only MSTO stars – are bright. For intermediate-age galaxies, that region is the mid-UV (2000-3000 Å), which at $z > 1$ is conveniently redshifted into the domain where large, groundbased telescopes are sensitive. By viewing galaxies in the restframe mid-UV, we can cleanly isolate the MSTO stars, which are the timekeepers of stellar evolution. The restframe mid-UV spectrum of such a stellar population looks like the spectrum of a single F-type star, so estimating the age of that population should be no harder than analyzing the spectrum of a single star.

Of course, galaxies are not single stellar populations. Particularly under the hypothesis of hierarchical galaxy formation, massive galaxies are composite: they consist of many generations of stars with the oldest ones generally being the most metal-poor. In this more complex scenario, the MSTO stars producing the mid-UV spectrum belong to the youngest stellar population of the galaxy. In other words, the mid-UV spectrum gives the age since the most recent episode of star-formation, not the age of the galaxy.

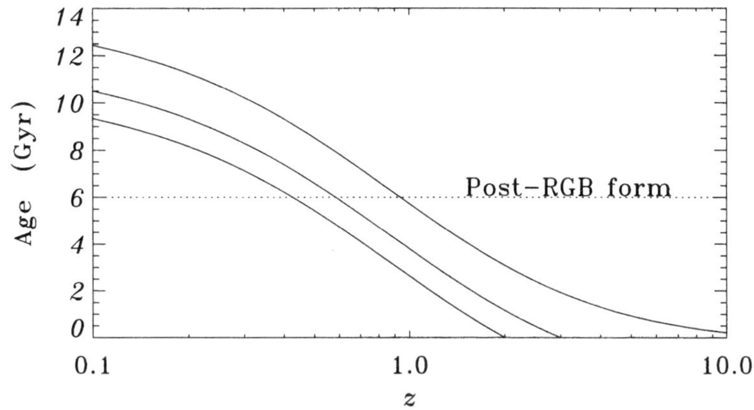

Fig. 1. Galaxy ages as a function of redshift for three different formation redshifts: $z_f =$ 2, 3, and 15. The concordance cosmology is assumed: $\Omega_M = 0.3$, $\Omega_\Lambda = 0.7$, $H_0 = 67$. The dotted line shows the age at which post-RGB stars form.

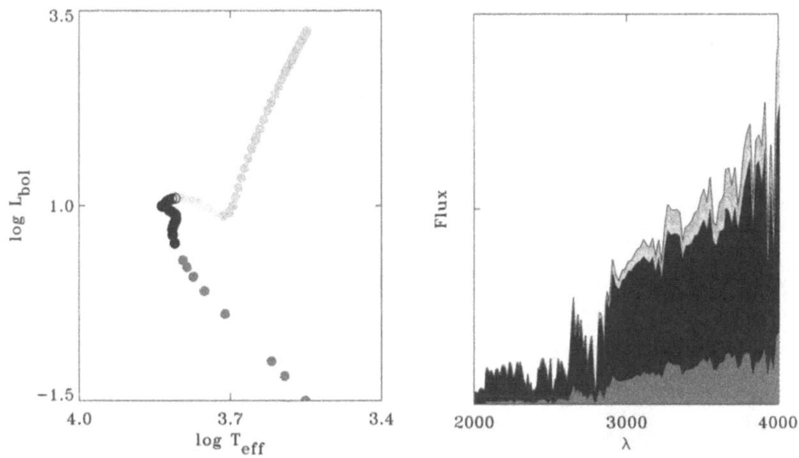

Fig. 2. Production of the mid-UV flux by MSTO stars. Left: isochrone of a 2 Gyr-old stellar population with a solar metallicity. Main sequence stars, stars on the MSTO (black), subgiants, and red-giants are shown in different shades of gray. Right: contributions to the SED by the four stellar components. Below 3000 Å, MSTO stars contribute nearly all the flux.

This more restricted meaning of age does not put us at a disadvantage. In fact, it helps provide a direct observational test of hierarchical galaxy formation. Please refer back to Figure 1. If we find giant ellipticals at $z \sim 1$ with mid-UV ages of 5-6 Gyr, then we will have proof that at least some massive galaxies were fully formed at a very early epoch. If instead we find that $z \sim 1$ galaxies have ages less than ~ 3.5 Gyr, then we can be confident that hierarchical galaxy formation is at work, and that merger-induced star-formation occurred as recently as $z_f \sim 3$.

The specific challenge before us boils down to determining the ages of $z \cong 1-2.5$ galaxies with an uncertainty of only about a Gyr. How hard is that? Estimating the ages from spectra of cleanly isolated MSTO stars would be straightforward were it not for the age-metallicity degeneracy, as we discuss below.

2 The Age-Metallicity Degeneracy

Figure 3 shows two sets of isochrones on the HR diagram to illustrate the age-metallicity degeneracy. The important parameter is the effective temperature of the MSTO. (The luminosity of MSTO stars is not relevant to our problem, since it only indicates the level of star-formation in the youngest generation of stars.) We are all familiar with the "banana-peel diagram," which shows how the MSTO moves to lower temperatures with increasing age, so we will not dwell on it here. We are less familiar with the "artichoke diagram," which shows that the MSTO also moves to lower temperatures with increasing metallicity. Hence, a young, metal-rich population can have the same MSTO temperature as an old, metal-poor population. Because of this degeneracy, we have to determine both temperature and metallicity of the MSTO stars to determine the time since the most recent star-formation episode. How can we do that? And can we infer T_{eff} and $\log Z$ of the MSTO from the mid-UV spectrum alone?

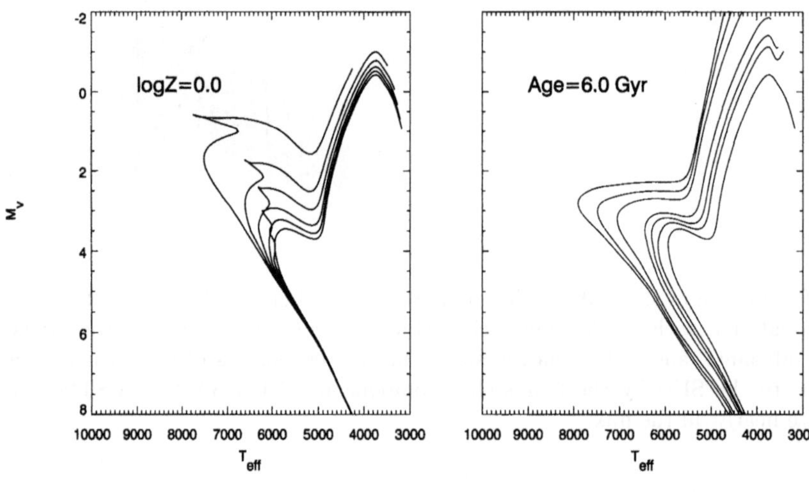

Fig. 3. Isochrones for intermediate-age galaxies, taken from Yi et al. (2003). Left: the Banana-Peel Diagram. The ages are (top to bottom): 1, 2, 3, 4, 5, 6 Gyr. All have a solar metallicity. Right: the Artichoke Diagram for stellar populations at an age of 6 Gyr. Values of metallicity are (left to right): Z=0.0001, 0.0004, 0.001, 0.004, 0.007, 0.01, and 0.02.

There have been a number of recent attempts to resolve the age-metallicity degeneracy. Some involve rather a rather complicated chain of reasoning, so in the interests of brevity, we will give only a summary list of approaches.

1. Dorman et al. (2003) argue that mid-UV plus UBV broadband colors can be used to separately determine age and abundance, or at least, narrow their uncertainties.
2. Lanz et al. (1999) found two spectral indices, the strength of MgI $\lambda 2852$ and the mid-UV color, can be used to separate effective temperature and metallicity, and hence to estimate age.
3. Lotz, Ferguson, and Bohlin (2000) found that simple model SED's (single age, single metallicity) can match the observed mid-UV features of globular clusters, but not most galaxies. The lack of agreement suggests that the star-formation histories of elliptical galaxies are complex.
4. Nolan et al. (2003) assert by that they can constrain the ages of elliptical galaxies through analysis of the entire mid-UV spectrum.

In our opinion, the best way to estimate age and metallicity of a galaxy is to follow the statistical approach of Kauffmann et al. (2003) by comparing the observed spectra of many $z \cong 1 - 2.5$ galaxies with spectral models of galaxies with different star-formation histories. These models would be similar to those generated by Kauffman et al. for the SDSS spectral survey of low-z galaxies except that (1) the spectral diagnostics would include features in the restframe mid-UV as well as Kauffman et al.'s spectral indices, $H\delta_A$ and D_{4000}; and (2) a galaxy model would include the metal-enrichment as well as star-formation history of the galaxy.

3 Mid-UV Spectral Analysis and Modeling

We now return to the practical problem of determining temperature and metallicity from the mid-UV spectrum, or equivalently, modeling the mid-UV spectrum of a star with a given $T_{\rm eff}$ and $\log Z$. This is not an easy undertaking - but it is not for lack of strong diagnostic features. As shown in Figure 4, the mid-UV is rich in prominent features that can be use for ionization balance as well as Mg/Fe abundance ratios. Rather, the problems have to do with the theoretical underpinnings of the model spectrum. Peterson et al. (2001) have documented these problems very well. In analyzing high-dispersion STIS spectra of several F-type MSTO stars, they found that they not only had to juggle basic stellar parameters and abundances, but they also had to contend with issues of convection, chromospheric emission, and inadequate UV atomic data, particularly for FeI. This last issue is especially serious, since it would take a long-term effort to develop the atomic data needed for analyzing the spectra of stars cool enough for FeI to be a major ion in the stellar atmosphere. Who would do the work? And who would fund the effort?

There is a way around this impasse, and that is to use the observed spectra of stars whose fundamental parameters (including abundances) are well known from

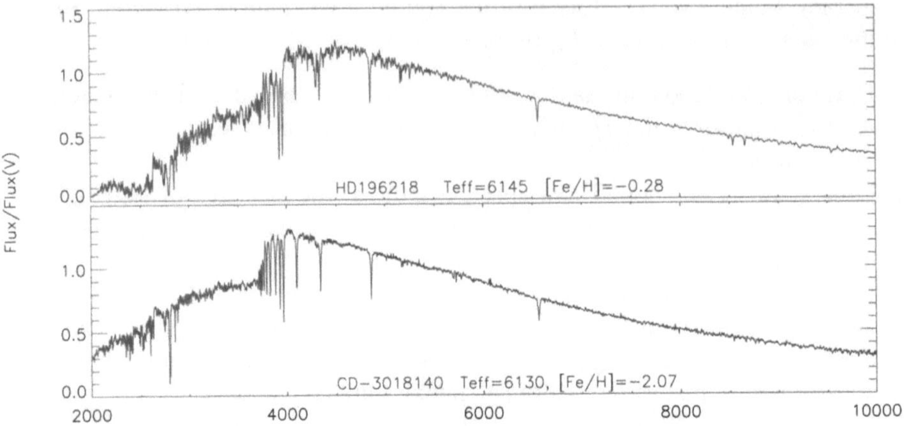

Fig. 4. Sample spectra of two F-type stars on the MSTO from the NGSL. The flux distributions have been normalized to 1 at 5500 Å. Note that the two stars have very different spectra especially in the mid-UV even though they have virtually the same effective temperature. The metal-poor star (bottom) looks bluer and hotter.

careful analysis of the *visual* spectrum. This is the tack we are taking. We are working with Hubble's New Generation Spectral Library (NGSL, PI=Michael Gregg), whose spectra cover the wavelength range from 2000 Å to 10,000 Å. Figure 4 shows some sample spectra from the NGSL. So far, there are over 200 stars in the NGSL, and observations are still being made. Most of the NGSL stars have been analyzed already (c.f. Cayrel et al. 2001), but to form a homogeneous library of high-dispersion visual spectra, Silva and colleagues at ESO are obtaining spectra of NGSL stars with UVES on VLT.

Our overall strategy is the following. We first form a library of stellar spectra using observed spectra where we can. However, since there are no metal-poor stars younger than 8 Gyr in the NGSL, we compute the spectra of these stars. (This is not as bad as it sounds. Metal-poor MSTO stars younger than 6 Gyr generally have effective temperatures above 7000 K. At these higher temperatures, convection is not a major problem; iron is mostly ionized, and the atomic data for FeII are pretty good.) We next compute a large library of Monte Carlo realizations of galaxies with different star-formation and metal-enrichment histories à la Kauffmann et al. We then compute the mid-UV spectra and spectral indices for each of these realizations and compare them to the observed spectra of the galaxy under study to derive the most probable history of that galaxy.

References

Cayrel, G., Soubiran, C., Ralite, N. 2001, A&A 373, 159
Dorman, B., O'Connell, R., Rood, R. 2003, ApJ, 591, 878

Kauffmann, G., Heckman, T., White, S. et al. 2003, MNRAS 341, 54
Lanz, T., Heap, S., Brown, T.M. et al., 1999, *Spectrophotometric Dating of Stars and Galaxies*, ed. Hubeny, I., ASP Conf. Proc., Vol. 192, p. 106
Lotz, J., Ferguson, H., & Bohlin, R.C. 2000, ApJ, 532, 830
Nolan, L., Jiminez, R., Dunlop, J., & Heavens, A. 2003, MNRAS 341, 464
Peterson, R., Dorman, B., & Rood, R. 2001, ApJ 559, 372
Yi, S., Kim, Y.-C., & Demarque, P. 2003, ApjS, 144, 259

Multiwavelength Surveys within the DEEP Fields

David C. Koo[1] and DEEP Team

[1] UCO/Lick Observatory & Department of Astronomy and Astrophysics, University of California, Santa Cruz, CA USA

Abstract. DEEP has launched DEEP2, a major new Keck spectroscopic survey of 50,000 galaxies at redshift $z \sim 1$ using a new spectrograph (DEIMOS). We give an overview of the project's science goals and why DEEP2 is especially attractive to support complementary coverage by a suite of faint multiwavelength imaging surveys. We highlight one such effort already underway in the DEEP2 field called the Extended Groth Strip - (EGS).

1 What is DEEP, DEEP1, DEEP2, & DEIMOS ?

DEEP (Deep Extragalactic Evolutionary Probe) was initiated over 10 years ago to use the Keck 10 m Telescopes for a major spectral survey of faint field galaxies. The use of DEIMOS (DEep Imaging Multi-Object Spectrograph) divides DEEP into two phases (see Table 1). The first (DEEP1) is comprised of a suite of pilot surveys using pre-DEIMOS spectrographs on Keck. DEEP1 was designed to establish the technical feasibility and scientific scope of the second phase (DEEP2) that does use DEIMOS.

DEEP2 is distinguished from prior redshift surveys, including DEEP1, by its large sample of 50,000 galaxies to $R_{AB} \sim 24$. The survey uses BRI two-color diagrams to preselect galaxies near $z \sim 1$ for three of its four fields. The exception is the EGS (see below). A competitive survey currently underway is the VMOS-VLT Deep Survey (see Le Fèvre in these proceedings).

DEEP2 is also distinguished by using spectral resolution high enough to yield rotation curves and linewidths. Such internal kinematics of galaxies provide a powerful new dimension related to masses of galaxies. Such masses are intimately tied to the dark matter halo masses which in turn are the fundamental components best modeled by theoretical simulations of galaxy formation.

The key to DEEP2's success is DEIMOS. This spectrograph allows about 120 targets per mask and uses a huge (8K x 8K) red sensitive CCD mosaic. Together with other design advances, DEIMOS achieves roughly a 7 fold increase in efficiency compared to LRIS. The reader can check papers [2] and [3] for more details on the reduction of data from and performance of DEIMOS, as well as to URL: http://www.ucolick.org/ ~loen/Deimos/deimos.html. For more information about the DEEP, DEEP1, and DEEP2 projects, the reader is referred to URL: http://deep.ucolick.org/ and http://deep.berkeley.edu.

Table 1. DEEP Survey Characteristics

-	DEEP1	DEEP2
Telescopes	Keck I & II; HST	Keck I & II
Instruments [a]	LRIS, ESI, HIRES, NIRSPEC	DEIMOS, LRIS-B
Survey Period	1995-2001 (30 Nights)	2002-2005 (120 Nights)
Fields [b] (FOV)	HDF-N & FF ($8' \times 8'$)	0230+00 ($30' \times 120'$)
-	GSS 1417+52 ($4' \times 42'$)	1417+52 ($16' \times 120'$) EGS
-	SA68 0017+15 (five $4' \times 7'$)	1652+35 ($30' \times 120'$)
-	-	2330+00 ($30' \times 120'$)
-	-	GOODS-N & S
No. Galaxies	1000 to $I \sim 23.5$ (1-4h)	1HS: 50,000 to $R_{AB} \sim 24$ (1h)
& Depth (exp)	-	3HS: few 1000 to $I \sim 24$ (3-10h)
Photometry	KPNO (*UBRI*); HST(*VI*)	UH (*BRI*); HST(TBD)
Science	Spheroid/Bulge Evol.	1HS: non-HST DEEP1 Science &
-	Disk Surface Brightness Evol.	Clustering Evol. at z $\sim$ 1
-	Compact & High z Galaxy Evol.	Lum. Funct. (z, color, etc.)
-	Tully Fisher & Fund. Plane Evol.	Volume Test (Dark Energy)
-	Red & Blue Galaxy Evol.	3HS: HST DEEP1 Science &
-	AGN/Variability/Lum. Funct.	"desert $z \sim 1.4 - 2.5$" galx Evol.
-	Star Form. & Metallicity Evol.	Red Galaxy Age & Metallicity
[a] ESI:	Echelle Spectrograph Imager	
LRIS:	Low Resolution Imaging Spectrog.	B - Blue side (UV sensitive)
HIRES:	High Resolution Echelle Spectrog.	
NIRSPEC:	Near Infrared Spectrog.	
DEIMOS:	DEep Imag. Multi-Obj. Spectrog.	
[b] HDF:	Hubble Deep Field;	N-North & FF-Flanking Fields
GOODS:	Great Obs. Origins Deep Surveys	N-North & S-South
GSS & EGS :	Groth Strip Survey	Extended Groth Strip
SA68:	Selected Area 68	-

2 Science Goals

The main theme that arises from our DEEP1 studies is that galaxy evolution is a complicated problem. Distant field galaxies are diverse in size, luminosity, structure; are composed of subcomponents which experience different star formation and dynamical histories and evolution; and reside in a wide range of environments that are likely to engage different physical mechanisms for their evolution. Results from our DEEP1 programs have established that kinematics are feasible with 10 m class telescopes for distant galaxies, and that such information provide new information not possible otherwise to extract from luminosity and colors alone [5].

DEEP2 will be pushing entirely new ground with its much larger sample spread over four fields. We hope to tackle several scientific puzzles in cosmology today: the nature of dark matter; the nature of dark energy; and the formation and evolution of structure, from dwarf galaxies to clustering and voids on 100 Mpc scales. DEEP's internal kinematics provide dynamical masses and thus dark as well as ordinary matter. Dark energy is traced through the classical volume test, which in turn measures the equation of state (w) of the Universe. By good fortune, the comoving volume density of halo masses of a given circular velocity appears to be virtually independent of cosmology at our redshifts near $z = 1$. Thus, like a standard candle in the supernovae experiments, DEEP is able to use galaxies of given internal kinematics as standard volume tracers [7]. For structure formation, again, internal kinematics will play an important role. Derived dynamical masses can be related to other galaxy properties such as stellar masses, luminosities, colors, metallicity, age, star formation rate, etc.

3 Advantages of DEEP2 for Multiwavelength Coverage

From its inception, DEEP has been designed to be a magnet for surveys at other wavelengths and for followup imaging by the Hubble Space Telescope. DEEP2 is particularly attractive for such intensive coverage because:

1) **Depth:** the DEEP2 survey reaches to $R \sim 24$ and thus beyond redshifts $z \sim 1$ when extensive star formation and AGN activity is expected. This limit nicely matches that from planned deep X-ray, far-UV, mid-IR, far-IR, and radio surveys.

2) **Size:** DEEP2 will have 50,000 spectra. Such large numbers are especially valuable to complement surveys of rarer targets from the X-ray to radio. Examples include AGN's, ultra-luminous infrared galaxies (ULIRGs), and luminous blue compact galaxies (LBCG). A big sample also enables studies as a function of environment, ranging from close pairs, groups, clusters, filaments, to voids. A larger survey is less susceptible to cosmic variance and is thus more reliable in deriving volume densities as a function of , e.g., luminosity, stellar mass, dynamical mass, and color. Finally, the sample size of DEEP2 enables very sensitive measurements of average properties by stacking data. An example of the value of this form of analysis might be the extraction of the average X-ray to optical luminosity ratio as a function of optical colors for non-AGN galaxies (i.e., the source of the X-rays are due to star formation), where even 1000's of sources may be summed in the X-ray maps to pull out the weak signals.

3) **Spectral Resolution:** the survey resolution is roughly 30 km/s. This is more than adequate to probe the internal kinematics of galaxy-scale masses (i.e., dynamical masses if inclinations and sizes are known). These can be directly compared to the stellar mass estimates possible with the IRAC instrument aboard SIRTF. The high resolution also enables the detection of merger signatures via kinematic distortions and asymmetries. Mergers and close pairs are expected to play an important role in the activation of dusty starbursts and AGNs that may dominate the sources at other wavelengths. And, of course, with the ability to detect small velocity dispersions, low mass structures such as groups can be detected and used to study their environments.

4) **Good Quality Spectra:** the large CCD detector used in DEIMOS allows both the high spectral resolution needed for internal kinematics as well as a decent spectral range. A wide range is important to capture a rich set of spectral features that complement key measures from other wavelength surveys. An important example is star formation rates. These can be probed from optical emission lines (Balmer and OII) in DEIMOS spectra and compared to measures of star formation from X-ray, restframe far UV continuum, far-infrared measures of dusty systems, submm signals, and ultra-low-fluxes in the radio. Depending on the redshift and thus specific set of lines that are accessible, metallicity of both the emission line gas (O/H) or stellar population (Fe/H) can be measured in the DEEP2 spectra and compared to SIRTF measurements of stellar mass from IRAC or amount of dust from MIPS. Even the ages of stellar populations or the starburst can be estimated from the optical lines and these are obviously important to be compared to, e.g., the dust estimated from SIRTF or the restframe far-UV upturn from the NUV band of GALEX, respectively.

4 Extended Groth Strip-EGS

While all four of our DEEP2 fields will have some near-IR imaging as well as several optical bands, one at 1417+52 (EGS) will receive the lion's share of very deep multiwavelength coverage. The heart of this field will be a $16' \times 120'$ area to be covered to $R \sim 24$ with DEEP2 spectra. No photometric redshift selection will be made so as to provide a wide range of redshifts and depth to optimize overlap with a diverse set of other deep surveys at other wavelengths. DEEP2 is expecting to complete the bulk of the 15,000 redshifts by end mid-2004, weather permitting.

At present, EGS is outstanding, perhaps unique, for its relatively large area with an extensive set of very deep surveys across the full spectrum. While the COSMOS field is larger (see article by Scoville), the depth of its complementary surveys are shallower, while the GOODS fields are smaller. Commitments in the EGS at other wavebands already include deep X-ray surveys with Chandra to a flux limit of 1.5 x E-16 ergs/s/cm^2 in the central region overlapping a $30'$ diameter circle as well as prior XMM exposures; near-UV imaging to 26 mag AB with GALEX in a roughly central $60'$ region (see articles by C. Martin and D. Schiminovich); a suite of optical imaging data from CFHT (Legacy Survey and Red Cluster Survey 2) and Subaru (R band) as well prior imaging by several groups that fully cover the entire EGS; coverage of a small portion (130 square arcmin) in the southwest with HST WFPC2 V and I (original Groth Survey) as well as some minor additional images with WFPC2 and NICMOS using HST; K band to about 22.5-23 AB for the whole EGS; total coverage in the central $10' \times 120'$ EGS with IRAC on SIRTF to 25 mag at 3.6um and to 22.5 mag at 8 um and with MIPS to 19 mag at 24 um and 13 mag at 160 um; and an existing deep radio survey at 5 Ghz to an rms noise of 2.5 uJy using the VLA over a $16'$ diameter field [4] which will be increased to the entire EGS field. Although the $R \sim 24$ limit of DEEP2's redshift survey is quite faint, the coverage of targets

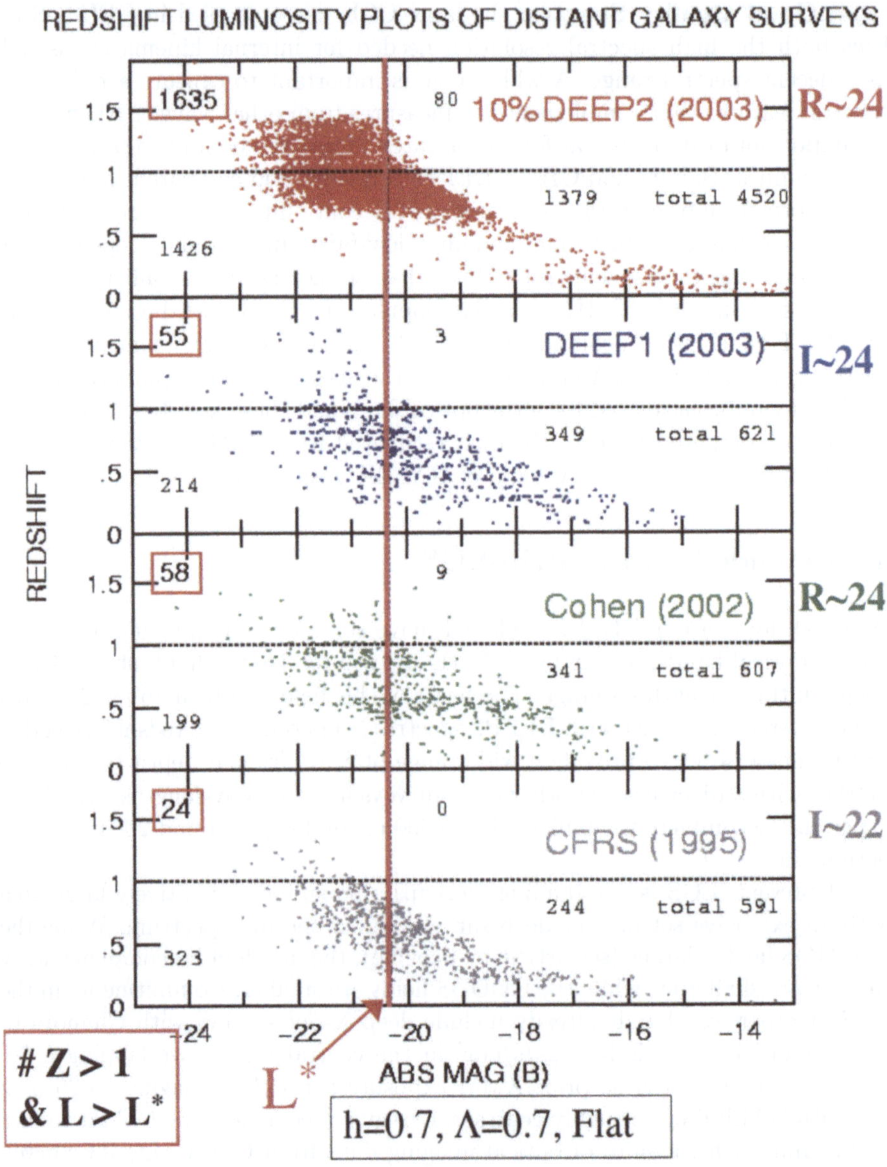

Fig. 1. Redshift vs luminosity (M_B assuming indicated cosmology) for four distant field galaxy surveys as indicated (CFRS: Canada France Redshift Survey [6]; Cohen: Caltech Faint Galaxy Redshift Survey [1]). The numbers on the right-hand side are the magnitude limits in the indicated passbands. Each of the panels is divided into quadrants by a vertical line at L^* of the local luminosity function of field galaxies and a horizontal line at redshift $z = 1$; the number of galaxies from each survey in each quadrant is noted, with the luminous, $z > 1$ sample size specified in a box at the upper left of each subpanel. Note that DEEP2, with only 10% of the survey completed, already has 1635 such galaxies, more than 10× the sum of the other three prior surveys (137).

will not be 100%. Moreover, several of the surveys at other wavelengths are so deep that a significant number of sources will not be within the range of DEEP2. For example, the GALEX team has already examined overlapping sources in a shallower NUV (near-UV) survey with those in the EGS and find a number of objects with extreme UV-optical colors. If such sources persist to GALEX's limit of 26 AB, these will require dedicated spectroscopic followup since they will be beyond the limit of DEEP2.

5 Early Science Results from DEEP2

The DEEP2 1HS program is now 40% complete. Figure 1 shows how DEEP2 compares to three previous faint field galaxy redshift surveys. With only 10% of the DEEP2 program, over 1600 galaxies with luminosities brighter than L^* and redshifts $z > 1$ have secure redshifts. This is well over 10 times larger than the number of such galaxies from the other three surveys *combined*. Figure 1 also shows how well we are able to select high redshift galaxies by using our BRI photometry. The vast fraction are at the desired redshifts $z > 0.7$. We do, however, find a small pool of extremely blue galaxies at redshifts less than 0.7; these contaminants are difficult to eliminate because they overlap blue galaxies at higher redshifts $z > 1$ in the BRI two color plots. This same 10% sample is now being actively worked on for a diverse range of early science (see J. Newman in these proceedings)

6 Summary and Future

DEEP1 is complete. DEIMOS has been fully commissioned and is performing superbly. The 1HS part of DEEP2, with 40% of the data already in hand, is on target for completion of 50,000 galaxies by 2005. The range of science enabled by DEEP, especially with measurements of internal kinematics of galaxies, is enormous, and includes tackling questions related to dark matter, dark energy, and the formation and evolution of galaxies and large-scale structure. With the recent successful launch of GALEX and SIRTF, and having a suite of other complementary multiwavelength data, the EGS promises to provide exciting new results for years to come.

References

1. J. G. Cohen et al.: ApJ, 538, 29 (2000)
2. M. Davis et al. : SPIE, **4834**, 161 (2003)
3. S. M. Faber et. al.: SPIE, **4841**, 1657 (2003)
4. E.B Fomalont, R. A. Windhorst, J. A. Kristian, and K. I. Kellerman: AJ, **102**, 1258 (1991)
5. D. C. Koo & DEEP Team: proc. of IAU 216, ASP Conf. Series, eds. M. Colless & L. Stavely-Smith, in press (2004)
6. S. J. Lilly et al. : ApJ, 455, 50 (1995)
7. J. A. Newman, M. Davis : ApJ, 564, 567 (2002)

The VVDS: Early Results on the Large Scale Structure Distribution of Galaxies out to $z \sim 1.5$

O. Le Fèvre[1], G. Vettolani[2], D. Maccagni[3], J.P. Picat[4], C. Adami[1], M. Arnaboldi[5,12], S. Arnouts[1], S. Bardelli[10], M. Bolzonella[10], M. Bondi[2], D. Bottini[3], G. Busarello[5], A. Cappi[2], P. Ciliegi[10], T. Contini[4], S. Charlot[7], S. Foucaud[3], P. Franzetti[3], B. Garilli[3], I. Gavignaud[4], L. Guzzo[8], O. Ilbert[1], A. Iovino[8], V. Le Brun[1], B. Marano[11], C. Marinoni[1], H.J. McCracken[2], G. Mathez[4], A. Mazure[1], Y. Mellier[6], B. Meneux[1], P. Merluzzi[5], R. Merighi[10], S. Paltani[1], R. Pellò[4], A. Pollo[8], L. Pozzetti[2], M. Radovich[5], D. Rizzo[8], R. Scaramella[9], M. Scodeggio[3], L. Tresse[1], G. Zamorani[10], A. Zanichelli[2], and E. Zucca[10]

[1] Laboratoire d'Astrophysique de Marseille, France
[2] Istituto di Radio-Astronomia - INAF, Bologna, Italy
[3] IASF - INAF, Milano, Italy
[4] Laboratoire d'Astrophysique - Observatoire Midi-Pyrénées, Toulouse, France
[5] Osservatorio Astronomico di Capodimonte, Naples, Italy
[6] Institut d'Astrophysique de Paris, France
[7] Max Planck fur Astrophysik, Garching, Germany
[8] Osservatorio Astronomico di Brera, Milano, Italy
[9] Osservatorio Astronomico di Roma, Italy
[10] Osservatorio Astronomico di Bologna - INAF, Italy
[11] Dipartimento di Astronomia - Universita di Bologna, Italy
[12] Osservatorio Astronomico di Torino, Italy

Abstract. The VIMOS VLT Deep Survey (VVDS) is an on-going program to map the evolution of galaxies, large scale structures and AGNs from the redshift measurement of more than 100000 objects down to a magnitude $I_{AB} = 24$, in combination with a multi-wavelength dataset from radio to X-rays. We present here the first results obtained from more than 20000 spectra. Dedicated effort has been invested to successfully enter the "redshift desert" $1.5 < z < 3$, by producing high S/N galaxy templates in the rest wavelength domain probed by VIMOS. We have produced the luminosity function (LF) of galaxies for the first time out to $I_{AB} = 24$, and the evolution of the LF for each morphological type is clearly established, indicating that most of the LF evolution is concentrated in the later types. We present for the first time the 3D spatial distribution of a well controlled sample of faint galaxies, and show that significant clustering is detected out to $z \sim 1.5$.

1 Introduction

The VIMOS VLT Deep Survey is dedicated to study galaxies, large scale structures and AGN evolution up to $z \sim 5$. To determine the evolutionary status of the galaxy and AGN population as a function of lookback time, it has been recognized early in the survey design phase that the large range in redshifts and the complex physical phenomena probed require to expand the core observations of the VVDS from the visible domain to a broad multi-wavelength spectrum. Observing the rest-frame SED of galaxies (e.g. in the B-band rest frame) at the same wavelength from $z \sim 0$ to $z \sim 5$, and having a sufficiently accurate estimate of the k-correction necessary to compute absolute magnitudes, requires to obtain deep photometric data from the U band to the K band. Furthermore, to get a complete picture of e.g. the star formation process, for galaxies with variable amount of dust extinction, or to study the evolution of large scale structures or AGNs, it is necessary to expand the wavelength baseline of observations from the far UV to the far IR, from the radio-waves to the X-rays.

We describe here the multi-wavelength approach followed by the VVDS, and present early results from the first epoch VVDS observations. We describe the effort to assemble for the first time an unbiased magnitude limited sample of galaxy redshifts and spectra $I_{AB} \leq 24$ up to $z \sim 5$, going through the "redshift desert" $1.5 < z < 2.5$, and present preliminary results on the evolution of galaxies and large scale structures up to $z \sim 1.5$.

2 The Multi-Wavelength VVDS

The VIMOS VLT Deep Survey has been designed from the start as an extensive survey with multi-color BVRI data obtained at CFHT [1], and deep spectroscopy obtained with VIMOS on the VLT [2], forming the core of the survey. The core observations are obtained on several fields, each $2 \times 2\text{deg}^2$: (i) multi-color photometry from B to I band, leading to clean I-band magnitude selected samples and broad band spectral energy distributions (SED), and (ii) multi-object spectra with a goal of ~ 100000 objects to $I_{AB} = 22.5$ and $I_{AB} = 24$.

In addition, our VVDS team has been conducting observations in other wavelength (U and K bands, radio), and has been connecting with other survey groups to stimulate other wavelengths observations in the VVDS fields (Galex-UV, X-rays with XMM and Chandra, SIRTF, see these proceedings).

We have also been advocating the need for the accurate knowledge of galaxy morphologies over large scales with the HST, and we are participating to the HST-COSMOS program set to observe 2deg^2 on the VVDS-10h field with the Advanced Camera for Surveys (ACS) [12].

The current spectroscopic VVDS data has been obtained in November- December 2002 in an observing campaign from which 18 nights were clear. More than 20000 spectra have been obtained, from which about 15000 secure galaxy and AGN redshifts are available in 3 different fields.

3 Accessing the "Redshift Desert" $1.5 < z < 2.5$

The VVDS is the first survey to assemble a complete sample of galaxies based on a simple I-band limit down to $I_{AB} = 24$. Magnitude limited samples have the advantage of a controlled bias in the selection of the target galaxies, which

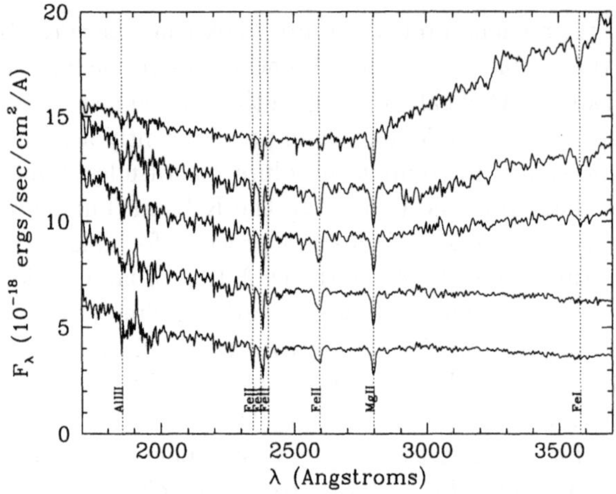

Fig. 1. Blowup of the VVDS galaxy templates in the range 1700 – 3700Å, produced from the combination of high redshift galaxies observed in the range 1.3 – 4 with VIMOS

can lead to a secure census of the galaxy population as seen at a given rest-frame wavelength (see e.g. [4]). The drawback is that, as the magnitudes get fainter, the redshift range gets larger, and identifying redshifts out of a very large range of possibilities becomes increasingly harder from a fixed set of observed wavelengths.

The VVDS default wavelength range for the VIMOS observations is 5500 – 9500Å, observed at a spectral resolution $R \sim 200$ to allow for a large multiplex gain during the multi-slit VIMOS observations. It allows a secure follow-up of the spectral signature of galaxies from [OII]3727Å to the 4000Å region and minimize the bias in the identification of galaxy redshifts up to $\sim$ 1.5. However, measuring the redshifts of galaxies with $1.5 < z < 3$ is quite challenging with generally faint features, and a lack of published observed galaxy templates in the range 1800 – 3000Å to be used in cross-correlation programs such as the KBRED environments developed for the VVDS [10]. Breaking up this "redshift desert" is critical to reduce the incompleteness of deep redshift surveys, and probe the galaxy population at an important time in the evolution.

In a first pass of redshift measurements of the first epoch spectra of the VVDS $I_{AB} = 24$ sample [3], the redshift of ~10% of the galaxies appeared hard to identify while a significant fraction of these spectra had a good continuum S/N. We have setup an iterative approach to identify these objects, based on the creation of rest-frame galaxy templates going as far as possible toward the UV, bridging the gap between the well observed wavelength range above [OII]3727Å and the range below 1800Å with the templates assembled by Shapley et al. [13]. These new templates (see Figure 1) have been included in our redshift measurement engine KBRED and tested against the parent samples of galaxies used to produce them, in the range up to redshift $z = 1.5$, then up to redshifts $z = 2$ and above as new redshifts were being identified.

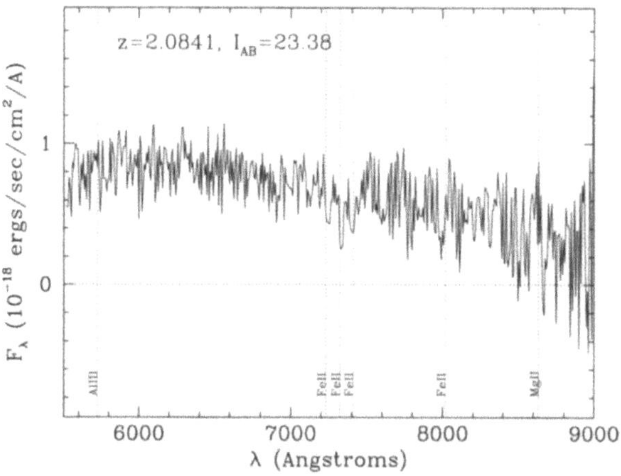

Fig. 2. VVDS galaxy identified in the "redshift desert" at a redshift z=2.0841

This approach has been quite successful on test samples, successfully turning up galaxies at redshifts $1.5 < z < 2.5$ as shown in Figure 2. We are currently going over our data with the expectation to significantly reduce the incompleteness which was observed at the end of our first pass.

4 Galaxy Evolution

A preliminary description of the evolution of the galaxy luminosity function out to $z = 1.5$ has been presented in Le Fèvre et al. [3]. The current VVDS data allows to build e.g. the luminosity function and trace its evolution across many time steps and, furthermore, to break up the population vs. spectral type, or vs. the local galaxy density.

In Figure 3, we present the evolution of the LF of galaxies out to redshift 1.5, for each of 4 spectral types from early to late types. Most of the LF evolution is concentrated in the later types, very strong for the type 4 presented in Figure3, while the LF of early types does not seem to change much over this period. These results, together with the Luminosity Density evolution will be discussed extensively in forthcoming papers [5],[6], [14]. The evolution of the star formation rate from U to K band data will be compared with the star formation indicators obtained at other wavelengths [14], e.g. in the radio [7], or in the far UV with Galex [8].

5 Large Scale Structure Evolution

The VVDS has been designed to trace large scale structures out to ~ 100Mpc at $z \sim 1$. Although our current data from the on-going survey does not yet reach these scales, we have assembled a set of redshifts in an area $\sim 30 \times 30$arcmin^2,

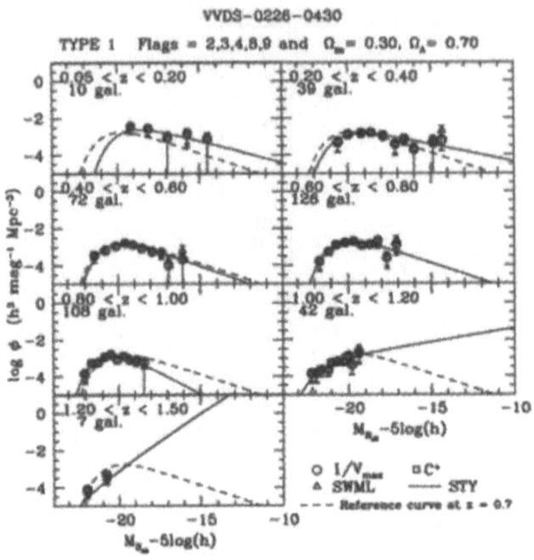

Fig. 3. Evolution of the galaxy luminosity function from the $I_{AB} \leq 24$ VVDS sample, separated in 4 spectral types from early types (upper left) to late types (bottom right). Most of the LF evolution is concentrated in the later type (preliminary, courtesy O. Ilbert).

with $\sim 1/3$ of all galaxies to $I_{AB} = 24$ with a measured redshift, which allows a first pass at tracing the large scale structures.

The preliminary redshift histogram of the VVDS-0226-04 field, shows a dense alternance of strong peaks in the 1D projection (Figure 4). To further detail the 3D distribution of galaxies, we have derived the galaxy density field from this data set using the Wiener filtering technique with a 3 Mpc smoothing scale. The deep probe in this field extends over more than 3000 Mpc, for a total volume sampled of $\sim 3 \times 10^6$Mpc. Several slices from this deep probe are shown in Figure 5. The density field appears to be already significantly clumped by $z \sim 1$ [11]. As an example, we show in Figure 6 a strong overdensity and very complex structure at $z \sim 0.962$ spreading across the full $30 \times 30\text{Mpc}^2$ of the field.

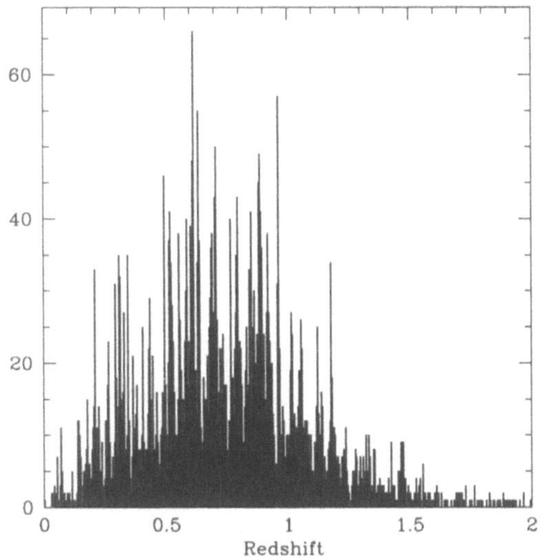

Fig. 4. Redshift distribution of galaxies in the field VVDS-0226-04. A total of 5010 galaxies are represented here from the $I_{AB} \leq 24$ magnitude limited sample. Strong galaxy peaks are traced all the way to $z \sim 1.5$.

The VVDS 0226-04 field has been observed by XMM [9], this will provide the opportunity to cross-correlate the X-ray flux from hot gas in clusters and groups with the structures identified from the redshift measurements.

The VVDS field at 1000+03 is now the focus of an extensive observational effort as part of the HST-COSMOS program [12], including the measurements of more than 50000 redshifts in the $2deg^2$ area to trace the galaxy density field. The other VVDS survey fields at 1400+00 and 2217+00 will continue to be the target of the VVDS team in future observational campaigns to trace the large scale structure distribution out to $z \sim 1.5$ in several independent fields.

6 Summary

The VVDS is providing for the first time a magnitude limited sample of ~ 15000 galaxy redshifts, with ~ 11000 redshift in the $I_{AB} \leq 24$ sample, and ~ 4000 in the $I_{AB} \leq 22.5$ sample, going successfully through the "redshift desert" $1.5 < z < 2.5$. The evolution of the luminosity function up to $z \sim 1.5$ is traced for several spectral type, with most of the evolution concentrated in the later types. Strong overdensities in the distribution of galaxies are visible over scales in excess of 30Mpc up to $z \sim 1.5$. This provides evidence that the galaxy distribution was already strongly structured when the universe was half of its present age.

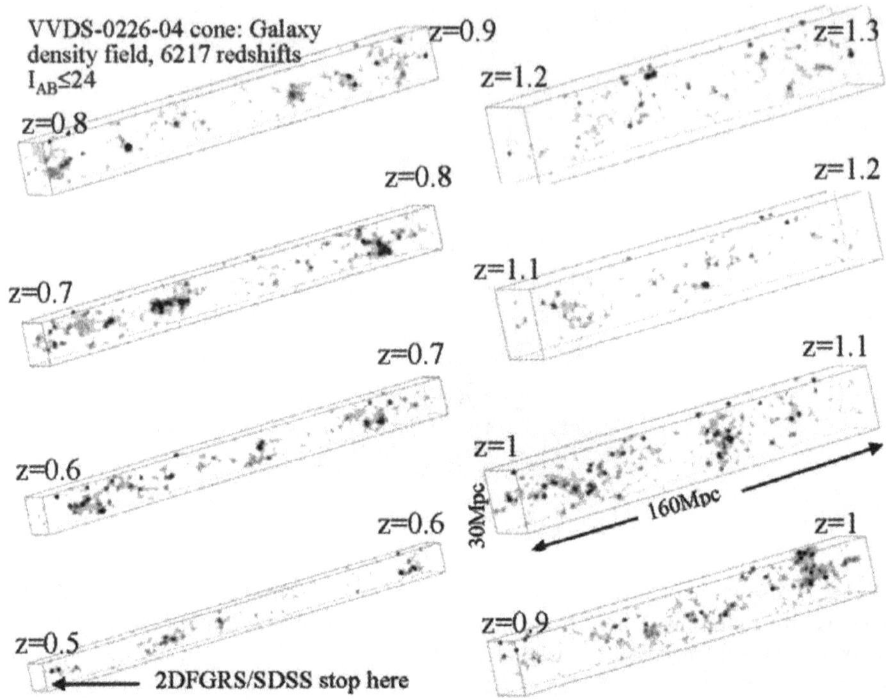

Fig. 5. Galaxy density field in the VVDS 0226-04 field in several redshift boxes up to $z = 1.3$ (preliminary, courtesy C. Marinoni).

Acknowledgements

We thank C. Cesarsky and ESO for the early allocation of VLT Guaranteed Observing Time, and A. Shapley who has provided fits files for her nice UV templates for $\lambda \leq 1700$Å.

References

1. O. Le Fèvre, O., Mellier, Y., et al., 2000, ASP Conf. Series, Clowes, Adamson and Bromage Eds, vol. 232, p.449
2. O. Le Fèvre, O., Vettolani, G., et al., 2001, proceeding of the ESO/ECF/STScI "Deep Field" workshop, Garching, Oct.2000, Renzini, Eds., astro-ph/0101034
3. O. Le Fèvre, O., Vettolani, G., et al., 2003, proceedings of IAU symposium 216 "Maps of the Cosmos", Sydney, July 2003, M. Colless & L. Staveley-Smith Eds., astro-ph/0311475
4. S.J. Lilly CFRS
5. O. Ilbert, L. Tresse, and the VVDS team, 2004, in preparation
6. E. Zucca, and the VVDS team, 2004, in preparation
7. Bondi, M., Ciliegi, P., et al., 2003, A&A, 403, 857
8. C. Martin, et al., 2003, these proceedings

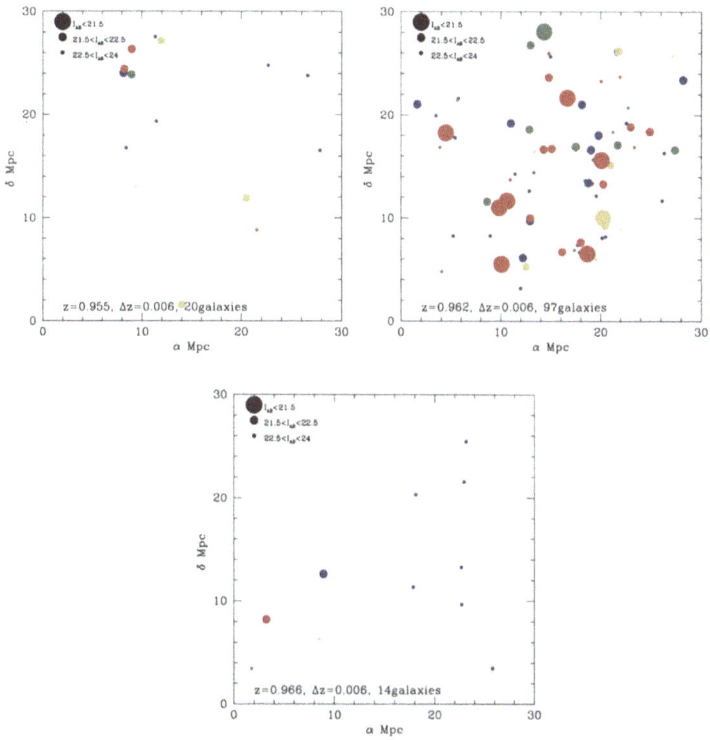

Fig. 6. Strong galaxy overdensity observed in the VVDS 0226-04 field, from left to right: galaxies observed in a $\Delta z = 0.006$ slice at $z = 0.955$, $z = 0.962$, and $z = 0.966$. The central slice has nearly 100 galaxies observed covering the full $30 \times 30\text{Mpc}^2$ in a web-like distribution, while the adjacent slices have only 14 and 20 galaxies respectively.

9. M. Pierre, et al., 2003, A&A, in press, astro-ph/0305191
10. R. Scaramella, and the VVDS team, in preparation
11. C. Marinoni, and the VVDS team, 2004, in preparation
12. N. Scoville, 2003, these proceedings
13. Shapley, et al., 2003, ApJ, in press, astro-ph/0301230
14. Tresse, L., and the VVDS team, 2004, in preparation

Directly Detecting the Evolution of Early-Type Galaxies

S.C. Trager[1], S.M. Faber[2], and A. Dressler[3]

[1] Kapteyn Astronomical Institute, Rijksuniversiteit Groningen, The Netherlands
[2] UCO/Lick Observatories and Department of Astronomy and Astrophysics, University of California, Santa Cruz, CA, USA
[3] The Observatories of the Carnegie Institution of Washington, Pasadena, CA, USA

Abstract. We describe observations focused on understanding the epochs and timescales of the formation and evolution of early-type galaxies, particularly those in clusters. We show that while early-type cluster galaxies are on average older and closer to coeval than their counterparts in the field, significant age spreads (a factor of two in massive ellipticals, more in low-mass galaxies, especially S0's) still exist in these objects. We then show that it is now possible to detect the evolution of early-type galaxies directly by comparing deep absorption-line spectroscopy of intermediate-redshift cluster galaxies with local cluster galaxies. We find that the stellar populations of early-type cluster galaxies do indeed evolve, at a rate that appears to be consistent with the cosmological lookback time.

1 Introduction

A fundamental question in modern astrophysics is understanding the epochs and timescales of galaxy formation and evolution. Spheroidal galaxies – elliptical and S0 galaxies and their close relatives, the bulges of spiral galaxies – constitute between 50% [25] and 75% [9] of the total *stellar* mass of the Universe. Observational evidence of when these galaxies form, how long this process takes, and how this process varies with different factors – specifically mass and environment – are crucial inputs for galaxy formation models.

Elliptical and S0 galaxies are predominantly found in high-density environments [1], with perhaps as many as 80% of ellipticals and 50% of S0's found in clusters [7]. Various lines of evidence suggest that cluster early-type galaxies formed the bulk of their stellar populations at high redshift, including the variation of the mass-to-light ratios [5,12,13,6] and optical and infrared colors [2,14,30] (just to name a few significant papers) of cluster galaxies with redshift. On the other hand, recent studies of the line strengths of early-type galaxies in the general field (primarily loose groups) [34,35] and the Fornax [15], Virgo [4], and possibly even Coma [22,17] clusters suggest that recent episodes of star formation are likely to have occurred, particularly in smaller galaxies. The resolution of this apparent disagreement requires a larger sample of local cluster galaxies with excellent line strength data, particularly in higher density regions like Coma, and line strengths of higher-redshift cluster galaxies.

A closely related problem is the consistency of astrophysical clocks. In essence, do the timescales of the "concordance" cosmological model [29] and those of modern stellar population evolution models [31] match? If we look back to $z = 0.4$, do the spectra of the oldest galaxies appear to be four billion years younger than those of the oldest galaxies today?

In this contribution, we describe observations focused on understanding the epochs and timescales of the formation of giant early-type galaxies, ellipticals and S0's, especially those found in clusters of galaxies. We begin a brief discussion of stellar population models, then discuss in turn the stellar populations of local and distant cluster galaxies. We conclude with some outstanding questions raised by this study and a hint of what is to come.

2 Models of Integrated Spectra: A Brief Overview

Unfortunately, we cannot study the vast majority of early-type galaxies star by star, and so we must be happy with the integrated spectra of these objects. In turn, the interpretation of integrated spectra requires stellar population models. This is an ancient subject [37] with a rich history and literature; a more complete description of stellar population models can be found in, e.g., [38,31,32].

The basic problem with interpreting the integrated spectra of galaxies (particularly those that are not currently forming stars) is that an increase in the stellar opacities of a stellar population by an increase in its metallicity is equivalent to a decreased mean temperature of that population [19]. In other words, a metal-rich population can be simulated by increasing the age – decreasing the mean temperature – of a more metal-poor population. This is the *age–metallicity* degeneracy.

Breaking the age–metallicity degeneracy requires finding a suitable tracer of temperature that is as independent of metallicity as possible. Two groups in the early 1980's [11,23] discovered that the Balmer absorption lines of hydrogen, in particular Hβ, were mostly temperature sensitive. This was reinforced by the comprehensive stellar population models of Worthey [38], which included metallicity effects for the first time. He found that breaking the age–metallicity degeneracy required both a Balmer absorption line and a metal absorption line. A modern version of this approach is shown in the left panel of Fig. 1 [31].

Modern stellar population models require either three or four ingredients. First, a set of stellar evolutionary models is needed to provide the basic astrophysical input to the models, in the form of a set of isochrones – different mass stars of the same age and abundance. Next, an initial mass function is chosen to populate the isochrones at a given age (time). Finally, either an extensive catalog of spectrophotometric stellar spectra or a combination of theoretical stellar fluxes and parameterized absorption-line strengths are needed to produce the integrated spectra or absorption-line strengths of simple (single-age, single-metallicity) stellar populations. Different approaches for each of these ingredients can be seen in, e.g., [38,31,3]. Calibration of stellar population models has re-

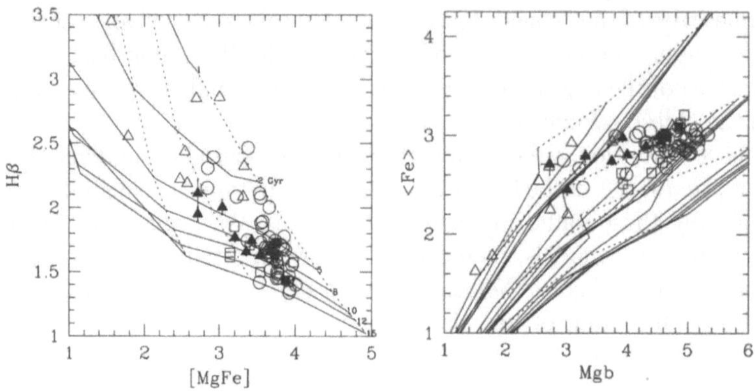

Fig. 1. The stellar populations of local early-type galaxies. (**Left**) Breaking the age–metallicity degeneracy with a Balmer line (here, Hβ) and a metal line (here, the [α/Fe]-insensitive composite index [MgFe]). Open circles are field ellipticals [10], open squares and triangles are ellipticals and S0's, respectively, in the Fornax Cluster [15], and solid squares and triangles are ellipticals and S0's in the Coma Cluster [36]. Solid lines are isochrones (constant ages: 1, 2, 5, 8, 10, 12, 15 Gyrs from top to bottom) and dashed lines are isofers (constant metallicity: [Z/H] = −0.33, 0., +0.67 dex from left to right) from [31]. (**Right**) The abundance ratios [α/Fe] of local early-type galaxies. Symbols as in the left panel; the models represent, from left to right, [α/Fe] = 0, +0.3, and +0.5 dex

cent been performed against both globular and open cluster color–magnitude diagrams, with comfortingly consistent results [27,28,26].

However, the right panel of Fig. 1 shows an additional problem, first noticed in the mid-1970's by a number of researchers [18,8,21] – early-type galaxies have much stronger Mg line strength than local disk stars with the same Fe line strengths. This has been found to be generally true of almost all α-elements [38], in the sense that [α/Fe] $\neq 0$ for giant early-type galaxies [21,34]. Hence, a correction for this deviation from the solar-neighborhood abundance pattern is required. A technique based on a few template synthetic spectra has been developed and applied to a few absorption-line strengths [34]; the right panel of Fig. 1 shows a set of models [31] using this method. A more general, self-consistent solution awaits a broader library of synthetic spectra or a library of suitable stars with similar abundance patterns to giant ellipticals (e.g., the Galactic Bulge).

Finally, there is the question of what is meant by the "ages" derived from these models. Because the absorption-line strengths of composite stellar population models add like vectors weighted inversely by the mass-to-light ratios of their component simple stellar populations [35], it is currently impossible to use line strengths alone to disentangle multiple-burst populations. Simple two-burst models suggest that "young" populations seen in early-type galaxies

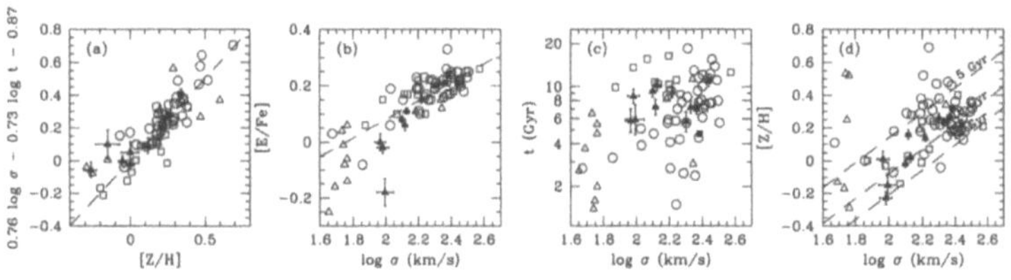

Fig. 2. The Z-plane and its projections. (**a**) The edge-on view of the plane (symbols as in Fig. 1) [35]. Note the slightly larger scatter of S0's compared to ellipticals around the plane. (**b**) The [E/Fe]–σ relation. The line is a fit to the field galaxies [10]; it is possible that low-mass clusters galaxies deviate from this relation (cf. [24]). (**c**) The age–σ plane. As these two parameters control the stellar populations of local early-type galaxies [35], all other two-dimensional projections will depend on the distribution of galaxies in this plane. (**d**) The metallicity–σ projection. While field ellipticals due not show a clear correlation in this projection due to frostings of recent star formation, massive early-type cluster galaxies have a clear velocity dispersion–metallicity relation. (All stellar population parameters derived using models from [38])

are likely to be small ($\ll$ 10% by mass) frostings of intermediate-aged stars ($t < 5\,$Gyr). Otherwise, hot blue stars that come from another stellar population (e.g., blue horizontal branch stars) must be invoked to account for the excess Balmer strengths [16]. This definition of "age" must be kept in mind whenever using stellar population models to analyze integrated spectra of "old" galaxies.

3 Stellar Populations of Early-Type Galaxies in Nearby Clusters

Figure 1 shows the application of stellar population models to the line strengths of early-type galaxies in the Fornax [15] and Coma [36] clusters, as well as a heterogeneous collection of isolated, loose and dense group, and Virgo cluster early-type galaxies called "field" galaxies [10] (we note here a number of other workers have observed both early-type galaxies in the Virgo and Coma clusters; [4,22,17] and others). Note that cluster *ellipticals* tend to be old galaxies, varying primarily in their metallicities, but note that the excellent Coma data suggests that a *factor of two* age spread exists even in massive Coma ellipticals (Fig. 2c). Cluster S0's, on the other hand, particularly in the Fornax cluster, have a very broad age spread, as do the "field" early-types. Thus, intermediate-aged stellar populations are present even in many early-type cluster galaxies at the present epoch.

Relationships between stellar population parameters and velocity dispersion are shown in Fig. 2. A principal component analysis [35] shows that early-type galaxies occupy a narrow plane in the 4-space of ($\log t$, [Z/H], [α/Fe], $\log\sigma$) that we refer to as the "Z-plane". The existence of this suggests that the stellar populations of giant early-type galaxies are controlled solely by two parameters,

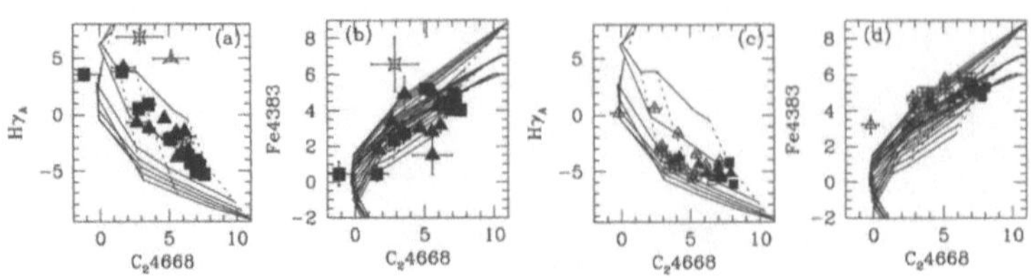

Fig. 3. The evolution of the stellar populations of cluster galaxies from $z \approx 0$ to $z = 0.41$ [33]. (**a**) The ages and metallicities of cluster galaxies in Abell 851, $z = 0.41$. Solid squares are ellipticals, solid triangles are S0's, open squares are early-type spirals, and open triangles are late-type spirals (model grids as in Fig. 1). (**b**) The abundance ratios of galaxies in Abell 851: from top to bottom, $[\alpha/\mathrm{Fe}] = 0, +0.3$, and $+0.5$ dex. (**c**) The ages and metallicities of early-type galaxies in the Fornax (open symbols) and Coma (filled symbols) Clusters. Aperture corrections to the same projected slits as if seen at $z = 0.41$ have been made. Note that even the oldest ellipticals in Abell 851 are younger than the majority of ellipticals in Fornax and Coma. Compared to the oldest galaxies in Fornax, the oldest galaxies in Abell 851 are about half as old. (**d**) The abundance ratios of early-type galaxies in the Fornax and Coma Clusters; the distributions do not appear to evolve significantly

which we associate with age and velocity dispersion (Fig. 2b). The abundance ratios $[\alpha/\mathrm{Fe}]$ are controlled solely by $\log\sigma$ (Fig. 2a), although there is a hint that there may be two relations, one for massive galaxies and another for lower-mass galaxies [24]. The existence of a mass–metallicity relation in early-type galaxies depends on the star formation history of the galaxies in question: cluster galaxies show a strong mass–metallicity relation, while field galaxies do not (Fig. 2d).

4 Stellar Populations of Early-Type Galaxies in Distant Clusters

Applying this method to early-type galaxies at even moderate redshifts ($z > 0.1$) requires significant time on the largest optical telescopes available today. We have been pursuing a campaign to acquire the required spectra at the Keck and VLT Telescopes; the first results, absorption-line strengths of galaxies in the cluster Abell 851 (CL0939+4713) at $z = 0.41$ are shown in Fig. 3. The typical exposure time was 14 to 21 hours of observation with the LRIS multislit spectrograph [20] to reach the required signal-to-noise ratios of $\sim 75/\mathrm{\AA}$.

Comparing Fig. 3a, the intermediate-redshift cluster galaxies, to Fig. 3c, the local cluster galaxies (projected to the same aperture size as in Abell 851), it is immediately apparent that the ages of *the oldest early-type galaxies in clusters have evolved* over the past 4 Gyr ($H_0 = 72\,\mathrm{km\,s^{-1}\,Mpc^{-1}}$, $\Omega_m = 0.3$, $\Omega_\Lambda = 0.7$). Note that relative ages are more reliable than absolute ages in stellar population models [34], and that the calibrations onto the Lick/IDS system for the line strengths of galaxies in Abell 851 are not yet complete at the time of writing (this can be alleviated using models by, e.g., [3]). We also note that $[\alpha/\mathrm{Fe}]$

variations strongly affect Hδ and Hγ at super-solar metallicities [36]; corrections for this effect for these lines are not yet available. We further note that the ages of ellipticals and S0s in Abell 851 spread by a factor of more than two, suggesting that small "frostings" of recent star formation may have happened in these galaxies as well, particularly in the S0 and spiral populations. (There appears however to be comfortingly little evolution in the metallicity and abundance ratio distributions, at least in the oldest galaxies.) Even with these cautions, we have clearly directly detected evolution of the oldest stellar populations within the last four billions years of cosmic time.

Although the oldest early-type galaxies have clearly evolved from $z = 0.4$ to the present, is there *too much* evolution? Preliminary age determinations suggest that the oldest galaxies in Abell 851 are half the age of the local cluster galaxies; given uncertainties in the models, this difference is between 4 and 6 Gyr. Within the uncertainties, then, this is equal to the lookback time to $z = 0.41$, and therefore stellar evolutionary and cosmological clocks are consistent.

5 Conclusions

Studies of the absorption-line strengths of early-type galaxies have revealed that their stellar populations are not completely old and static. This activity is likely to be recent star formation events consisting of a small fraction of the total mass. But while prevalent in field galaxies, even elliptical galaxies in the Coma Cluster are not immune: massive ellipticals in the core of that cluster show a spread of a factor of two in age, and early-types in the Virgo and Fornax Clusters show even larger age spreads. There is a hint that low-mass galaxies in these clusters may not have the same trends in parameter correlations as the giant galaxies; this will require deeper spectroscopy of larger samples to confirm.

With the availability of significant amounts of observing time with efficient spectrographs on the largest telescopes, it is now possible to detect the evolution of the stellar populations of early-type galaxies directly by measuring absorption-line strengths at a variety redshifts. We presently have such data in cluster galaxies out to $z = 0.75$, providing a baseline of lookback times of 6 Gyr, allowing us to probe directly the evolution of early-type galaxies for at least half the age of the Universe; higher redshifts are accessible with current generation of spectrographs and reasonable observing times. First results from the stellar populations of the oldest cluster galaxies at $z = 0.41$ prove that early-type galaxies evolve at the rate consistent with current cosmological parameters. It appears that we have (finally) reached a point where cosmological and stellar evolutionary clocks agree.

SCT would like to thank the conference organizers, particularly R. Bender, A. Renzini, C. Maraston, and M. Salvato, for a very enjoyable meeting in one of the world's most beautiful cities.

References

1. M. R. Blanton, D. J. Eisenstein, D. W. Hogg, D. J. Schlegel, J. Brinkmann: ApJ, in press
2. R. G. Bower, J. R. Lucey, R. S. Ellis: MNRAS, **254**, 601
3. G. Bruzual A., S. Charlot: MNRAS, **344**, 1000 (2003)
4. N. Caldwell, J. A. Rose, K. D. Concannon: AJ, **125**, 2891
5. P. G. van Dokkum, M. Franx: MNRAS, **281**, 985 (1996)
6. P. G. van Dokkum, S. A. Stanford: ApJ, **585**, 78 (2003)
7. A. Dressler: ApJ, **236**, 351 (1980)
8. S. M. Faber, R. E. Jackson: ApJ, **204**, 668 (1976)
9. M. Fukugita, C. J. Hogan, P. J. E. Peebles: ApJ, **503**, 518 (1998)
10. J. J. González: Line Strength Gradients and Kinematic Profiles in Elliptical Galaxies. PhD Thesis, University of California, Santa Cruz (1993)
11. J. E. Gunn, L. L. Stryker, B. M. Tinsley: ApJ, **249**, 48 (1981)
12. D. D. Kelson, P. G. van Dokkum, M. Franx, G. D. Illingworth, D. Fabricant: ApJL, **487**, L13 (1997)
13. D. D. Kelson, G. D. Illingworth, P. G. van Dokkum, M. Franx: ApJ, **531**, 184 (2000)
14. T. Kodama, N. Arimoto, A. J. Barger, A. Aragón-Salamanca: A&A, **334,** 99 (1998)
15. H. Kuntschner: MNRAS, **315**, 184 (2000)
16. C. Maraston, D. Thomas: ApJ, **541**, 126 (2000)
17. S. A. W. Moore, J. R. Lucey, H. Kuntschner, M. Colless: MNRAS, **336**, 382
18. R. W. O'Connell: ApJ, **206**, 370 (1976)
19. R. W. O'Connell: 'Population Synthesis of Early Type Galaxies'. In *Stellar populations; Proceedings of the First Space Telescope Science Institute Symposium, Baltimore, MD, May 20–22, 1986*, ed. by C. A. Norman, A. Renzini, M. Tosi (Cambridge Univ. Press, Cambridge 1986) pp. 167–189
20. J. B. Oke, et al.: PASP, **107**, 375 (1995)
21. R. C. Peterson: ApJL, **210**, L123 (1976)
22. B. M. Poggianti et al.: ApJ, **562**, 689 (2001)
23. D. Rabin: ApJ, **261**, 85 (1982)
24. P. Sánchez-Blázquez, J. Gorgas, R. Peletier: private communication (2003)
25. P. L. Schechter, A. Dressler: AJ, **94**, 563 (1987)
26. R. P. Schiavon, N. Caldwell, J. A. Rose: ApJ, in press
27. R. P. Schiavon, S. M. Faber, B. V. Castilho, J. A. Rose: ApJ, **580**, 850 (2002)
28. R. P. Schiavon, S. M. Faber, J. A. Rose, B. V. Castilho: ApJ, **580**, 873 (2002)
29. D. N. Spergel et al.: ApJS, **148**, 175
30. S. A. Stanford, P. R. Eisenhardt, M. Dickinson, B. P. Holden, R. De Propris: ApJS, **142**, 153
31. D. Thomas, C. Maraston, R. Bender: MNRAS, **339**, 897 (2003)
32. S. C. Trager: 'Abundances from the Integrated Light of Globular Clusters and Galaxies'. In: *Origin and Evolution of the Elements, Carnegie Observatories Astrophysics Series, Vol. 4,* ed. by A. McWilliam, M. Rauch (Cambridge Univ. Press, Cambridge 2004), in press
33. S. C. Trager, A. Dressler, S. M. Faber: in preparation
34. S. C. Trager, S. M. Faber, G. Worthey, J. J. González: AJ, **119**, 1645 (2000)
35. S. C. Trager, S. M. Faber, G. Worthey, J. J. González: AJ, **120**, 165 (2000)
36. S. C. Trager, S. M. Faber, A. Dressler: in preparation
37. F. L. Whipple: Harvard Coll. Obs. Circ., **404**, 1 (1935)
38. G. Worthey: ApJS, **95**, 107

The Near-Infrared View of Galaxy Evolution

Andrea Cimatti

INAF - Osservatorio Astrofisico di Arcetri, Largo E. Fermi 5, I-50125, Firenze, Italy

Abstract. Near-infrared surveys provide one of the best opportunities to investigate the cosmic evolution of galaxies and their mass assembly. We briefly review the main results obtained so far with the K20 and other recent near-IR surveys on the redshift distribution, the evolution of the luminosity function and luminosity density, the nature of old and dusty EROs, the evolution of the galaxy stellar mass function, the properties of the galaxies in the "redshift desert" and the nature of luminous starbursts at $z \sim 2$.

1 Introduction

Despite the detection of objects up to $z \sim 6.5$ and the impressive success of the ΛCDM scenario to account for the properties of the cosmic microwave background, one of the main and still controversial questions remains how and when galaxies assembled their mass as a function of the cosmic time. The hierarchical scenario predicts that galaxies built up their present-day mass through a progressive assembly of smaller sub-systems driven by the merging of dark matter halos.

With ground-based observations, one of the most solid approaches to address the problems of galaxy formation and evolution is to study samples of field faint galaxies selected in the near-infrared, particularly in the K-band (2.2μm) [3,23]. Firstly, since the rest-frame near-IR luminosity is a good tracer of the galaxy stellar mass [5], K-band surveys allow to select galaxies according to their mass up to $z \sim 1.5$ ($\lambda_{rest} \sim 0.9 - 1.0\mu$m). At higher redshifts, the K-band starts to sample the rest-frame optical and UV regions, and space-based observations at $\lambda_{obs} > 2\mu$m are needed to cover the rest-frame near-infrared (e.g. SIRTF, ASTRO-F). Secondly, the similarity of the spectral shapes of different galaxy types in the rest-frame near-IR makes the K-band selection free from strong biases against or in favor of particular classes of galaxies. In contrast, the selection of high-z galaxies in the observed optical bands is more sensitive to the star formation activity than to the stellar mass because it samples the rest-frame UV light and makes optical samples biased against old passive or weakly star-forming galaxies. Last but not least, near-infrared surveys are less affected by dust extinction than optical surveys.

Motivated by the above considerations, several near-infrared surveys have been undertaken during the last decade [12,10,6,18,24,17,15,19] (Fig. 1). Spectroscopic surveys are particularly relevant thanks to their capability not only to derive the redshifts, but also to unveil the nature and the spectral properties

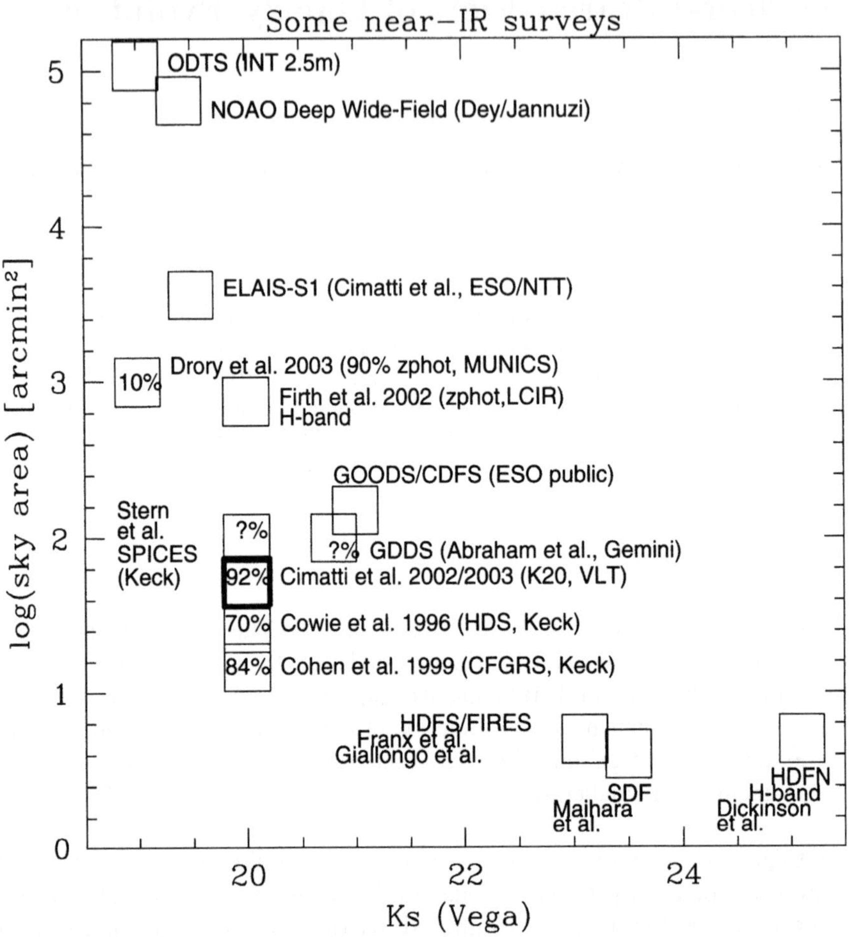

Fig. 1. Some imaging and spectroscopic near-infrared surveys. The redshift completeness is shown for spectroscopic surveys.

of the targeted galaxies. Since near-IR multi-object spectrographs are not fully available on 8-10m-class telescopes, most spectroscopy of K-selected galaxies has been done in the optical. The most crucial probes of massive galaxy evolution are galaxies with the very red colors expected in case of passive evolution at $z > 1$ (e.g. $R - K_s > 5$, usually called Extremely Red Objects, EROs). However, for the typical limiting fluxes of near-IR surveys ($K_s < 20 - 21$), they have very faint optical magnitudes ($R \sim 25 - 26$) already close to the spectroscopic limits of 8-10m-class telescopes. In addition, for $z > 1$ their main spectral features (e.g. the 4000 Å break and H&K absorptions) fall in the very red part of the observed spectra, where the strong OH sky lines and the CCD fringing and low quantum efficiency make spectroscopy even more demanding.

2 The K20 Survey

The K20 survey is a project aimed at investigating galaxy evolution through deep ESO VLT spectroscopy of a sample of 546 objects with $K_s < 20$ selected from two fields covering a total area of 52 arcmin^2. The two fields are completely independent, and one of them is a sub-area of the Chandra Deep Field South. Most spectroscopy was done in the optical with FORS1 and FORS2 and designed to reach the highest possible signal-to-noise ratio in the red. ISAAC near-IR spectroscopy was also done for a small fraction of the sample. The multi-band ($UBVRIzJK_s$) imaging available for both fields was used to derive high-quality photometric redshifts "trained" with the spectroscopic redshifts. The final sample covers a redshift range of $0 < z < 2.5$, and the current redshift completeness is 92% (spectroscopic) and 100% (spectroscopic + photometric redshifts).

The K20 survey represents a significant improvement with respect to other surveys for faint K-selected galaxies thanks to its high spectroscopic redshift completeness (the highest to date) extended to very faint red objects, the larger sample, the coverage of two fields and the availability of optimized photometric redshifts. We note here that the distribution of the sample over two independent sky fields is a significant advantage in reducing the field-to-field variation effects. For more details on the K20 survey see [6] and *http://www.arcetri.astro.it/~k20/*.

3 Established Results: Galaxy Evolution to $z \sim 1$

Overall, the K20 and other recent near-IR surveys show that galaxies are characterized by little evolution to $z \sim 1$, so that the observed properties can be mimicked by a pure luminosity evolution (PLE)-like scenario. This is in contrast with the current ΛCDM hierarchical merging models where the assembly of galaxies occurs later than what is actually observed. In particular:

• For $K_s < 20$, $N(z)$ has a median redshift of $z_{med} \sim 0.8$ and a high-z tail extended beyond $z \sim 2$. Current hierarchical models do not match the observed redshift distribution [8,18] (Fig.2), but an improved treatment of the merging processes is providing encouraging results [29,25].

• The rest-frame K_s-band luminosity function shows a mild luminosity evolution up to $z \sim 1$, with a brightening of about 0.5 magnitudes. Significant density evolution is ruled out up to $z \sim 1$ [11,26,17]. The high-luminosity tail at $z \sim 1$ is dominated by red early-type galaxies [26]. Current hierarchical models fail in reproducing the shape and the evolution of the LF as they over-predict the number of sub-L^* galaxies, under-predict that of luminous galaxies, and predict a strong density evolution.

• The rest-frame K_s-band luminosity density evolves slowly to $z \sim 1$ with $\rho_{K_s}(z) \propto (1+z)^{0.37}$, and much slower than the UV luminosity density [26].

• The properties of "old" EROs with $K_s < 20$ at $z \sim 1$ (morphology, spectra, luminosities, ages, stellar masses, clustering) imply the existence of a substantial population of old (a few Gyr), passively evolving and fully assembled massive spheroids which requires that major episodes of massive galaxy formation occurred at least at $z_{form} \sim 2$ [7,9]. Their number density at $z \sim 1$ is consistent

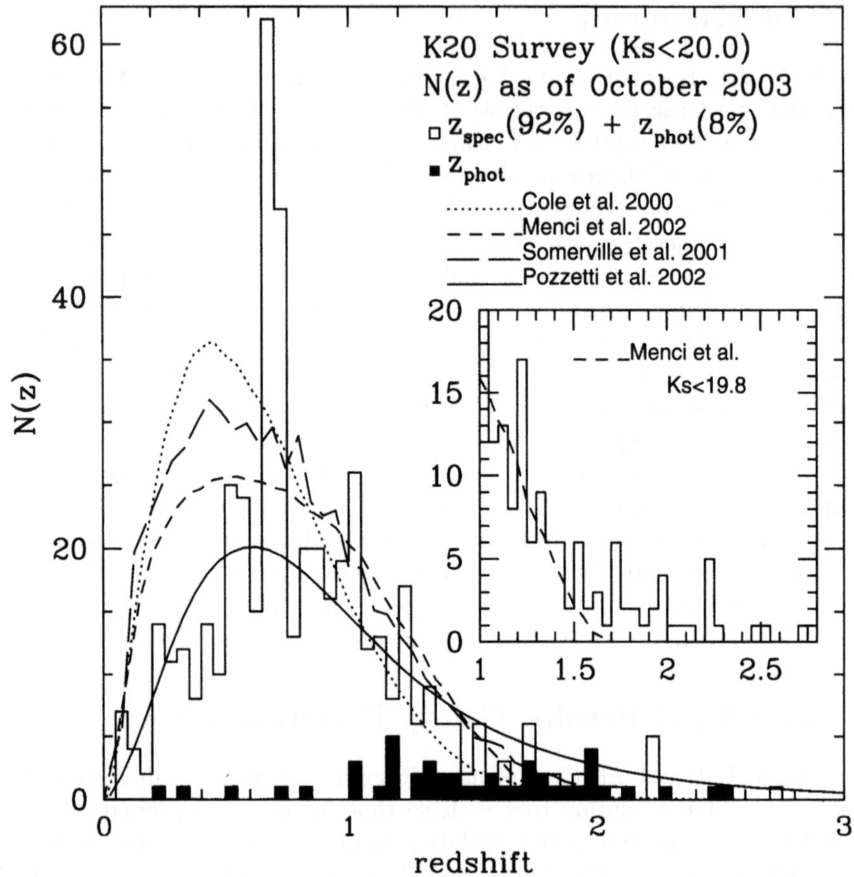

Fig. 2. The K20 survey $N(z)$ compared with the predictions of three hierarchical merging model (dotted and dashed lines) and a pure luminosity evolution (PLE) model (solid line). The models refer to *total* magnitudes $K_s < 20$ and are not corrected for the average under-estimate of 0.2 magnitudes due to the photometric selection effects evaluated in the K20 sample (see [6]). Including this bias in the models results in improving the agreement with the PLE models and increasing the discrepancy with the hierarchical models, as shown in the inserted plot where the model of Menci et al. (2002) for $K_s < 19.8$ is compared to the observed $z > 1$ tail of $N(z)$ (see [8]).

(within 2σ) with that of local luminous E/S0 galaxies. Current hierarchical models severely under-predict the number of "old" EROs [7,9].

• A numerous population of dusty star-forming EROs with disk-like and irregular morphologies emerges at $0.7 < z < 1.7$ from near-IR surveys [7,28,9]. These objects are often too faint to be detected in submm surveys due to their inferred far-IR luminosities $< 10^{12}$ $L_\odot$ and represent an ensemble of galaxies important to complement other star-forming systems selected with different techniques. They are also expected to be important contributors to the cosmic star-formation density [7,9,4,28]. Also in this case, current hierarchical models under-estimate the number of "dusty" EROs.

• "Old" EROs seem to have much stronger clustering than "dusty" EROs, with a comoving r_0 similar to that of present-day luminous ellipticals [13].

• The number density of massive galaxies and the cosmic stellar mass density Ω_* show a slow decrease from $z \sim 0$ to $z \sim 1$ qualitatively consistent with a hierarchical scenario, but much slower than what the current models predict [20,2,16,11,15,19].

4 Beyond $z \sim 1$: Galaxies in the "Redshift Desert"

At higher redshifts, the picture becomes more controversial: at $z > 1$ the evolution of the near-IR luminosity function seems to depart from a PLE-like pattern [26], it is still not clear if the number density of old spheroids drops at $z > 1.3$ [1,27], and unexpected populations of massive star-forming galaxies are being found in the range of $1.5 < z < 3$ thanks to near-IR surveys [21,31,14].

Moreover, the galaxy stellar mass function and $\Omega_*(z)$ display a fast evolution at $z > 1$ [15,19], and the K20 survey results suggest a rapid increase of the stellar mass density from $20^{+20}_{-10}\%$ of the local value at $2 < z < 3$ to about 100% at $z \leq 1$ [20].

These results suggest that $1.5 < z < 2.5$ may be the critical cosmic epoch during which most star formation activity and galaxy mass assembly took place. However, probing the $1.5 < z < 2.5$ range is difficult due to the lack of strong spectral features and emission lines redshifted in the observed optical and to the strong OH emission sky lines severely affecting the spectra beyond ~8000 Å. Due to the limited number of spectroscopically identified galaxies, this redshift range has been traditionally called "redshift desert".

Thanks to the very deep red-optimized optical spectroscopy and to near-IR spectroscopy, the K20 survey allowed us to populate the "redshift desert" and to unveil the nature of K-selected galaxies in this critical redshift range. By combining deep HST+ACS imaging with K20 spectroscopy, we found that a variety of galaxy types are present at $1.5 < z < 2.5$: spheroids at $z \sim 1.5$, disk- and spiral-like systems, and many irregular merging-like objects.

Particularly relevant is a sub-sample of luminous galaxies at $1.7 < z < 2.3$ spectroscopically identified in the K20 survey [14]. Their inferred star formation rates of $\sim 100 - 500\ M_\odot yr^{-1}$, stellar masses up to $10^{11}\ M_\odot$ derived from the fitting of their multi-color spectral energy distributions, and their merging-like optical morphologies becoming more compact in the near-IR, suggest that these galaxies may be massive galaxies caught during their major episode of mass assembly and star-formation, possibly being the progenitors of the present-day massive spheroidal systems. Current semi-analytical models under-predict by a factor of ~ 30 the number density of such galaxies [14,29]. Other near-IR surveys are independently finding candidate massive systems at $z \sim 2 - 3$ [21,31].

In order to draw a coherent picture of galaxy evolution at $1.5 < z < 2.5$, it will be crucial to combine K-selected samples (e.g. K20, FIRES [21], GDDS

[22]) with UV- and optically-selected samples more biased toward star-forming objects with little dust extinction such as those resulting from the Lyman-break selection at $z > 1.4$ [30].

Acknowledgements

The K20 survey team includes: T. Broadhurst (HUJ), S. Cristiani (INAF-Trieste), S. D'Odorico (ESO), E. Daddi (ESO), A. Fontana (INAF-Roma), E. Giallongo (INAF-Roma), R. Gilmozzi (ESO), N. Menci (INAF-Roma), M. Mignoli (INAF-Bologna), F. Poli (University of Rome), L. Pozzetti (INAF-Bologna), A. Renzini (ESO), P. Saracco (INAF-Brera), and G. Zamorani (INAF-Bologna).

References

1. Benitez N. et al. 1999, ApJ, 515, L65
2. Brinchmann J., Ellis R.S. 2000, ApJ, 536, L77
3. Broadhurst, T., Ellis, R.S., & Grazebrook, K. 1992, Nature, 355, 55
4. Brusa M. et al. 2002, ApJ, 581, L89
5. Gavazzi G., Pierini D. & Boselli A. 1996, A&A, 312, 397
6. Cimatti A., Mignoli M., Daddi E. et al. 2002c, A&A, 392, 395
7. Cimatti A., Daddi E., Mignoli M. et al. 2002a, A&A, 381, L68
8. Cimatti A., Pozzetti L., Mignoli M., et al. 2002b, A&A, 391, L1
9. Cimatti A., Daddi E., Cassata P. et al. 2003, A&A, in press, astro-ph/0310742
10. Cohen J.G., Blandford R., Hogg D.W. et al. 1999, ApJ, 512, 30
11. Cohen J.G. 2002, ApJ, 567, 672
12. Cowie, L.L., Songaila A., Hu E.M. & Cohen J.G. 1996, AJ,112,839
13. Daddi E., Cimatti A., Broadhurst T. et al. 2002, A&A, 384, L1
14. Daddi E., Cimatti A., Renzini A. et al. 2004, ApJL, in press, astro-ph/0308456
15. Dickinson M., Papovich C., Ferguson H.C., Budavari T. 2003, ApJ, 587, 25
16. Drory N., Bender R., Snigula J. et al. 2001, ApJ, 562, L111
17. Drory N., Bender R., Feulner G. et al. 2003, ApJ, 595, 698
18. Firth A.E., Somerville R.S., McMahon R.G. et al. 2002, MNRAS, 332, 617
19. Fontana A., Donnarumma, I. Vanzella E. 2003, ApJ, 594, L9
20. Fontana A. et al. 2003, A&A, submitted
21. Franx M., Labbé I., Rudnick G. et al. 2003, ApJ, 587, L79
22. Glazebrook K. et al. 2003, astro-ph/0311045
23. Kauffmann G.,Charlot S., White S.D.M. 1996, MNRAS, 283, L117
24. Labbé I., Franx M., Rudnick G. et al. 2003, AJ, 125, 1107
25. Menci N. et al. 2004, ApJ, in press, astro-ph/0311496
26. Pozzetti L. et al. 2003, A&A, 402, 837
27. Rodighiero G., Franceschini A., Fasano G. 2001, MNRAS, 324, 491
28. Smail I. et al. 2002, ApJ, 581, 844
29. Somerville R.S. et al. 2004, ApJL, in press, astro-ph/0309067
30. Steidel C.S. et al., this volume
31. van Dokkum P.G., Förster Schreiber N.M., Franx M. et al. 2003, ApJ, 587, L83

Rapid Growth of Massive Galaxies: A Paradox for Hierarchical Formation Models

Hsiao-Wen Chen[1] & David Crampton[2], and the LCIRS[3] & GDDS[4] Teams

[1] MIT Center for Space Research, Cambridge, MA 02139, USA
[2] Dominion Astrophysical Observatory, HIA, Victoria, BC V9E 2E7, Canada
[3] http://www.ociw.edu/lcirs/lcir.html
[4] http://www.ociw.edu/lcirs/gdds.html

Abstract. On behalf of the survey teams I summarize the designs and results of the Las Campanas Infrared Survey and Gemini Deep Deep Survey, both of which were initiated to understand the nature of red galaxies and to study the history of stellar mass assembly. Our results from luminosity function analysis, ISM absorption line measurements, and spectral synthesis modeling show that near-infrared selected galaxies at $1 < z < 2$ are not only massive and abundant but also old and metal enriched, indicating rapid formation of massive systems at higher redshifts.

1 Background

Various galaxy surveys in the optical, near-infrared, and sub-mm have uncovered different galaxy populations at redshift $z > 2$ (see Steidel, Cimatti, and Chapman in these proceedings for discussions), but whether these galaxies are representative of the galaxy population at high redshifts or how they are related to the present-day population is not clear. For example, rest-frame UV-selected samples are presumably sensitive to active star-forming galaxies and biased against evolved, quiescent systems, while sub-mm samples select mostly dusty star-forming galaxies. In contrast, near-infrared based surveys are sensitive to early-type galaxies at $z = 0 - 3$ that are known to dominate the total stellar mass at $z = 0$. A complete sample of massive systems at different epoch allows us to study the history of stellar mass assembly and offers important clues for discriminating between different galaxy formation scenarios (e.g. [1]).

But past studies based on near-infrared surveys have yielded inconsistent space density measurements of red galaxies at $z \geq 1$ (see [2] for a list of references). Early-type galaxies are strongly clustered and have no prominent narrow-band spectral features at UV wavelengths. If these red galaxies are progenitors of early-type galaxies we see in the local universe (rather than dusty star-forming galaxies), then a wide-field infrared survey is required to minimize the effect of surface density variation between fields due to strong clustering (e.g.[3,4]). Furthermore, deep galaxy spectroscopy is required to obtain a complete sample of these massive systems for spectral diagnostics of their ISM and stellar content.

The Las Campanas Infrared Survey (LCIRS) and Gemini Deep Deep Survey (GDDS) are complementary studies designed to probe the nature of red galaxies and to determine mass assembly history using near-infrared selected galaxies. In the following section I summarize the goals and results of the surveys.

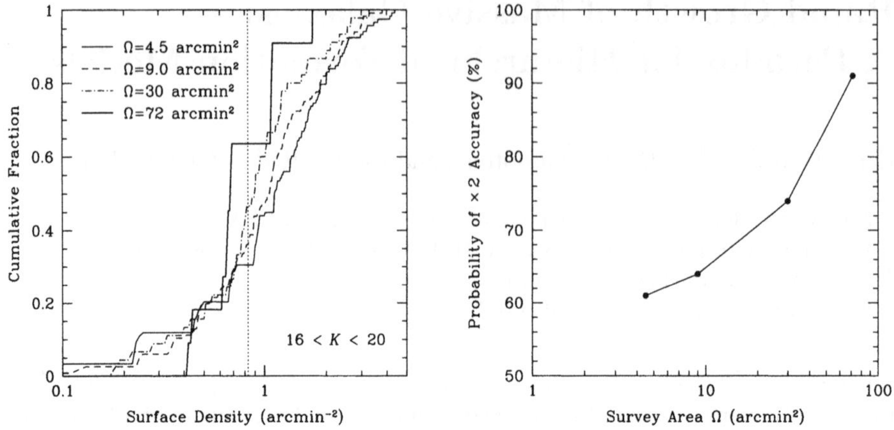

Fig. 1. Uncertainty of surface density measurements of red $I - K \geq 4$ galaxies with $16 < K < 20$ in different survey areas Ω. The surface densities were determined empirically within separate sub-areas randomly selected from the LCIRS fields. The left panel shows the cumulative probability of different surface density measurements for $\Omega = 4.5 - 72\,\mathrm{arcmin}^2$. The vertical dotted line indicates the mean surface density of $0.83\,\mathrm{arcmin}^{-2}$ determined over the entire LCIRS K-band selected red galaxies. Based on these curves, we show in the right panel the probability of obtaining a measurement within a factor of two accuracy of the nominal value versus Ω.

2 The Las Campanas Infrared Survey

The LCIRS is a deep, wide-field near-infrared and optical imaging survey, designed to identify a large number of red galaxies at $1 < z < 2$, while securing a uniform sample of galaxies of all types to $z \sim 2$ using photometric redshift techniques ([2,4,5]). The primary objectives are: (1) to examine the nature of the red galaxy population and identify evolved galaxies at redshifts $z > 1$; (2) to study the space density and luminosity evolution of early-type galaxies at redshifts $z \leq 2$; and (3) to measure spatial clustering of massive galaxies, thereby inferring merger rates of these galaxies for constraining theoretical models.

We have completed the phase-I H-band survey to $H = 20.5$ over 1.1 deg^2 and are completing the phase-II K-band survey to $K = 20.6$ over 0.75 deg^2. The size of our survey area was determined in order to obtain a representative measure of the surface density of red galaxies and a significant signal in clustering analysis [4]. Five random fields have been chosen and each field covers $\approx 26 \times 26\,\mathrm{arcmin}^2$, corresponding to a projected co-moving distance 20 h^{-1} Mpc. This is twice the correlation length observed for early-type galaxies at $z = 0$ (see [4] for references). The large survey area of the LCIRS allows us to estimate empirically the uncertainties in the surface density measurements of red galaxies identified in smaller-area surveys, assuming that the LCIRS galaxy sample are representative of these systems at $z \geq 1$. Figure 1 shows the results based on red galaxies with $I - K \geq 4$ found in non-overlapping, random sub-areas within the LCIRS fields,

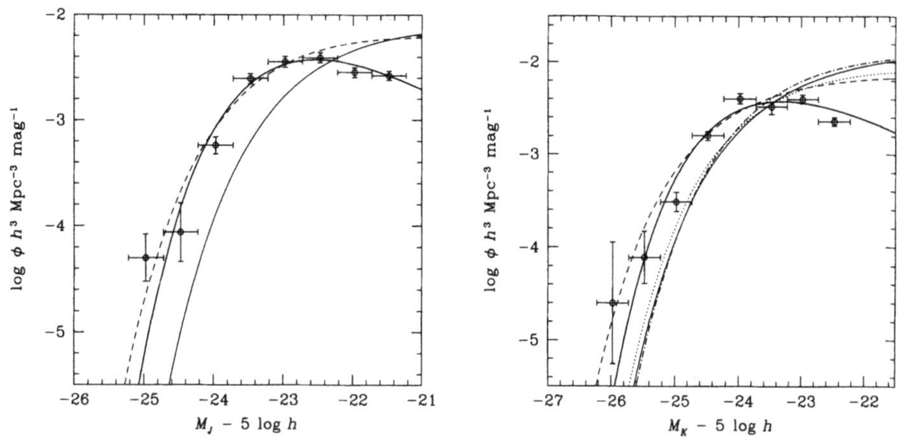

Fig. 2. Rest-frame J (left) and K (right) galaxy luminosity function of > 4000 galaxies identified at $1 \leq z \leq 1.5$ in the LCIRS using photometric redshift techniques. The points were evaluated using a modified stepwise maximum likelihood analysis, explicitly incorporating redshift uncertainties from photometric redshift analysis [6]. Error bars were estimated using a bootstrap technique that takes into account photometric uncertainties, redshift errors, and sampling biases. The thick solid curve in each panel shows our best-fit Schechter luminosity function. For comparison, we present measurements for K-selected galaxies at $0.75 < z < 1.3$ from the K20 survey team [7] (dashed curves). We also plot local measurements from the 2MASS survey [8,9] (thin curves).

in comparison to 0.83 arcmin^{-2} as the nominal surface density of $I - K \geq 4$ galaxies determined based on the entire LCIRS red sample. We find based on the curves that a survey area of at least 70 arcmin^2 is necessary to identify a statistically representative sample of red galaxies. Our initial clustering analysis based on an H-band selected sample over 0.62 deg^2 of sky shows that red galaxies with $I - H \geq 3$ have a co-moving correlation length of $9 - 10\ h^{-1}$ Mpc at $z \simeq 1$, comparable to that of present-day early-type galaxies [4].

Photometric redshifts have been determined for the entire LCIRS galaxy sample based on a combination of available optical $BVRIz'$ and near-infrared JHK photometry [6], making the LCIRS a *complete* redshift survey of faint galaxies at $z \geq 1$. But because of relatively large redshift uncertainties, accounting for these errors in all analyses that make use of photometric redshift measurements is necessary to minimize systematic errors. Adopting an H-band selected sample and using a technique that explicitly accounts for the *non-gaussian characteristics* of photometric redshift uncertainties, we have shown that the evolution of the rest-frame R-band co-moving luminosity density ℓ_R is characterized by $\Delta \log \ell / \Delta \log(1 + z) = 0.6 \pm 0.1$ and that luminous early-type galaxies exhibit only moderate evolution over $z = 0.3 - 1.5$ [6].

Rest-frame J- and K-band luminosity functions have also been calculated using > 4000 K-band selected galaxies at $1 \leq z \leq 1.5$. Figure 2 shows the best-fit luminosity functions in J (left panel) and K (right panel) using modified stepwise maximum likelihood method and STY approach as described in Chen et al. [6]. The best-fit Schechter parameters are $M_{J_*} - 5\log h = -22.6, \alpha_J = -0.3, \phi_{J_*} = 0.0144\,h^3\,\mathrm{Mpc}^{-3}$, and $M_{K_*} - 5\log h = -23.4, \alpha_K = -0.2, \phi_{K_*} = 0.0142\,h^3\,\mathrm{Mpc}^{-3}$. Figure 2 also shows that our results are consistent at the bright end with the estimates presented by the K20 team for 170 galaxies found at $0.75 < z < 1.3$ over a total sky area of 52 arcmin^2 [7] (dashed curves), but exhibit a hint of a turn-over at the faint-end. Comparisons with existing measurements for local galaxies using 2MASS data [2,9] indicate that there is approximately $0.4 - 0.6$ mag fading in luminous ($> L_*$) near-infrared selected galaxies from $z = 1.2$ to $z = 0$, the amount of which may be explained by stellar evolution over the same period of time. These results together suggest little/no evolution in the space density of massive early-type galaxies since $z = 1.5$ and support the hypothesis that these systems were formed at redshift much beyond $z = 1.5$.

3 The Gemini Deep Deep Survey

The GDDS is an ultra-deep spectroscopic survey of near-infrared selected galaxies at $0.7 < z < 1.8$ [10,11]. The primary objectives are: (1) to obtain spectral diagnostics of the ISM and stellar content of massive galaxies at high redshift; (2) to construct the stellar mass function over the target redshift range; and (3) to determine ages and star-formation histories of evolved systems at $z > 1$.

The survey makes use of nod & shuffle observing mode [12] for accurate sky subtraction and fringe removal at wavelength beyond 7500 Å. It is aimed at identifying near-infrared selected faint galaxies at $z > 0.8$ to a high completeness level. We have observed 301 spectra in four fields using GMOS on Gemini North. Each of the four fields covers 30 arcmin^2 field of view and is located in a separate LCIRS field. The pointing and target selection of the GDDS sample is guided by known information in the wider-field LCIRS sample to maximize the efficiency of identifying a representative sample of faint galaxies at $z > 0.8$. For example, each GDDS pointing is selected from within an LCIRS field to have the number of red galaxies comparable to the global mean, which is important for reliable volume density analyses of early-type galaxies. In addition, we exclude foreground contaminating sources at $z < 0.8$ using photometric redshifts available for all LCIRS sources. Figure 3(a) shows a comparison of the observed $V - I$ color versus z for CFRS [13], GDDS [10], and UV-selected galaxies [14]. It demonstrates that the GDDS is probing a representative galaxy population concentrated at $0.8 \leq z < 2$ with a broad-range of star formation history, from quiescent through active star-forming systems. Conversely, the GDDS results offer necessary tests and verification of the accuracy of LCIRS photometric redshifts. Figure 3(b) shows that the rms scatter between z_{phot} and z_{spec} is $\sigma_{\Delta z/(1+z)} \approx 0.1$, consistent with previous results based on a smaller number of galaxies at $0 \leq z < 1$ [6].

The GDDS data have revealed a wealth of information regarding the ISM and stellar content of massive galaxies at $z > 1$. Sample spectra are presented in Figure 4, where we show various GDDS composite spectra as well as individual sample spectra to demonstrate the data quality. The full sample of GDDS

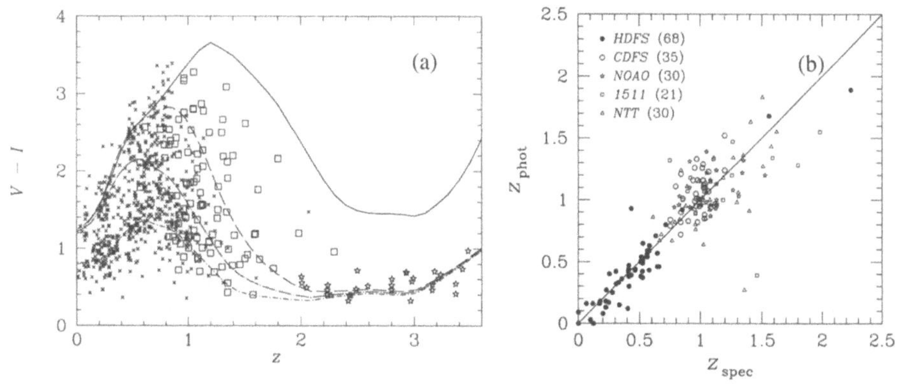

Fig. 3. (a) The observed $V-I$ color versus redshift for the GDDS sample (squares), in comparison to the CFRS sample [13] (crosses), UV-selected galaxies [14] (stars), and model predictions for a non-evolving elliptical galaxy (solid curve), to an exponentially declining star formation rate (SFR) with e-folding time of $\tau = 1$ (dashed curve) and 2 Gyr (long dash-dotted curve), and a constant SFR (short dash-dotted curve). (b) Comparison of photometric and spectroscopic redshifts for ~ 200 galaxies identified in the LCIRS. Spectroscopic redshifts for galaxies in the HDFS and CDFS were collected in part from the literature and in part from our own observations. Spectroscopic redshifts for galaxies in the NOAO, 1511, and NTT fields were obtained from the GDDS.

spectra is presented in Abraham et al. [10]. Specifically, we show in panel (a) the composite spectrum of early-type (absorption-line dominated) galaxies, panel (b) intermediate-type galaxies, and in panel (c) late-type (emission-line dominated) galaxies. For comparison, we include in panel (a) the SDSS composite for luminous red galaxies (LRG) at $\langle z \rangle = 0.3$ [15] (red curve). Our composite spectrum for early-type galaxies at $z = 1-2$ is strikingly similar to the LRG composite spectrum at $\langle z \rangle = 0.3$, showing little evolution in the stellar content of massive and quiescent systems over the redshift interval spanned by the two samples. We also show the spectrum of a galaxy at $z = 1.34$ in panel (d) together with the spectrum of the host galaxy of 53W091 at $z = 1.55$ [16] (red curve) and a starburst galaxy at $z = 1.67$ in panel (e). These spectra demonstrate that we have good signal-to-noise ratios to obtain accurate and precise redshifts of individual galaxies using both rest-frame ultraviolet metal absorption features and weak continuum bumps originating in evolved stellar populations.

In addition, we have constructed a composite spectrum for 13 K-band selected galaxies at $1.3 < z < 2$ from the GDDS sample that are also luminous in the UV with $M_{2000} < -19.5$ [11]. A normalized version is shown in panel (f) of Figure 4, together with a normalized composite spectrum of local starburst galaxies from Tremonti et al. [17]. The composite spectrum shows prominent absorption features from Mg II, Fe II together with Mg I and Mn II, all of which are stronger than those observed in local starburst systems. It indicates that the ISM of these massive star-forming galaxies has been heavily

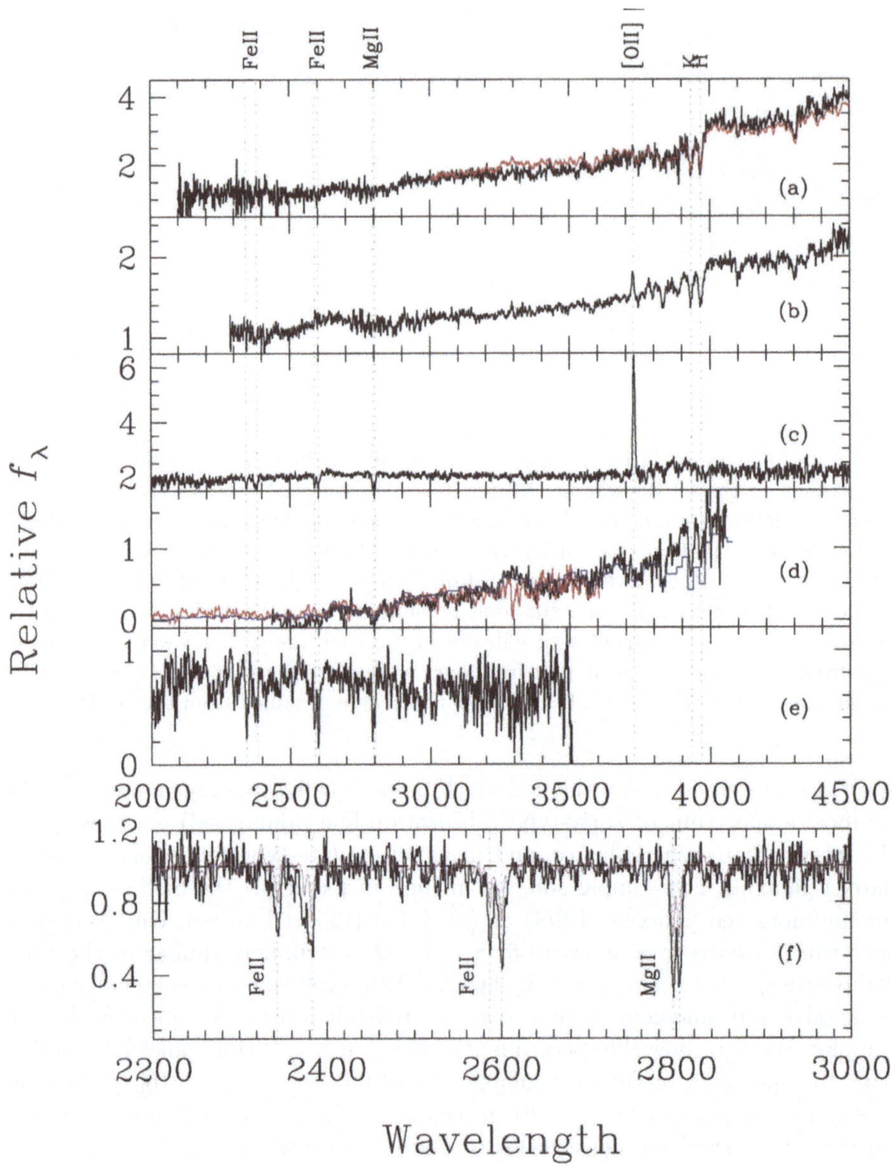

Fig. 4. (a)-(c) GDDS composite spectra for respectively early-type, intermediate, and late-type galaxies at $z = 1 - 2$. (d) The spectrum of a galaxy at $z = 1.34$ from the GDDS, together with the spectrum of the host galaxy of 53W091 at $z = 1.55$ [16] in red, and the model spectrum of a single burst system formed at $z_f = 6$ in blue. (e) The spectrum of a starburst galaxy at $z = 1.67$, showing strong metal absorption features. (f) A normalized composite spectrum of 13 massive star-forming galaxies identified at $\langle z \rangle = 1.6$ from the GDDS, together with a normalized composite spectrum of local starburst galaxies [17].

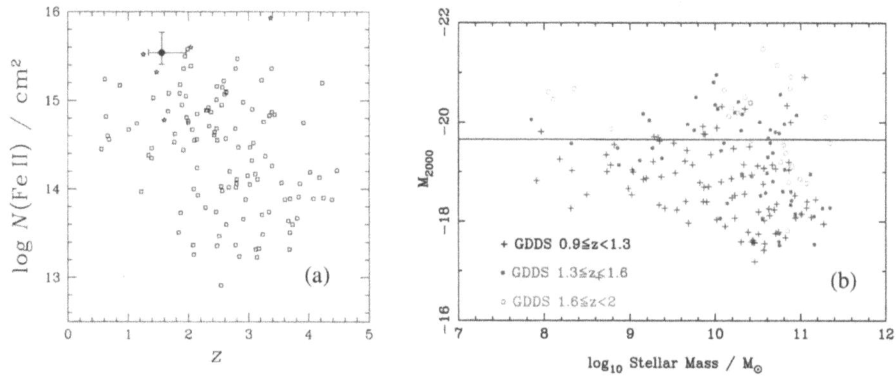

Fig. 5. (a) Fe II column density vs. redshift in different systems: solid point with error bars represents the GDDS measurement; open squares are for damped Lyα absorption systems compiled from the literature; and stars are for the ISM of gamma-ray burst host galaxies. (b) Rest-frame UV magnitude versus inferred stellar mass for galaxies in three redshift intervals. The solid line indicates the minimum UV flux of the galaxies included for the composite spectrum presented in panel (f) of Figure 4.

enriched with metals by $\langle z \rangle = 1.6$. Based on a curve-of-growth analysis of all available transitions, Savaglio et al. ([11]) find a mean Fe II column density of $\log N_{\mathrm{Fe\,II}}/(\mathrm{cm}^2) = 15.54^{+0.23}_{-0.13}$ that is among the highest observed in either known damped Lyα absorption systems or in the ISM of host galaxies of gamma-ray bursts (Figure 5a). While the metallicity is uncertain due to the unknown neutral hydrogen content, it is clear that the ISM of these galaxies contains abundant heavy elements. Because of the unknown nature of the galaxies that give rise to damped Lyα systems, direct column density measurements of the ISM of luminous galaxies at high redshift offer a promising alternative for understanding cosmic chemical enrichment history.

We also determine the formation epoch of the stars seen in quiescent systems that show no signs of recent star formation (lack of emission features in the rest-frame UV). Using a likelihood analysis, we derive a minimum age for each system by comparing predicted spectral energy distributions (SEDs) from different stellar synthesis models with the observed ones established from the GDDS spectra and available LCIRS optical and near-infrared photometry. An example is shown in panel (d) of Figure 4 for a red galaxy at $z = 1.34$. The model (blue curve) was obtained for a single burst occurring at $z_f = 6$ and passively evolving to lower redshift. The agreement between the observed spectrum and models suggests that this galaxy is at least 3.5 Gyr old, pushing the formation epoch of this galaxy to beyond $z_f = 6$ (McCarthy et al. 2004).

Finally, we estimate the total stellar mass of every galaxy in the GDDS sample using a K-band mass-to-light ratio determined from comparing the observed multi-color data with a grid of model SEDs with varying dust extinction, metallicity, and star formation history. Figure 5(b) shows rest-frame UV flux (expressed in M_{2000}) versus the inferred stellar masses of individual galaxies in three redshift bins spanned by the GDDS sample. We see that a large fraction of the GDDS galaxies have masses $> 10^{10}\, M_{\odot}$, and these systems exhibit on average increasing rest-frame UV fluxes (and therefore increasing star formation rate) with increasing redshift. In addition, while few systems with stellar masses $< 10^{10}\, M_{\odot}$ are seen at $z > 1.3$ in the GDDS sample due to the K limit of the survey, a tendency for the most massive galaxies to be UV quiescent is found in the low-redshift sample (crosses)–a trend which is similar to what is seen at $z = 0$. This result supports the view that most star formation takes place in lower mass systems even at $z > 1$ (Glazebrook et al. 2004).

In summary, the wide-field multi-color imaging data from the LCIRS and the high-quality nearly complete spectral sample from the GDDS together allow us to conduct a comprehensive study of massive early-type galaxies at $z = 1 - 2$. Our luminosity function analysis indicates that most and perhaps all of the present-day infrared luminous galaxies were in place by $z \approx 1.2$. The rest-frame UV spectra of massive star-forming systems show abundant metals that arise in the ISM of these galaxies, indicating an early heavy element enrichment by $z \approx 1.6$. Finally, spectral synthesis modelling showed that near-infrared selected galaxies at $z > 1$ are massive and that the quiescent ones with no evidence of recent star formation are genuinely old and likely to form at $z > 4$. Together these results suggest a rapid and early formation process of massive galaxies, presenting difficulties for conventional hierarchical formation scenarios.

References

1. G. Kauffmann, S. Charlot: MNRAS **297**, L23 (1998)
2. H.-W. Chen, P. J. McCarthy, R. O. Marzke et al.: ApJ **570**, 54 (2002)
3. E. Daddi et al.: A&A **361**, 535 (2000)
4. P. J. McCarthy, R. G. Carlberg, H.-W. Chen et al.: ApJ **560**, L131 (2001)
5. A. E. Firth, R. Somerville, R. G. McMahon et al.: MNRAS, **332**, 617 (2002)
6. H.-W. Chen, R. O. Marzke, P. J. McCarthy et al.: ApJ **586**, 745 (2003)
7. L. Pozzetti et al.: A&A, **402**, 837 (2003)
8. S. Cole et al.: MNRAS, **326**, 255 (2001)
9. C. S. Kochanek et al.: ApJ, **560**, 566 (2001)
10. R. G. Abraham et al.: AJ submitted
11. S. Savaglio et al.: ApJ in press (astro-ph/0310437)
12. K. Glazebrook and J. Bland-Hawthorn: PASP, **113**, 197 (2001)
13. S. J. Lilly et al.: ApJ, **455**, 108
14. M. Dickinson et al.: ApJL in press (astro-ph/0309070)
15. D. J. Eisenstein et al.: ApJ, **585**, 694 (2003)
16. J. Dunlop et al.: Nature, **381**, 581 (1996)
17. C. A. Tremonti, C. Leitherer, T. M. Heckamn, & D. Calzetti. 2003, ApJ submitted

The MUNICS Project: Galaxy Assembly at $0 < z < 1$

Niv Drory[1], Ralf Bender[2,3], Georg Feulner[3], Gary J. Hill[1], Ulrich Hopp[3], Claudia Maraston[2], and Jan Snigula[3]

[1] University of Texas at Austin, Austin, Texas 78712
[2] Max–Planck Institut für extraterrestrische Physik, Giessenbachstraße, Garching, Germany
[3] Universitäts–Sternwarte München, Scheinerstraße 1, D-81679 München, Germany

1 The MUNICS Survey and its Results

MUNICS is a wide-area, medium-deep, photometric and spectroscopic survey selected in the K band, targeting randomly-selected high Galactic latitude fields. It covers an area of roughly one square degree in the K and J bands with complementary optical follow-up imaging in the I, R, V, and B bands in 0.5 square degrees.

The limiting magnitudes of this main part of the survey are 19.5 in K, 21.5 in J, 22.5 in I, and 23.5 in R (50 % completeness for point-like sources). This multicolor catalog probes field galaxies in a large volume out to redshifts of roughly 1.5 (for massive galaxies) and is by far the largest catalog of near-infrared selected distant galaxies published so far. It thus comprises a suitable and highly competitive multi-color field galaxy survey. The photometric survey is described and characterized in [11] and [21]. The survey spans the redshift range $0 < z < 1.5$ and selects typically L^* and brighter objects.

The MUNICS photometric survey is complemented by spectroscopic follow-up observations at 4m-class telescopes of all galaxies down to $K \leq 17.5$ in 0.25 square degrees. This survey is complete down to $K \leq 16.5$ and 80% complete at $16.5 < K < 17.5$ to the present date. Furthermore, a sparsely selected deeper sample down to $K \leq 19$ was observed with the ESO VLT, covering a much smaller area of 100 square arcmin. The whole spectroscopic sample contains 593 secured redshifts thus far. The spectra cover a wide wavelength range 4000 – 8500Å at 13.2Å resolution, and sample galaxies at $0 < z < 1$. Details of the spectroscopic survey are given in [12].

Since it is impossible to obtain spectroscopic redshifts for the whole MUNICS sample (most galaxies in the sample are too faint to be observed currently with optical spectrographs) one must rely on photometric redshift techniques to obtain distance estimates for the complete sample of almost 6000 galaxies. Comparing to > 500 spectroscopic redshifts, the scatter in the relative redshift error $\Delta z/(1+z)$ is 0.055. The distribution of photometric redshifts peaks around $z \approx 0.5$. and has a tail extending to $z \approx 3$.

Given these limits, MUNICS contains mostly massive (stellar mass $M > 10^{10} M_\odot$) field galaxies, spanning a significant fraction of cosmic time and is ideal for studying their formation and evolution, specifically their mass assembly history.

2 The Rest-Frame K-Band Luminosity Function to z $\sim$ 1.2

Fig. 1 shows the rest-frame K–band luminosity function (LF) derived from the MUNICS data in four redshift bins spanning $0.4 < z < 1.2$ [8]. Absolute magnitudes were derived using the photometric redshifts, extrapolating the best fitting SED to the rest-frame K band. Errors due to this extrapolation are expected to be small in the redshift range we probe, since the near-IR slopes of the SEDs differ only very little over the galaxy types probed by our survey (the K–band k-corrections are small and almost type-independent).

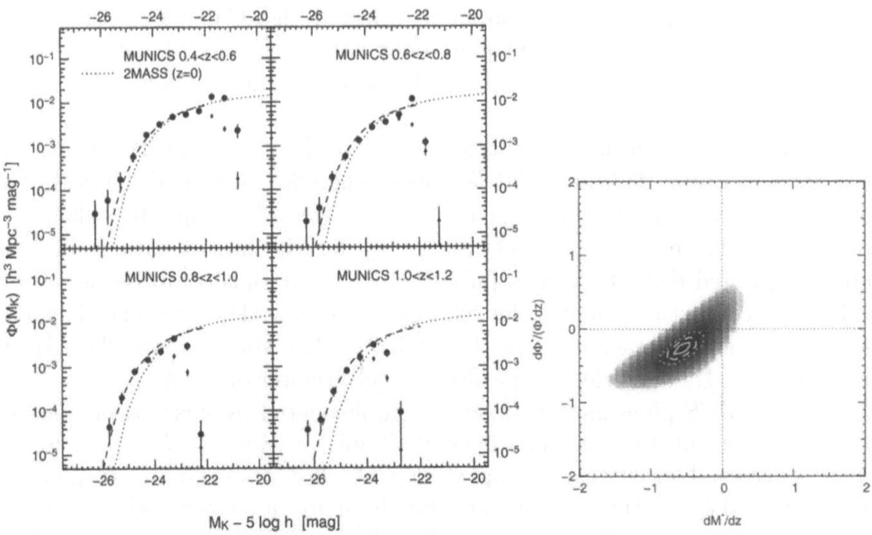

Fig. 1. Left panel: The rest-frame K–band luminosity function derived from the MUNICS data in four redshift bins spanning $0.4 < z < 1.2$. The dotted curve is the $z = 0$ LF [16]. The dashed lines are parameterized fits to the data. **Right panel:** Likelihood map for the change in the Schechter parameters, $d\Phi^*/(\Phi^*\,dz)$ and dM^*/dz, with redshift.

Quantitative analysis of the evolution of the LF yields a mild decrease in number density by $\sim 25\%$ to $z = 1$ accompanied by brightening of the galaxy population by ~ 0.5 mag. These results are fully consistent with an analogous analysis using only the spectroscopic MUNICS sample [12], which is shown in Fig. 2.

To interpret these results in terms of a picture of galaxy evolution is not straight–forward, though. Since the K–band light traces stellar mass, we may

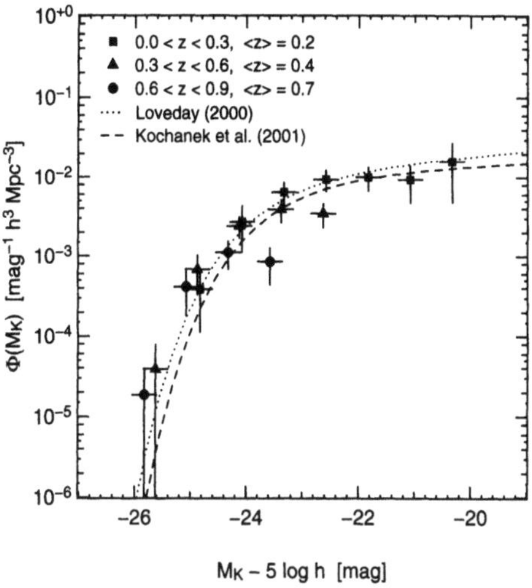

Fig. 2. The K band luminosity function of galaxies from spectroscopic observations of MUNICS galaxies. The dotted and dashed curves are the $z = 0$ LFs [18,16].

attribute a change in K band luminosity function either to a change in the mass–to–light ratio alone (passive evolution) or to a combination of a change in M/L and in stellar mass. A change in stellar mass can be due to star formation and/or due to merging and accretion. These models cannot be discriminated on basis of the luminosity function alone. In fact, if the increase seen in the global star formation rate to $z \sim 1$ (e.g. [17,14,15,6]) is attributable to normal field galaxies, the stellar mass of these systems is expected to evolve by roughly a factor of 1.5 to 2 over the redshift range $0 < z < 1$. Therefore, if the K band light does reflect stellar mass, number density evolution in the K band is indeed expected.

3 The Stellar Mass of Field Galaxies to z ∼ 1.2

The multicolor data in MUNICS with their wide range in wavelength allow us to investigate properties of the stellar populations of individual objects in greater detail. We can therefore use the MUNICS data to derive the stellar mass function in the redshift range $0.4 < z < 1.2$ [9,10] by modeling and fitting the stellar mass-to-light ratios in the NIR.

We parameterize star formation histories as $\psi(t) \propto \exp(-t/\tau)$, with $\tau \in \{0.1, 0.2, 0.4, 1.0, 2.0, 3.0, 5.0, 8.0, 10.0, 13.0\}$ Gyr. We extract spectra at 28 ages between 0.001 and 14 Gyr and allow A_v to vary between 0 and 3 mag using a Calzetti [3] extinction law. The models use solar metallicity and are based on the Simple Stellar Population models by C. Maraston [20]. We assume a value of 3.33 for the absolute K–band magnitude of the Sun. We use a Salpeter IMF with lower and upper mass cutoffs of 0.1 and 100 $M_\odot$. This choice allows us to compare our results directly with the literature. The use of an IMF with a flatter slope at the low mass end will not affect the shape of the mass function, it will only change its overall normalization. If, however, the IMF depends on the mode of star formation, e.g. being top–heavy in starbursts, our results will be affected. We convert the absolute K–band magnitude, M_K, into stellar mass by using the K–band mass–to–light ratio, M/L_K, of the best fitting CSP model in a χ^2 sense.

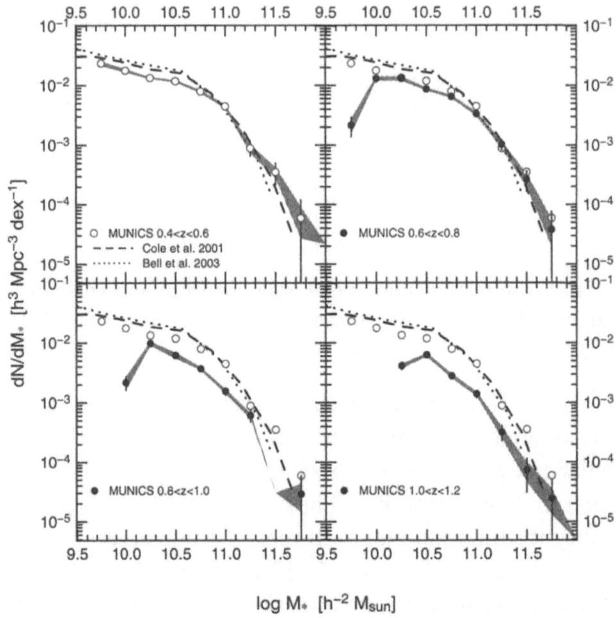

Fig. 3. The evolution of the stellar mass function with redshift. The open symbols are the MUNICS values at $0.4 < z < 0.6$, the closed symbols are the MUNICS values at higher redshifts. The lowest z values are shown in all panels for comparison. Error bars denote the uncertainty due to Poisson statistics. The shaded areas show the 1 σ range of variation in the mass function given the total systematic uncertainty in M/L_K. The dotted and dashed lines show the $z = 0$ stellar mass function derived similarly to our methods using SDSS, 2dF, and 2MASS data [1,5].

Fig. 3 shows the mass function of galaxies with $10^{10} < M/(h^{-2}\,M_\odot) < 10^{12}$ in four redshift bins centered at $z = 0.5$, $z = 0.7$, $z = 0.9$, and $z = 1.1$. The results on the local stellar mass function [5,1] derived by fitting stellar population models to multicolor photometry and deriving NIR mass–to–light ratios similarly to the method employed here are also shown for comparison as a dashed and dotted line, respectively. The data from the lowest redshift bin, $0.4 < z < 0.6$ are shown alongside the higher–redshift data for easier comparison as open symbols. Error bars denote the uncertainty due to Poisson statistics. The shaded areas show the 1 σ range of variation in the mass function from Monte–Carlo simulations given our estimate of the total systematic uncertainty in M/L_K.

The lowest redshift bin shows remarkable agreement with the $z = 0$ values, despite the different selection at low versus high redshift and the different model grids used, although we obtain slightly lower number densities at $\log M/(h^{-2}\, M_{\odot}) \leq 10.5$. Therefore, there seems to be not much evolution in stellar mass at $z < 0.5$. The general trend at higher redshift, is for the total normalization of the stellar mass function to goes down and for the knee to move towards lower masses. This causes the higher masses to evolve faster in number density than lower masses and is well visible in Fig. 3 at $10.5 < \log M/(h^{-2}\, M_{\odot}) < 11.5$.

4 The Total Stellar Mass Density

The total stellar mass density of the universe – the integral of the star formation history of the universe (e.g. [19,22]) and observationally its complement – is shown in Fig. 4. We also plot the local value from [5], values from the Hubble Deep Fields [7,13] covering $z > 1$ and values from [4,2] at $z < 1$. Additionally, we integrate the star formation history curve (including extinction correction) from [22] for comparison.

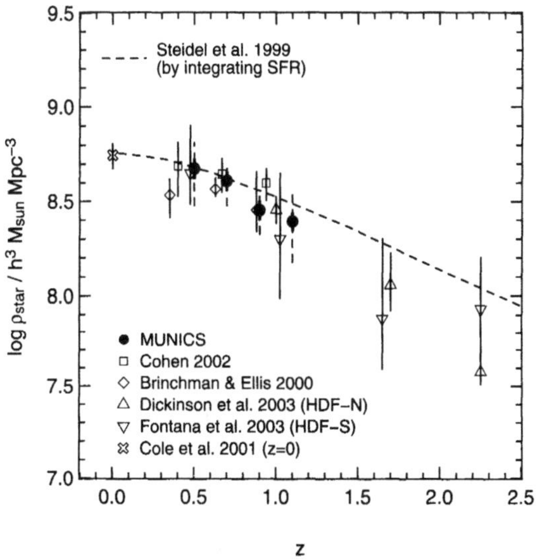

Fig. 4. The evolution of the total stellar mass density in the universe. The closed circles are the MUNICS values, open symbols are values from the literature. The integrated star formation rate (dashed curve) is shown for comparison. The thick error bars on the MUNICS values (solid dots) are the statistical errors associated with the data. The dashed error bars show the variance we get from the MUNICS data in GOODS size patches.

From our data, as well as from the HDFs and the integrated star formation rate from the UV luminosity density, it appears that 50% of the local mass in stars has formed since $z \sim 1$. The data from [2] are consistent with ours at $z \sim 1$ but are lower at $z \sim 0.4$ and seem to under–predict the local value if extrapolated. The values obtained by [4] are higher at $z \sim 1$ which we think is attributable to their method of deriving M/L_K.

It is worth noting that the results from the HDF–N differ by a factor ~ 2 from the HDF–S, attributable to cosmic variance. The MUNICS values in Fig. 4 are shown with their statistical errors (thick error bars), which amount to roughly 10%. We also show the variance we get in our sample divided into GOODS size patches of 150 square arcminutes (dashed error bars), showing that at these redshifts, even surveys like GOODS are expected to be dominated by cosmic variance. We expect differences of around 50% between the two GOODS areas.

References

1. Bell, E. F., McIntosh, D. H., Katz, N., & Weinberg, M. D. 2003, ApJ, submitted
2. Brinchmann, J., & Ellis, R. S. 2000, ApJ, 536, L77
3. Calzetti, D., Armus, L., Bohlin, R. C., Kinney, A. L., Koornneef, J., & Storchi-Bergmann, T. 2000, ApJ, 533, 682
4. Cohen, J. G. 2002, ApJ, 567, 672
5. Cole, S., et al. 2001, MNRAS, 326, 255
6. Cowie, L. L., Songaila, A., & Barger, A. J. 1999, AJ, 118, 603
7. Dickinson, M., Papovich, C., Ferguson, H. C., & Budavári, T. 2003, ApJ, 587, 25
8. Drory, N., Bender, R., Feulner, G., Hopp, U., Maraston, C., Snigula, J., & Hill, G. J. 2003a, ApJ, 595, 698
9. Drory, N., Bender, R., Feulner, G., Hopp, U., Maraston, C., Snigula, J., & Hill, G. J. 2003b, ApJ, submitted
10. Drory, N., Bender, R., Snigula, J., Feulner, G., Hopp, U., Maraston, C., Hill, G. J., & de Oliveira, C. M. 2001a, ApJ, 562, L111
11. Drory, N., Feulner, G., Bender, R., Botzler, C. S., Hopp, U., Maraston, C., Mendes de Oliveira, C., & Snigula, J. 2001b, MNRAS, 325, 550
12. Feulner, G., Bender, R., Drory, N., Hopp, U., Snigula, J., & Hill, G. J. 2003, MNRAS, 342, 605
13. Fontana, A., et al. 2003, ApJ, 594, L9
14. Hammer, F., et al. 1997, ApJ, 481, 49
15. Hogg, D. W., Cohen, J. G., Blandford, R., & Pahre, M. A. 1998, ApJ, 504, 622
16. Kochanek, C. S., et al. 2001, ApJ, 560, 566
17. Lilly, S. J., Le Fèvre, O., Hammer, F., & Crampton, D. 1996, ApJ, 460, L1
18. Loveday, J. 2000, MNRAS, 312, 557
19. Madau, P., Ferguson, H. C., Dickinson, M. E., Giavalisco, M., Steidel, C. C., & Fruchter, A. 1996, MNRAS, 283, 1388
20. Maraston, C. 1998, MNRAS, 300, 872
21. Snigula, J., Drory, N., Bender, R., Botzler, C. S., Feulner, G., & Hopp, U. 2002, MNRAS, 336, 1329
22. Steidel, C. C., Adelberger, K. L., Giavalisco, M., Dickinson, M., & Pettini, M. 1999, ApJ, 519, 1

Tracing the Formation of Massive Spheroids from High-z Galaxy Clustering

Emanuele Daddi

European Southern Observatory, Karl-Schwarzschild-Str. 2,
D-85748 Garching bei München, Germany

Abstract. The high-z progenitors of local massive early-type galaxies should be characterized by a strong level of clustering, larger than that measured for $z = 3$ Lyman Break Galaxies and comparable to that of $z \sim 1$ EROs. First possible evidences for such strongly clustered objects at $z \gtrsim 2$ were found by the FIRES and K20 surveys, that have identified new classes of faint high-z K-selected galaxies. Some details are given here for the new population of massive star-forming galaxies at $z \sim 2$, found by the K20 survey in the GOODS-South area. Because of their much redder UV continuum, most of these galaxies would not be selected by the Lyman Break criterion. Such objects are good candidates for the precursors of local ellipticals caught in their formation phase. We have calibrated a two color criterion to allow the identification of these highest redshift galaxies in bright K-limited samples.

1 Introduction

Understanding the evolution, star-formation and assembly histories of early type galaxies up to their highest redshifts of formation is a crucial observational question that remains unsolved. Passively evolving, massive spheroidal galaxies appear to be common up to $z \approx 1$ with little evolution, if any, from present days [6,22,2]. It is not well constrained yet if and how many passive spheroids exist at higher redshifts, $z > 1.5$–2. Moreover, it is unclear up to what redshift most precursors of spheroids maintain the low-redshift signatures of morphological and stellar population properties. Approaching their formation epoch, spheroid progenitors will appear as star-forming and possibly morphologically irregular galaxies. Therefore, identifying the precursors of todays ellipticals close to their most interesting formation or assembly phases requires to rely on independent and alternative signatures than their morphologies or old stellar population properties.

2 Expected Clustering of Spheroid Precursors

Spheroids inhabit preferentially the densest environment in the local universe. They have the largest correlation lengths among galaxies in the present day universe and up to $z \sim 1$ [19,7]. *Searching for the most clustered population at any redshift is therefore a natural way to locate such progenitors and to investigate their properties.*

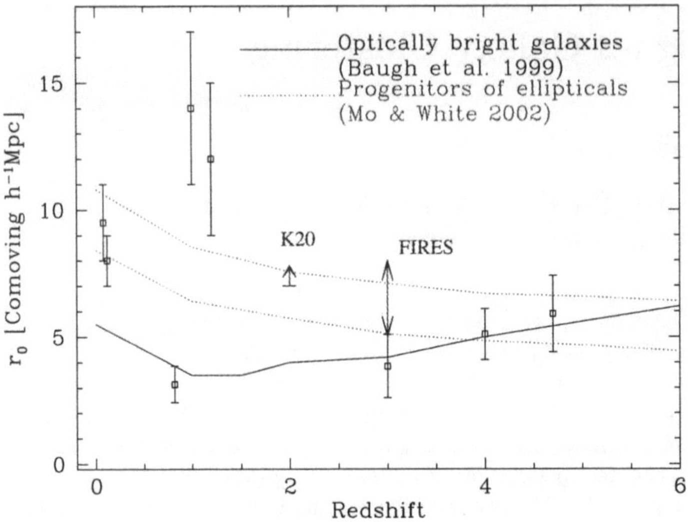

Fig. 1. Predicted evolution for the correlation length of halos that will evolve into local massive ellipticals (dotted lines, from [17]) and of optically selected galaxies (solid line, from [1]). Together with the new constraints coming from K-selected populations at $z \gtrsim 2$ (heavy symbols), we show relevant measurements taken from the literature: ellipticals at $z \sim 0$ from [19], optically selected galaxies at $z \sim 0.8$ from [5], EROs and radio galaxies at $z \sim 1$ from [7,21], $z = 3$ LBGs from [13] and $z = 4$–5 LBGs from [20]

This view is supported by model predictions for the redshift evolution of the correlation length of halos hosting today massive ellipticals (Fig. 1, based on [17]). Halos clustered with $r_0 \sim 8$–10 in the local universe, are predicted to have $r_0 \sim 5$–8 h^{-1} Mpc even at $2 < z < 4$. Being referred to halos, these predictions could represent only lower limits for galaxies that, at the relatively small scales probed by high-z measurements, can have enhanced correlations due to large halo occupation numbers (e.g. [18]).

In practice, it is impossible to observationally preselect a population by its clustering properties and one has to look at the clustering properties of the available classes of high-z galaxies to find putative progenitors.

Extremely red objects (EROs) and giant $z \sim 1$ radio galaxies, as well known, have an high clustering and therefore are good candidate progenitors of local ellipticals [6,16,7,21], in agreement with some significant fraction of these objects being passive spheroids (e.g. [4]).

Measurements for the highest redshifts ($z \sim 4$–5) star-forming galaxies [20] of $r_0 \sim 5$–6 may be consistent with the predictions for forming spheroids. This is also in qualitative agreement with the observation that the oldest stars (thus the ones formed at the highest redshifts) are nowadays in spheroids.

Typical $z = 3$ Lyman Break Galaxies (LBGs) appear instead to start falling short of the expected clustering level of ellipticals progenitors, with the best current estimates of $r_0 \lesssim 4$ h^{-1} Mpc [24] and with fainter LBGs having even much weaker clustering [13]. In fact the clustering of observed LBGs at $z = 3$ is much lower than that of $z \sim 1$ EROs, suggesting that possibly a large fraction of them do not evolve into EROs and ellipticals. Overall, the observed clustering of optically selected (i.e. actively star forming) galaxies from $z = 5$ to $z = 1$ [5] is in very good agreement with the model predictions of [1]. The comparison of the two set of predictions of Fig. 1, diverging below $z \sim 4$, also confirm that by $z \lesssim 3$ ordinary star formation has began producing stars that have mainly ended in today later type or less massive galaxies.

Should it be possible to isolate the progenitors of local spheroids as a class of galaxies at $1.5 < z < 3.5$, their clustering would be expected to be fairly strong, in the cited range $r_0 \sim$ 5–8 h^{-1} Mpc or even more. New populations of K-selected, $z \sim$ 2–3 galaxies possibly satisfying those clustering requirements were recently reported by the FIRES and K20 surveys [9,10]. Along with SCUBA sources, these galaxy populations represent the best candidates so far identified for the precursors of massive spheroids.

3 $J - K$ Red FIRES Galaxies at $2 < z < 4$

The FIRES survey of the HDF-South [15] has produced the first large sample of about 120 K-selected galaxies in the (photometric) redshift range $2 < z < 4$. From the analysis of their correlation properties [9], these $z = 3$ K-selected galaxies appear more strongly clustered than LBGs (at a fixed number density), although the comparison suffers from the small area of HDFS. A significant result was the detection in the FIRES sample of a strong color dependence of clustering, in the usual way that redder galaxies appear more clustered (but a previously unobserved effect at any $z > 1$). The clustering depends primarily on the $J - K$ color, which brackets the Balmer (or 4000Å) break at $z \approx 3$, suggesting the effect could be due to age. Intriguingly, the existence of populations older and more clustered than LBGs could solve the paradox that $z = 3$ LBGs are too young to be direct progeny of $z >$ 4–6 star forming galaxies [12].

The galaxies with the reddest ($J - K > 1.7$) colors at $z \approx 3$ have an angular two-point correlation function consistent with $r_0 \sim 8\ h^{-1}$ Mpc (using a typical slope $\delta = 0.8$), although a weaker $r_0 \gtrsim 5\ h^{-1}$ Mpc could be made consistent with the data assuming enhanced small scale correlations in the "halo occupation distribution" framework [26,14]. Overall, these clustering strengths of 5–8 h^{-1} Mpc are at the level expected for the progenitors of ellipticals seen at $z \sim 3$ (see Fig. 1). The first handful spectroscopic redshifts for this class of $J - K$ red galaxies [25] are grouped in a single redshift spike, a feature that support the presence of strong clustering.

4 Near-IR Bright Galaxies at $z \sim 2$ from the K20 Survey

The K20 survey resulted in the observation of a significant high redshift tail of $1.7 < z < 2.3$ galaxies with bright near-IR magnitudes ($K_{Vega} < 20$, [3]).

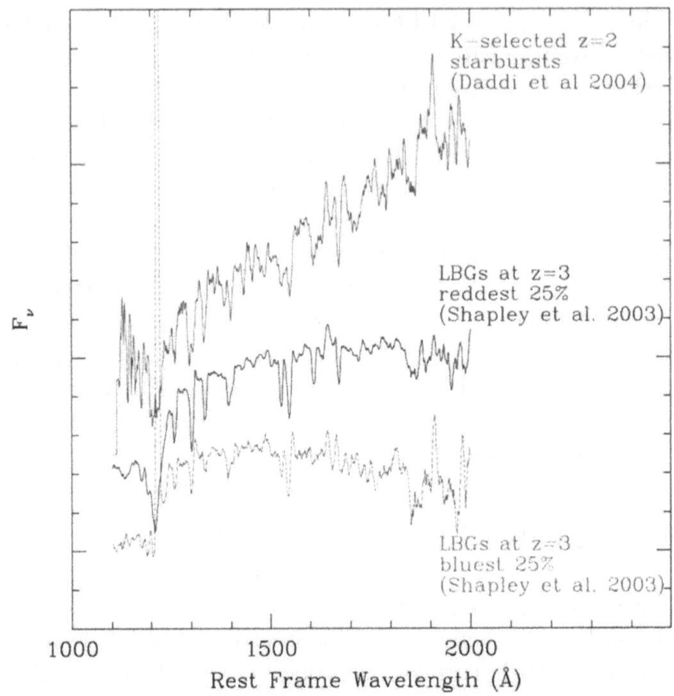

Fig. 2. The coadded spectrum of the $z = 2$ K-selected galaxies (top, from [10]) is compared to the reddest and bluest composite spectrum of LBGs (center and bottom, from [23]). The strong Lyα emission is from the bluest LBG spectrum.

The vast majority of such galaxies appear to be actively star-forming, reddened, morphologically irregular and massive objects [10].

Remarkably, 6 out of the 9 galaxies with measured spectroscopic redshift belong to spikes with $\lesssim 1500$ km/s of redshift separation. Assuming a random redshift selection function within the $1.7 < z < 2.3$ range such a level of pairing has a probability of about 10^{-5}, a clear evidence of clustering. It is intriguing that even for EROs a similarly strong level of redshift pairing is not observed. We performed Monte Carlo simulations building samples of clustered populations (using the method presented in [8]) in order to see what correlation length is necessary to produce a similar redshift pairing in our small sample. It is found that in order to have a 5% or more probability to find as many pairs as in the observed sample, $r_0 \gtrsim 7\ h^{-1}$ Mpc is required. The fraction of close pairs in a survey is directly related to the integral of the real space two-point correlation function, and has little dependence on the correlation slope δ. The pairing evidence thus suggest that these near-IR bright $z \sim 2$ galaxies are consistent with having a high-level of clustering, like the one expected for the progenitors of local ellipticals seen at $z \sim 2$. It would be extremely interesting to spectroscopically

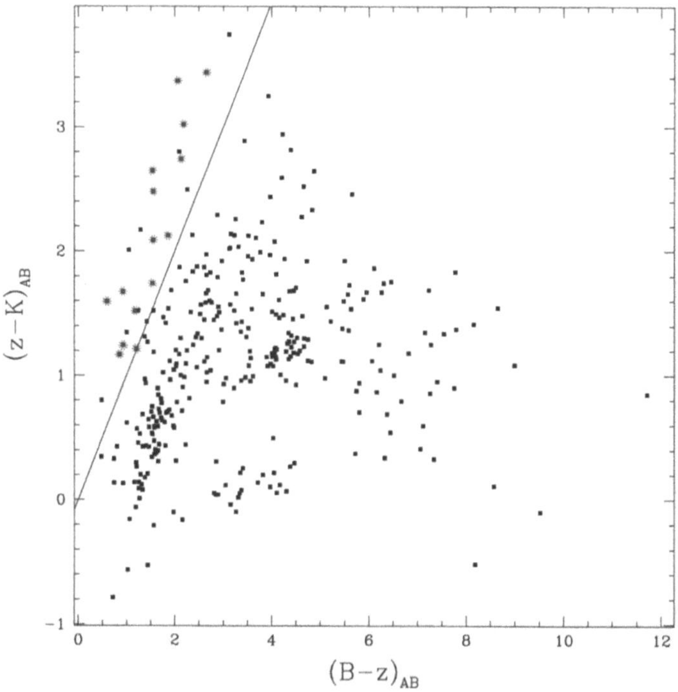

Fig. 3. Two color diagrams showing galaxies from the K20 survey in the CDFS field with the 9 near-IR bright galaxies at $z \sim 2$ shown as asterisks, including also a few redder objects with reliable photometric redshift $z > 2$. The straight line has $(z - K)_{AB} = (B - z)_{AB}$.

confirm a much larger samples, in order to obtain improved estimates for the clustering of this galaxy population.

Galaxies at $z \sim 2$ with bright K-band magnitudes were unknown before [10] and [25], and actually most theoretical models had predicted that they should not exist at all (see e.g. [3] for relative references). As they appear to be actively star-forming, just like LBGs, it is interesting to assess whether these K-selected galaxies are just a special subsample (the brightest tip of the iceberg) of LBGs, or a different population. This is especially important as now large samples of $\sim$ 1000 LBGs have been assembled down to $1.5 < z < 2.5$ [24]. Fig. 2 shows that the coadded spectrum of $z = 2$ K-selected galaxies is remarkably redder than typical LBGs at $z = 3$, even for the reddest among them. For a more direct comparison to LBGs, we used the available multicolor photometry in the CDFS/GOODS area to determine the synthetic G and R magnitudes of our 9 $z = 2$ objects, matching the $z = 2$ selection criterion of [24] (the available U-band data closely match the system of [24]). Although the procedure is a bit uncertain, the $G-R$ colors of near-IR bright $z \sim 2$ objects appear again generally much redder than what required by the LBG selection criterion in the same redshift range [11].

These near-IR bright $z = 2$ galaxies are thus a new population of high-z galaxies, that cannot be selected with the usual LBG criterion. As they are only $\approx 5\%$ of the population in a K-limited sample at $K < 20$, it would be very useful to have an alternative method to distinguish them from lower redshift field galaxies. Using the highly complete spectroscopic sample of the K20 survey we have calibrated a very efficient two color criterion: galaxies with $(z - K)_{AB} \gtrsim (B - z)_{AB}$ have an high probability to be star-forming objects at $z \sim 1.5$–2.5 (Fig. 3). We are currently using this BzK criterion to build up a much larger sample of spectroscopically identified near-IR bright $z = 2$ galaxies. This will be crucial in order to understand their properties and the role played in the formation history of galaxies and to confirm their nature of possible star forming progenitors of local spheroids.

Acknowledgements

The K20 and FIRES teams are warmly thanked. The work presented here would have not been possible without the dedicated efforts of all the people involved in those projects.

References

1. Baugh C. M., et al., 1999, MNRAS, 305, L21
2. Bell E., et al. 2004 (astro-ph/0303394, see also these proceedings)
3. Cimatti A., et al., 2002, 391, L1
4. Cimatti A., et al., 2003, A&A, 412, L1
5. Coil A.L., et al., 2003 (astro-ph/0305586)
6. Daddi E., et al., 2000, A&A 361, 535
7. Daddi E., et al., 2001, A&A, 376, 825
8. Daddi E., et al., 2002, A&A, 384, L1
9. Daddi E., et al., 2003, ApJ, 588, 50
10. Daddi E., et al., 2004, ApJ, 600, L127
11. Erb D.K., et al., 2003, ApJ, 591, 101
12. Ferguson H., Dickinson M., Papovich C., 2002, ApJ, 569, L65
13. Giavalisco M. & Dickinson M., 2001, ApJ, 550, 177
14. Kravtsov A.V., et al., 2004, (astro-ph/0308519)
15. Labbé I., et al., 2003, AJ, 125, 1107
16. McCarthy P. J. et al., 2001, ApJ, 560, L131
17. Mo H. J., White S. D. M., 2002, MNRAS, 336, 112
18. Moustakas L. A., Somerville R. S., 2002, ApJ, 577, 1
19. Norberg P. et al., 2002, MNRAS, 332, 827
20. Ouchi M., et al., 2004 (astro-ph/0309657)
21. Overzier R. A., et al., 2003, A&A, 405, 53
22. Pozzetti L., et al., 2003, 402, 837
23. Shapley A., et al., 2003, ApJ, 588, 65
24. Steidel C., et al., these proceedings
25. van Dokkum P., et al., 2003, ApJ, 587, L83
26. Zheng Z., 2004 (astro-ph/0307030)

Galaxy Formation and Evolution since z = 1

François Hammer[1], Hector Flores[1], YanChun Liang[1], XianZhong Zheng[1], David Elbaz[2], and Catherine Cesarsky[3]

[1] GEPI, Observatoire de Paris, 92195 Meudon Principal, France
[2] CEA, Saclay-SAp, Ormes des Merisiers, 91191 Gif sur Yvette, France
[3] ESO, Karl-Schwarzschild-Strasse 2, D-85748 Garching bei München, Germany

Abstract. Determination of the star formation rate can be done using mid-IR photometry or Balmer line luminosity after a proper correction for extinction effects. Both methods show convergent results while those based on UV or on [OII]λ3727 luminosities underestimate the SFR by factors ranging from 5 to 40 for starbursts and for luminous IR galaxies, respectively. Most of the evolution of the cosmic star formation density is related to the evolution of luminous compact galaxies and to luminous IR galaxies. Because they were metal deficient and were forming stars at very high rates (40 to 100 $M_{\odot} yr^{-1}$), it is probable that these (massive) galaxies were actively forming the bulk of their stellar/metal content at z $\leq$ 1.

1 Introduction

Determining how galaxies formed and grew is one of the outstanding problems of modern astrophysics. The epoch at which half the stellar mass was formed has been estimated to be z=1-1.5 from simple integrations of the global star formation history of the Universe ([1]; [2]). Similar results have been found by estimating the growth of galaxies by calculating the stellar mass from SEDs with broad wavelength coverage ([3]; [4]; [5]). However, this method depends strongly on the estimated M/L of individual galaxies which is sensitive to the assumptions about their star formation history, dust distribution, metallicity, and IMF – all of which are usually poorly constrained. Many efforts have been made to study galaxies at redshifts higher than 1, in attempts to track the formation of half the present day stellar mass. However the situation at z> 1 is difficult, because of their faintness and because most of the important lines for diagnostics are redshifted to the near IR. Here we choose to investigate the properties of z $\leq$ 1 galaxies, for which the current generation of telescopes is able to provide accurate measurements of their properties (star formation rates, masses, metal abundances). In a cosmological model with H_0 =70km s^{-1} Mpc^{-3}, Ω_{Λ}=0.7 and Ω_M=0.3, z=1 corresponds to 58% of the Universe age. At z> 0.4, the emergence of two galaxy populations have been reported, namely the luminous infrared galaxies (LIRGs, [1]; [6]) and the luminous compact galaxies (LCGs, [7]; [8]). These two populations are responsible of most of the evolution of the IR and UV luminosity density evolutions, and then of a significant fraction of the cosmic star formation density at z$\sim$ 1.

In the following, we present a summary of results based on follow-up studies of the Canada France Redshift Survey (CFRS) using the Very Large Telescope

(VLT), the Infrared Space Observatory (ISO) and the Hubble Space Telescope (HST).

2 Estimating Extinction and SFR at z $\sim$ 1

A prerequisite for estimating star formation from emission is to properly estimate the extinction. $H\alpha$ luminosity is one of the best indicator of the instantaneous SFR. However low resolution spectroscopy (R $<$ 500) often produces misleading results [9]. Only good S/N spectroscopy with moderate spectral resolution (R>600) allows a proper estimate of the extinction from the $H\alpha/H\beta$ ratio after accounting for underlying stellar absorption. SFRs can be otherwise underestimated or overestimated by factors reaching 10, even if one accounts for an *ad hoc* extinction correction. These effects are prominent for a large fraction of evolved massive galaxies especially those experiencing successive bursts (A and F stars dominating their absorption spectra). Further estimates of the cosmic star formation density at all redshifts mandatorily requires moderate resolution spectroscopy to avoid severe and uncontrolled biases.

Liang et al ([9]) have given an obvious warning for the studies based on low resolution spectroscopy aimed at measuring individual galaxy properties (gas chemical abundances, interstellar extinction, stellar population, ages as well as star formation rates and history), particularly for dusty spiral galaxies. For example, it has been shown ([10]) that 1/4 of the Lilly et al sample ([11]) of z $\sim$ 0.7 galaxies was made of LIRGs which are dust enshrouded systems (average A_V= 2.4). Assuming a constant extinction of A_V=1 for these LIRGs, leads to underestimate their [OII]λ3727/$H\beta$ ratio, providing an overestimate of their metallicity by 0.3 to 0.5 dex.

ISOCAM observations give us a unique opportunity to test the validity of our SFR estimates. It has been shown ([6], see also David Elbaz's contribution) that the mid-IR and radio luminosities correlate well up to z=1. Bolometric IR measurements derived from mid-IR are validated, unless if both the radio-FIR and the MIR-FIR correlations are invalid at high redshifts. Mid-IR photometry provides an unique tool to estimate properly the SFR of LIRGs (defined as $L_{IR} > 2\ 10^{11}\ L_\odot$) up to z=1.2, and of starbursts ($L_{IR} < 2\ 10^{11}\ L_\odot$) up to z=0.4. For a sample of 16 ISO galaxies at 0 $<$ z $<$ 1, SFRs have been derived from the extinction corrected $H\alpha$ luminosities ([12]). These values agree within a factor 2 with mid-IR estimates (median value of $SFR_{IR}/SFR_{H\alpha}$=1.05). Using moderate spectral resolution (R = 1200) with good S/N of 90 ISO galaxies (0.2 $<$ z $<$ 1), Liang et al [10] have shown that the extinction can be properly derived from the $H\beta/H\gamma$ ratio. Such measurements can be performed for all galaxies up to z=1 within an accuracy of 0.6 mag for A_V values, and then for SFR values.

It has been argued ([13]) that the [OII]λ3727 luminosity can provide an efficient way to derive SFRs of high redshift galaxies using spectrographs in the visible range. However preliminary results from [14] and [15] have shown that the correlation between $H\alpha$ and [OII]λ3727 is rather poor, because of extinction effects and also because it considerably depends on the galaxy metallicity,

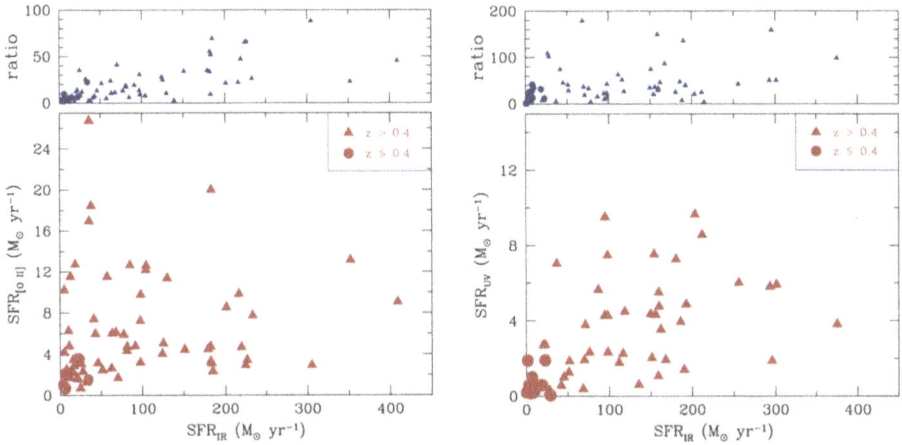

Fig. 1. Comparison of SFRs calculated using the formalism of [13]. (*Left panel*): [OII]λ3727 estimates are compared to IR estimates for a sample of 70 galaxies from the CFRS and Marano fields (see [10]). [OII]λ3727 estimates have been done using the formulae from [7] which have been adequately multiplied by 2.46 to account for the fact that they have used a different IMF than [13]. Upper left panel provides the $SFR_{IR}/SFR_{[OII]}$ ratio. Median values of the ratio are 5 and 22 for starbursts (SFR $<$ 40 $M_{\odot}yr^{-1}$ and LIRGs respectively. (*Right panel*): same comparison for UV (2800Å) luminosity estimated for 61 CFRS galaxies. Median values of the SFR_{IR}/SFR_{UV} ratio are 13 and 36 for starbursts (SFR $<$ 40 $M_{\odot}yr^{-1}$ and LIRGs, respectively.

luminosity and spectrophotometric type. How [OII]λ3727 estimates compare to IR (or $H\alpha$) estimates of the SFR ? Figure 1 shows that the former strongly underestimates the SFR by factor 5 for starbursts and factor 22 for LIRGs. A similar result is found by comparing SFR estimates from UV to those from IR, and has been already discussed by [8] for a sample of distant LCGs (average SFR= 40 $M_{\odot}yr^{-1}$). The fact that UV fluxes provide even lower SFR values than [OII]λ3727 fluxes might be related to the expected increase of [OII]λ3727/$H\beta$ ratio with decreasing metal abundance (high z systems being expected to be less metal abundant than local ones).

3 Were All Massive Galaxies Already Formed at z=1?

It has been claimed ([16]) that the stellar mass built up of massive spirals was mostly done at z$\sim$ 1. This claim was based on the very low values of the $SFR/M_{stellar}$ for CFRS galaxies, yielding very low stellar mass increases (ranging from 1 to 10%) for massive galaxies ($M_{stellar} > 10^{11}M_{\odot}$) since z=1. If true, this result would imply that most of the star formation at z$<$ 1 occurred in dwarf galaxies. However, present day dwarfs contain only a marginal fraction of the stars or metals, and the Brinchman and Ellis 's result ([16]) is somewhat at odd with integrations of the star formation history and with stellar mass density at different lookback times. We notice that SFRs calculated by [16] were based on the formulae from [7], and provide values 54 and 12 times lower than our estimates for LIRGs and starbursts, respectively. This potentially leads to a severe revision of the conclusions from [16].

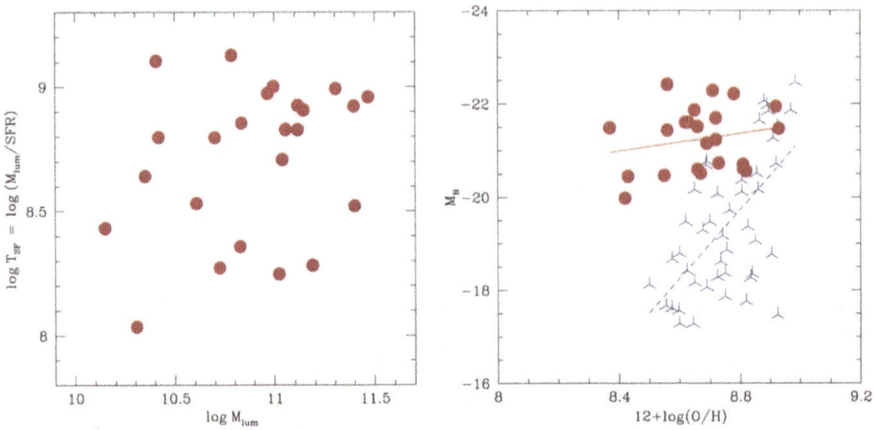

Fig. 2. (*Left panel*): $M_{stellar}/SFR$ against $M_{stellar}$ for distant LIRGs (from [17]). Because of their very high SFRs, LIRGs can build up of significant part of their masses in relatively short times.(*Right panel*): M_B versus O/H abundance ratio of distant LIRGs (from [10], full dots) compared to those of local spirals (from [18] and [15], skeletal symbols). Full line and dotted lines report the regression law for LIRGs and local spirals, respectively. At z$\sim$ 0.75, LIRGs have log(O/H) values 0.3 dex lower than those of local spirals.

A pioneering study of 14 distant LCGs ($M_B <$ -20; $r_{half} \leq 3.5\ h_{70}^{-1}$ kpc) have revealed that they are dust enshrouded starbursts superimposed to an old and evolved stellar population. It has lead us ([8]) to suggest that LCGs are the progenitors of present-day spiral bulges, prior to the star formation in the disk component. Because LCGs correspond to 23% of the luminous galaxy population at z $\sim$ 0.75, they could be the progenitors of a similar fraction of present-day luminous spirals.

Recently, important progress has been made in our understanding of distant LIRGS ([10]; [17]). They are forming rapidly new generations of stars and possess stellar masses comparable or slightly smaller than that of present-day massive spirals (Figure 2; [17]). Galaxies experiencing such strong star forming events could form a significant fraction of their masses during the last 8 Gyr. Their metal content have been found to be on average half that of today massive spirals (Figure 2; [10]): at such rates of star/metal production, they can be the progenitors of massive spirals within timescales much smaller than a Hubble time. Indeed LIRG morphologies are intimately related to giant disks (Figure 3;

[17]), and, incidentally include a significant fraction of compact galaxies. Because LIRGs and LCGs are much more frequent at z> 0.4 than today, a substantial fraction of massive galaxies could be actively forming their stellar content at z < 1. It does not necessarily contradict previous studies (e.g. [19]) which have shown that the number density of large spirals ($r_{disk} > 2.8\ h_{70}^{-1}$ kpc) was similar at z=0.75 than today. In fact we realize that one third of the large disks in the sample of [19] were indeed LIRGs, strengthening our prediction that many large spirals were actively forming the bulk of their stellar content at z < 1.

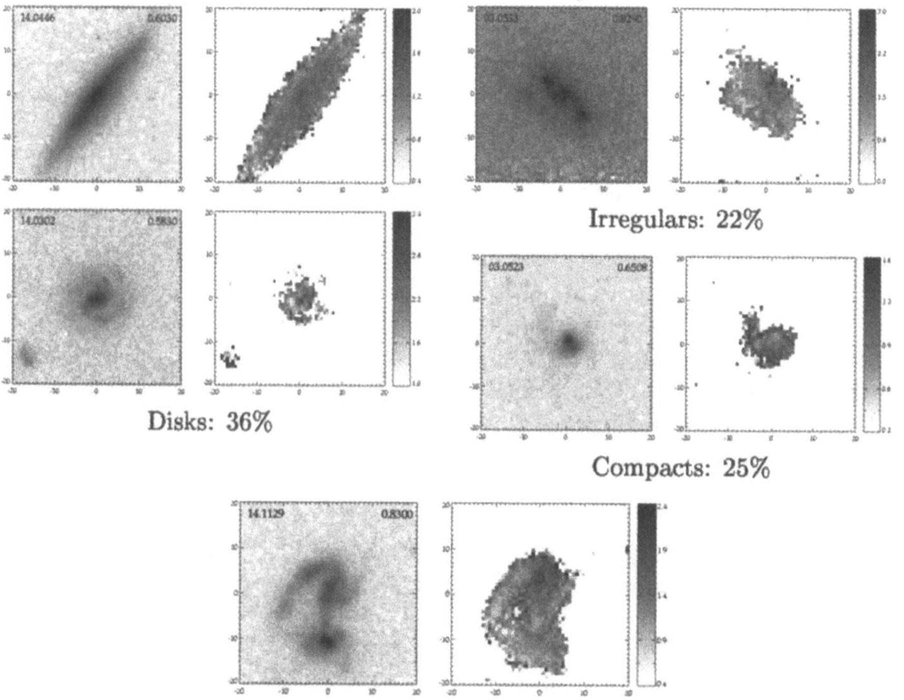

Fig. 3. Morphological classifications using color maps from a study of 36 LIRGs by [17]. Using color map and deconvolution software (GIM2D), it has succeeded to classify all the galaxies (fractions of each type are given). For each galaxy, the left stamp (40x40 kpc^2 area) shows the F814W WFPC2 image, and the right stamp shows the color map (F606W-F814W) following the prescription of Zheng et al ([17]). Color map have been derived after a very accurate alignment of V and I images down to 0.015 arcsec, which is a prerequisite for galaxies with some irregularities in their morphologies. Notice the fact that most distant galaxies show small blue and very red regions, interpreted as HII regions and dusty enshrouded regions, respectively. Notice also the very blue color in the center of compact galaxies.

Extragalactic studies might appear inaccurate and affected by numerous biases and uncertainties when compared to WMAP results (see Licia Verde's contribution): they have been nicknamed "dirty physics" during the first talk of the Conference. Indeed, the extensive use of photometric redshifts or of low resolution spectroscopy of z> 1 galaxies can cast some doubts about our knowledge of their properties. On the other hand, physical parameters of z≤ 1 galaxies can be determined rather accurately, because these sources are enough bright to be at the reach of current telescopes. This requires to apply methodologies (for example high S/N and medium resolution spectroscopy) which have been successful for analysing local galaxy properties. In this paper, we also systematically compare two independent estimates of each individual parameter: SFR calculations have been done by comparing IR and Balmer line luminosities and stellar production in LIRGs and in LCGs have been compared to their metal content relative to that of local spirals.

We intend to pursue our analyses towards the general population of field galaxies up to z=1, including by studying their morphologies, metal content and SFRs (Hammer et al, 2004, in preparation). We believe that establishing a new classification sequence for all galaxies at z < 1 is at the reach of the present generation of telescopes. An important goal will be to study their dynamics using the multi-integral field unit mode of FLAMES/GIRAFFE at VLT. It would provide an important test for the merging scenario, as well as a solid estimate of the evolution of the Tully-Fisher relation up to z=1. Dynamical masses of galaxies could be then compared to their stellar masses.

References

1. Flores, H., Hammer, F., Thuan, T.X. et al.: Astrophys. J. **517**, 148 (1999)
2. Madau, P. and Pozzetti, L. : MNRAS **312**, 9 (2000)
3. Pozzetti, L., Cimatti, A., Zamorani, G. et al: Astron. Astrophys. **402**, 66 (2003)
4. Fontana, A., Donnarumma, I., Vanzella, E. et al.: Astrophys. J. **594**, 9 (2003)
5. Dickinson, M., Papovitch, C., Ferguson, H. and Budavari, T.: Astrophys. J. **587**, 25 (2003)
6. Elbaz, D., Cesarsky, C. J., Chanial, P. et al.: Astron. Astrophys. **384**, 848 (2002)
7. Guzman, R., Gallego, J., Koo, D. et al.: Astrophys. J. **489**, 559 (1997)
8. Hammer, F., Gruel, N., Thuan, T.X. et al.: Astrophys. J. **550**, 570 (2001)
9. Liang, Y.C., Hammer, F., Flores, H. et al.: Astron. Astrophys., in press (2003a,astro-ph/0312341)
10. Liang, Y.C., Hammer, F., Flores, H. et al.: Astron. Astrophys., submitted (2003b)
11. Lilly, S., Carollo, C. and Stockton, A.: Astrophys. J. **597**, 730 (2003)
12. Flores, H., Hammer, F., Elbaz, D. et al.: Astron. Astrophys., in press (2003, astro-ph/0311113)
13. Kennicutt, R.: Ann. Rev. Astron. Astrophys. **36**, 189 (1998)
14. Hammer, F., Flores, H., Lilly, S. et al.: Astrophys. J. **481**, 49 (1997)
15. Jansen, R.., Franx, M. and Fabricant, D..: Astrophys. J. **551**, 825 (2001)
16. Brinchman, J. and Ellis, R. : Astrophys. J. **536**, L77 (2000)
17. Zheng, X.Z., Hammer, F., Flores, H. et al.: Astron. Astrophys., submitted (2003)
18. Kennicutt, R. C.: Astrophys. J. **388**, 310 (1992)
19. Lilly, S., Schade, D., Ellis, R. et al.: Astrophys. J. **500**, 75 (1998)

Characteristic Scales in Galaxy Formation

Avishai Dekel[1,2]

[1] The Hebrew University of Jerusalem, Israel
[2] Institut d'Astrophysique and Observatoire de Paris, France

Abstract. Recent data, e.g. from SDSS and 2dF, reveal a robust *bi-modality* in the distribution of galaxy properties, with a characteristic transition scale at stellar mass $M_* \sim 3 \times 10^{10} M_\odot$ (near L_*), corresponding to virial velocity $V \sim 100\,\mathrm{km\,s^{-1}}$. Smaller galaxies tend to be blue disks of young populations. They define a "fundamental line" of decreasing surface brightness, metallicity and velocity with decreasing M_*, which extends to the smallest dwarf galaxies. Galaxies above the critical scale are dominated by red spheroids of old populations, with roughly constant high surface brightens and metallicity, and they tend to host AGNs. A minimum in the virial M/L is obtained at the same magic scale. This bi-modality can be the combined imprint of several different physical processes. On smaller scales, disks are built by *cold flows*, and *supernova feedback* is effective in regulating star formation. On larger scales, the infalling gas is *heated by a virial shock* and star formation can be suppressed by *AGN feedback*. Another feedback mechanism – gas evaporation due to photo-ionization – may explain the existence of totally dark halos below $V \sim 30\,\mathrm{km\,s^{-1}}$. The standard cooling barriers are responsible for the loose upper and lower bounds for galaxies: $10 < V < 300\,\mathrm{km\,s^{-1}}$.

1 Introduction

Observations indicate four characteristic scales in the global properties of galaxies, expressed either in terms of virial velocity V, or stellar mass M_*:

- An upper limit at $V \sim \mathbf{300}\,\mathrm{km\,s^{-1}}$, which is the lower bound for clusters.
- A robust bi-modality scale at $M_* \simeq 3 \times 10^{10} M_\odot$ [15], corresponding to $V \sim \mathbf{100}\,\mathrm{km\,s^{-1}}$. This is comparable to L_* of the Schechter luminosity function, marking the exponential upper bound for disk galaxies. Smaller galaxies tend to be blue, star-forming disks with the surface brightness and metallicity decreasing with decreasing M_* (LSB). More massive galaxies are dominated by red, old-population spheroids with high surface brightness and metallicity insensitive to M_* (HSB) [15]. AGNs typically reside in halos above the bi-modality scale [16]. The halo M/L shows a minimum at the same scale [35].
- Dark-dark halos (DDH) with no luminous components seem to dominate the halo distribution below $V \sim \mathbf{30}\,\mathrm{km\,s^{-1}}$. The relatively flat faint-end luminosity function and the steeper mass function predicted by the standard ΛCDM cosmology cannot be reconciled with the virial theorem and the Tully-Fisher relation obeyed by the luminous galaxies (§5). Also, most dwarf spheroidals in the Local Group lie in this range, while dwarf irregulars are typically $V > 30\,\mathrm{km\,s^{-1}}$ [22] [32] [13] [9] [34].
- A lower bound at $V \sim \mathbf{10}\,\mathrm{km\,s^{-1}}$ for dwarf galaxies (see [9]).

We summarize here how *theory* predicts the following characteristic scales:

- A loose upper bound for cooling (by bremsstrahlung) on a dynamical time at $M \sim 10^{12-13} M_\odot$, corresponding to $V \sim 300\,\mathrm{km\,s^{-1}}$.
- A virial shock-heating scale at $M_* \sim 3 \times 10^{10} M_\odot$, or $V \sim 100\,\mathrm{km\,s^{-1}}$. In less massive halos virial shocks cannot form and the gas flows cold into the disk vicinity (§2).
- A supernova-feedback scale at $V \sim 100\,\mathrm{km\,s^{-1}}$, below which the potential wells are shallow enough for the energy fed to the ISM by supernova to significantly heat the halo gas, or even blow it out (§3).
- A photo-ionization scale at $V \sim 30\,\mathrm{km\,s^{-1}}$. Once the gas is continuously ionized by the background UV flux, IGM gas cannot accrete into smaller halos and gas already in such halos evaporates via a steady thermal wind (§5).
- A lower bound for cooling (by atomic hydrogen) at $V \sim 10\,\mathrm{km\,s^{-1}}$.

We briefly address the origin of these scales, and try to associate them with the observed scales.

2 Shock-Heating Scale

The standard picture of disk formation is that after the dark halo virializes, the infalling gas is shock heated near the virial radius. It then cools radiatively and, if the cooling time is shorter than the dynamical time, the gas accretes onto a central disk and forms stars with the associated feedback. The upper-limit halo mass for efficient cooling is $M_{\rm cool} \sim 10^{12-13} M_\odot$ [27] [33]. The common wisdom used to be that this explains the upper bound for disk galaxies, but it has been realized that the predicted scale is somewhat higher than the observed scale.

Fig. 1 shows the time evolution of the radii of Lagrangian gas shells in a spherical system consisting of gas (with primordial composition) and dark matter, as simulated by Birnboim & Dekel [5] using a 1D hydro code. Not shown are the dark-matter shells, which collapse, oscillate and tend to increase the gravitational attraction exerted on the gas shells. The left panel focuses on massive halos of $\sim 10^{12} M_\odot$. Indeed, a shock exists near the virial radius at all times; it gradually propagates outward, encompassing more mass in time. The hot post-shock gas is in a quasi-static equilibrium, pressure supported against gravity. The right panel focuses on halo masses smaller by an order of magnitude. An interesting new phenomenon is seen here, where a stable shock forms and propagates from the disk toward the virial radius only after a mass of $\sim 10^{11} M_\odot$ has virialized. In less massive systems, the cooling rate is faster than the infall time, making the "post-shock" gas unstable against gravitational infall and thus not being able to support an extended shock.

This behavior can be understood via the following instability analysis of the gas just behind the shock [5]. The standard equation of state is written as $P = (\gamma - 1)\rho e$ where ρ and e are the gas density and internal energy, and $\gamma = 5/3$ for a mono-atomic gas. In the no-cooling case, the adiabatic index is defined as

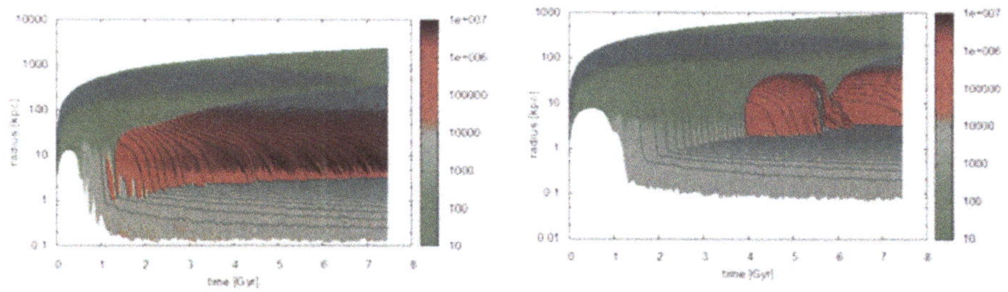

Fig. 1. Time evolution of the radii of Lagrangian gas shells (lines) in a spherical simulation of a protogalaxy consisting of primordial gas and dark matter. Temperature is marked by color. A shock shows up as a sharp break in the flow lines and as an abrupt change in temperature. The lower discontinuity marks the "disk" radius, formed due to an artificial centrifugal force. (**a**) A massive system, where the virialized masses are in the range $10^{11} - 10^{13} M_{\odot}$. (**a**) A less massive system, $10^{10} - 10^{12} M_{\odot}$. A virial shock exists only in systems more massive than $> 10^{11} M_{\odot}$.

$\gamma \equiv (\partial \ln P / \partial \ln \rho)_{\rm ad}$, and the system is known to be gravitationally stable once $\gamma > 4/3$. When there is radiative cooling at a rate q, we define a new quantity:

$$\gamma_{\rm eff} \equiv \left(\frac{d \ln P}{d \ln \rho} \right) = \gamma - \frac{\rho}{\dot{\rho}} \frac{q}{e} = \gamma - \frac{t_{\rm dyn}}{t_{\rm cool}} \,. \qquad (1)$$

The second equality follows from energy conservation, $\dot{e} = -P\dot{V} - q$, plugged into the equation of state. Note that $\gamma_{\rm eff} = \gamma$ when $q = 0$, and that the difference between the two is nothing but the ratio of characteristic times $t_{\rm dyn}/t_{\rm cool}$, where $t_{\rm dyn} \equiv \rho/\dot{\rho} \simeq r_{\rm s}/u$, with $r_{\rm s}$ the shock radius and u the infall velocity into it. Using the jump conditions across the shock, one can express $\gamma_{\rm eff}$ in terms of the pre-shock gas quantities:

$$\gamma_{\rm eff} = \gamma - \frac{(\gamma+1)^4}{6(\gamma-1)^2} \frac{\rho r_{\rm s} \Lambda(T)}{|u^3|} \,, \qquad T = \frac{\mu}{k_{\rm B} N_{\rm A}} \frac{2\gamma - 1}{(\gamma+1)^2} u^2 \,, \qquad (2)$$

where ρ is the pre-shock density.

In order to test for stability, we perturb a shell by $r \to r + \delta r$ and compute the sign of the restoring force: $\ddot{\delta r}/\delta r$. We assume a homologous infall $\delta r = u \delta t$ (supported by the finding in the simulations above), and balancing forces $\ddot{r} = -\rho^{-1} \nabla P - GM/r^3 = 0$. This leads to

$$\ddot{\delta r}/\delta r \propto \gamma - 4/3 - 2\gamma_{\rm eff}^{-1} (\gamma - \gamma_{\rm eff}) \,. \qquad (3)$$

The system is stable when this expression is positive, namely for

$$\gamma_{\rm eff} > \gamma_{\rm crit} = \frac{2\gamma}{\gamma + 2/3}, \quad \text{or} \quad \frac{t_{\rm cool}}{t_{\rm dyn}} > \frac{\gamma + 2/3}{\gamma^2 - 4\gamma/3} \,. \qquad (4)$$

For $\gamma = 5/3$, we get $\gamma_{\rm crit} = 10/7$ replacing the $\gamma_{\rm crit} = 4/3$ of the adiabatic case. The critical scale is then characterized by $t_{\rm cool}/t_{\rm dyn} = 21/5$ at the disk vicinity, which translates to $t_{\rm cool}/t_{\rm dyn} \simeq 1$ at the virial radius. Note that in our notation $t_{\rm dyn} \simeq R/V$ at the virial radius, while the dynamical time used in the order-of-magnitude arguments of [27] and [33] is several times larger.

When $\gamma_{\rm eff}$ is computed using eq. (2) just above the disk radius at every time of the spherical simulation, we find that the criterion of eq. (4) indeed works very well. Before the shock forms, it is well below $\gamma_{\rm crit}$ and gradually rising, reaching $\gamma_{\rm crit}$ almost exactly when the shock starts propagating outward. It then oscillates about $\gamma_{\rm crit}$ with a decreasing amplitude, following the oscillations in the shock radius seen in Fig. 1.

When put in a cosmological setting, we find that the critical halo mass for virial-shock heating is

$$M_{\rm shock} \simeq (1-6) \times 10^{11} M_\odot \,. \tag{5}$$

The higher estimate is obtained for near solar metallicity. This scale is comparable to the observed bi-modality scale, and is quite insensitive to redshift. To see this, we use the spherical-collapse virial relations at z: $M \propto (1+z)^{-3/2} V^3 \propto (1+z)^3 R^3$, and the general trend in the relevant range $\Lambda \propto T^{-1/2}$, to obtain $t_{\rm cool}/t_{\rm dyn} \propto M$ independent of z. Since at $z > 2$ most halos are of $M < M_{\rm shock}$, we conclude that in most forming disks the gas never heats to the virial temperature – it flows cold all the way to the disk vicinity.

Although we are back to a condition involving $t_{\rm cool}/t_{\rm dyn}$, which seems to remind us of the original criterion by Rees & Ostriker, it is now based on analyzing the actual physical process that could be responsible for the heating to the virial temperature, namely the existence of a virial shock. Because the dynamical time involved in the current analysis is smaller than the one used in the original estimate by a factor of a few, our derived upper bound for halo masses in which cold flows prevail is smaller by a similar factor.

Preliminary results from cosmological hydro simulations indicate that the phenomenon is not restricted to spherical systems, and that our stability criterion may in fact be approximately valid in the general case where the inflow is along filaments. Fig. 2 shows snapshots from Eulerian simulations by A. Kravtsov [19] [20], showing the gas temperature and flow fields in two protogalaxies: one of $M \simeq 3 \times 10^{11} M_\odot$ at $z \simeq 4$, and the other of $M \simeq 2 \times 10^{10} M_\odot$ at $z \simeq 9$. While the more massive galaxy shows a hot gas component at the virial temperature behind a virial shock, the smaller galaxy has mostly cold gas inside the virial radius. Also seen in the massive galaxy are cold streams penetrating the hot medium toward the central disk. Similar results have been obtained by Fradal et al. [11], who emphasis the feeding of galaxies by cold flows preferentially at early epochs, corresponding to less massive galaxies. We are in a process of testing our shock-stability criterion using cosmological simulations, and generalizing it to the realistic case of elongated streams.

Beyond introducing a new scale in galaxy formation, the cold streaming toward the disk, as opposed to the commonly assumed shock heating and slow accretion, may have several interesting implications. For example:

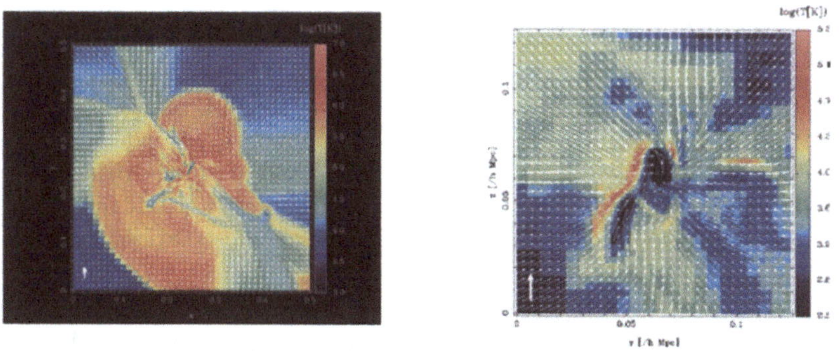

Fig. 2. Snapshots from cosmological hydro simulations (courtesy of Andrey Kravtsov), showing the gas temperature and flow fields in two protogalaxies. (**a**) A halo of $M \simeq 3\times 10^{11} M_\odot$ at $z \simeq 4$, showing a virial shock and heated gas penetrated by cold streams. (**b**) A halo of $M \simeq 2 \times 10^{10} M_\odot$ at $z \simeq 9$, showing cold infall with little shock heating inside the virial radius of $\sim 7\,h^{-1}$kpc.

- It may modify the star formation rate and the associated supernova feedback. If the result of the cold stream hitting the disk is a burst of stars at high z, this may help the semi-analytic models (SAMs) explain better the presence of luminous red galaxies at $z \sim 1$. This process should be properly incorporated in the SAMs.
- It may reduce the soft X-ray flux from halos. This is consistent with observed individual galaxies [6], and may explain why the observed X-ray background is lower than the predictions based on shock heating in all halos [26].
- The cold gas may emit a significant fraction of its infall energy in Ly-α, either if the streams cool and merge into the rotating disk gradually without shocking [17], or after shocking at the disk vicinity. In the latter, the X-rays produced are likely to be captured inside a dense Strongrem sphere of a few kpc and the energy may eventually be released by Ly-α. This may explain the observed Ly-α emitters [21].

3 Supernova Feedback Scale

Supernova feedback, which is a very different physical process, gives rise to a characteristic scale in the same ballpark. The energy fed to the ISM by supernovae can be written as

$$E_{\rm SN} \simeq \nu\epsilon \dot{M}_* t_{\rm rad} \propto M_*(t_{\rm cool}/t_{\rm dyn}) \,, \tag{6}$$

where ν is the number of supernovae per unit mass of forming stars, ϵ is the typical supernova energy, and $t_{\rm cool}$ is the time available for the supernova remnant to share its energy with the medium during its adiabatic phase, before a significant fraction of it is radiated away. The second proportionality follows from the crude assumption $\dot{M}_* \sim M_*/t_{\rm dyn}$. Dekel & Silk [8] pointed out that for halos in the relevant range, where the virial temperature is $T \sim 10^5$K, the cooling rate behaves roughly like $\Lambda \propto T^{-1}$, and then $t_{\rm cool}/t_{\rm dyn} \simeq 0.01$ for all galaxies. This means that, despite the significant radiative losses, the energy fed to the ISM is $E_{\rm SN} \propto M_*$. When this energy input is compared to the energy required for significantly heating the gas or blowing it out, $E_{\rm binding} \simeq M_{\rm gas} V^2$, one ends up with the critical scale for effective feedback,

$$V_{\rm SN} \simeq 100\,{\rm km\,s^{-1}}\,. \tag{7}$$

In shallower potential wells, the feedback from a burst of stars can significantly suppress further star formation, and produce low-surface-brightness (LSB) and dwarf galaxies. We note that $V_{\rm SN}$ corresponds to $M_* \simeq 3\times10^{10} M_\odot$, comparable to the observed bi-modality scale.

Tight correlations between the properties of galaxies below $3\times10^{10} M_\odot$ define a "*fundamental line*" (FL) which stretches across five decades in luminosity down to the smallest dwarfs. The mean scaling relations, involving M_*, V, the metallicity Z and the surface brightness μ, are approximately [15] [9] [34] [31]

$$V \propto M_*^{0.2}, \quad Z \propto M_*^{0.4}, \quad \mu \propto M_*^{0.6}\,. \tag{8}$$

Note that the familiar Tully-Fisher relation at the bright end, $V \propto M_*^{0.3}$ [7], seems to flatten in the dwarf regime. Supernova feedback can explain the origin of the FL in simple terms [9]. The above energy criterion, $E_{\rm SN} \propto M_* \propto M_{\rm gas} V^2$, with the assumption that the relevant gas mass is a constant fraction of the halo mass, $M_{\rm gas} \propto M$, immediately implies the relation driving the FL:

$$M_*/M \propto V^2\,. \tag{9}$$

Assuming spherical collapse to virial equilibrium, the halo virial quantities are related via, $M \propto V^3 \propto R^3$. In the instantaneous recycling approximation, we approximate $Z \propto M_*/M_{\rm gas}$. For the stellar radius we adopt the standard assumption of angular-momentum conservation [10] [24], $R_* \propto \lambda R$, with λ a constant spin parameter, and use it in $\mu \propto M_*/R_*^2$. Combining the relations from eq. (9) and on, we recover the scaling relations, eq. (8). This success of this simplest possible toy model indicates that supernova feedback could indeed be the primary driver of the FL below the bi-modality scale, and it strengths the association of this process with the origin of the transition scale itself.

4 The Origin of Bi-Modality

In Fig. 3, we schematically put together the physical scales discussed so far in the classic density-temperature diagram. Galaxies are loosely confined to the region where $t_{\rm cool} < t_{\rm ff}$. The line $M_{*,\rm shock} \sim 3\times10^{10} M_\odot$ distinguishes between massive galaxies which suffer virial shock heating and lower mass galaxies where the gas flows cold toward the disk. The line $V \sim 100\,{\rm km\,s^{-1}}$ marks the transition from deep potential wells where HSBs form to shallow potential wells where supernova feedback is effective and leads to LSBs and dwarfs along the FL.

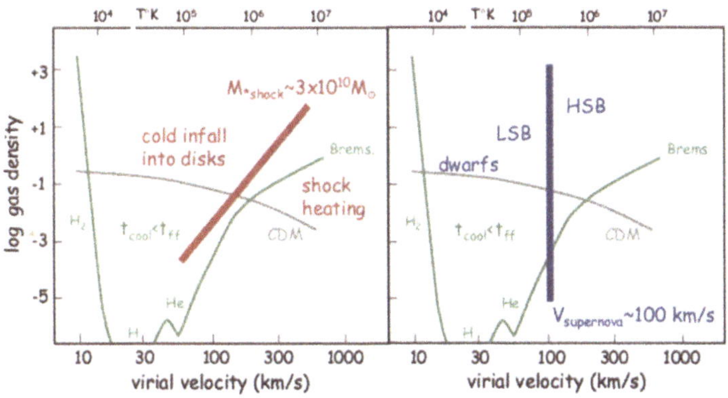

Fig. 3. The characteristic scales in the classic diagram of gas density versus virial velocity. The horizontal (gray) curve marks 2-σ halos in ΛCDM. The V-shaped (green) curve refers to $t_{\rm cool} = t_{\rm dyn}$; galaxies can form only above this line. (**a**) The shock-heating scale. (**b**) The supernova feedback scale. The two scales are comparable, at $V \sim 100\,{\rm km\,s^{-1}}$ or $M_* \sim 3 \times 10^{10} M_\odot$.

The two different physical processes happen to yield a similar characteristic scale, and they thus combine to produce the bi-modality. A possible scenario is as follows:

- Halos below $M_{\rm shock}$ had cold infall to disks, possibly associated with bursts of star formation. Galaxies above this threshold are built from progenitors that were below the threshold at high z and thus formed early disks, which later merged to spheroids of red, old stars. Any halo gas in such massive halos consists of a hot medium and possibly cold streams.
- In halos below $V_{\rm SN}$, the SN feedback regulates the star-formation rate, which leads to blue, young stars and the FL, $M_*/M \propto V^2$, with M/L rising toward smaller halos. The SN feedback may also suppress any AGN activity in these halos.
- In more massive halos, the gas, which has been heated by virial shocks, may be prevented from cooling by AGN feedback, causing M/L to rise toward larger halos in proportion to the total-to-cold gas ratio. There seems to be enough energy in AGNs, though the actual feedback mechanism is not properly understood yet.
- Protogalaxies below $M_{\rm shock}$ may be seen as Ly-α emitters rather then as X-ray sources, while the hot gas in more massive halos still produces X-rays.

5 Photo-Evaporation Scale

Some of the halos below $V \sim 30\,{\rm km\,s^{-1}}$ must be *totally dark*, with no luminous component. Marginal evidence is starting to emerge from the statistics of image multiplicities in gravitational lenses [18], and independently from the following comparison between theory and the observed frequency of dwarf galaxies. While the faint-end galaxy luminosity function is roughly $\phi(L) \propto L^{-1}$ (or steeper), the halo mass function predicted in ΛCDM is roughly $\psi(M) \propto M^{-2}$ (or flatter). These cannot be reconciled with the virial relation $M \propto V^3$ and the Tully-Fisher relation in the dwarf regime, $L \propto V^5$ (or flatter). If, however, only a fraction $f_{\rm L}(V)$ of the small halos have a luminous component (with $L/M \propto V^2$ as discussed above), then the observations can be reconciled once $f_{\rm L} \propto V^2$. Thus, if the dark-dark halos appear at $V \sim 30\,{\rm km\,s^{-1}}$, we expect $\sim 90\%$ of the halos to be totally dark at the faint end, $V \sim 10\,{\rm km\,s^{-1}}$.

Supernova outflows alone are not likely to produce DDHs, both because they must leave a trace of stars behind, and because they tend not to be equally effective in all directions [23]. Ram pressure due to outflows from nearby galaxies may remove gas in some cases [28], but probably not enough for the purpose. On the other hand, the origin of DDHs could be explained by *radiative feedback*. The first stars and AGNs produce UV flux which ionizes most of the gas in the universe by $z_{\rm ion} \sim 10$, and keeps it at $T \sim (1-2) \times 10^4$K until $z_{\rm end} \sim 2$ or so [2]. Further infall of gas is suppressed in halos below the Jeans scale, which is crudely estimated to be about $V \sim 30\,{\rm km\,s^{-1}}$ [12], but this does not explain the complete removal of the gas already residing in those halos. A significant fraction of $T \simeq 10^4$K gas is expected to escape instantaneously from halos of a correspondingly shallower potential, $V < 10\,{\rm km\,s^{-1}}$ [1], but not from halos of $10 < V < 30\,{\rm km\,s^{-1}}$.

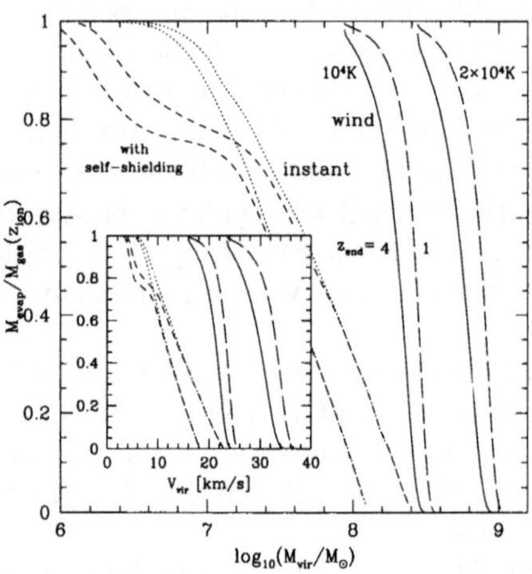

Fig. 4. The evaporated mass fraction as a function of virial mass and velocity (insert). The curves on the left correspond to instantaneous expulsion, and the curves on the right refer to steady evaporation, where the ionization is assumed to be efficient from $z = 8$ till either $z = 4$ (solid) or $z = 1$ (dashed). The gas temperature is either 10^4K (left) or 2×10^4K (right). While half the gas is lost instantaneously from $V \sim 10\,{\rm km\,s^{-1}}$ halos, a steady wind can remove a similar fraction of the gas from $V \sim 30\,{\rm km\,s^{-1}}$ halos.

However, if a continuous input of energy keeps the gas ionized for many dynamical times, it can evaporate via a *steady thermal wind* even from deeper potential wells. Once the gas is lost from the shallow edge of the potential well, it is continuously replenished by interior gas, and so on. In Shaviv & Dekel [29] we analyzed this process following the classic solar-wind analysis [25], and found that a significant fraction of the gas eventually evaporates from halos as large as $V \sim 30\,\mathrm{km\,s^{-1}}$. Fig. 4 summarizes the results of this analysis.

Thus, halos in the range $10 < V < 30\,\mathrm{km\,s^{-1}}$ lose most of their gas after the cosmological reionization. They may end up as gas-poor dwarf spheroidals with an old stellar population [9] [29], or as DDHs (see also [30]).

Halos of $V < 10\,\mathrm{km\,s^{-1}}$ are not expected to form stars at all because of the cooling barrier at $T \sim 10^4$K (Fig. 3). Atomic line cooling is very efficient just above this temperature, but the only cooling agent below 10^4K is molecular hydrogen, which cannot survive the UV background and is anyway very inefficient [14].

6 Conclusion

The following table summarizes the characteristic scales originating from the physical processes discussed above and their possible association with the observed scales. The classic argument of cooling on a vaguely-defined dynamical time provides relatively loose upper and lower bounds for luminous galaxies. The combination of (a) the newly discovered scale distinguishing between virial shock heating and cold flows and (b) the familiar supernova feedback scale seems to produce the robust bi-modality imprinted on the observed galaxy properties, and in particular the characteristic upper limit for disk galaxies near L_*. This is provided that AGN feedback, or another process such as thermal conductivity, indeed prevents the shock-heated gas in massive halos from ever forming stars. The details of the proposed scenario are yet to be worked out (see also [3] [4]). While supernova feedback can be the primary process responsible for the "fundamental line" of LSB and dwarf galaxies, the radiative feedback due to cosmological reionization may be responsible for totally evacuating small halos from gas and producing dwarf spheroidals and dark-dark halos. Once these physical

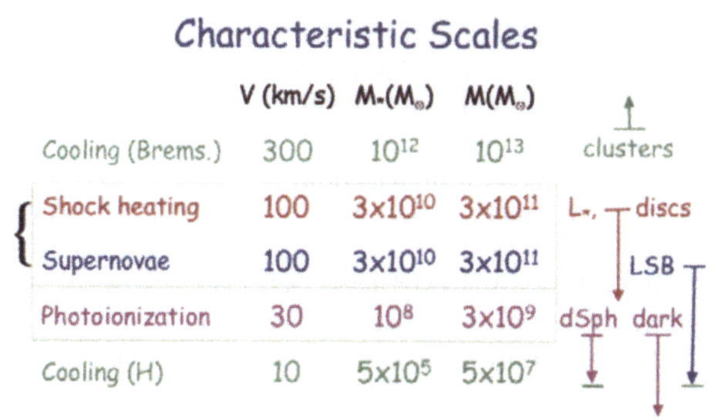

process are properly incorporated in semi-analytic models of galaxy formation, they may help solving the apparent conflicts between theory and observation.

Acknowledgment: I thank my collaborators Yuval Birinboimi and Nir Shaviv, and those responsible for the 3D hydro simulations, R. Dave, N. Katz, D. Keres, A. Kravtsov, & D. Weinberg. This research has been supported by ISF 213/02 and NASA ATP NAG5-8218.

References

1. Barkana R., Loeb A., 1999, ApJ, 523, 54
2. Barkana R., Loeb A., 2001, PhR, 3
3. Benson A.J., Bower R.G., Frenk C.S., Lacey C.G., Baugh C.M., Cole S., ApJ, 599, 38
4. Binney J., 2004, MNRAS, astro-ph/0308172
5. Birnboim, Y., Dekel, A., 2003, MNRAS, 345, 349
6. Ciotti L., Pellegrini S., Renzini A., D'Ercole A., 1991 ApJ, 376, 380
7. Courteau, S., MacArthur, L.A., Dekel, A. et al., 2004, astro-ph/0310440
8. Dekel, A., Silk, J., 1986, ApJ, 303, 39.
9. Dekel, A., Woo, J., 2003, MNRAS, 344, 1131
10. Fall M., Efstathiou G., 1980, MNRAS, xxx
11. Fardal, M.A., Katz, N., Gardner, J.P., Hernquist, L., Weinberg, D.H., Dave, R., 2001, ApJ, 562, 605
12. Gnedin N.Y., 2000, ApJ, 542, 535
13. Grebel E.K., Gallagher J.S., Harbeck D., 2003, AJ, 125, 1926
14. Haiman Z., Rees M.J., Loeb A., 1996, ApJ, 467, 522
15. Kauffmann et al. 2003, MNRAS, 341, 54
16. Kauffmann et al. 2004, in preparation
17. Keres D., Dave R., Katz, N., Weinberg, D.H. et al., 2004, in preparation
18. Cohn J.D., Kochanek C.S., 2004, astro-ph/0306171
19. Kravtsov A.V., 2003, ApJL, 590, L1
20. Kravtsov A.V., Gnedin, O.Y., 2003, ApJ submitted (astro-ph/0305199)
21. Manning C., Stern D., Spinrad H., Bunker A.J., 2000, ApJ 537, 65
22. Mateo M., 1998, ARA&A, 36, 435
23. Mac low M.M., Ferrara A., 1999, ApJ, 513, 142
24. Mo H.J., Mao S., White S.D.M., 1998, MNRAS, 295, 319
25. Parker E.N. 1960, ApJ, 132, 821
26. Pen U. L., 1999, ApJL 510, L1
27. Rees M.J., Ostriker J.P., 1977, MNRAS, 179, 541
28. Scannapieco E. Ferrara A., Broadhurst T., 2000, ApJL, 536, L11
29. Shaviv, N., Dekel, A., 2004, astro-ph/0305527
30. Somerville, R.S. 2002, ApJL, 572, L23
31. Tremonti et al. (SDSS), 2004, in preparation
32. van den Bergh S., 2000, The Galaxies of the Local Group (Cambridge University Press)
33. White, S.D.M., Rees, M.J., 1978, MNRAS, 183, 341.
34. Woo, J., Dekel, A., Courteau, S., 2004, in preparation
35. Yang X., Mo H.J., van den Bosch F.C., 2003, MNRAS, 339, 1057

Galaxy Evolution in Three Dimensions: Time, Space and Mass

Tadayuki Kodama[1], Richard Bower[2], Philip Best[3], Patrick Hall[4], Toru Yamada[1], and Masayuki Tanaka[5]

[1] National Astronomical Observatory of Japan, 2–21–1 Osawa, Mitaka, Tokyo, 181–8588, Japan
[2] Department of Physics, University of Durham, South Road, Durham DH1 3LE, UK
[3] Institute for Astronomy, Royal Observatory Edinburgh, Blackford Hill, Edinburgh EH9 3HJ
[4] Department of Astrophysical Sciences, Princeton University, Peyton Hall, Ivy Lane, Princeton, NJ 08544–1001, USA
[5] Department of Astronomy, School of Science, University of Tokyo, Bunkyo-ku, Tokyo 113–0033, Japan

Abstract. There are three major axes to describe the evolution of galaxies, namely, time (redshift), space (environment) and mass (stellar mass). In this article, one topic each will be presented along these axes. (1) Based on the Subaru wide-field (30′) optical imaging of two distant clusters (z=0.55 and 0.83), we show the large scale structures in galaxy distributions on a scale greater than 10 Mpc, which serve as an evidence for hierarchical growth of rich clusters of galaxies through assembly of surrounding groups. (2) Based on the deep NIR imaging of high-z clusters at $z \sim$1 and 1.5, we show massive galaxies in clusters have assembled most of their mass by $z = 1.5$, which is earlier than the hierarchical model predictions. (3) Based on the Subaru deep and wide optical imaging of Subaru/XMM-Newton Deep Survey Field, we show a deficit of red and faint galaxies and a lack of blue massive galaxies in the high density regions at $z \sim 1$, which suggest down-sizing in galaxy formation as apparently opposed to the CDM-based bottom-up scenario.

1 *Space*: Panoramic Imaging and Spectroscopy of Cluster Evolution with Subaru (PISCES)

First of all, along the space axis, we present panoramic imaging of two distant clusters taken with Subaru wide-field optical camera Suprime-Cam which provides $34' \times 27'$ field of view corresponding to 15×12 Mpc^2 at $z = 0.8$. This project called PISCES (Panoramic Imaging and Spectroscopy of Cluster Evolution with Subaru) has started since 2003, and we aim to target 10-15 X-ray selected distant clusters in total at $0.4 \lesssim z \lesssim 1.3$, in good coordination with *ACS/HST*, *XMM*, and *Chandra* observations. The goals of this programme is to map out the large scale structures around clusters to trace the cluster assembly history and then to look into the galaxy properties as a function of environment along the structures to directly identify the environmental effects acting on galaxies during their assembly to higher density regions. This unique project is currently underway and just a preliminary result on the large scale structures is shown

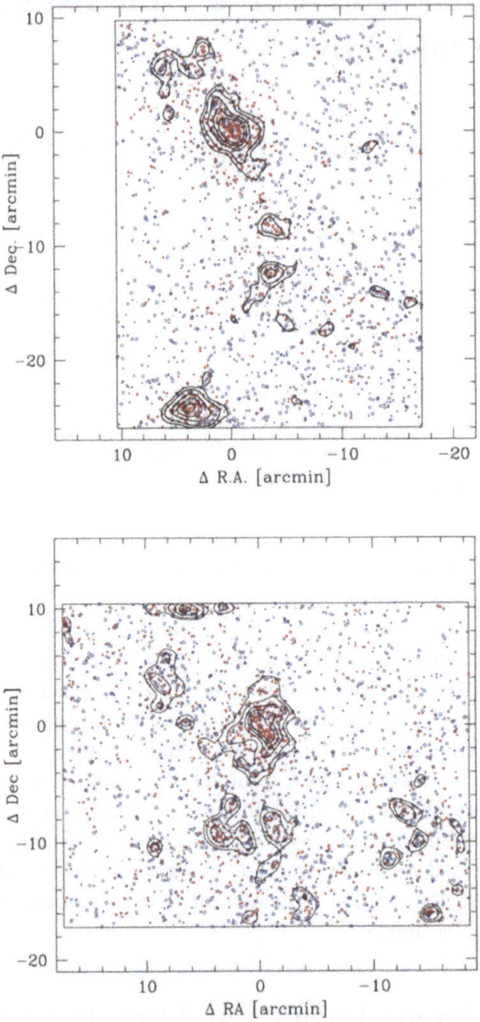

Fig. 1. The panoramic maps of CL0016+16 cluster (z = 0.55) and RXJ0152.7–1357 (z=0.83) are presented on the top and the bottom panels (10 arcminutes corresponds to 3.8 Mpc and 4.6 Mpc, respectively). Using photometric redshift technique based on multi-colour data ($BVRi'z'$ and $VRi'z'$, respectively), only photometric member candidates are plotted selected with redshift cuts of $0.47 \leq z \leq 0.6$ and $0.75 \leq z \leq 0.9$, respectively. Contours show local 2-D number density of galaxies at 1.5, 2, 3, 4, 5 σ above the mean density. Coordinates are shown relative to the centre of the main cluster. Large scale filamentary structures (>10Mpc) are seen in both clusters.

in Fig. 1. These two rich clusters at z=0.55 and 0.83 are imaged in multi optical bands with Suprime-Cam, and photometric redshifts [13] have been applied to efficiently remove foreground/background contaminations and to isolate the cluster member candidates (cf., [14]). Many substructures are now clearly seen around the main body of the clusters which tend to be aligned in filamentary structures extending to >10 Mpc scale across. Although these structures should be confirmed spectroscopically later on, these already provide good evidence for cluster assembly in the hierarchical Universe.

[14] have presented the environmental dependence of galaxy colours along the filamentary structures around the A851 cluster (z=0.41), and have shown that the galaxy colour changes rather sharply at relatively low density regions like galaxy groups along the filaments well outside of the cluster core. Together with the similar findings in the local Universe [18][9]. the environmental effects that truncate star formation are not the cluster specific phenomena such as ram-pressure stripping [1] but are found to be much wider spread into low density regions. It is important to extend this analysis to higher redshifts as the galaxy environment is expected to change dramatically during the course of vigorous assembly, which is probably related to the appearance of morphology-density relation [8].

2 *Time:* Stellar Mass Assembly of Massive Galaxies in High-z Clusters

Secondly, along the time axis, we present the stellar mass functions of galaxies in high-z clusters constructed from deep near-infrared imaging (J and K_s) (Fig. 2). The stellar mass function of galaxies derived from the K-band observations is a good tracer of mass assembly history of galaxies (eg., [10][3]). We have combined two $z \sim 1$ clusters (3C336 and Q1335+28) [16], and five $z \sim 1.5$ clusters (Q0835+580, Q1126+101, Q1258+404, Q0139–273, and Q2025–155) [20][21][4] to increase statistics. We have subtracted the control field counts taken from the literature [23][24][4]. Applying the same technique described in [16], we construct the field-subtracted stellar mass functions of galaxies in high-z clusters, primarily using K_s-band flux and also using $J - K_s$ colour as a measure of the M/L ratio. As shown, little evolution is observed since $z = 1.5$ to the present-day (2MASS clusters [2]), indicating that the mass assembly on galaxy scale is largely completed by $z \sim 1.5$ in the cluster environment. This epoch of mass assembly of massive galaxies is earlier than the prediction of the hierarchical models as shown for comparison [19].

It is also interesting to trace back the stellar mass to even higher redshift to firstly identify the epoch of assembly of massive galaxies where they start to break down into pieces. Recent deep NIR observations in fact start to enter such formation epoch (eg., [7][22][6]), and this aspect will be further extended by on-going and future space missions such as *SIRTF* and *Astro-F*.

3 *Mass:* Down-Sizing in Galaxy Formation Seen at $z \sim 1$

Finally, along the mass axis, we present how the star formation history depends on the mass of the galaxies at $z \sim 1$ based on the unique Suprime-Cam imaging

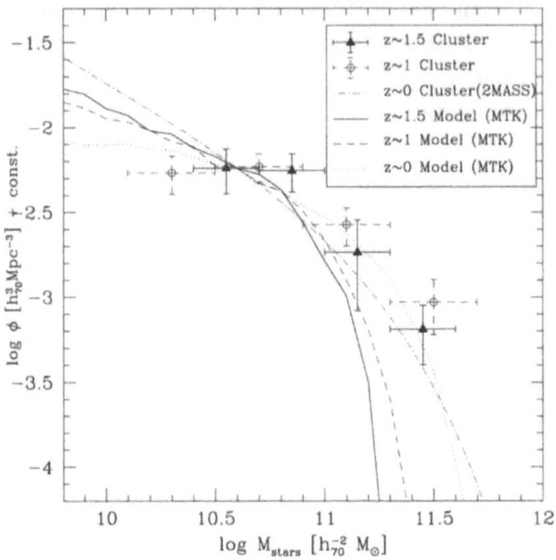

Fig. 2. Stellar mass functions of galaxies in clusters as a function of redshift. The Kennicutt's initial mass function [11] are used to scale the stellar mass. The open diamonds and the filled triangles show the stellar mass functions for $z \sim 1$ and $z \sim 1.5$ clusters, respectively, which are compared to the local counterpart from the 2MASS survey (dot-dashed curve). All the curves and the data points are normalised at $5{\times}10^{10}$M$_\odot$ so as to have the same amplitude. The theoretical predictions from a semi-analytic model [19] are presented for comparison, which are made for galaxies in the haloes whose circular velocities are greater than 1000 km/s at each epoch. The mass assembly of massive galaxies in the real Universe is much faster than the hierarchical model prediction.

data ($BRi'z'$) on the Subaru/XMM-Newton Deep Survey (SXDS) Field. Our data are both sufficiently deep (z'_{AB}=25, 6-10σ) and wide (1.2 deg^2), which enable us for the first time to investigate the photometric properties of statistical sample of galaxies at $z \sim 1$ down to $\sim M^*+3$ with respect to the passive evolution. We first identify five $z \sim 1$ high density regions by applying colour cuts at $1.7<R-z'<2.0$ and $0.8<i'-z'<1.1$, which correspond to the colours of passively evolving galaxies at $z \sim 1$ [12]. We then combine these five regions (amounting to 141 arcmin^2 in total) and subtract off the low density regions at $z \sim 1$ (scaled to the same area) on the colour-magnitude diagram to isolate only $z \sim 1$ galaxies in a statistical sense, since the both regions are expected to have the same amount of foreground/background contaminations.

Thus constructed field-corrected colour-magnitude diagram for $z \sim 1$ galaxies in high density regions is shown in Fig. 3. Most striking feature of this diagram is that there are two distinct populations, 'bright+red' and 'faint+blue'. More precisely, we show a deficit of red and faint galaxies below M^*+2 or 10^{10}M$_\odot$ in stellar mass and a lack of blue massive galaxies beyond $M^*-0.5$ or $8{\times}10^{10}$M$_\odot$ in stellar mass. This bimodality in colour-magnitude distributions can be also

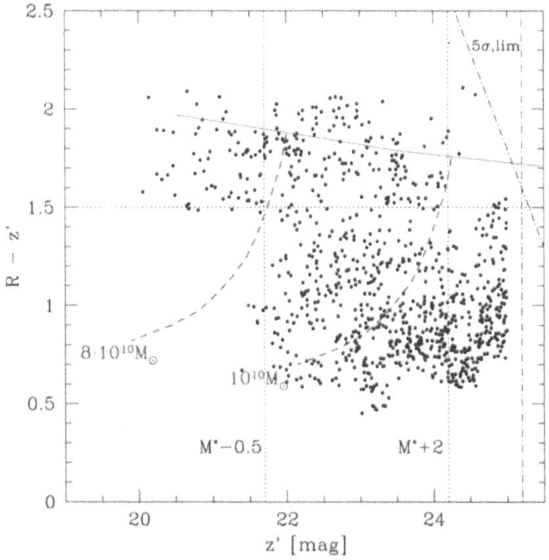

Fig. 3. Field-corrected colour-magnitude diagram for the $z \sim 1$ galaxies in high density regions [17]. The solid line show the expected location of colour-magnitude sequence at $z \sim 1$ assuming a passive evolution with $z_{\rm form}$=5 [12]. The deficit of both blue galaxies at the bright/massive end and the deficit of red galaxies at the faint/less-massive end are both clearly identified.

seen in the colour-dependent stellar mass function shown in Fig. 4. Red galaxies tend to be bright and show a bell-like shape, while the blue galaxies tend to be faint and show a much steeper faint-end slope. These results indicate the down-sizing in galaxy formation [5], where star formation and mass assembly in massive galaxies take place early in the Universe, while the star formation seems to be progressively shifted towards less massive systems as the Universe ages. Next obvious important step is to directly detect the brightening of such break luminosity or mass with increasing redshift [15].

We have assumed the cosmological parameters of (H_0, Ω_m, Ω_Λ)=(70, 0.3, 0.7), throughout this paper.

Acknowledgements

We thank Drs. Ian Smail and Eric Bell for useful discussion. We also acknowledge to the SXDS team (Kaz Sekiguchi et al.). This work was financially supported in part by a Grant-in-Aid for the Scientific Research (No. 15740126) by the Japanese Ministry of Education, Culture, Sports and Science.

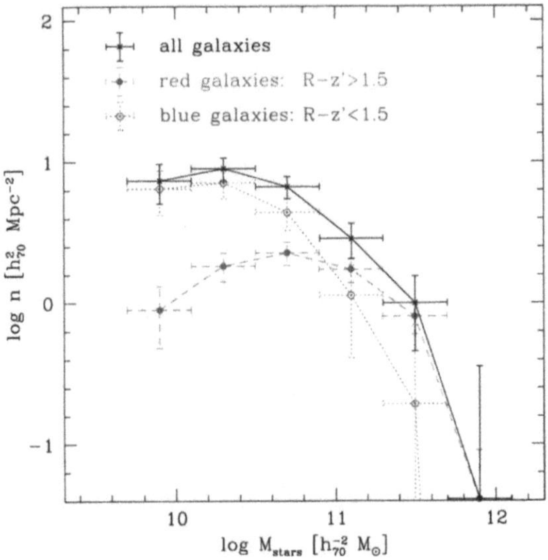

Fig. 4. Colour-dependent stellar mass functions of the $z \sim 1$ galaxies in high density regions [17]. The stellar masses are scaled using the Kennicutt's IMF [11].

References

1. Abadi, M.G., Moore, B., Bower, R.G.: MNRAS, **308**, 947 (1999)
2. Balogh, M.L., Christlein, D., Zabludoff, A.I., Zarisky, D.: ApJ, **557**, 117 (2001)
3. Baugh, C.M., Benson, A.J., Cole, S., Frenk, C.S., Lacey, C.: astro-ph/0203051 (2002)
4. Best, P.N., Lehnert, M.D., Miley, G.K., Röttgering, H.J.A.: MNRAS, **343**, 1 (2003)
5. Cowie, L.L., Songaila, A., Hu, E.M., Cohen, J.G.: AJ, **112**, 839 (1996)
6. Daddi, E., et al.: astro-ph/0308456 (2003)
7. Dickinson, M. Papovich, C., Ferguson, H.C., Budavari, T.: ApJ, **587**, 25 (2003)
8. Dressler, A.: ApJ, **236**, 351 (1980)
9. Gómez, P.L., et al.: ApJ, **584**, 210 (2003)
10. Kauffmann, G., Charlot, C.: MNRAS, **297**, 23 (1998)
11. Kennicutt, R.C. Jr.: ApJ, **272**, 54 (1983)
12. Kodama, T., Arimoto, N., Barger, A.J., Aragón-Salamanca, A.: A&A, **334**, 99 (1998)
13. Kodama, T., Bell, E.F., Bower, R.G.: MNRAS, **302**, 152 (1999)
14. Kodama, T., Smail, I., Nakata, F., Okamura, S., Bower, R.G.: ApJ, **562**, L9 (2001)
15. Kodama, T., Bower, R.G.: MNRAS, **321**, 18 (2001)
16. Kodama, T., Bower, R.G.: MNRAS, **346**, 1 (2003)
17. Kodama, T., Yamada, T., et al.: submitted (2003)
18. Lewis, I., et al.: MNRAS, **334**, 673 (2002)
19. Nagashima, M., Yoshii, Y., Totani, T., Gouda, N.: ApJ, **578**, 675 (2002)
20. Hall, P.B., Green, R.F., Cohen, M.: ApJS, **119**, 1 (1998)
21. Hall, P.B., et al.: AJ, **121**, 1840 (2001)
22. Rudnick, G., et al.: astro-ph/0307149 (2003)
23. Saracco, P., et al.: A&A, **349**, 751 (1999)
24. Saracco, P., et al.: A&A, **375**, 1 (2001)

The Evolution of Evolved Galaxies

Giuseppe Gavazzi

Università degli Studi di Milano - Bicocca, P.zza delle scienze 3, 20126 Milano, Italy

1 Introduction

The plethora of high redshift multifrequency surveys currently under way, that were extensively illustrated at this meeting will shortly provide us with a sequence of "fossil" galaxies, eventually disclosing the secret of their evolution, much as fossil organisms guided paleontologists tracing the evolution of species. Meanwhile we wish to remind to both theorists and to observers that the characterization of local galaxies, representing the boundary condition at $z = 0$ of any evolutionary model, is not yet fully achieved. With this purpose we conceived an extensive observational campaign aimed at providing the phenomenology of local galaxies in the broadest possible frequency range. We took observations and collected data from the literature for over 3600 local ($z < 0.03$) galaxies, mainly members to rich clusters, spending a large effort in making the literature data as homogeneous as possible with our own. The data cover the range from 2000 Å(UV) to the centimetric radio domain. The Web site "GOLDmine" (Galaxy On Line Database Milano Network) [10] is designed to provide world-wide access to this massive data-set on local galaxies.

2 GOLDmine

GOLDmine (http://goldmine.mib.infn.it) is focused on 9 local clusters of galaxies: A262 (Perseus-Pisces), Cancer, A1367, A1656 (Coma), Virgo, A2147, A2151, A2197, A2199 (Hercules). In addition it contains a filament of nearly isolated galaxies, the so called "Great Wall", thus providing the ideal laboratory for comparative analyses of galaxies in different environments, spanning a factor of 20-100 in local galaxy density. Objects are selected in the above regions with strictly optical completeness criteria. Galaxies brighter than $m_p = 15.7$ are taken from the Catalogue of Galaxies and of Clusters of Galaxies (CGCG) by [14] in all clusters except Virgo where objects brighter than $m_p = 20.0$ are taken from the Virgo Cluster Catalogue (VCC) by [3]. Obviously, due to the factor of ~ 5 difference in distance between Virgo and the other clusters, this selection limit results in dwarf galaxies being included in our database only for the Virgo cluster. However globally GOLDmine covers the whole range (4 orders of magnitude) of luminosities spanned by real galaxies. GOLDmine contains 3649 galaxies.

Extensive campaigns were carried out to observe as many as possible of the 3649 target galaxies through all possible observational windows, a task that we did not complete yet.

The parameters listed in the GOLDmine database are divided into 5 categories: General, Continuum and Line photometry, Dynamical and Structural. They can be obtained from GOLDmine by querying the database for an individual galaxy name or "by parameters", "by near name or position" or "by available images". In this case all galaxies in a given range of photographic magnitude, and morphological type can be selected.
General parameters include Catalogue designations, (J2000) celestial coordinates, optical diameters, photographic magnitude, redshift, distance, morphological type.
Continuum parameters include: UV, U, B, V, J, H, K magnitudes computed at the optical radius (25th mag arcsec^{-2}) (see [8]); IRAS 60 and 100 micron fluxes; radio continuum fluxes densities at 0.6 and 1.5 GHz.
Line photometry includes: the atomic (HI) and molecular (H_2) hydrogen mass; the Hα+[NII] line equivalent width and flux.
Dynamical parameters include: the width of the HI line, with a quality flag; the width of the Hα line and the central velocity dispersion.
Structural parameters include: the light concentration index (C31); the effective radius R_e; the effective surface brightness μ_e; the total asymptotic magnitude. These quantities (see [13]) are given separately for the H, V and B bands.

The novelty of GOLDmine consists of its image section, where images can be downloaded in JPG and FITS format. Images include:
Finding Charts from the Digitized Palomar Sky Survey for all galaxies.
Broad band images obtained in the B, V, H and K bands.
Narrow band images in the light of Hα and a red image of the underlying stellar continuum near Hα.
RGB images. For some galaxies we combined several images to obtain "true" color pictures.
Radial profiles of the light distribution as obtained on the available (B, V, H) images (see [9]). When at least two radial profiles are available the color radial profile is also shown.
Optical spectra integrated over the whole surface of the galaxy, obtained in drift-scan mode, i.e. by drifting the spectrograph slit over the galaxy extension (see [11]).
Spectral Energy Distributions (SEDs) from the UV to the centimetric radio continuum obtained from broad-band photometry. The plotted data are total fluxes (extrapolated to the optical radii), unlike the individual aperture data given by NED. However they are given as observed, i.e. uncorrected for extinction from our Galaxy and for internal extinction (see [5]).

It is our goal to provide a homogeneous set of keywords in all FITS header to characterize the data, including: effective integration time, filter, telescope, WCS parameters, photometric effective zero point. This homogenization is not yet complete. Its progress will be detailed in the online "GOLDmine news" section.

3 An Application

To give just one example of the potential use of GOLDmine for constraining the properties of local galaxies, we briefly discuss an application aimed at deriving the star formation history (SFH) of galaxies in nearby clusters. Combining the spectro-photometric survey of [11] with photometry taken over the broadest possible wavelength baseline, including UV (2000 Å), optical (UBV) and infrared (JHK) photometry, we construct spectral energy distributions covering the domain 2000–22000 Å. The (extinction corrected) SEDs are fitted with Bruzual & Charlot [6] population synthesis models assuming a Salpeter IMF, an age T=13 Gyr and letting the metallicity and the time scale τ of the assumed SFH as free parameters. We adopt a star formation history with delayed-exponential form "a la Sandage" which mimics the SFH first proposed by [12] (see Fig.1). Fig.2 shows the fit of the B&C models to templates SEDs obtained averaging the data of several galaxies in 10 bins of Hubble type. The whole Hubble sequence can be modeled assuming increasingly delayed star formation histories of longer duration (τ) along with decreasing metallicity.

B&C models were also fitted to the SEDs of the individual galaxies and τ is found to decrease with increasing H luminosity (e.g. dynamical mass). For L_H spanning 4 decades, τ varies between 3 and 15 Gyr. The most luminous galaxies

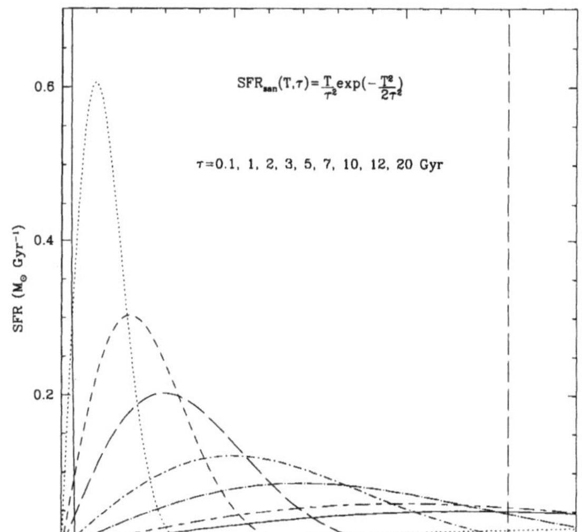

Fig. 1. A family of star formation histories with delayed-exponential form "a la Sandage" is here shown assuming that galaxies are coeval (T=13 Gyr), for several τ. We find that $\tau \leq$ 3 Gyr well represent giant galaxies, whereas longer τ, up to 15 Gyr represent dwarf galaxies.

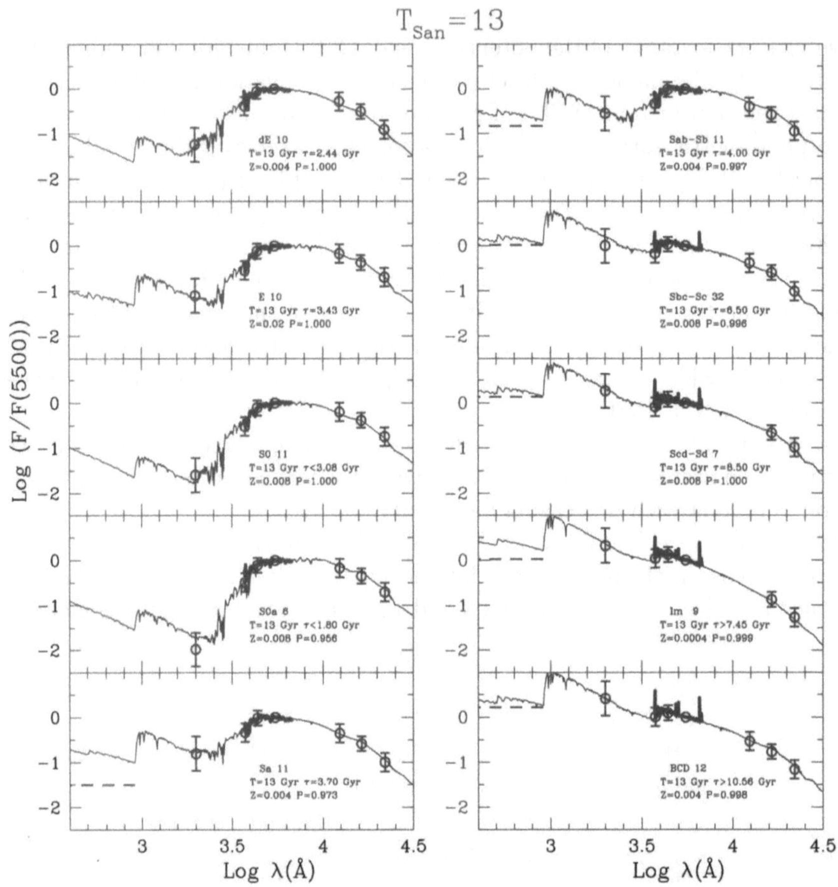

Fig. 2. The observed template SEDs (points and thick line) in ten bins of Hubble type are fitted with Bruzual & Charlot population synthesis models (thin line) assuming a SFH as in Fig.1 and a common age T=13 Gyr. The derived metallicity and τ are listed in each panel.

have experienced a "short" burst of star formation in their early history, whereas the less luminous systems have shallower star formation histories, consistent with the ongoing star formation observed at the present epoch.

4 Conclusions

We have assembled the GOLDmine database containing data for 3649 galaxies in 9 nearby clusters and made it available to the community via the WWW. This is a suitable tool for studying the phenomenology of evolved galaxies, in particular of the members to rich clusters.

As an application of GOLDmine we showed how the star formation history depends on the systemic mass: big galaxies (both ellipticals and bulge-dominated spirals) have a star formation history consistent with an early (10 Gyr ago) intense burst. Dwarfish spirals experienced a shallower star formation activity of longer duration, such that even at the present epoch they are currently transforming gas into stars. Such a mass dependence is one of the many similar evidences we found (see [7]). Other are the color-magnitude relation ([13]), todays re-discovered by SDSS [2]. In ellipticals this is most likely the result of a metallicity sequence ([1]), whereas in spirals it reflects a genuine "ageing" of the stellar population with increasing mass ([4]). Another mass dependence is the significant increase of the light concentration index with mass, such that the light profiles become progressively more "cuspy" (old star are more concentrated toward the center producing significant bulges) with increasing mass ([13]). Again this hods true for elliptical and spiral galaxies.

Altogether it appears that massive galaxies are "older" systems than their corresponding low mass counterparts. This evidence is beginning to be taken into serious considerations by theories and semi analytic models of galaxy evolution, as this meeting contributed showing.

Acknowledgements

Alessandro Boselli, Luca Cortese, Alessandro Donati, Paolo Franzetti, Anna Gallazzi, Marco Scodeggio, Antonio Zaccardo and Stefano Zibetti are acknowledged for their precious contributions to GOLDmine.

References

1. N. Arimoto, Y. Yoshii, A&A, 173, 23 (1987)
2. I. K. Baldry, K. Glazebrook, J. Brinkmann, Z. Ivezic, R. H. Lupton, R. C. Nichol, A. S. Szalay, 2003, ApJ, (in press, astro-ph/0309710)
3. B. Binggeli, A. Sandage, & G. Tammann, AJ, 90, 1681 (1985)
4. A. Boselli, G. Gavazzi, J. Donas, M. Scodeggio, AJ, 121, 753 (2001)
5. A. Boselli, G. Gavazzi, Sanvito, G., A&A, 402, 37 (2003)
6. G. Bruzual & S. Charlot, ApJ, 405, 538 (1993)
7. G. Gavazzi, D. Pierini, A. Boselli, A&A, 312, 397 (1996)
8. G. Gavazzi, A. Boselli, ApL&C, 35, 1 (1996)
9. G. Gavazzi, P. Franzetti, M. Scodeggio, A. Boselli, D. Pierini, A&A, 361, 863 (2000)
10. G. Gavazzi, A. Boselli, A. Donati, P. Franzetti, M. Scodeggio, A&A, 400, 451 (2003)
11. G. Gavazzi, C. Bonfanti, G. Sanvito, A. Boselli, M. Scodeggio, ApJ, 576, 135 (2002)
12. A. Sandage, A&A, 161, 89 (1986)
13. M. Scodeggio, G. Gavazzi, A. Boselli, P. Franzetti, S. Zibetti, D. Pierini, A&A, 384, 812 (2002)
14. F. Zwicky, & et al., in California Inst. Techn., 1-6 (1961-1968)

TP-AGB Stars to Date High-Redshift Galaxies with the Spitzer Space Telescope

Claudia Maraston

Max Planck Institut für Extraterrestrische Physik, Giessenbachstr. 1, 85748 Garching bei München, Germany

Abstract. We present new stellar population models that include the contribution of the Thermally Pulsing Asymptotic Giant Branch (TP-AGB) phase also in the synthetic spectral energy distribution (SED). The TP-AGB phase is essential for a correct modeling of intermediate-age ($0.2 \lesssim t/\mathrm{Gyr} \lesssim 1 \div 2$) stellar populations, because it provides $\sim$ 40 % of the bolometric contribution, and up to $\sim$ 80 % of that in the K-band. These models are obtained by coupling the energetic of the TP-AGB phase as calibrated with data of Magellanic Clouds star clusters ([9]), with empirical spectra of TP-AGB stars ([3]). Now that the Spitzer Space Telescope (SST) allows the sigth of the rest-frame IR at high redshifts, these models provide the opportunity to use the TP-AGB phase as an age indicator also for high-redshift stellar populations. Here we focus on redshift $\sim$ 3 and provide predictions of the colours of various galaxy models as will be measured by means of the IRAC imaging instrument on board the SST. We find a sizable magnitude difference between TP-AGB-dominated high-redshift stellar populations and those being older or younger. The first releases of GOODS data should allow a check of these predictions.

1 The Dating of Galaxies and the TP-AGB Phase

The epoch(s) of galaxy formation is constrained by dating the stellar populations at zero as well as at high redshift, because the timescales of stellar evolution are independent of cosmological models. Such a constraint provides an important check of current models of hierarchical galaxy formation, in the framework of which the assembly of massive galaxies appears to occur over a rather extended redshift range, with a substantial amount of star formation at redshift lower than 1 (see reviews by S. White and R. Sommerville, *this volume*). Such prediction appears to be at odd with the old average age, and the small spread in ages, derived for local massive ellipticals (Es) using optical absorption features and chemical evolution arguments (Thomas, Maraston, Bender, *this volume*; see also G. Gavazzi, *this volume*; [8]). Also, the finding of galaxies already massive at high redshifts (see contributions by A. Cimatti; R. Genzel, *this volume*; [16]) is difficult to accommodate in such models. On the other hand, there are also galaxies whose average ages appear to be rather young and could indeed be consistent with small formation redshifts ($z \lesssim 1$), like the so-called k+a galaxies ([13]), or lenticulars and some low-mass Es in the field (see Thomas, Maraston, Bender, *this volume*).

However, the dating of unresolved stellar populations by means of spectrophotometric indicators in the optical is limited by the well-known *age/metallicity degeneracy* ([4], [17], [11]), i.e. the phenomenon that a low metallicity can mimick a low age, and vice versa. When a stellar population ages above 1÷2 Gyr, the Red Giant Branch and the Main Sequence share almost equally the energy production (see e.g. Figure 3 in [9]), and their contributions evolve very smoothly with age. At the same time there are no other stellar phases of short duration and relevant energetics that become important and could be used as age indicators. Therefore, at old ages the age/metallicity degeneracy works at its best in confusing the age determination. To trace back the beginning of the formation of a stellar system it would be ideal to recognize it before it becomes a few Gyr old.

A clear signpost of intermediate age ($t \sim 1$ Gyr) stellar populations are Thermally-Pulsing Asymptotic Giant Branch stars (TP-AGB; [15], [9]). According to stellar evolution, the TP-AGB stellar phase, the brightest and the coolest on the HR diagram, becomes fully developed in stars with degenerate carbon-oxygen cores. The onset of such event in the life of a stellar population has been called the AGB-*phase transition* ([15]). The observational evidence of the onset of the TP-AGB is a sizable jump in the V−K colour that increases from $\sim$ 1.4 to $\sim$ 3.2, as observed among the Magellanic Clouds globular clusters (see Section 2). The narrow interval in evolutionary mass ($1.5 - 3\ M_{\odot}$) constrains the whole duration of the TP-AGB dominance to be $\sim$ 1 Gyr ([9]). Therefore, picking up the TP-AGB is a powerful way of dating a stellar population, and this technique has been applied with success to local stellar populations ([12]).

How can we extend this approach of age dating to $z > 0$?

An early suggestion in this direction is due to [14], who indicated that the AGB phase transition potentially is an effective tool to date the high redshift formation of Es. Two factors has hampered the exploitation of this idea until now. First, the rest-frame IR at redshift 2÷3 is sampled by the observed frame around $8 - 10\ \mu$m. This window is only now available thanks to the advent of the S(pitzer)S(pace)T(elescope). Second, synthetic spectral energy distributions (SEDs) including the TP-AGB phase are clearly required, but usually evolutionary population synthesis models include only the early part of the AGB phase (so-called the Early-AGB), thereby missing the TP-AGB that is the one energetically important (see Section 2).

In this contribution we introduce model SEDs that include the TP-AGB phase (Section 3) and show the substantial effect on the integrated SEDs of intermediate-age stellar populations. In Section 4 we construct diagnostic colour-colour diagrams for the imaging instrument IRAC on board the SST for the illustrative redshift of 3. We further discuss the use of these models at every redshift.

2 TP-AGB in SSP Models: State of the Art

The TP-AGB is a critical stellar phase to be accounted for in a Stellar Population (SP) model, because its energetic and duration are affected by mass-loss and nuclear burning in the envelope, both phenomena requiring parametrizations to be calibrated with data ([7]). However, the TP-AGB phase is the dominant phase in intermediate-age stellar populations (0.2 Gyr$\lesssim t \lesssim$ 1 ÷ 2 Gyr), contributing $\sim$ 40% to the bolometric light, as observed among the globular clusters (GCs) in the Magellanic Clouds ([6]). Maraston (1998, [9]) uses the *fuel consumption* theorem ([15]) to include the TP-AGB phase in SP models in a semi-empirical way, by calibrating the energetic of the TP-AGB phase with data of intermediate-age Magellanic Clouds GCs. The calibration of the bolometric contribution, and the

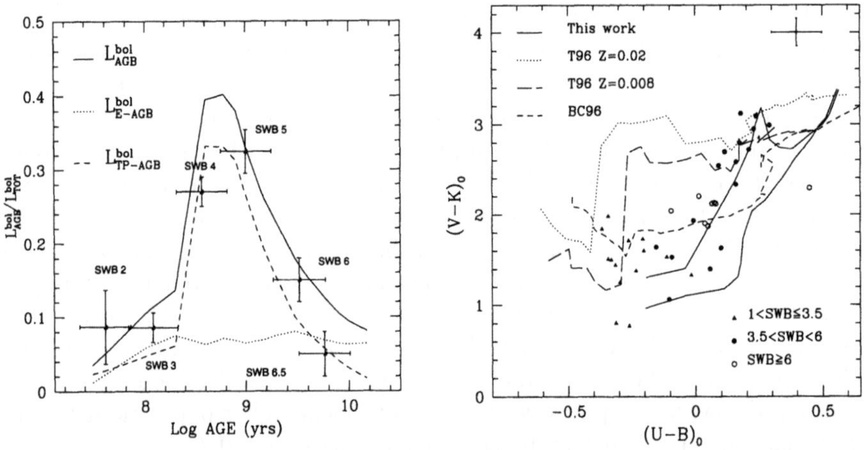

Fig. 1. From [9]. *Left-hand panel.* Calibration of the bolometric contribution of the TP-AGB phase of SP models as a function of age, with data of Magellanic Clouds GCs. *Right-hand panel.* Calibration of the synthetic V−K vs. U−B (higher solid line) with the same data. The intermediate-age objects are the filled circles. The lower solid line shows the same SPs but without the TP-AGB phase, to appreciate the deficit of IR flux in the latter case. The other line styles show SPs from other authors (see [9].)

comparison of the synthetic broad-band colours with the same data of Magellanic Clouds GCs, is shown in Figure 1. The AGB phase transition among the intermediate-age GCs (filled circles) appears as an enhacement of the IR luminosity with respect to the optical one and with the V−K colour reaching values larger than 3. The figure shows that the inclusion of the TP-AGB phase (models as thick solid lines) is crucial to match the integrated near-IR colours of intermediate-age stellar populations, as emphasized by a model (solid thin line) in which the TP-AGB contribution has been subtracted on purpouse and only the E-AGB considered. The SP models shown in Figure 1 were restricted at the broad-band colours due to the unavailability at the time of spectra, either theoretical or empirical, appropriate to TP-AGB stars. However, in order to use the TP-AGB-induced jump of the near-IR flux as an age indicator also at $z > 0$, this phase has to be included in the synthetic spectral energy distribution. Such improvements of the models will be described in the next Section.

3 TP-AGB in SSP Models: Extension to SEDs

Recently, a library of observed spectra of carbon-rich and oxygen-rich type TP-AGB stars in the wavelength range 0.5 ÷ 2.5 μm has become available ([3]). We have used these empirical spectra and the calibrated TP-AGB fuel ([9]) to include the TP-AGB phase in the synthetic SEDs of SP models. The SEDs are constrained to match our previously published model broad-band colours, because these match the observed colours of the Magellanic Clouds GCs (Figure 1). Figure 2 shows as an example the 1 Gyr SED (thick solid line). The inclusion of

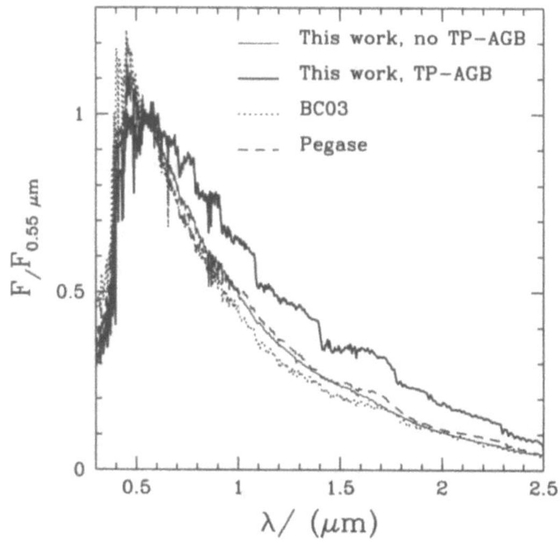

Fig. 2. The effect of the TP-AGB on the spectral energy distribution of a 1 Gyr old stellar population. The solid thick and thin lines show our models with and without TP-AGB. Dotted and dashed lines are models from the literature ([1], [5]).

the TP-AGB changes substantially the spectrum at wavelengths $\gtrsim 0.7\mu$m, while leaving unchanged the optical side. Also shown are the SEDs of other models ([1], [5]. Since the latter do not include the TP-AGB phase in the synthesis, the detection of intermediate age SP in galaxies based on the near-IR predictions of these models should be taken with caution.

The models presented here are on course of publication, and we refer to the article for more details ([10]). In this proceeding we use them to compute the observed-frame colours of SP models at high-redshift in the imaging bands of IRAC-SST to illustrate the power of the TP-AGB-based diagnostic of SP ages.

4 How to Find the TP-AGB at $z > 0$

At high redshifts when galaxies are dominated by $\sim$ 1 Gyr old populations the TP-AGB signature in the rest-frame near-IR must show up, with e.g. the rest-frame V$-$K colour mapping into the observed K$-$10 μm at $z = 3$ ([14]). The conclusion of Chiosi et al. ([2]) that the jump in the mid-IR colours due to the AGB is "wiped out by cosmological effects" is due to their consideration of the *observed* V$-$K *at every redshift* instead of the rest-frame one, as appropriate.

The fingerprint of the TP-AGB starts to be appreciable at rest-wavelength $\sim 0.7\mu$m (Figure 2). Already existing K-selected galaxy surveys like MUNICS and the K20 (see Drory; Cimatti, *this volume*) could be used to search for intermediate-age SP using these models in galaxies at moderate redshifts ($z \sim 1$), where the observed K samples the rest J. This is the subject of a forthcoming paper.

Here we focus on redshift 3 to search for the TP-AGB signature in the progenitors of massive Es, for which the α-enhancement of their stellar populations constrain the star formation history to be short (e.g. 1-2 Gyr, see D. Thomas, *this volume*). This maximizes the relative fraction of stars that spend the TP-AGB phase *simultaneously*. At $z = 3$ the rest-frame near-IR maps into the observed mid-IR ($3 \lesssim \lambda/\mu m \lesssim 10$) and therefore SST is required. Figure 3 shows colour-

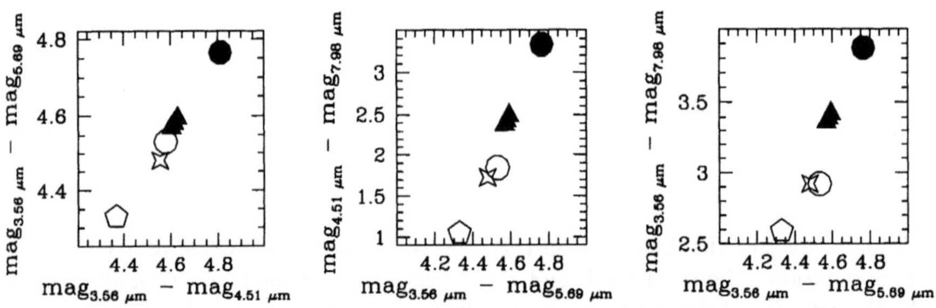

Fig. 3. Diagnostic colour-colour diagrams for the IRAC-SST filters to detect intermediate-age SP at the illustrative redshift of 3. Various synthetic stellar populations with solar metallicity are shown for which the stars started to form 0.8 Gyr before $z = 3$: a single burst (SSP, filled circle) and models with exponentially declining star formation with e-folding times of 1, 2 and 10 Gyr (filled triangles). For comparison, an SSP in which the TP-AGB phase is not included is shown as an open circle. Also shown is a younger and metal-poor and an older and metal-rich SSP (pentagon and star, respectively)

colour diagrams in the four filters available on the imaging instrument IRAC on board the SST, covering the region $3.6 \div 8$ μm. The central λ of each filter is indicated in the axis labels. Note that the bluest available IRAC filter does not map into the rest-frame V at $z = 3$, but into the rest-frame 0.89 μm. While such longer wavelength does not maximize the signature of the TP-AGB, it has the advantage of being better sheltered than optical bands from other effects, such as dust attenuation or newly born hot stars. These effects will be explored in a forthcoming paper.

Various SP models are shown in Figure 3 in which the stars started to form 0.8 Gyr before redshift 3. The filled circle is a single burst, or Simple Stellar Population (SSP), that includes the TP-AGB. The empty circle shows the same model, but in which the TP-AGB phase has not been included.

Figure 3 displays the same effect as we saw in Figure 1 using the familiar Johnson bands: the presence of TP-AGB stars enhances the rest-frame mid-IR flux. The single burst maximizes the effect because of the single age of the stars. However, the TP-AGB signature is still visible when extended star formation histories are considered (filled triangles). As mentioned above, this is due to the fact that at this redshift the three reddest IRAC filters still sample the mid-IR region, where the impact of the very young stars is minimized. Also shown in Figure 3 is a young, pre TP-AGB model with lower metallicity (pentagon), and an old, post TP-AGB, but metal-enriched model (star). As it can be seen, the larger metallicity is not able to mimick the red colours due to TP-AGB. From Figure 3 we infer a sizable magnitude difference as the marker of the transition between young and intermediate-age SP that could be effectively used as age indicator for high-redshift galaxies. The data collected with SST will allow the TP-AGB-based dating of stellar populations at large look-back times.

References

1. G. Bruzual, S. Charlot: MNRAS **344**, 1000 (2003)
2. C. Chiosi, G. Bertelli, A. Bressan: "Models of population synthesis". In: *The Stellar Populations of Galaxies. IAU Symp. 149*, ed. by B. Barbuy and A. Renzini (Dordrecht, Kluwer, 1992), pp. 321-324
3. A. Lancçn, M. Mouchine: A&A **393**, 167 (2002)
4. S. Faber: A&A **393**, 167 (1972)
5. M. Fioc, B. Rocca-Volmerange: A&A **326**, 950 (1997)
6. J.A. Frogel, J.R. Mould, V.M. Blanco: ApJ **352**, 96 (1990)
7. I. Iben, Jr. & A. Renzini: ARA&A **21**, 271 (1983)
8. G. Kauffmann, T.M. Herckman, S.D.M. White, et al.: MNRAS **341**, 54 (2003)
9. C. Maraston: MNRAS **300**, 872 (1998)
10. C. Maraston: MNRAS, *submitted* (2004)
11. C. Maraston, D. Thomas: ApJ **541**, 126 (2000)
12. C. Maraston, M. Kissler-Patig, J.P. Brodie, et al.: A&A **370**, 176 (2001)
13. B.M. Poggianti, I. Smail, A. Dressler, et al.: ApJ **518**, 576 (1999)
14. A. Renzini: 'The Age Ladder from low to High-Redshift Populations'. In: *The Stellar Populations of Galaxies. IAU Symp. 149*, ed. by B. Barbuy and A. Renzini (Dordrecht, Kluwer, 1992) pp.325–336
15. A. Renzini, A. Buzzoni: "Global properties of stellar populations and the spectral evolution of galaxies". In *Spectral evolution of galaxies*, Proceedings of the Fourth Workshop, Erice, Italy, March 12-22, 1985, ed. by C. Chiosi, A. Renzini (Dordrecht: Reidel 1986), pp. 195-231
16. P. Saracco, M. Longhetti, et al.: A&A **370**, 176 (2003)
17. G. Worthey: ApJS **95**, 107 (1994)

The Epochs of Early-Type Galaxy Formation in Clusters and in the Field

Daniel Thomas[1], Claudia Maraston[1], and Ralf Bender[1,2]

[1] MPI für extraterrestrische Physik, Giessenbachstraße, D-85748 Garching, Germany
[2] Universitäts-Sternwarte München, Scheinerstr. 1, D-81676 München, Germany

Abstract. Using our models of absorption line indices that account for variable abundance ratios, we derive ages, total metallicities, and element abundance ratios of 126 early-type galaxies in various environments. The data are analyzed by comparison with mock galaxy samples created through Monte Carlo simulations taking the typical average observational errors into account, in order to eliminate artifacts caused by correlated errors. We find that all three parameters age, metallicity, and α/Fe ratio are correlated with velocity dispersion. We further find evidence for an influence of the environment on the stellar population properties. Massive early-type galaxies in low-density environments appear on average several Gyrs younger and ~ 0.1 dex more metal-rich than their counterparts in clusters. No offsets in the α/Fe ratios, instead, are detected. With the aid of a simple chemical evolution model, we translate the derived ages and α/Fe ratios into star formation histories. We show that most star formation activity in early-type galaxies is expected to have happened between redshifts ~ 3 and 5 in high density and between redshifts 2 and 3 in low density environments.

1 Introduction

The most direct way to constrain the formation and evolution of galaxies certainly is to trace back their evolution with redshift [1,3,40,14,38,27]. The price to be paid, however, is that high-redshift data naturally have lower quality and are therefore more difficult to interpret. A clear complication is the so-called progenitor bias, which implies that galaxies observed at low and high redshift are not necessarily drawn from the same sample [38]. The alternative approach is the detailed analysis of the stellar populations in local galaxies. By determining their average ages and element ratios, it is possible to unravel their (star) formation histories and to constrain the main epoch of galaxy formation. The confrontation with predictions from models of galaxy formation is certainly most meaningful, when the two approaches, the archeology of local galaxies and the mining of the high-redshift universe set consistent constraints.

In this paper we follow the former approach and analyze a homogeneous, high-quality data sample of early-type galaxies in various environmental densities. With our new stellar population models ([34], hereafter TMB03) we determine, besides ages and total metallicities, α/Fe ratios from absorption line indices. The main aim is to reveal possible correlations of these parameters with velocity dispersion, hence, galactic mass, and to derive star formation histories as a function of redshift, in order to constrain the main epochs of early-type galaxy formation.

2 Data and Models

We analyze a sample of 126 early-type galaxies, 71 of which are field and 55 cluster objects, containing roughly equal fractions of elliptical and lenticular (S0) galaxies. The sample is constructed from the following sources: 41 Virgo cluster and field galaxies [11], 32 Coma cluster galaxies [22,23], and 53 mostly field galaxies (highest quality data from [5]) selected from the ESO–LV catalog [17]. In the latter sample, objects with a local galaxy surface density $NG_T > 9$ are assumed to be cluster galaxies. NG_T is given in [17] and is the number of galaxies per square degree inside a radius of one degree around the considered galaxy. Line indices are measured as functions of galaxy radius. We adopt the central indices determined within 1/10 of the effective radius, so that the analysis presented here is free from aperture effects. The average $1-\sigma$ errors in Hβ, Mg b, and $\langle\mathrm{Fe}\rangle$ are 0.09, 0.06, 0.07 Å, respectively.

In TMB03 new stellar population models with different chemical mixtures and element abundance ratios are presented. All optical Lick indices from CN_1 to TiO_2 [39] in the wavelength range $4000\,\text{Å} \leq \lambda \leq 6500\,\text{Å}$ are computed for various chemical mixtures modifying the abundance ratios between α-elements (i.e. O, Mg, Ca, Na, Si, Ti), and iron-peak elements (Fe, Cr). The models are based on the evolutionary population synthesis of [19,20]. The impact from element ratio changes is computed with the help of the [37] model atmosphere calculations, using an extension of the method introduced by [35]. We refer to TMB03 for more details. These models are used here to derive the stellar population parameters age, metallicity, and α/Fe ratio from the line indices Hβ, Mg b, and $\langle\mathrm{Fe}\rangle$.

3 Results

An uncomfortable effect of the degeneracy between age and metallicity, is that the errors of these stellar population parameters are not independent. A possibility to get a handle on correlated errors is the performance of Monte Carlo simulations taking into account the observational uncertainties of the line indices from which the stellar population parameters are derived [36,16]. In this paper we extend the approach of [16] to a 3-dimensional parameter space including also α/Fe ratios besides age and total metallicity. We produce artificial samples of galaxies with given ages, metallicities, and α/Fe ratios, from which we compute through our stellar population models the line indices Hβ, Mg b, and $\langle\mathrm{Fe}\rangle$. By means of Monte Carlo simulations we perturb these 'exact' values with the $1-\sigma$ errors of the observational sample, assuming a Gaussian error distribution. The outcome is then compared with the observational data.

The result is shown in Fig. 1, in which the α/Fe-independent index [MgFe]$'$ (see TMB03) vs. the Balmer line index Hβ is plotted. Grey symbols are the observational data, while the filled symbols are the Monte Carlo realizations. Left and right panels refer to high and low density environments, respectively. The sample splits in two subclasses divided by the Balmer absorption index at H$\beta \approx 2\,\text{Å}$. The main part of the data ($\sim$ 75 per cent) is centered around

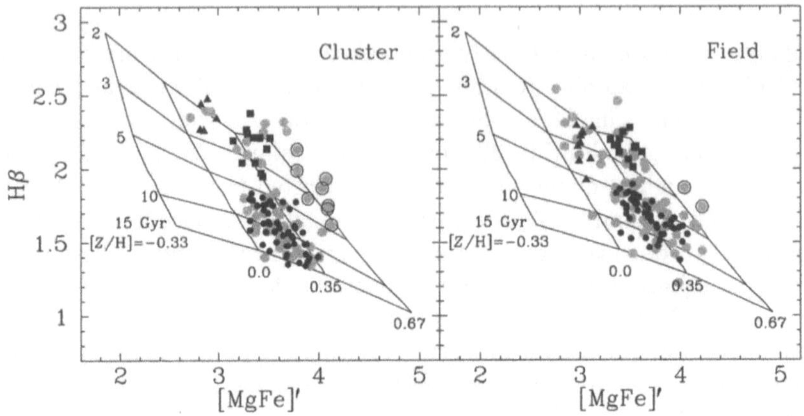

Fig. 1. Indices [MgFe]$'$ vs. Hβ. The grey circles are the observed values of early-type galaxies [11,5,23]. Objects in high density and low density environments are shown in the left and right panels, respectively. Black symbols are Monte Carlo realizations (see text). Index values are measured within $1/10\ r_e$. The solid lines are stellar population models grids (TMB03) with $[\alpha/\text{Fe}] = 0.0$ for various ages and metallicities (see labels).

H$\beta = 1.6$ Å, while one quarter of the objects have relatively high Hβ absorption, which implies the presence of either young stellar populations or blue horizontal branch morphologies. In the cluster environment, all objects with high Hβ are lenticular galaxies (see also [23]), while in the low density environments also some ellipticals exhibit strong Hβ. Fig. 2 shows the indices Mg b vs. $\langle\text{Fe}\rangle$, and can be used to get estimates of the α/Fe ratios of the sample galaxies.

In both figures the distributions of the observational data are well reproduced by the Monte Carlo realizations. In the latter we have assumed linear correlations of the parameters age, metallicity and α/Fe ratio with velocity dispersion. We have not assumed any scatter around these relationships. The observed scatter in Fig. 1 is fully compatible with the observational errors, while some intrinsic scatter in the α/Fe ratio might be required to fully reproduce the data Fig. 2. We find the following dependencies (brackets indicate low density environment):

$$\begin{aligned} t\ (\text{Gyr}) &= 8.4\ (5.4) + 0.01\ \sigma \\ [Z/\text{H}] &= -0.02\ (0.08) + 0.0012\ \sigma \\ [\alpha/\text{Fe}] &= 0.058\ (0.058) + 0.0008\ \sigma \end{aligned} \qquad (1)$$

All three parameters, age, metallicity, and α/Fe ratio increase with increasing galaxy mass. As noted in previous studies [36,16], age seems to be the weakest variable, although still we find an anti-hierarchical relationship, such that the stellar populations in the most massive early-type galaxies are the oldest. At the same time they are also more enhanced in α/Fe, and have higher total metallicities. The above relationships imply that the Mg-σ relation [2] is driven mainly by metallicity, like also the color-magnitude relation [15], followed by α/Fe ra-

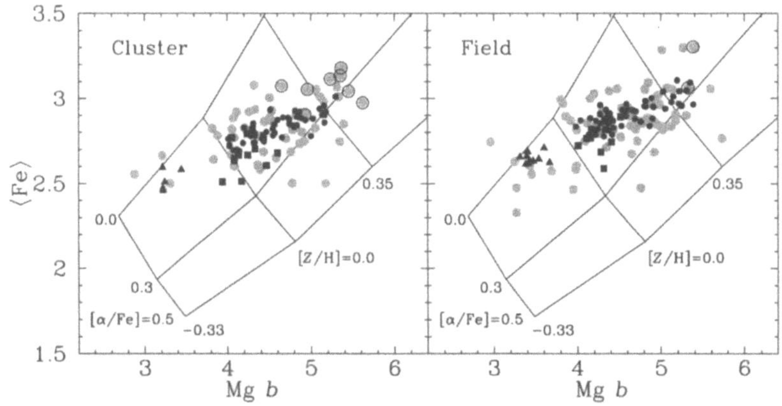

Fig. 2. Mg b vs. $\langle \mathrm{Fe} \rangle$. Symbols see Fig. 1. Solid lines are stellar population models (TMB03) with age $t = 12$ Gyr for various metallicities and α/Fe ratios.

tio ($\sim$ 30 per cent share), and only slightly by age. There is a clear difference between cluster and field objects, in the sense that galaxies in low density environments are several Gyrs younger and have slightly higher metallicities. The α/Fe ratios, instead, are independent of the environmental density. They depend only on the galaxy's velocity dispersion, independent of the environment in which the object resides (see also [33]). Because of this, and because the age and metallicity effects compensate, the zero-points of the Mg-σ relations change very little as a function of environment [4].

To explain the high Hβ indices in term of a young population, a very high (50 to 100 per cent in mass) fraction of fairly young ($\sim 1-2$ Gyr) populations on top of the relations of Eqn. 1 is required (triangles, squares and open circles in Figs. 1 and 2). This solution is not unique, however, and is degenerate with the presence of old populations with blue horizontal branch morphologies [21]. A future work is devoted to a detailed investigation of this problem.

4 Star Formation Histories

The relations given in Eqn. 1 can be used to constrain the epochs of the main star formation episode and the star formation timescales as a function of velocity dispersion. Assuming a Gaussian distribution for the star formation rate, we calculate the chemical enrichment of α and Fe-peak elements for an initially primordial gas cloud [32], varying the star formation timescale (width) and the lookback time of the maximum star formation (position of the peak).

The resulting star formation rates as a function of lookback time for various velocity dispersions in both dense and loose environments are shown in Fig. 3. In general, the more massive the galaxy, the shorter is its formation timescale and the higher is its formation redshift, which is in good agreement with the star

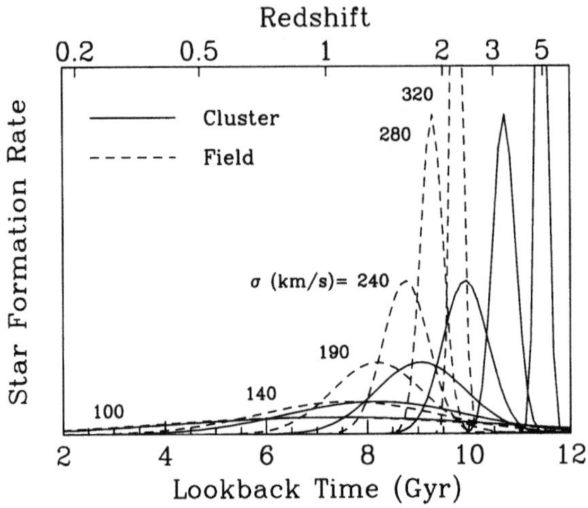

Fig. 3. Star formation rates as functions of lookback time for various velocity dispersions σ (see labels). High density and low density environments are indicated by solid and dashed lines, respectively. They are derived from the relationships in Eqn. 1. Redshifts assume the 'concordance model' $\Omega_m = 0.3,\ \Omega_\Lambda = 0.7,\ H_0 = 75$ km/s/Mpc.

formation histories found by [9]. Low-mass galaxies have extended star formation histories, and only they exhibit significant star formation at redshifts $z < 1$. The formation of massive galaxies in less dense environments appears delayed by a few Gyr with respect to the cluster environment (in qualitative agreement with [12]). The star formation histories derived here predict that very massive objects with high star formation rates should be observed at redshifts between 2 and 5.

Such objects have indeed been recently found. At least part of the SCUBA sources at $2 \leq z \leq 3$ turn out to be massive star forming galaxies with extremely high star formation rates up to 1000 $M_\odot$/yr [30,10]. These objects are likely to be the precursors of the most massive ellipticals ($\sigma \approx 300$ km/s) forming in a violent star formation episode at high redshift. Ly-break galaxies, instead, exhibit more moderate star formation rates of the order a few 10 $M_\odot$/yr [26], and may therefore be the precursors of less massive ellipticals in Fig. 3 ($\sigma \approx 200$ km/s). [25] find cB58 (a lensed galaxy at redshift $z = 2.73$ [29]) to have a significantly α-enhanced *interstellar medium*, which is in good agreement with the α/Fe-enhancement derived in this paper for the *stellar population* of local ellipticals.

The prediction made here for the formation epochs of early-type galaxies fits well to the detection of red, old galaxies at redshift 1 [7,28]. Deeper redshift surveys are now finding evolved objects even at redshifts around 2 [8] and massive star forming objects up to redshifts around 5 [18]. It seems the two approaches, archeology of stellar populations in local galaxies and redshift surveys, are converging to a consistent picture pushing the epoch of early-type galaxy formation to fairly high redshifts. A modification of current hierarchical models of galaxy formation ([13,31,6,24]) has become unavoidable.

References

1. A. Aragón-Salamanca, R.S. Ellis, W.J. Couch, D. Carter: MNRAS 262, 764 (1993)
2. R. Bender, D. Burstein, S.M. Faber: ApJ 411, 153 (1993)
3. R. Bender, B.L. Ziegler, G. Bruzual: ApJ 463, L51 (1996)
4. M. Bernardi et al.: ApJ 508, 143 (1998)
5. J. Beuing, R. Bender, C. Mendes de Oliveira, D. Thomas, C. Maraston: A&A 395, 431 (2002)
6. S. Cole, C.G. Lacey, C.M. Baugh, C.S. Frenk: MNRAS 319, 168 (2000)
7. E. Daddi et al.: A&A 361, 535 (2000)
8. M. Franx et al.: ApJ 587, L79 (2003)
9. G. Gavazzi, C. Bonfanti, G. Sanvito, A. Boselli, M Scodeggio: ApJ 576, 135 (2002)
10. R. Genzel, A.J. Baker, L.J. Tacconi, D. Lutz, P. Cox, S. Guilloteau, A. Omont: ApJ 584, 633 (2003)
11. J. González: Phd thesis, University of California, Santa Cruz (1993)
12. K. Gebhardt, S.M. Faber, D.C. Koo, et al.: ApJ 597, 239 (2003)
13. G. Kauffmann, S.D.M. White, B. Guiderdoni: MNRAS 264, 201 (1993)
14. D.D. Kelson, G.D. Illingworth, P.G. van Dokkum, M. Franx: ApJ 531, 137 (2000)
15. T. Kodama, N. Arimoto: A&A 320, 41 (1997)
16. H. Kuntschner, J.R. Lucey, R.J. Smith, M.J. Hudson, and R.L. Davies: MNRAS 323, 615 (2001)
17. A. Lauberts, E.A. Valentijn: The Surface Photometry Catalogue of the ESO–Uppsala Galaxies. ESO, Garching (1989)
18. M.D. Lehnert, M. Bremer: ApJ 593, 630 (2003)
19. C. Maraston: MNRAS 300, 872 (1998)
20. C. Maraston: MNRAS, submitted (2004)
21. C. Maraston, D. Thomas: ApJ 541, 126 (2000)
22. D. Mehlert, R.P. Saglia, R. Bender, G. Wegner: A&AS 141, 449 (2000)
23. D. Mehlert, D. Thomas, R.P. Saglia, R. Bender, G. Wegner: A&A 407, 423 (2003)
24. N. Menci, A. Cavaliere, A. Fontana, E. Giallongo, F. Poli, V. Vittorini: ApJ, astro-ph/0311496 (2003)
25. M. Pettini, S.A. Rix, C.C. Steidel, K.L. Adelberger, M.P. Hunt, A.E. Shapley: ApJ 569, 742 (2002)
26. M. Pettini, A.E. Shapley, C.C. Steidel, J. Cuby, M. Dickinson, A.F.M. Moorwood, K.L. Adelberger, M. Giavalisco: ApJ 554, 981 (2001)
27. R.P. Saglia, C. Maraston, L. Greggio, R. Bender, B. Ziegler: A&A 360, 911 (2000)
28. P. Saracco et al.: A&A 398, 127 (2003)
29. S. Seitz, R.P. Saglia, R. Bender, U. Hopp, P. Belloni, B. Ziegler: MNRAS 298, 945 (1998)
30. I. Smail, R.J. Ivison, A.W. Blain, J.-P. Kneib: MNRAS 331, 495 (2002)
31. R.S. Somerville, J.R. Primack: MNRAS 310, 1087 (1999)
32. D. Thomas, L. Greggio, R. Bender: MNRAS 302, 537 (1999)
33. D. Thomas, C. Maraston, R. Bender: Ap&SS 281, 371 (2002)
34. D. Thomas, C. Maraston, R. Bender: MNRAS 339, 897 (2003)
35. S.C. Trager, S.M. Faber, G. Worthey, J.J. González: AJ 119, 164 (2000)
36. S.C. Trager, S.M. Faber, G. Worthey, J.J. González: AJ 120, 165 (2000)
37. M.J. Tripicco, R.A. Bell: AJ 110, 3035 (1995)
38. P.G. van Dokkum, M. Franx, D. Fabricant, G.D. Illingworth, D. Kelson: ApJ 541, 95 (2000)
39. G. Worthey, S.M. Faber, J.J. González, D. Burstein: ApJS 94, 687 (1994)
40. B.L. Ziegler, R.P. Saglia, R. Bender, P. Belloni, L. Greggio, S. Seitz: A&A 346, 13 (1999)

Spectroscopic Ages of Elliptical Galaxies – Subaru Observation

Nobuo Arimoto[1], Yoshihiko Yamada[1,2], Alexandre Vazdekis[3], and Reynier Peletier[4]

[1] National Astronomical Observatory of Japan, Mitaka, Tokyo 181-8588, Japan
[2] Institute of Astronomy, University of Tokyo, Mitaka, Tokyo 181-0015, Japan
[3] Instituto de Astrofisica de Canarias, Via Lactea, E-38200 La Laguna, Tenerife, Spain
[4] Groningen Kapteyn Laboratorium Rijksuniversiteit, Landleven 12, Postbus 800, 9700 AV Groningen, Netherlands

1 Introduction

Breaking age/metallicity degeneracy ([19], [1]) is crucial for understanding stellar populations in elliptical galaxies. Spectroscopic properties, such as line strength, can partially break age/metallicity degeneracy. Reference [19] found that Hβ index, defined in [4] (Lick indices), is relatively sensitive to luminosity-weighted age of galaxies. But emission line makes Hβ shallower, therefore correction of emission line effect is necessary. Hγ and Hδ indices, such as Hγ_A, Hδ_A, Hγ_F and Hδ_F, which are less affected by emission than Hβ, are defined in [8] and [20]. However, those indices are still affected by age/metallicity degeneracy, since they are measured on low resolution spectra (9–11 Å). Reference [8] defined Hγ_{HR} to find it strongly sensitive to age. But Hγ_{HR} is much affected by velocity dispersion (σ) of galaxies, since the bandpasses of feature and pseudocontinua are very narrow.

Recently, [18] introduced the new age indicator, Hγ_σ, which they show is very successful in breaking the age/metallicity degeneracy and usable for wide range of σ. While the presence of non-solar abundance ratio in most early-type galaxies ([7], [10], [15] etc.) introduces systematical errors of ages if using Lick indices, errors from Hγ_σ come almost only from poisson noise. We have newly measured Hγ_σ of 27 elliptical galaxies by using Subaru and other telescopes to determine the ages and metallicities, from which we can discuss a possible star formation history in ellipticals and their origin.

2 Observation, Sample Selection, and Data Reduction

We performed our observations with three telescopes, WHT 4.2m, NTT 3.6m, and Subaru 8.2m. The details of our observations are listed in Table 1. Spectral resolution of the observations is 2.4–3.1Å, which coincident with 70–90 km s^{-1} of σ, and dispersion ranges from 0.44–0.92 Åpix^{-1}. Spectral coverage is set to include Hγ feature at λ ~4340Å. Multiple exposures of 15-35 minutes were acquired for each galaxy, as well as CuArNe (WHT), ThAr (NTT, Subaru) arc lamp exposures for λ calibration at each galaxy pointing. For WHT observation, we took Tungsten lamp flat exposures to remove wiggles in the spectra. We also observed a spectrophotometric standard star, G191B2B (WHT), LTT4364 (NTT), Feige 34, and G191B2B (Subaru).

Table 1. Overview of Our Observations

Telescope	WHT 4.2m	NTT 3.6m	Subaru 8.2m
Date	1997 Oct. 28–29	2002 May 19-21	2002 Apr. 17–18
	1999 Apr. 21–22		2003 Jan. 27
			2003 Jun. 26
Instruments	ISIS blue	EMMI	FOCAS
	R600B grating	Grating #12	300R grism, L600 filter
	EEV12 CCD	Blue #31 CCD	SITe CCD
Gain	1.01 e^-/ADU	1.44 e^-/ADU	1.99 e^-/ADU
Readout Noise	3.9 e^-	4.9 e^-	8–10 e^-
Dispersion	0.44 Åpix^{-1}	0.92 Åpix^{-1}	0.67 Åpix^{-1}
Resolution	2.4 Å	2.5 Å	3.1 Å
Wavelength Range	3800–5250 Å	3900–4750 Å	3800–5800 Å
	4000–5500 Å		
Spatial Scale	0".57pix^{-1}	0.74 pix^{-1}	0".21pix^{-1}
	(binning 2 pixels)	(binning 2 pixels)	(binning 2 pixels)

We observed 14 and 13 *genuine* elliptical galaxies in the Virgo cluster and in the field (group), respectively. Table 2 lists the basic parameters of the observed galaxies, such as radial velocity (V_r), effective radii (r_e), velocity dispersions (σ), visual magnitude (m_V), colour of $U-V$, $B-V$, telescope, exposure time, used slit width, and signal to noise ratio around 4340Å. Our Virgo sample were mainly taken from [3] along a well defined colour-magnitude relation. Our field (group) sample were selected from LEDA database ([12], [13], [14]) with the following criteria: 1) nearby sample, i.e., $V_R < 3500$ km/s; 2) neither very faint nor small galaxies, $\log r_{25} < 0.14$, $M_B < -18.5$; 3) morphological type index t < -2.9; 4) no strong emission line; 5) low extinction, i.e., $|b| > 20$ deg. In both Virgo and field, we selected galaxies to cover a wide range of σ, 70–300 km s^{-1} and luminosity.

Data reduction was done with the standard IRAF packages in the standard way; i.e. overscan correction, bias subtraction, flat fielding, λ calibration, sky subtraction and flux calibration. We achieved $\Delta\lambda$ (error of λ) smaller than 0.07Å in the λ calibration, which is one of the most important process in the age determination by $H\gamma_\sigma$, since even small λ shift affects the $H\gamma_\sigma$ measurement significantly ([18]). Cosmic rays were removed using the "cleanest" task in the REDUCEME package. Exposures whose spectra around $H\gamma$ feature were hit by cosmic ray were excluded, since even 'cleanest' task could not remove them in a perfect way. Flux calibration was done with the observed flux standard stars.

Table 2. Properties and Observational Parameters of our Sample Ellipticals

NGC	V_r	R_e	σ	m_V	$U-V$	$B-V$	Obs.	Exp.Time	Slit	S/N
	km s^{-1}	arcsec	km s^{-1}	mag	mag	mag		min.		Å^{-1}
	(1)	(2)	(3)	(4)	(5)	(6)	(7)	(8)	(9)	(10)
NGC4239	921	15	63	12.81	-	0.89	S1	100	0".4	115
NGC4339	1281	32	114	11.73	1.45	0.92	S2	50	0".6	150
NGC4365	1227	80	261	9.67	1.59	0.99	W	70	1".5	250
NGC4387	561	16	112	12.29	1.45	0.91	W	105	1".5	130
NGC4458	668	28	106	12.33	1.32	0.88	S1	70	0".4	145
NGC4464	1255	3	121	12.93	1.37	0.86	W	105	1".5	160
NGC4467	1426	6	83	14.33	1.46	0.86	S3	150	0".6	115
NGC4472	983	107	303	8.55	1.62	0.98	S1	20	0".4	170
NGC4473	2237	42	193	10.33	1.49	1.02	W	70	1".5	280
NGC4478	1382	13	143	11.46	1.32	0.93	W	105	1".5	135
NGC4489	957	28	62	12.38	1.21	0.89	S2	40	0".6	120
NGC4551	1189	18	113	12.14	1.47	0.92	S2	80	0".6	160
NGC4621	430	44	230	9.81	1.55	0.96	W	140	1".5	490
NGC4697	1236	72	181	10.56	1.57	0.89	S1	30	0".4	255
NGC0221	-145	36	78	7.84	1.28	0.88	S1	10	0".6	285
NGC0584	1875	27	217	10.30	1.38	0.91	W	60	1".0	200
NGC0720	1716	40	247	10.17	1.44	0.96	W	90	1".0	180
NGC0821	1716	45	199	10.79	-	0.93	W	60	1".0	150
NGC1700	3881	24	233	11.10	1.4	0.91	W	60	1".0	200
NGC3379	922	35	201	9.24	1.46	0.94	W	20	1".0	185
NGC3617	1744	12	88	12.44	1.12	0.82	N	180	1".5	120
NGC5638	1832	28	159	11.15	1.34	0.91	N	150	1".5	150
NGC5831	1820	27	166	11.39	1.47	0.92	W	35	1".5	160
NGC5831	1820	27	166	11.39	1.47	0.92	S2	60	0".6	190
NGC5898	2083	22	218	11.00	1.41	0.92	N	180	1".5	150
NGC5903	2125	35	235	10.87	-	0.87	N	180	1".5	185
NGC7144	1923	41	185	10.78	1.33	0.90	N	120	1".5	175
NGC7454	2007	24	112	11.81	1.19	0.89	W	105	1".0	115

3 Breaking Age/Metallicity Degeneracy

To derive the ages and metallicities of the galaxies, we use the revised evolutionary stellar population synthesis model of [17], which has been made from the empirical stellar spectral library of [6]. This model has been updated for new Padova isochrones ([6]). The model predicts spectral energy distributions (SEDs) in the optical wavelength range, λλ3976Å–4456Å and λλ4764Å–5460Å, for simple stellar populations at a resolution of 1.8Å (FWHM). The new model provides flux-calibrated, high-resolution spectra, therefore allowing an analysis of galaxy spectra at the natural resolution given by their internal velocity broadening and instrumental resolution. For example, no correction for velocity broadening has to be applied as made in the correction to Lick/IDS system.

4 Age and Metallicity of Elliptical Galaxies

Figure 1 presents age – velocity dispersion relations of elliptical galaxies in the Virgo cluster and in the field (poor groups). Virgo ellipticals with $\sigma \geq 120$ km s^{-1} show uniformly old ages ($\geq$ 10 Gyr; see also Vazdekis et al. 2001), while significant fraction of field ellipticals are as young as $\sim 6-10$ Gyr. On the other

Table 3. Age and Metallicity of Elliptical Galaxies: Virgo (left) and Field (right)

NGC	Age(Gyr)	Age Error	[Fe/H]	[Mg/Fe]	NGC	Age(Gyr)	Age Error	[Fe/H]	[Mg/Fe]
4239	5.2	(−1.6,+1.9)	−0.20	+0.05	221	3.4	(−0.2,+0.3)	+0.17	−0.25
4339	10.3	(−3.1,+7.5)	−0.04	+0.11	584	8.4	(−1.3,+1.9)	−0.01	+0.34
4365	20.5	(−6.5,+4.5)	−0.25	+0.47	720	11.2	(−3.0,+5.2)	−0.14	+0.74
4387	12.9	(−2.9,+6.1)	−0.20	+0.17	821	11.6	(−4.5,+10.4)	−0.10	+0.41
4458	13.5	(−3.0,+5.5)	−0.40	+0.39	1700	7.2	(−1.5,+2.4)	+0.04	+0.34
4464	18.7	(−6.1,+3.3)	−0.41	+0.39	3379	16.4	(−6.0,+6.6)	−0.30	+0.48
4467	16.0	(−5.3,+5.2)	−0.35	+0.27	3617	10.0	(−2.5,+6.8)	−0.31	
4472	10.5	(−4.2,+9.7)	−0.15	+0.56	5638	13.5	(−6.5,+10.1)	−0.45	
4473	9.8	(−1.6,+3.4)	−0.04	+0.54	5831	6.5	(−1.6,+1.5)	+0.19	+0.32
4478	8.6	(−1.7,+3.3)	−0.04	+0.32	5831	6.8	(−3.5,+4.7)	+0.15	+0.42
4489	3.0	(−0.7,+0.7)	+0.25	+0.05	5898	18.2	(−9.9,+11.8)	−0.28	
4551	6.9	(−1.0,+1.4)	+0.10	+0.32	5903	11.8	(−5.2,+15.2)	−0.08	
4621	10.6	(−2.1,+2.1)	−0.05	+0.63	7144	9.9	(−3.4,+8.4)	−0.14	
4697	10.6	(−2.2,+4.4)	−0.09	+0.39	7454	6.9	(−2.4,+2.4)	−0.20	+0.13

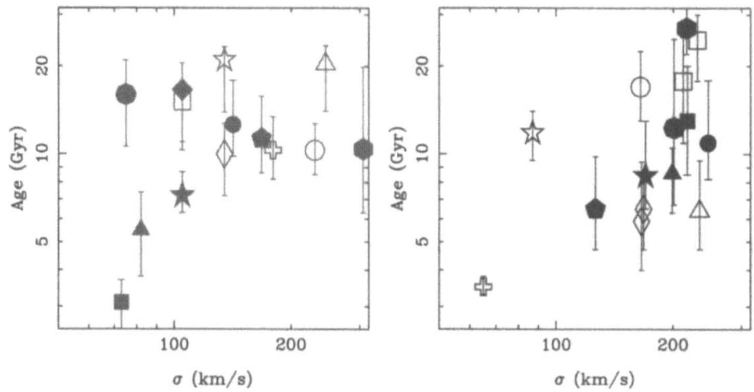

Fig. 1. Age-σ relation of elliptical galaxies in the Virgo cluster (left) and in the field (right)

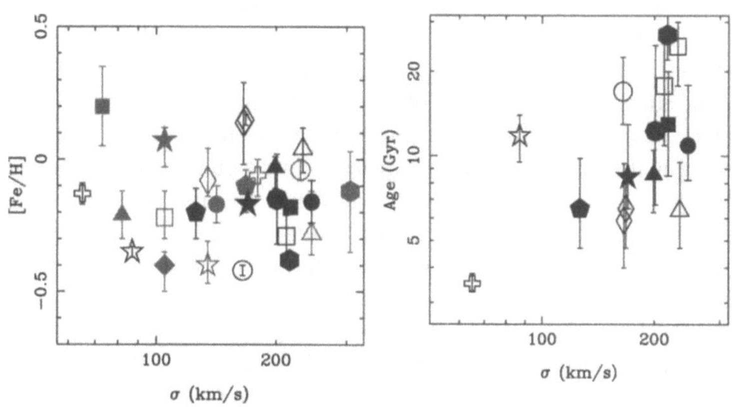

Fig. 2. [Fe/H] *vs* σ (left) and [Mg/H] *vs* σ (right) diagram of Virgo and field ellipticals

hand, in both cluster and field, ages of less massive elliptical galaxies (σ<120 km s^{-1}) range from 3 Gyrs to $\sim$ 16 Gyrs, showing a large scatter.

Our results are consistent with [16] who claimed that field ellipticals cover large age range, while Virgo ellipticals are generally old. Reference [11] also found that the age spread among field ellipticals is larger than cluster ones (Fornax), and the mean age is younger than cluster. We note, however, that these previous claims based on indices analyses that were not entirely free from age/metallicity degeneracy.

Reference [16] noted a weak trend suggesting that low σ galaxies tend to be younger. Reference [5] also show age spread for great samples of field ellipticals by using Hn/Fe−Fe4383 diagram. We find, however, no clear trend of progressively younger age towards less massive galaxies. There are certainly some dwarf ellipticals that are as old as massive ellipticals ($\sim$ 15 Gyr).

Figure 2 shows [Fe/H] (left) and [Mg/H] (right) of both Virgo and field ellipticals as a function of σ. Surprisingly, the iron abundances of elliptical galaxies are rather uniform, irrespective of their environment and velocity dispersion (mass), showing [Fe/H] $\simeq -0.2$. If one takes into account [Fe/H] gradient in individual galaxies, the mean stellar iron abundance of our sample ellipticals is roughly $\overline{[Fe/H]} \simeq -0.5$. This new result may solve the so-called iron-abundance discrepancy of elliptical galaxies (see [2]). [Mg/H] shows a clear trend with σ – massive ellipticals have higher [Mg/Fe] ratios. This implies that star formation stopped earlier in massive ellipticals, while that star formation lasted much longer in dwarf galaxies, suggesting somewhat different origin for dwarf and giant ellipticals.

5 Conclusions

We have obtained high S/N spectra of 27 elliptical galaxies in low (field and poor groups) and high (Virgo cluster) density environments with WHT, NTT, and Subaru Telescope, and determined their ages and metallicities by using $H\gamma_\sigma$ index that is free from age/metallicity degeneracy ([18]). We find that massive ellipticals in cluster are uniformly old, but those in the field show considerably large scatter in age (3 – 16 Gyr). Some of dwarf ellipticals ($\sigma \leq 120$ km s^{-1}) are as young as the youngest field ellipticals, but some dwarfs are as old as the oldest cluster ellipticals, showing no clear trend of galaxy age with the velocity dispersion. [Fe/H] is almost constant, while [Mg/H] increases towards more massive galaxies. Our results suggest that massive ellipticals in dense environment formed in rather short period at $z > 2$, while some of massive ellipticals in the field formed as late as $z \sim 0.6$. Dwarf ellipticals show wide variety of star formation history, perhaps depending on initial conditions of galaxy formation.

This work was financially supported in part by a Grant-in-Aid for the Scientific Research (no. 13640230) by the Japanese Ministry of Education, Culture, Sports and Science.

References

1. N. Arimoto: 'Stellar Population Synthesis Models for Elliptical Galaxies'. In: *From Stars to Galaxies: The Impact of Stellar Physics on Galaxy Formation, proceeding of a conference held at Porto Elounda Mare, Crete (Greece) 9-13 October 1995*, ASP Conference Series, Vol 98, 1996, ed. C. Leitherer, U. Fritze-von-Alvensleben, and J. Huchra (1996), pp.287-298
2. N. Arimoto, K. Matsushita, Y. Ishimaru, T.Ohashi, A. Renzini: ApJ **477**, 128 (1997)
3. R.G. Bower, J.R. Lucey, R.S. Ellis: MNRAS **254**, 589 (1992)
4. D. Burstein, S.M. Faber, C.M. Gaskell, N. Krumm: ApJ **287**, 586 (1984)
5. N. Caldwell, J.A. Rose, K.D. Concannon: AJ **125**, 2891 (2003)
6. L. Girardi, A. Bressan, G. Bertelli, C. Chiosi: A&AS **141**, 371 (2000)
7. J.J. González: Line Strength Gradients and Kinematic Profiles in Elliptical Galaxies. Ph.D. Thesis, University of California, Santa Cruz (1993)
8. L.A. Jones, G. Worthey: ApJL **446**, 31 (1995)
9. L.A. Jones: Differential Spectral Synthesis of Low-Luminosity Elliptical Galaxies. Ph.D. Thesis, The University of North Caroline, Chapel Hill (1998)
10. H. Kuntschner, J.L. Lucey, R.J. Smith, M.J. Hudson, R.L. Davies: MNRAS **323**, 615 (2001)
11. H. Kuntschner, R.J. Smith, M. Colless, R.L. Davies, R. Kaldare, A. Vazdekis: MNRAS **337**, 172 (2002)
12. G. Paturel, L. Bottinelli, N. Gouguenheim: A&A **286**, 768 (1994)
13. G. Paturel et al.: A&AS **124**, 109 (1997)
14. G. Paturel, G. Theureau, L. Bottinelli, Gouguenheim, N. Coudreau-Durand, N. Hallet, C. Petit: A&A **412**, 57 (2003)
15. B.M. Poggianti et al.: ApJ **563**, 118 (2001)
16. S.C. Trager, S.M. Faber, G. Worthey, J.J. Gonzalez: AJ **120**, 165 (2000)
17. A. Vazdekis: ApJ **513**, 224 (1999)
18. A. Vazdekis, N. Arimoto: ApJ **525**, 144 (1999)
19. G. Worthey: ApJS **95**, 107 (1994)
20. G. Worthey, D.L. Ottaviani: ApJS **111**, 377 (1997)

Evolution and Environment of Early-Type Galaxies

Mariangela Bernardi

Department of Physics, Carnegie Mellon University, Pittsburgh, PA 15213

Abstract. The properties of early-type galaxies drawn from the SDSS database depend weakly, but significantly, on environment. Objects in regions of lower density tend to have slightly lower luminosities, velocity dispersions, bluer colors, and to be less homogeneous than objects in denser cluster-like regions. The chemical abundances of early-type galaxies drawn from the SDSS database also depend weakly on environment. Since chemical abundances also evolve with redshift, the observed trends constrain how strongly the luminosity weighted age depends on environment, provided the low redshift population is simply a passively aged version of the more distant population. The constraint does not depend on the detailed properties of stellar population synthesis models.

1 Introduction

There are two competing scenarios for the formation of giant early-type galaxies. Either they formed from a monolithic collapse, or they formed from the stochastic mergers of smaller systems. The differences between these scenarios are small if the stellar populations in these galaxies formed at high redshift. In the merger model, stars are likely to have formed earlier in the densest regions, so early-type galaxy observables are expected to correlate with environment, as is the evolution of the population with redshift. In what follows, I describe the results of a search for trends with environment and redshift. I select a sample of early-type galaxies from the SDSS database, and assign them to one of three types of environment: high density, low density, and in between. Very briefly, all objects with concentration index in the i^* band $\geq 2.75 - 2.5$ (depending on redshift), and `eclass < 0`, for which the spectroscopic pipeline had measured velocity dispersions, are selected (59405 objects). The subset with $14.5 \leq r_{\rm Pet} \leq 17.6$ and inside the SDSS "masked" region contains 39320 objects. Of these, 39251 have $z \leq 0.30$, and 25423 have $z \leq 0.14$. These are assigned to one of three different environments as shown in Figure 1 (see Bernardi et al. 2004 for details). Only objects in the densest (bottom left) and least dense (top right) environments are used in the analysis which follows: in all there were 3685 and 6258 early-types in the two environments at $z \leq 0.14$.

The typical S/N of the galaxy spectra in my catalog (~18) is considerably smaller than the value (~50) required to make reliable estimates of line-strengths. Therefore, for each environment, I construct high S/N composite spectra, suitable for line-index measurements, by co-adding the spectra of similar

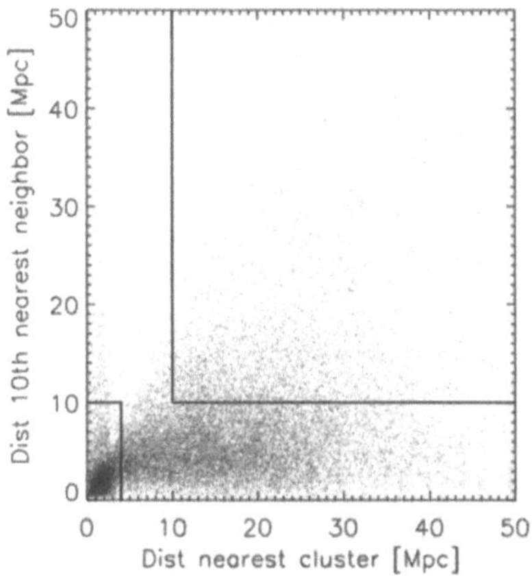

Fig. 1. Classification into high density (lower left) and low density environments (upper right) is based on the comoving distance to the tenth nearest luminous ($L > 3L_*$) neighbour, and on the comoving distance to the nearest massive cluster in the C4 catalog (Miller et al. 2004; each cluster is more luminous than $1.75 \times 10^{11}h^{-2}L_\odot \sim 10L_*$).

objects (I use narrow bins in luminosity, size, velocity dispersion, and redshift; see Bernardi et al. 2003 for details).

2 Structural Parameters, Chemical Abundances and Environment

At any given redshift, the objects in less dense regions tend to be slightly less luminous, to have smaller velocity dispersions, masses and sizes, and to be bluer, although these differences tend to be smaller than ten percent of the rms variation across the entire sample. The left panel of Figure 2 illustrates these trends using objects at $z \sim 0.1$; results at other redshifts are similar. (Of course, because of the magnitude limit, the higher redshift samples contain objects which are more luminous, have higher velocity dispersions, larger masses and sizes, and are redder.) In all cases, the population in dense regions is slightly more homogeneous.

The colors and chemical abundance indicators (Bernardi et al. 2003) of early-type galaxies correlate primarily with velocity dispersion. To see if the environment plays an additional role in determining chemical abundances we have selected objects in narrow magnitude and redshift bins in which the median velocity dispersion is the same in both environments. The panel on the right of Figure 2 shows that objects in less dense regions tend to be bluer, even though the luminosities and velocity dispersions are the same in both environments. This

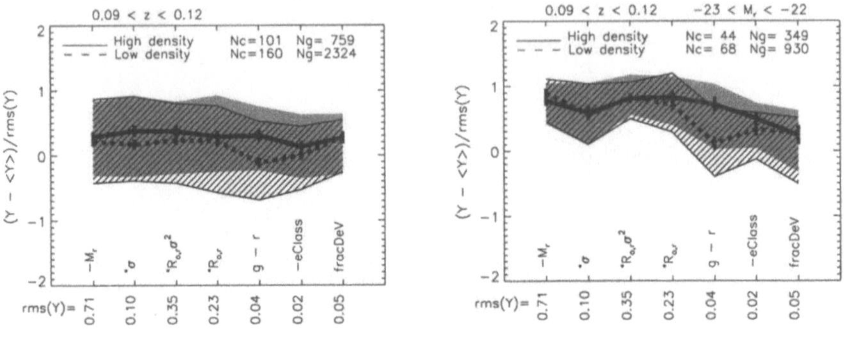

Fig. 2. Median values of luminosity, velocity dispersion, mass, size, color, PCA classification (i.e. spectral type) and light-profile shape (i.e. morphological type) for objects in dense and less dense regions, for a few narrow bins in redshift and luminosity. To facilitate comparison with one another, all observables have been rescaled by subtracting the mean and then dividing by the rms spread across the entire sample (i.e. across all redshifts and luminosities). The x-axis lists the observable and the value of the rms. An asterisk denotes that the quoted rms is for $\log_{10}$ of the parameter. Shaded regions show the 25th and 75th percentile values. Text in top right of each panel indicates the total number of composites in each redshift bin, and the total number of galaxies which made-up those composites. Results are similar for other choices of redshift and luminosity.

illustrates clearly that the environment plays an important role in determining galaxy colors.

The top panel of Figure 3 shows correlations between various emission and absorption lines which are not part of the Lick system (the Balmer absorption line indices H_δ, H_γ, which are sensitive to star formation at lookback times smaller than about 2 Gyr), the strength, D4000, of the break in the spectrum at 4000Å (which is also an indicator of age), the combinations [MgFe]$'$ and Mg/$\langle$Fe$\rangle$), and the environment. Mg/$\langle$Fe$\rangle$ is an indicator of the α-element abundance ratios (in early-type galaxies, this ratio is enhanced relative to the solar value), and [MgFe]$'$ is an index which is an indicator of metallicity that is not sensitive to the α-element abundance ratios.

The bottom panel shows the Lick indices (Trager et al. 1998) for these objects, arranged in order of increasing wavelength, so the figure can be thought of as illustrating how (little) the spectrum depends on environment. Small but significant trends are evident. For example, the Mg line strengths tend to be slightly weaker in the field (about 33% of the rms across the full sample, making Mg_2 weaker by ~ 0.01), consistent with Bernardi et al. (1998). In general, galaxies in the field tend to have stronger emission lines, stronger Balmer lines, weaker 4000Å breaks, lower metallicities and smaller α-element abundance ratios. In addition, notice that the scatter appears to be slightly larger for the population in less dense regions.

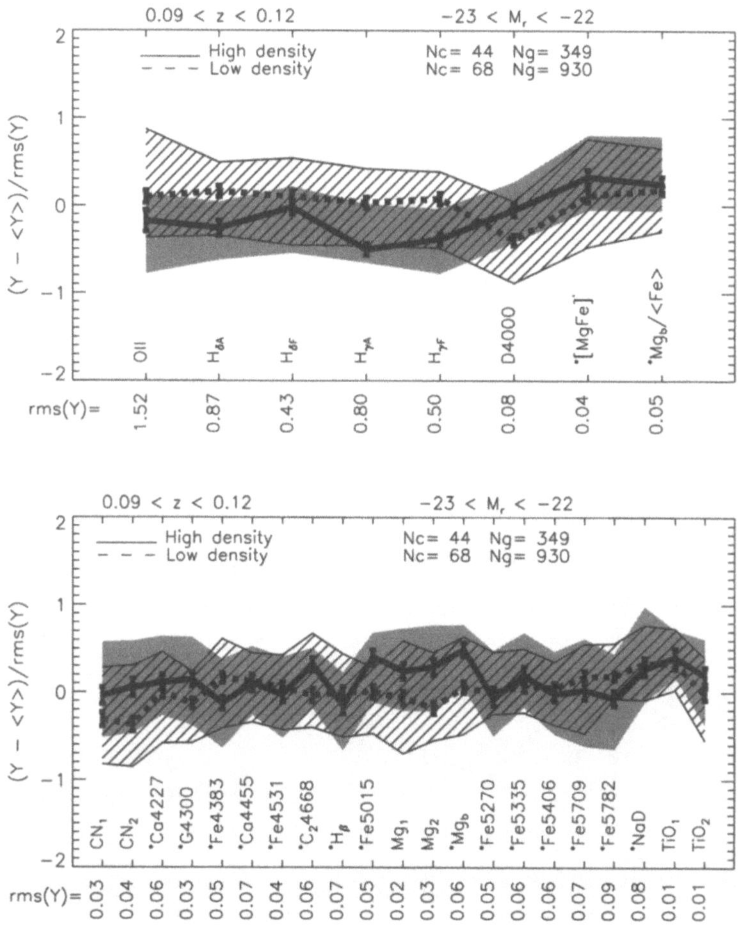

Fig. 3. Top: Emission line strengths (OII), recent (~2 Gyr) star-formation indicators (H_δ, H_γ), the strength of the break at 4000Å (D4000), and indicators of metallicity and α-element abundance ratios ([MgFe]$'$ and $\mathrm{Mg}/\langle\mathrm{Fe}\rangle$), for objects in a narrow range of redshifts and luminosities (as indicated). Bottom: Lick indices for objects with the same narrow range of redshifts and luminosities. (An asterisk signifies that the quoted rms is for $\log_{10}$ of the index.) In most observables in the top panel, the cluster and field samples differ by about 20% of the rms spread across the entire sample. Recall that the mean velocity dispersions are the same in dense and less dense environments, so trends with environment are not caused by correlations with σ. Note also that the scatter around the median is slightly smaller in the denser environments.

3 Evolution as a Model-Independent Clock

The previous section showed that, at any given redshift, most chemical abundances depend weakly, but significantly, on environment. On the other hand, many of the line strengths have evolved significantly between $z = 0.05$ and $z = 0.15$ (Bernardi et al. 2003, 2004).

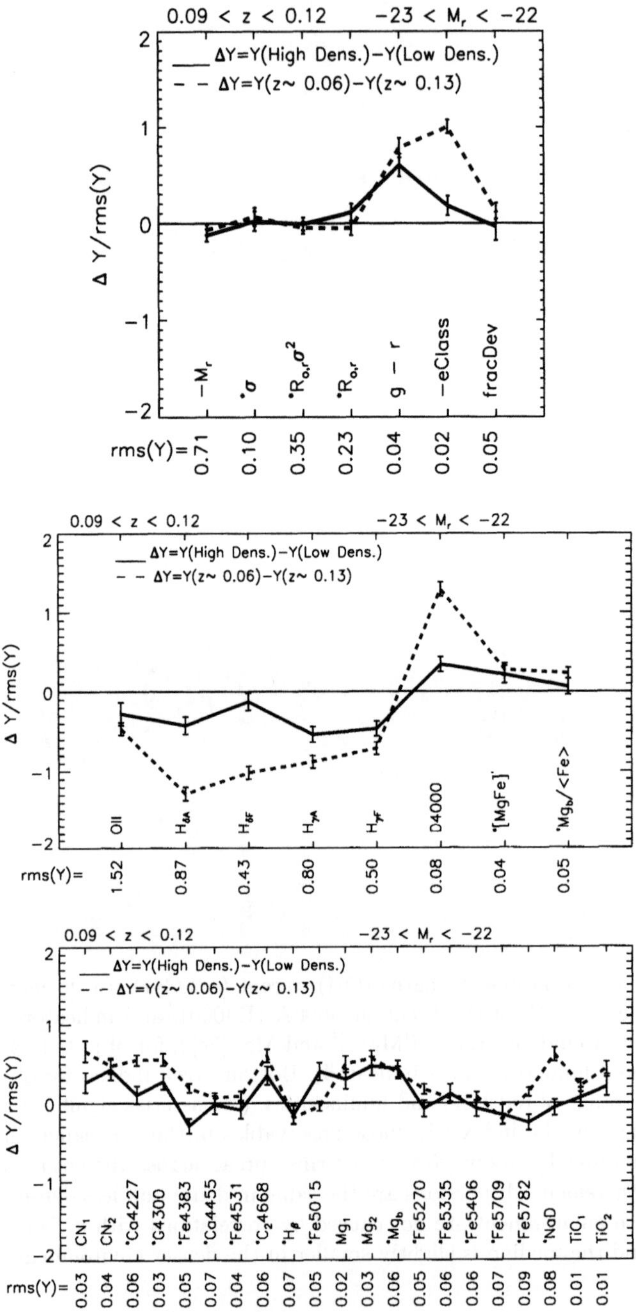

Fig. 4. Evolution of observed parameter (dashed), compared to dependence on environment (solid). The time difference between $z \sim 0.13$ and $z \sim 0.06$ is 0.85 Gyrs. The Balmer lines vary more in this short time interval than they do between dense and less dense environments. Similarly the evolution of most Lick indices is similar to or greater than the difference between environments.

To see what this means, I have attempted to compare these results with predictions from stellar population synthesis models. Unfortunately, most models to date do not yet account for the fact that the stellar populations of early-types have non-solar α-element abundances and provide poor fits to my data. The recent work of Thomas, Maraston & Bender (2003) provides α-enhanced models of the Lick indices, but does not provide corrected models for color, D4000 or higher order Balmer lines. Due to these problems, I have chosen to do the following.

The luminosity function of this sample evolves slightly with redshift (Bernardi et al. 2003). The observed evolution can be accounted-for if one assumes that the number densities are not changing, but the higher redshift population is slightly more luminous than that at lower redshifts. Since a pure luminosity evolution model is consistent with the data, it is plausible that the population at lower redshifts is a passively evolved sample of the population at higher redshifts. If so, then the observed evolution with redshift can be used as a clock. The argument is as follows.

The dashed and solid lines in Figure 4 have approximately the same values for color as for Mg, suggesting that color and Mg respond similarly to changes in age and environment. Presumably, metallicity and α-abundance differences play some role in the dependence on environment. A constraint on the importance of these effects can be obtained if one believes that the trend with redshift is primarily a reflection of age effects. For color, at least, the observation that the slope of the color-magnitude relation does not evolve out to redshifts of order unity suggests that residuals from the color magnitude relation are indicators of age. Thus, the dashed line in the top panel of Figure 4 shows the change in color associated with $\sim$ 1 Gyr in age. Since this change in color is similar to the variation between low and high density environments, the age difference between these two environments must be $\sim$1 Gyr. Moreover, the fact that Mg behaves similarly suggests that the entire environmental variation of Mg can be accounted for by age effects: metallicity effects are not very important, and the age difference really is 1 Gyr. Metallicity effects may be important for the indices which are not well fit by the dashed line: in principle, the argument above allows one to calibrate the relative roles of age and metallicity on the different indices. This leads me to a slightly provocative conclusion: claims that metallicity effects are important for the Lick indices like Mg are incompatible with currently popular interpretations of the color-magnitude relation.

References

1. Bernardi, M., Renzini, A., da Costa, L. N. et al. 1998, ApJL, 508, 143
2. Bernardi, M., Sheth, R. K., Annis J. et al. 2003, AJ, 125, 1882
3. Bernardi, M., Nichol, R. C., Sheth, R. K. et al. 2004, AJ, *submitted*
4. Miller, C. J., Nichol, R. C., Reichart, D. et al. 2004, AJ, *submitted*
5. Thomas, D., Maraston, C., & Bender, R. 2003, MNRAS, 339, 897
6. Trager, S. C., Worthey, G., Faber, S. M., Burstein, D., & González, J. J. 1998, ApJS, 116, 1

Local Redshift Surveys and Galaxy Evolution

Roberto De Propris[1], Matthew Colless[1], and Darren Croton[2]

[1] Research School of Astronomy and Astrophysics,
Australian National University,
Weston, ACT, 2611,
Australia
[2] Max-Planck Institut für Astrophysik,
Garching, D-85740, Germany

Abstract. We present observations of galaxy environmental dependencies using data from the 2dF Galaxy Redshift Survey. From a combined analysis of the luminosity function, Butcher-Oemler effect and trends in Hα line strengths we find support for a model where galaxy properties are mainly set by initial conditions at the time of their formation.

1 Introduction

Local redshift surveys, undertaken to map the three-dimensional distribution of galaxies and therefore address topics in cosmology, are also useful to derive a sample of local galaxies and analyze their properties, such as luminosity functions and star formation rates. Although the current generation of surveys is too local to be useful for galaxy evolution studies (but see descriptions of the DEEP2 and VIMOS surveys in the present volume), the 2dF and SDSS are adequate to obtain an in-depth description of the local world of galaxies.

2 The Galaxy Luminosity Function

This is sometimes regarded as a 0th order statistics, the simplest characterization of galaxy populations. It is essential to reproduce its shape as a first step towards a consistent model of galaxy formation. A review of this topic is presented in Driver & De Propris (2003). Here we compare luminosity functions for the field, clusters and as a function of local density from the 2dF galaxy redshift survey: since these luminosity functions are derived from the same catalogue of redshifts and photometry, they should share most of the selection biases and therefore should be fairly comparable to gain a picture of environmental effects on galaxy luminosities.

Figure 1 shows a schematic view of the behaviour of the two relevant parameters of the luminosity function as a function of density, δ_8, which is the density, for each galaxy, measured within an 8 Mpc sphere. We see that:

- M^* becomes brighter in higher density regions
- α becomes steeper in these regions

- This is essentially due to evolution in the M^* and α of early-type galaxies (i.e. spectroscopically quiescent objects)
- There is little or no evolution among late-type galaxies

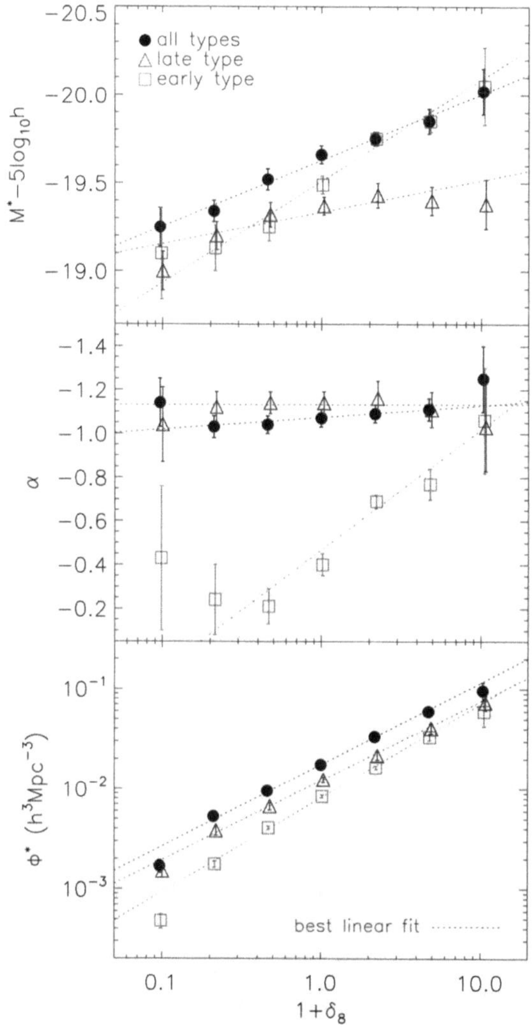

Fig. 1. The best Schechter fit parameters to the luminosity functions in the density regimes considered by Croton et al. (2003) for early-type galaxies (squares) and late-type galaxies (triangles),

This is reminiscent of the simple model for galaxy evolution presented in De Propris et al. (2003), where galaxies were simply assumed to move from star-forming to quiescent with little luminosity or density evolution. Although this is a very simplistic scenario, it appears to reproduce the observations to a good degree. If this is correct, (i) mergers are not necessary to form galaxies; suppression of star formation suffices and (ii) even the blue luminosity of most massive galaxies is dominated by the underlying old population.

3 The Butcher-Oemler Effect

A traditional measure of star formation and its variation in clusters is the blue fraction as defined by Butcher & Oemler (1984). Our sample of clusters lies at $z < 0.11$ and is therefore unsuitable to study how the blue fraction evolves, but it allows us to explore how the blue fraction depends on cluster properties and therefore identify the mechanisms responsible, as well as control the selection effects present in cluster samples at higher redshift.

Figure 2 shows that the blue fraction does not strongly depend on cluster properties such as richness, velocity dispersion and Bautz-Morgan type. With proper accounting for errors, the blue fraction may not be significantly different from cluster to cluster. The only dependencies are with luminosity (the blue fraction is higher for lower luminosity limits), which argues that the blue galaxies are intrinsically faint, and with radius.

The above implies that the original claims for a trend in blue fraction with redshift may have resulted from a selection effect and an underestimate of the actual errors and that the cause for the suppression of star formation responsible for the trend in the luminosity function (and the Hα relations shown below) may not be associated with the effects of the cluster environment (e.g. ram stripping, tides, harassment).

We have explored the Butcher-Oemler effect as a function of the more general mean density and show that there is a clear relation between lower density and higher blue fraction, of which the clusters represent an extreme. Clusters are simply a continuation of field trends (Figure 3).

However, as shown in Figure 4, this is not due to star formation being suppressed, but simply to a changing fraction of red and blue galaxies, consistent with our simple model for instantaneous suppression of star formation with conservation of numbers and luminosity.

4 Galaxy Ecology: The Hα-Environment Relation

Balogh et al. (2003) have carried this analysis a step further in their study of trends in Hα emission with local density from a combined 2dF and SDSS sample. Consistent with the previous findings, they show that:

- The population of galaxies is bimodal: one population shows star formation and the other is quiescent
- The star formation rate is largely independent of density: the numbers of star forming and quiescent galaxies vary smoothly and continuously with density.

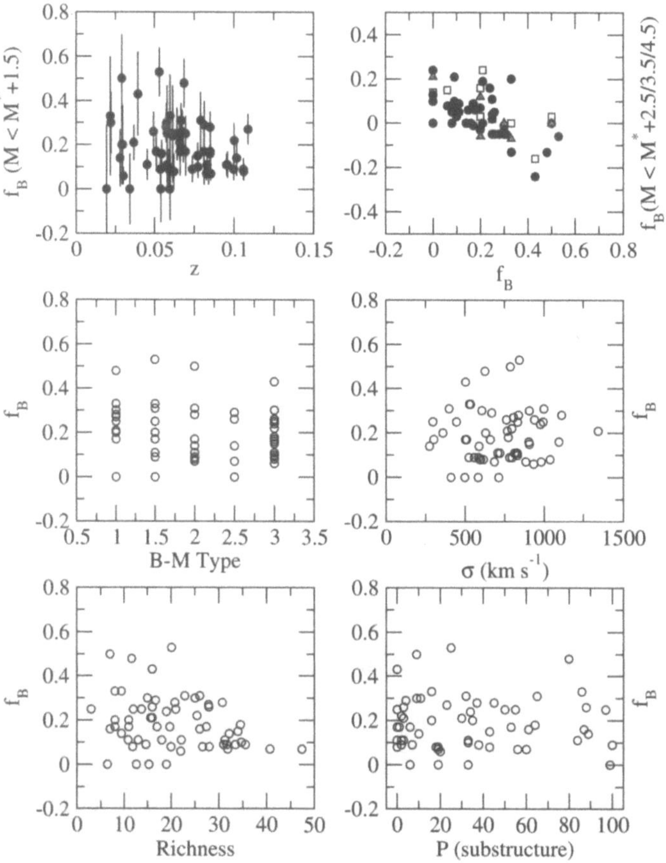

Fig. 2. Dependence of the blue fraction in clusters on redshift, luminosity and cluster properties. The top panel on the right shows the difference between the blue fraction measured to $M^* + 2.5$ (filled circles), +3.5 (open squares) and +4.5 (triangles) with respect to the blue fraction measured to $M^* + 1.5$ (which is the one used in all other panels. We omit error bars for clarity, but show them in the top left panel.

- The fraction of star forming galaxies is sensitive primarily to density over large scales

These results argue that the local environment does not influence galaxy star formation rates, and therefore that galaxy properties are set at high redshift via processes involving relatively short timescales. This obviously accounts for the lack of any correlation between the blue fraction and cluster properties.

5 A 'Predestination' Model of Galaxy Evolution

We present here a speculative scenario for galaxy formation and evolution, which attempts to account for the observations described above.

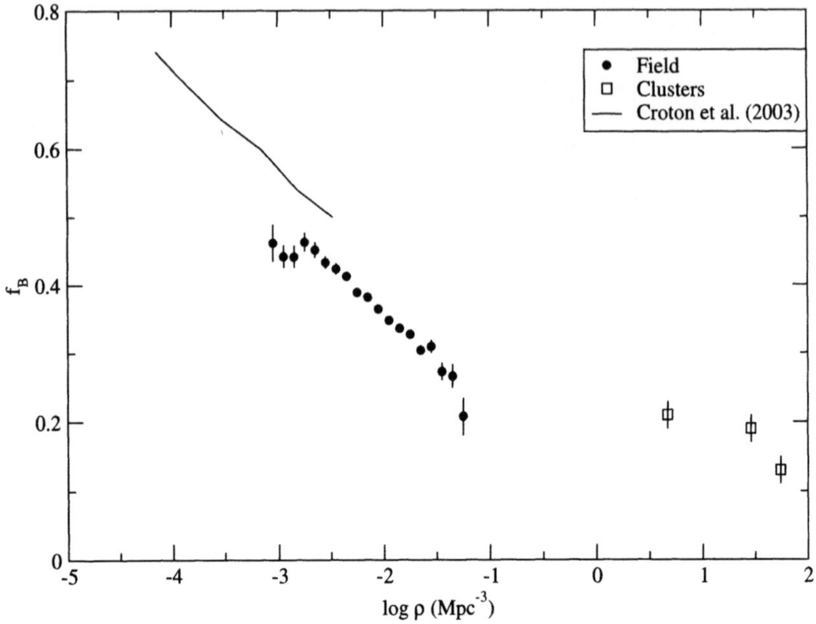

Fig. 3. Variation of the blue fraction with local density.

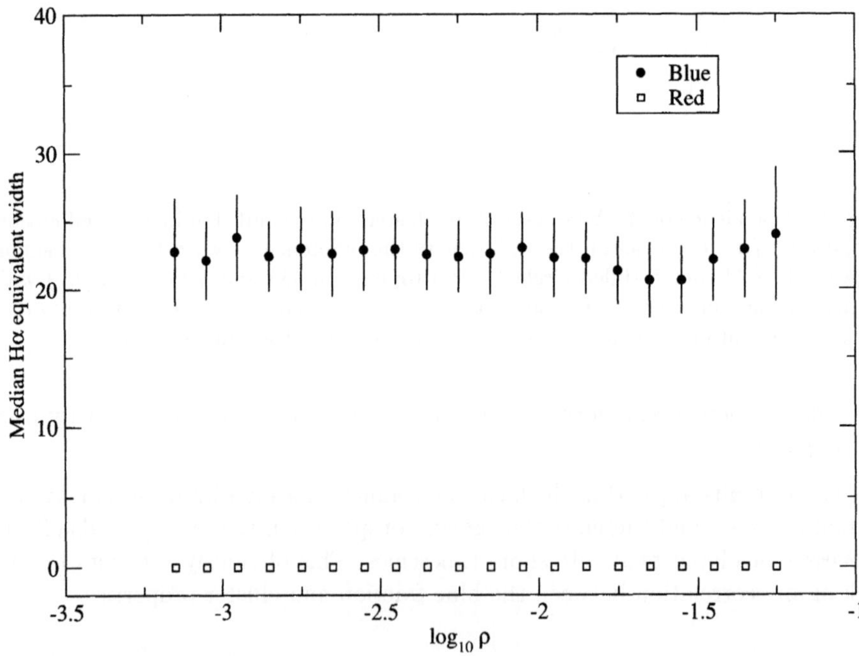

Fig. 4. Median Hα equivalent width for red and blue galaxies as a function of density.

In our model, the properties of galaxies are set by their initial conditions. Galaxies that form in the highest density regions evolve faster and burn all their gas quickly, therefore becoming ellipticals. Blue galaxies are a residual population of star-forming objects, which are therefore more prevalent in lower density regions. The fact that the U (Cortese et al. 2003) and B band luminosity functions for star-forming galaxies are identical in all environments (De Propris et al. 2003) is a necessary consequence of this mechanism.

Star formation may be triggered at high redshift in small groups, which evolve into the dense regions of today. The likely trigger is close interactions, which have been shown to induce star formation (Barton et al. 2003), or mergers between gas-rich galaxies. It is unlikely that these mergers have been important in forming galaxies to $z < 2$, as also shown by the results of the K20 survey described elsewhere in these proceedings.

The above scenario is remarkably similar to the original scenario of galaxy formation in Gaussian random peaks presented by Bardeen et al. (1986), where the main galaxy properties are set by the initial conditions at the time of their formation. The slow evolution detected in nearly all high redshift surveys (with the possible exception of the Combo-17 results presented elsewhere in this volume) is a strong pointer to this scenario.

The 2dF Galaxy Redshift Survey Team: I. K. Baldry (JHU), Carlton M. Baugh (Durham), Joss Bland-Hawthorn (AAO), Terry Bridges (AAO), Russell Cannon (AAO), Matthew Colless^{PI} (RSAA), Chris Collins (LJMU), Shaun Cole (Durham), Warrick Couch (UNSW), Nicholas Cross (STScI), Gavin Dalton (Didcot), Roberto De Propris (RSAA), George Efstathiou (Cambridge), Richard S. Ellis (Caltech), Carlos Frenk (Durham), Karl Glazebrook (JHU), Edward Hawkins (Nottingham), Carole Jackson (ATNF), Ofer Lahav (Oxford), Ian Lewis (Oxford), Stuart Lumsden (Leeds), Steve Maddox^{PI} (Nottingham), Darren Madgwick (UCSC), Peder Norberg (ETHZ), John Peacock^{PI} (Edinburgh), Will Percival (Edinburgh), Bruce A. Peterson (RSAA), Will Sutherland (AAO), Keith Taylor (Caltech)

References

1. M. Balogh et al.: Mon. Not. Roy. Ast. Soc. in press (2003)
2. J. Bardeen, J. Bond, N. Kaiser, A. Szalay: Astrophys. J. **304**, 15 (1986)
3. E. Barton Gillespie, M, Geller, S. Kenyon: Astrophys. J. **582**, 688 (2003)
4. H. Butcher, A. Oemler: Astrophys. J. **285**, 426 (1984)
5. L. Cortese et al.: Astron. Astrophys. **410**, 25 (2003)
6. D. Croton et al.: Mon. Not. Roy. Ast. Soc. submitted (2003)
7. R. De Propris et al: Mon. Not. Roy. Ast. Soc. **342**, 725 (2003)
8. S. Driver, R. De Propris: Astrophys. Sp. Sci. **285**, 175 (2003)

Star-Forming Galaxies in the Sloan Digital Sky Survey – The View from Pittsburgh

Regina E. Schulte-Ladbeck[1], Christopher J. Miller[2], Ulrich Hopp[3], Andrew Hopkins[1], Robert C. Nichol[2], Wolfgang Voges[4], and Taotao Fang[5]

[1] University of Pittsburgh (Pitt), Pittsburgh, USA
[2] Carnegie Mellon University (CMU), Pittsburgh, USA
[3] Munich Observatory & MPE, Munich, Germany
[4] Max-Planck Institut für Extraterrestrische Physik (MPE), Munich, Germany
[5] CMU, now at University of California, Berkeley, USA

Abstract. We used data from the Data Release 1 (DR1) of the Sloan Digital Sky Survey (SDSS) in order to define a catalog of about 13,000 star-forming galaxies (SFG). We discuss the results of two projects. First, we matched our catalog against the ROSAT All-Sky Survey (RASS) and catalogs of pointed ROSAT observations. We identify eight X-ray emitting star-forming galaxies; four were known previously, e.g., the famous "most metal-poor" dwarf galaxy I Zw 18, but another four are new identifications. The data confirm the calibration of X-ray luminosity to a star-formation rate (SFR) by Ranalli et al. (2003), and are used to derive SFRs for these SDSS SFGs. We also suggest two new candidate galaxy clusters. Second, after carefully eliminating all cases of "shredded" SFGs, we derive total absolute blue magnitudes and ionized gas metallicities (O/H ratios). We discuss the luminosity-metallicity (L-Z) relation for galaxies of different physical size (compact, small and large depending on Petrosian half-light radius, R_e). We report the discovery of evolution in the L-Z relation for the redshift range from 0 to 0.3, in the sense that galaxies at higher redshifts tend to have lower O/H ratios at a given luminosity. This evolution is strongest for large galaxies, and weakest for compact galaxies.

1 Introduction

The CMU-Pitt Value Added Catalog (VAC) database (http://astrophysics.phys.cmu.edu/vac/) allows users to conduct research on SDSS spectroscopic galaxy data. The current public VAC contains 133,440 SDSS DR1 galaxies, of which 124,411 are unique. This enables the pursuit of projects which require very large galaxy samples. We here describe two such projects; first, the search for rare, X-ray emitting normal SFGs, and second, an investigation of the L-Z relation for galaxies of different sizes – compact to large – and its evolution with redshift.

2 Sample Definition

The sample was drawn from all objects spectrally classified as galaxies. We required that the Hβ rest equivalent width of an object be larger than 5Å. The threshold of 5Å was set because the SDSS pipeline does not account for underlying Balmer absorption when calculating equivalent widths, and was motivated by the work of Kobulnicky et al. (1999), who recommend a statistical correction of 3±2Å for underlying Hβ absorption to integrated spectra of late-type galaxies. We adopted a conservative criterion for separating SFGs and AGN. We required that the galaxies fell into the SFG regime in all three diagnostic diagrams of Kewley et al. (2001). The SDSS SFGs/AGN separation is also described in Miller et al. (2003). In addition, we checked that all galaxies classified as SFGs conform to the same relation of Hβ versus [OIII]$\lambda\lambda$5007 line width. The SDSS wavelength range and the SFGs/AGN separation together, limit the redshift range of the sample from about 0.02 to 0.3. We use a concordance cosmology with $\Omega_0 = 0.3$, $\Omega_\Lambda = 0.7$, H_0=70 km s^{-1} Mpc^{-1}.

3 SDSS/ROSAT SFGs

We searched for positional matches to our objects with RASS Position Sensitive Proportional Counter (PSPC) data, with catalogs of pointed PSPC observations, and with catalogs of pointed High Resolution Imager (HRI) observations, maintained at MPE. The search radius was 60" for PSPC, and 20" for HRI catalogs. This resulted in a list of 146 unique SDSS DR1 SFGs which have positional coincidence with a ROSAT source (or source list). The ROSAT error circle may include a cluster of galaxies, an AGN or an X-ray bright star, as well as the SFG, but we are interested only in secure X-ray/SFG detections. AGN are X-ray bright and fairly numerous, and a match of our list against NED and SDSS QSO catalogs (Richards et al. 2002) immediately revealed that 1/4 could be QSOs, while

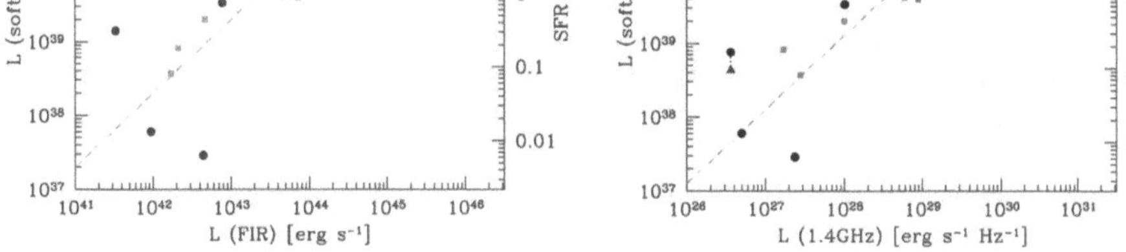

Fig. 1. *Left.* The soft–X-ray luminosity is shown as a function of FIR luminosity. The Ranalli et al. (2003) ASCA & BeppoSAX data are shown by the squares; their fit, by the dashed line, and their SFR calibration, on the right-hand axis. Our ROSAT data (for the 0.5-2.0keV range) are shown by circles. *Right.* Same for soft–X-ray luminosity as a function of 1.4 GHz luminosity density. The triangle indicates the Chandra 0.5-2.0keV luminosity of I Zw 18. It is lower than the ROSAT point, which had a confusing source within the extraction radius resolved by Chandra.

another 1/10 could be clusters. We used a soft-band luminosity of 10^{41} erg s^{-1} to separate SFGs from luminous, AGN and clusters in the remaining 2/3 of sample. In order to transform the ROSAT count rates into an unabsorbed flux, we used the Galactic HI column densities of Dickey & Lockman (1990), and assumed an intrinsic powerlaw spectrum with a photon index of 2.0. We converted redshifts to luminosity distances and used these to convert fluxes to luminosities. There are twelve objects with soft X-ray luminosities $\leq 10^{41}$ erg s^{-1}. In addition, the data suggest two possible matches based on small f_X-over-f_R ratios. While these have $L_X > 10^{41}$ erg s^{-1}, they were added to our candidate sample, bringing it up to 14 objects. We investigated the properties of all 14 cases in great detail (but did not pursue the nature of the rest of the sources).

Four SFGs were previously known to be soft X-ray emitters, namely the edge-on SB galaxy IC 2233, the Blue Compact Dwarf (BCD) I Zw 18, the SA galaxy NGC 4030, and the edge-on SA galaxy NGC 5907. There are four newly identified X-ray emitting SFGs: SDSS J014143.17+134033.16, a face-on Spiral galaxy also known as NPM1G +13.0066, SDSS J081313.18+455940.7, the BCD Mrk 86, SDSS J130039.24+023002.81, the face-on Virgo-cluster SB galaxy NGC 4900, and SDSS J165817.7+644220.11, an inclined Spiral galaxy. The fluxes in the 0.1–2.4 keV band range from about 0.5 to 1.4×10^{-14} erg s^{-1} cm^{-2}; fairly deep because of pointed ROSAT observations. The logarithmic luminosities range from about 37.8 to 41.5. Matches coinciding with the known clusters Abell 119, Abell 318 and RDCS J0910+5422 were not attributed to SFGs. We suggest that there may be galaxy groups or clusters at J0233+0041, with a redshift of about 0.022, and at J1717+6420, with a redshift of about 0.035.

Table 1. Optical Data

SDSS name	Other name	Type	z	d [Mpc]	r [mag][a]
SDSS J014143.17+134033.16	NPM1G+13.0066	Spiral	0.0453	201.2	14.84
SDSS J081313.18+455940.7	Mrk 86	BCD	0.0015	6.3	13.31
SDSS J081358.8+454440.61	IC 2233	SB	0.0019	8.2	13.56
SDSS J093402.03+551427.88	I Zw 18	BCD	0.0026	11.0	16.44
SDSS J120026.27-010607.95	NGC 4030	SA	0.0051	21.6	15.88
SDSS J130039.24+023002.81	NGC 4900	SB	0.0032	13.5	11.81
SDSS J151602.21+561733.02	NGC 5907	SA	0.0015	6.3	13.86
SDSS J165817.7+644220.11	...	Spiral	0.0160	69.2	15.81

[a] Petrosian magnitudes; beware, I Zw 18 is "shredded" & magnitude under-estimated

Table 1 lists the basic data for the X-ray emitting SFGs identified in this work. Table 2 gives fluxes and luminosities which can be calibrated to a SFR. Our objects follow the soft-X-ray—FIR and soft-X-ray—radio relations derived by Ranalli et al. (2003). They propose a soft-X-ray—SFR calibration based on 23 galaxies, which is nicely confirmed by our added sample of 8 SDSS/ROSAT/SFGs, as shown by Fig. 1. The last column of Table 2 gives the X-ray SFRs of our sample based on equation (14) of Ranalli et al. (2003).

- Our results underscore the utility of soft X-ray fluxes to derive SFRs.

Table 2. FIR, Radio and X-Ray Data

SDSS name	$S_{60\mu m}$ [Jy]	$S_{100\mu m}$ [Jy]	$\log L_{FIR}$ [erg s^{-1}]	$S_{1.4GHz}$ [mJy]	$\log L_{1.4GHz}$ [erg $s^{-1}Hz^{-1}$]	$\log F_X$[a] [erg $s^{-1}cm^{-2}$]	$\log L_X$[a] [erg s^{-1}]	SFR[a] [$M_\odot yr^{-1}$]
S.J01..+13..	0.78	1.44	44.32	3.5	29.23	-13.50	41.19	33.70
S.J08..+455.	3.46	6.78	41.97	10.5	26.70	-13.90	37.78	0.01
S.J08..+454.	0.81	1.19	41.52	...	...	-12.78	39.15	0.29
S.J09..+55..	...	...	...	2.5	26.56	-13.28	38.88	0.17
[b]	"	"	"	"	"	-13.54	38.63	0.09
S.J12..-01..	17.24	45.50	43.80	153.9	28.93	-12.65	40.10	2.75
S.J13..+02..	5.52	14.00	42.89	46.7	28.01	-12.81	39.53	0.74
S.J15..+56..	11.52	44.17	42.65	50.5	27.38	-14.22	37.46	0.01
S.J16..+64..	0.14	0.36	42.72	...	...	-13.30	40.46	6.32

[a] The energy range is from 0.5–2.0 keV.
[b] *Chandra* data of I Zw 18, fitted with a power law & Raymond-Smith model.

4 SDSS L-Z Relation

Local galaxies are observed to conform to a luminosity-metallicity relation: luminous galaxies exhibit higher metallicities than do dwarfs (Melbourne & Salzer 2002, and references therein). The L-Z relation is usually shown as total absolute B magnitude versus O/H ratio. The L-Z relation is increasingly being used to study galaxy evolution at intermediate (e.g. Lilly et al. 2003) and high (e.g. Pettini et al. 2001) redshifts, when galaxies were younger and smaller.

In order to study the dependence of the L-Z relation on galaxy size as well as on redshift, we divided our SDSS DR1 SFG sample into three groups depending on *r*-band size: compact galaxies with $0<R_e[\mathrm{kpc}]\leq 2$, small galaxies with $2<R_e[\mathrm{kpc}]\leq 5$; this is the largest group and includes Milky-Way–sized objects, and large galaxies having $5<R_e[\mathrm{kpc}]\leq 100$. Photometry in *g* and *r* was transformed to the B band following Smith et al. (2002); galactic foreground reddening was removed and K-corrections were applied (Blanton et al. 2003). In many cases, the SDSS pipeline magnitudes referred to HII regions within galaxies, rather than to entire galaxies. This phenomenon is also known as "shredding". Of course, the compact galaxy sample was most severely affected by shredding, and we used in our analysis only "parents" which had not been deblended into "children". This has eliminated a number of real compact galaxies, e.g., the two HII regions of I Zw 18 are listed as separated galaxies in the DR1. We used the N2 parameter (Denicoló et al. 2002) in order to derive O/H ratios. The N2 parameter has two advantages compared with the R23 index (Pagel et al. 1979): it is not affected by reddening, and its calibration to O abundance is single-valued. The SDSS fibers have a size of 3" and the O/H ratios usually refer to the inner regions of the galaxies. The L-Z relation for about 5,000 SDSS SFGs with high signal-to-noise data is shown in Fig. 2.

The results of the KISS survey for ≈500 local SFGs (Melbourne & Salzer 2002) indicate a flattening of the L-Z relation for dwarfs. Fig. 2 suggests that

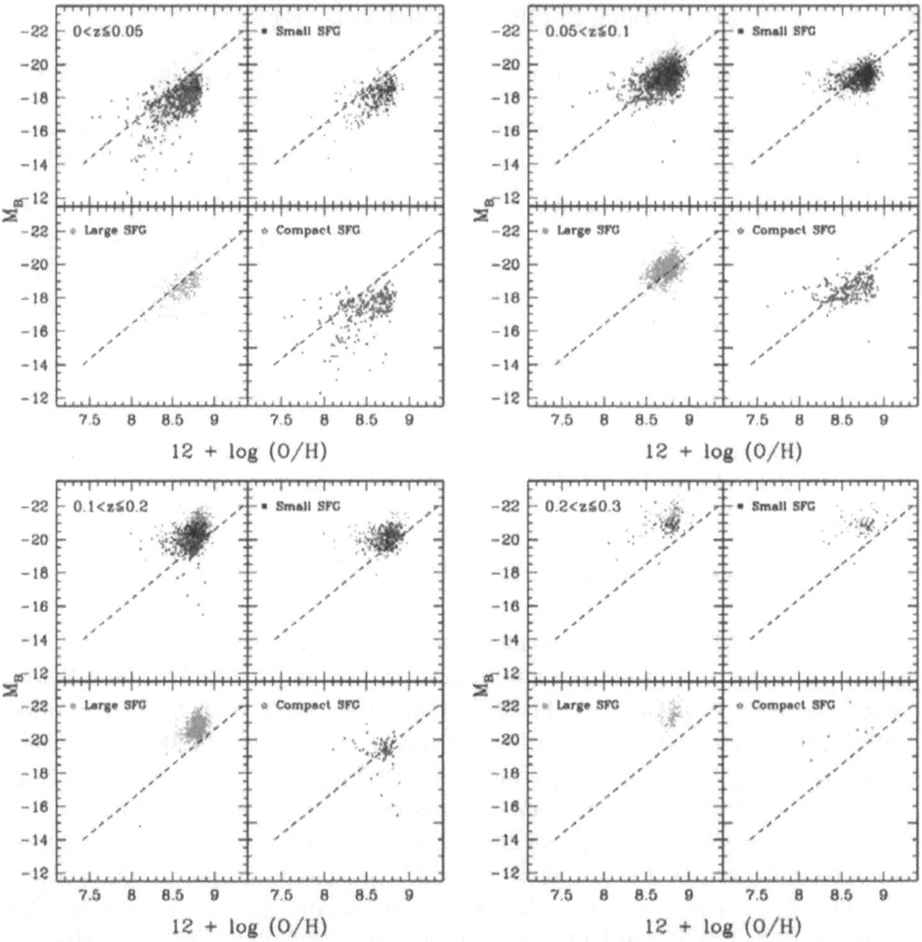

Fig. 2. The L-Z relation for SDSS SFGs in four redshift bins. The dashed line shows the relation in the KISS sample, which includes galaxies with redshifts of up to 0.095.

small SDSS SFGs have a flatter L-Z relations than large ones. We also see a large amount of scatter among compact, dwarf galaxies. We interpret this to indicate that these galaxies are affected more by feedback than luminous/large galaxies.

The top left quadrant of each panel in Fig. 2 shows where SDSS SFGs (of all sizes) in four redshift slices lie relative to the KISS L-Z relation (Melbourne & Salzer 2002). We see a strong evolution, in the sense that the SDSS SFGs systematically shift across the KISS L-Z relation as redshift increases. Specifically, the data suggest that at a given metallicity (say, 12+log(O/H) = 8.7), galaxies are less luminous now than they were just 2-3 Gyrs ago; or, that at a given L (say, M_B = -19.5), galaxy metallicity declines with increasing redshift.

The data in each of the redshift slices, do not extend over the luminosity range of the entire sample: in the lowest-z bin the volume is too small to include luminous galaxies, while in the highest-z bin, the limited depth of the survey precludes the detection of dwarfs. Therefore, the main concern is that luminosity introduces some kind of bias. In order to start to address this question, we used a magnitude limited sample with $r \leq 17.77$, as in the SDSS main galaxy sample (Strauss et al. 2002). This has the effect of removing galaxies at low O/H and decreasing the scatter among compact galaxies, but also, of cutting galaxies from the two highest-z bins. In principle, the L-Z relation could vary in both, slope and zero point, but because our galaxies cover only a small fraction of the entire L range in each redshift slice, we assumed a constant slope is retained and tested only for variation in zero point with redshift. The difference in zero point of the L-Z relation when comparing the lowest with the now highest, $0.1<z\leq 0.2$ bin, is significant, and amounts to over 1.5 magnitudes.

• SDSS data indicate a very rapid evolution in the L-Z relation for late-type galaxies since $z< 0.3$.

Acknowledgments Drs. M. Bernardi and D. Vanden Berk helped us understand SDSS parameters. RS-L thanks the MPE for the invitation to Munich in fall of 2002. AMH gratefully acknowledges support provided by NASA through Hubble Fellowship grant HST-HF-01140.01-A awarded by the Space Telescope Science Institute (contract NAS 5-26555). Funding for the creation and distribution of the SDSS Archive has been provided by the Alfred P. Sloan Foundation, the Participating Institutions, the National Aeronautics and Space Administration, the National Science Foundation, the U.S. Department of Energy, the Japanese Monbukagakusho, and the Max Planck Society. The SDSS Web site is http://www.sdss.org/. We made extensive use of NED.

References

1. Blanton, M.R., et al. 2003, in preparation
2. Denicoló, G., Terlevich, R., Terlevich, E. 2002, MNRAS, 330, 69
3. Dickey, J.M., Lockman, F.J. 1990, ARA&A, 28, 215
4. Kewley, L. J., Dopita, M. A., Sutherland, R. S., Heisler, C. A., Trevena, J. 2001, ApJ, 556, 121
5. Kobulnicky, H.A., Kennicutt, R.C., Jr., Pizagno, J.L. 1999, ApJ, 514, 544
6. Lilly, S. J., Carollo, M., Stockton, A.N. 2003, astro-ph/0209243
7. Melbourne, J., Slazer, J.J. 2002, AJ, 123, 2302
8. Miller, C.J., Nichol, R.C., Go;mez, P.L., Hopkins, A.M., Bernardi, M. 2003, ApJ, 597, 142
9. Pagel, B. E. J., Edmunds, M. G., Blackwell, D. E., Chun, M. S., Smith, G. 1979, MNRAS, 198, 95
10. Pettini, M., Shapley, A.E., Steidel, C.C., Cuby, J.G., Dickinson, M., Moorwood, A.F.M., Adelberger, K.L., Giavalisco, M. 2001, ApJ, 554, 981
11. Ranalli, P., Comastri, A., Setti, G. 2003, A&A, 339, 39
12. Richards, G.R., Fan, X., Newberg, H., et al. 2002, AJ, 123, 2945
13. Smith, J. A., et al. 2002, AJ, 123, 2121
14. Strauss, M.A., Weinberg, D.H., Lupton, R.H. et al. 2002, AJ, 124, 1810

Exploring the Reionization Epoch with HST and JWST

S. Michael Fall[1], Massimo Stiavelli[1], and Nino Panagia[1,2]

[1] Space Telescope Science Institute, 3700 San Martin Drive, Baltimore, MD 21218, USA

[2] Affiliated with the Space Telescope Division of the European Space Agency, ESTEC, Noordwijk, Netherlands

Abstract. We discuss the detectability of the sources responsible for reionization with existing and planned telescopes. Our approach is based on the number of UV photons required for reionization and depends on relatively few undetermined parameters. We conclude that if the reionization sources are UV-efficient, minimum-luminosity sources, it may be difficult to detect them before the launch of the James Webb Space Telescope. Until then, the best approach may be either to exploit gravitational lensing by clusters of galaxies, or to search for strong Lyα emission by narrow-band imaging or slitless grism spectroscopy.

1 Introduction

Motivated by recent evidence that the epoch of hydrogen reionization may have ended at a redshift as low as $z \approx 6$ [1,2], we have considered the detectability of the sources responsible for this reionization. The main idea is that reionization places limits on the mean surface brightness of the population of reionization sources. We have defined a family of models characterized by two parameters: the fraction f_c of Lyman-continuum photons that escape from the sources and the clumpiness parameter C of the intergalactic medium. The minimum required surface brightness corresponds to a value of unity for both of these parameters. The maximum surface brightness is set by the condition that the reionization sources not produce too many heavy elements. Our general approach is applicable to most types of reionization sources, but in specific numerical examples, we focus on those composed of Population III stars, which have high effective temperatures and therefore produce ionizing photons with high efficiency (e.g. [3]). Our predictions are compared to the parameter space that can be probed by existing and future telescopes, primarily the Hubble and James Webb Space Telescopes (HST and JWST) in order to help plan the most effective searches. A full account of our work has been published recently in the Astrophysical Journal [4].

2 Continuum Searches

In Figure 1, we show the expected mean surface density of the reionization sources as a function of their apparent AB magnitude in the non-ionizing UV

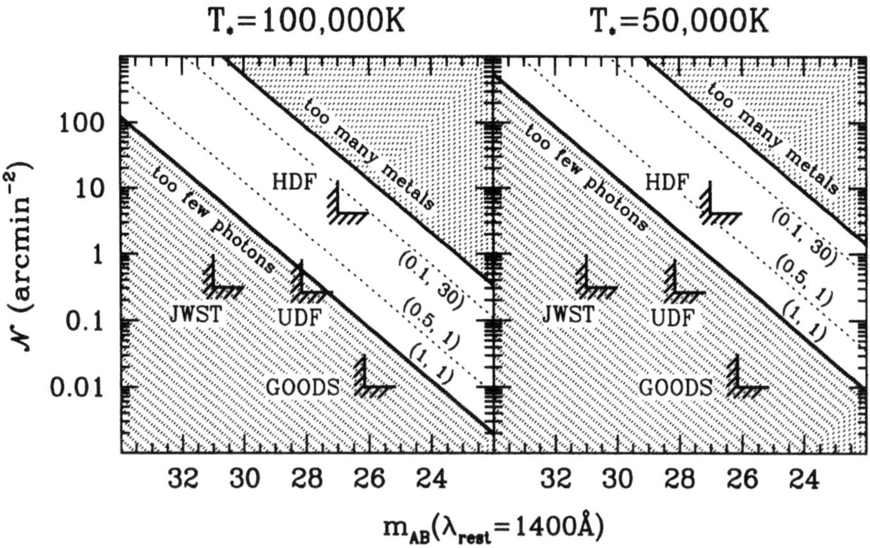

Fig. 1. Surface density vs. apparent magnitude of identical reionization sources composed either of Population III stars (left hand panel) or Population II stars (right hand panel)

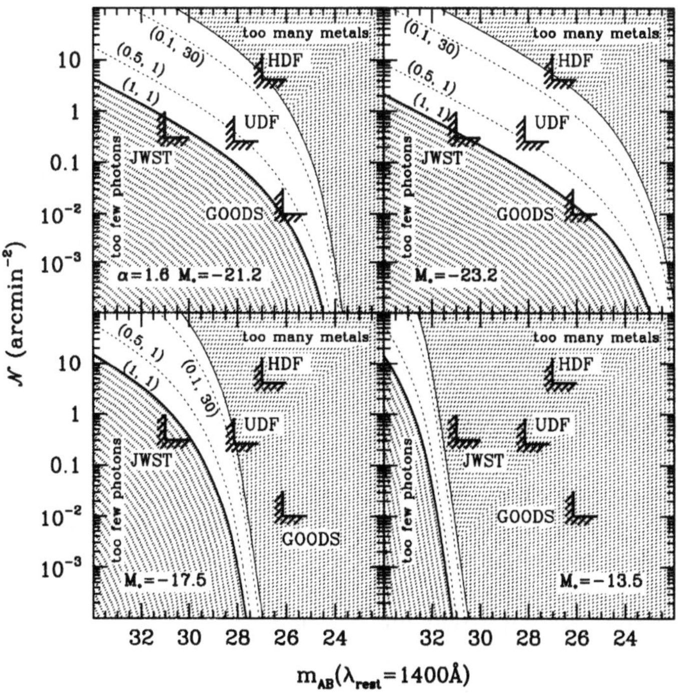

Fig. 2. Cumulative surface density vs. apparent magnitude of non-identical reionization sources with different luminosity functions (different M_* but same α)

continuum at $\lambda_{\rm rest} = 1400$ Å. In this case, the sources are assumed for simplicity to be identical to each other. The left panel refers to sources with Population III stars, with an effective temperature of 10^5 K; the right panel refers to those with Population II stars, with an effective temperature of 5×10^4 K. In both panels, the lower solid line represents the minimum surface brightness model, with $(f_c, C) = (1, 1)$, while the upper solid line represents the global metallicity constraint $Z \leq 0.01\ Z_\odot$ at $z = 6$. The thin dotted lines represent models with $(f_c, C) = (0.5, 1)$ and $(0.1, 30)$. Populations of stellar sources can be excluded from the upper shaded regions on the grounds that they would have produced too many heavy elements. The L-shaped markers delimit the quadrants (i.e. the areas upward and rightward) probed by the GOODS/ACS survey [5], the HDF/NICMOS and HDFS/NICMOS surveys [6,7], the planned HST Ultra-Deep Field (UDF), and a hypothetical ultra-deep survey with JWST.

In Figure 2, we show the expected cumulative surface density of non-identical reionization sources with a broad range of luminosities as a function of their apparent AB magnitude. We assume that the luminosity function of the sources has the Schechter form, parameterized by the knee M_* and slope α. This is likely to be more realistic than the idealized situation depicted in Figure 1. The luminosity function adopted for the upper-left panel of Figure 2 is similar to that of Lyman-break galaxies at $z = 3$ ($M_{*,1400} = -21.2$ and $\alpha = 1.6$); the knee of the function is brighter or fainter in the other panels. As before, the upper solid line in each panel represents the global metallicity constraint, while the lower solid line represents the minimum surface brightness model, with $(f_c, C) = (1, 1)$. The thin dotted lines show results for models with $(f_c, C) = (0.5, 1)$ and (0.1, 30). Again, the L-shaped markers delimit the quadrants probed by the GOODS/ACS survey, the HDF/NICMOS and HDFS/NICMOS surveys, the UDF, and an ultra-deep survey with JWST, respectively.

From these results, it appears that if reionization is caused by UV-efficient, minimum-luminosity sources with Population III stars, then the non-ionizing continuum emission from these sources will be difficult to detect before the launch of JWST. However, if the sources are composed instead of cooler Population II stars or AGNs, they will be brighter by 1–2 magnitudes and thus easier to detect.

3 Emission-Line Searches

Figure 3 presents the expected surface density of reionization sources as a function of their Lyα line flux. The left panel shows models with identical sources and different values of the parameters f_c and C. As before, the top solid line indicates the global metallicity constraint. The right panel illustrates results for non-identical sources with different luminosity functions. The solid line refers to a luminosity function similar to that of the Lyman-break galaxies at $z = 3$ ($M_{*,1400} = -21.2$ and $\alpha = 1.6$), while the dashed line refers to a luminosity function with a fainter knee and a shallower slope ($M_{*,1400} = -17.5$ and $\alpha = 1.1$). In each panel, the L-shaped marker indicates the quadrant probed by

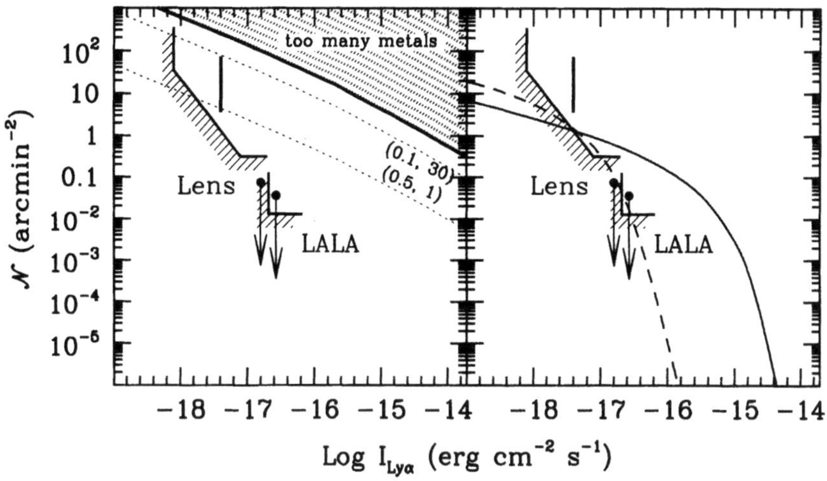

Fig. 3. Surface density vs. Lyα line flux of reionization sources

a narrow-band survey at $z \geq 6$ (LALA survey [8]). The oblique marker labeled *Lens* represents a hypothetical 100-orbit survey with the ACS grism on a cluster of galaxies to exploit gravitational amplification. The solid bar represents the surface density estimated from the single source at $z = 6.56$ detected by Hu et al. [9], while the two points with downward arrows represent their upper limits.

Evidently, emission-line searches based on narrow-band imaging or slitless grism spectroscopy are promising and might lead to the first detection of the reionization sources.

References

1. R.H. Becker, et al.: AJ **122**, 2850 (2001)
2. X. Fan, et al.: AJ **123**, 1247 (2002)
3. N. Panagia, M. Stiavelli, H. Ferguson, H.S. Stockman: in preparation (2004) [see also astro-ph/0209278 and /0209346]
4. M. Stiavelli, S.M. Fall, N. Panagia: ApJ **600**, 508 (2004)
5. M. Dickinson, M. Giavalisco: 'The Great Observatories Origins Deep Survey'. In: *The Mass of Galaxies at Low and High Redshift*, ed. by R. Bender, A. Renzini, (Springer-Verlag, 2003) pp. 324–331
6. R. Thompson, L. Storrie-Lombardi, R.J. Weymann, M.J. Rieke, G. Schneider, E. Stobie, D. Lytle: AJ **117**, 17 (1999)
7. R.E. Williams, et al.: AJ **120**, 2735 (2000)
8. J.E. Rhoads, S. Malhotra: ApJ **563**, L5 (2001)
9. E.M. Hu, et al.: ApJ **568**, L75 (2002)

COSMOS 2□° Survey

Nick Scoville

Caltech, Pasadena, CA 90065, USA

Abstract. The Cosmic Evolution Survey (COSMOS) will provide complete multi-wavelength imaging and spectroscopy of a 2□° equatorial field aimed at sampling the full range of cosmological environments from voids to massive clusters to understand the coupled evolution of dark matter, luminous galaxies and clusters and AGN in the epoch of maximum formation activity from $z = 0.5$ to 3. The COSMOS survey has several major primary components : 1) HST-ACS imaging covering the entire field in I-band (8100Å); 2) ground-based broadband imaging (Subaru, NOAO and CFHT) providing photometric redshifts (typical accuracy 10%); 3) low resolution optical spectra (R = 200 with VLT-VIMOS) for 90,000 galaxies with I < 25 mag. Over 60 astronomers from Europe, North America and Japan are actively involved in COSMOS multi-λ observations and science.

1 Overview

The COSMOS HST survey is the largest ACS survey ever undertaken – imaging an equatorial, 2□° field ($E_{B-V} \simeq 0.02$ mag) with single-orbit I-band exposures to a depth of 27 mag (10σ). Over 2 million galaxies (of all types) are expected with I < 27 mag (AB, 10 σ), sampling a volume in the high redshift universe approaching that sampled locally by the Sloan survey. Extensive multi-λ ground and space-based observations of this field have also been committed. Spectroscopic redshifts will be obtained for 90,000 galaxies with I < 25 mag (mostly using VLT-VIMOS) and photometric redshifts for the remainder (using Subaru, NOAO and CFHT).

2 COSMOS Science

The principle science goals for COSMOS are: quantifying large scale structure in dark matter through weak lensing and mapping the evolution of galaxies in the context of large scale structure. The expected number of objects in a 2□° field is given in Table 1. With more than 70,000 galaxies having measured redshifts (VIMOS survey) in the range $z = 0.5 - 2$ and $0.05''$ HST imaging, 7 redshift bins can be constructed, each with $> 10^4$ galaxies, to probe evolution of the morphological distribution (E, Sp, Irr, etc.) as a function of both LSS and time. Evolution of the luminosity and spatial correlation functions for type-selected galaxies will be analyzed with unprecedented statistical accuracy. The growth of galaxies, AGN and dark matter structure will be accurately traced over a period corresponding to ~75% of the age of the universe.

Table 1. Expected Numbers of Objects in COSMOS 2□° Field

Class	#	I_{AB} (10σ))	Reference
All objects	1.9×10^6	< 27	Metcalfe *et al.* 2001
All objects	300,000	< 25	Metcalfe *et al.* 2001
XMM-AGN	3400	27	c.f. Lockman Hole
XMM-clusters	~120	5×10^{-17} cgs	COSMOS XMM
strong lens systems	60–80		Fassnacht *et al.* 2004
VIMOS Gal.w/ Spectra	$\sim10^5$	I≤25	Kneib *et al.*
QSOs	600(100)	24(21)	Croom *et al.* 2001
z > 4 QSOs	50	25	Cristiani *et al.* 2004
ULIRGs	3,000	26	Smail *et al.* 2002
ExtremelyRedObjects	25,000	25	Daddi 2000
LymanBreakGalaxies (z≤2)	65,000	25.5	Steidel *et al.* 2004
LymanBreakGalaxies (z~3)	10,000	25.5	Shapley *et al.* 2001
Red high-z Galaxies (z > 2)	10,000	25.5	Labbé *et al.* 2002
SN I&II	~100	26.5	GOODS
L,T Dwarfs	300(<200 pc)	$28(4\sigma)$	Burgasser *et al.* 2002
KuiperBeltObjects	100-250	27	Jewitt (priv. com.)

Galaxies in the early universe are built up by two major processes: dissipational collapse and merging of lower mass protogalactic and galactic components. Their intrinsic evolution is then driven by the conversion of primordial and interstellar gas into stars, with galactic merging and interactions triggering star formation and starbursts. These activities are believed to be strongly correlated and occur first in regions of the LSS with highest overdensity.

COSMOS will be fundamental to nearly every area of observational and theoretical cosmology, yielding:

- **the assembly of galaxies, clusters and dark matter on scales up to $\geq 2\times10^{14}$ $M_\odot$, well sampled as a function of redshift;**
- **full reconstruction of the dark matter distributions and characteristics out to z ~1 using weak gravitational lensing;**
- **the evolution of galaxy morphology, galactic merging and star formation as a function of LSS environment and redshift;**
- **evolution of AGN and the dependence of black hole growth on galaxy morphology and environment.**

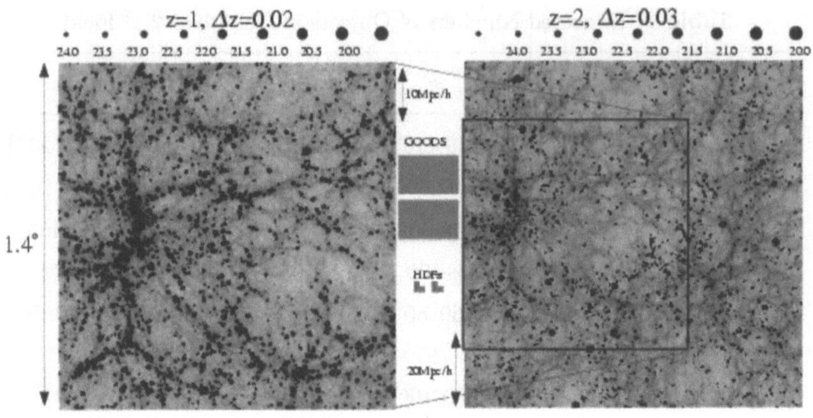

Fig. 1. Λ-CDM simulation results for 2□° at z = 1 and 2, illustrating the evolution of galaxies and clusters and the enormous contrast between void and wall regions at both redshifts (Frenk *et al.* 2002). The gray-scale indicates the dark matter distribution and the symbols show magnitudes of galaxies from semi-analytic models computed for F814W. At each redshift, the slice depth is 50 Mpc corresponding to $\Delta z = 0.02$ (at z = 1). Also shown are the HDF and GOODS field sizes ; the GEMS field size is 1/4□ °. Spectroscopic redshifts provided by VIMOS are an essential feature of COSMOS since photometric redshifts typically have 10% accuracy and thus don't adequately resolve the LSS.

2.1 Large Scale Structure

Figure 1 shows the results of a LSS Λ-CDM simulation for two characteristic redshifts (Virgo Consortium [7]). Observational studies of Lyman break galaxies and deep X-ray imaging with CXO/XMM are broadly consistent with these models with respect to the LSS ([8],[1],[9]).

The need to sample very large scales arises from the fact that structure occurs on mass scales up to $\geq 10^{14} M_{\odot}$ and existing smaller surveys (Figure 1) are likely to be unrepresentative at z $\sim$ 1. This is illustrated in Figure 2, which shows the probability of enclosing a given mass as a function of field size. GOODS and GEMS, adequately sample masses up to $3 \times 10^{13} M_{\odot}$, whereas COSMOS samples the largest known structures $\sim 2 \times 10^{14} M_{\odot}$. The evolution of the halo and cluster mass distribution is shown in the right panel of Figure 2 – dramatically demonstrating the evolution of the dark matter on scales probed by COSMOS. Evolution of the luminous-galaxy occupation number in halos as a function of both redshift and halo mass, will provide stringent tests of LSS models.

3 COSMOS Field

The COSMOS field is located near the equator to ensure coverage by all astronomical facilities. The time requirements for HST-ACS imaging over a total area of 2□° and ground-based spectra of 90,000 galaxies at I $\leq$ 25 mag makes

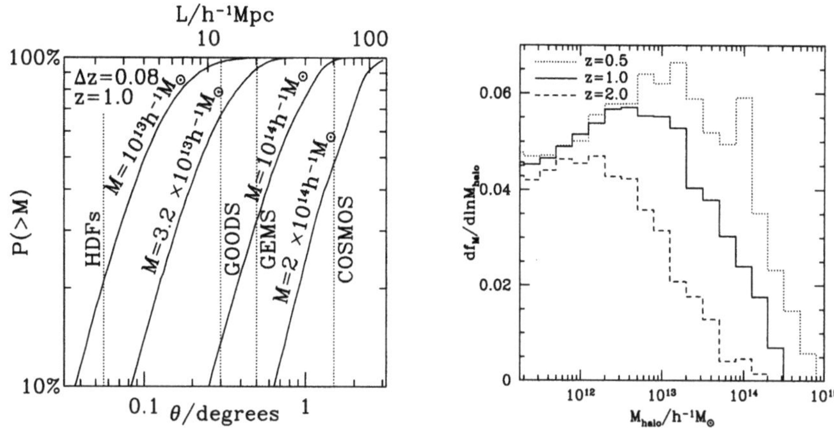

Fig. 2. On the left, the probability of enclosing at least one structure of the specified mass is shown as a function of the field size at z = 1 from the Virgo consortium Λ-CDM simulation [7]. The masses shown correspond approximately to : the Local group, a 'poor' cluster, Virgo and about 30% of Coma. **On the right**, the distribution of halo masses is shown for z = 0.5, 1 and 2, illustrating the expected (but not yet verified) evolution in the halo and cluster mass distributions from the LSS simulations.

it strategically imperative that the field be readily observable by all large optical/IR telescopes and unique non-optical instruments. The selected COSMOS field is a 1.4° ×1.4° square, aligned E-W, N-S, centered at RA = 10:00:28.6 ,

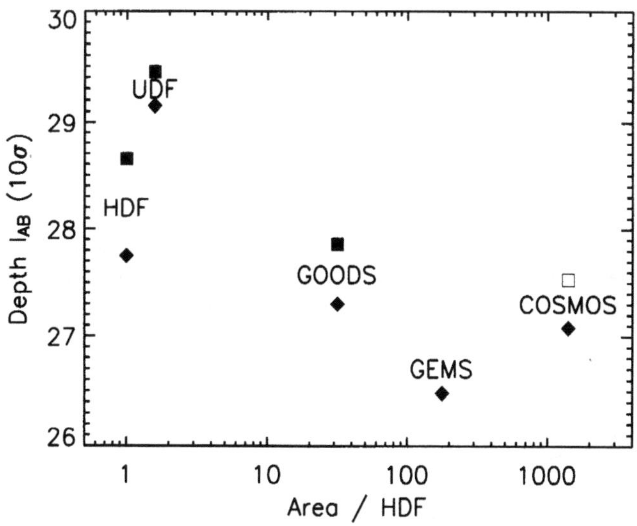

Fig. 3. Relative field areas and sensitivities of major HST surveys at ~8000Å(◇) and 4500Å(□). COSMOS has 9× the area of GEMS (the next largest survey) with sensitivity just 1.25× less than GOODS (in 20% the time due to the higher throughput of F814W vs F850LP – GOODS).

DEC = +02:12:21.0 (J2000). This is close to but shifted slightly from the VIRMOS 10 hr field, in order to minimize backgrounds. The COSMOS field has exceptionally low Galactic extinction (mean $E_{B-V} \simeq 0.2$ mag), low FIR background (a mean FIR background of 0.85 MJy sr^{-1}, compared to ~0.5 MJy sr^{-1} in the very best, non-equatorial fields such as Lockman Hole) and no bright X-ray, radio or stellar sources).

4 Comparison with Other HST Surveys

The relative sensitivities and areas of major HST surveys are compared in Figure 3 and Table 2. As seen in Figure 3, the sensitivity of COSMOS at ~8000Å is only 25% worse than GOODS.

Table 2. Survey Pt. Source Sensitivities

Filter	Survey	10σ AB	Orbits
F435	UDF	29.45	56
..	GOODS	27.86	3
F475	**future COSMOS**	**27.53**	**1**
F606	UDF	29.91	56
..	GOODS	28.22	2.5
F775	UDF	29.71	150
..	GOODS	27.49	2.5
F814	**COSMOS**	**27.08**	**1**
F850	GOODS	27.30	5
..	UDF	29.15	150

Sensitivities for optimally-extracted point sources; for a source uniformly extended over ~0.25″ diameter the limiting magnitudes are ~1 mag greater. (Assumes: 2700s per orbit, λ^{-1} power law spectrum, 4 cr-split, no reddening)

5 Multi-λ COSMOS

The COSMOS field is equatorial and thus is accessible to essentially all astronomical facilities, enabling complete, multi-λ datasets, in addition to the ACS imaging. Their status is summarized in Table 3.

Additional information on COSMOS may be found at the public web-site (**www.astro.caltech.edu/~cosmos**).

Table 3. Multi-λ COSMOS Data

Data	Bands,λ,Res.	AB mag 5σ pt. src	status	PI
HST-ACS	814I	27	U	Scoville
HST-ACS	475g	27	U	Scoville
HST-NIC3	160W	22.9(5% area)	U	Scoville
HST-WFPC2	300W	26.4	U	Scoville
Subaru-SCam	B, r', z'	27, 27, 25	U	Taniguchi
Subaru-SCam	NB816	25	U	Taniguchi
CFHT-Megacam	u^*	27	U	Sanders
CFHT-Megacam	u,i^*	26	U	LeFevre
CFHT-LS	u-z		U	DLSS
NOAO	K_s	21	U	Mobasher
CFHT	K	23 ($9\times9'$)	U	Sanders
UH-88	J	21	U	Sanders
GALEX	FUV,NUV	26.1,25.8	U	Schminovich
XMM-EPIC	0.5 – 10 keV	10^{-15} cgs	U	Hasinger
CXO	0.5 – 7 keV	pointed	F	Elvis
VLT-VIMOS sp.	(R=200)	#=18000 ($I_{AB}<23$)	U	Kneib
VLT-VIMOS sp.	(R=200)	#=72000 ($I_{AB}<25$, $z\geq 0.8$)	F	Kneib
Mag.-IMAX sp.	(R=3000)	#=2000	U	McCarthy
Keck/GEMINI sp.	(R=5,000)	#=4000 (I<24)	F	
Spitzer-MIPS	160,70,24μm	17,1,0.15 mJy(5σ)	F	Sanders
Spitzer-IRAC	8,6,4.5,3μm	11,9,3,2 μJy(5σ)	F	Sanders
IRAM-30m	1.2 mm	1 mJy ($20\times20'$)	U	Bertoldi
CSO-Bolocam	1.1 mm	3 mJy	U	Blain
VLA-A	20cm	24μJy(1σ)	U	Schinnerer
VLA-A/C	20cm	8μJy(1σ)	F	Schinnerer
SZA(full field)	9 mm	S-Z to $2\times10^{14}M_\odot$	F	Carlstrom

Status: U $\rightarrow$ underway (done or scheduled time); F $\rightarrow$ future proposal

References

1. Adelberger, K. *et al.* : ApJ, **505**,18 (1998)
2. Burgasser, A. *et al.* : 2002, ApJ, **564**, 421 (2002)
3. Cristiani, S. *et al.* : ApJL, **(Jan 20)**, 14 (2004)
4. Croom, S.M., Wareen, S.J. & Glazebrook,K. : MNRAS, **328**, 150 (2001)
5. Daddi, E.*et al.* : A&A, 361, 535; 2003 ApJ, **203**, 588 (2000)
6. Fassnacht, C. *et al.* : APJL, **600**, L155 (2004)
7. Frenk, C. *et al.* : astro-ph/0007362 (2002)
8. Giavalisco, M *et al.* : ApJ, **503**, 543 (1998)
9. Gilli, B. *et al.* : ApJ, **592**, 721 (2003)
10. Labbé *et al.* : astro-ph/0212236 (2002)
11. Metcalfe, T.S. *et al.* : MNRAS, **323**, 795 (2001)
12. Shapley, A. *et al.* : 2001 ApJ, **562**, 95 (2001)
13. Steidel *et al.* : astro-ph/0401061 (2004)
14. Smail, I. *et al.* : MNRAS, **331**, 795 (2002)

The UKIDSS Ultradeep Survey – Mapping the Early Stages of Galaxy Formation

Omar Almaini

Institute for Astronomy, University of Edinburgh, Royal Observatory, Blackford Hill, Edinburgh EH9 3HJ, UK

Abstract. The UKIRT Infrared Deep Sky Survey (UKIDSS) is scheduled to commence in mid 2004, using the new Wide-Field Camera (WFCAM) to carry out a major series of long-term public surveys. Data will become public to the entire ESO community immediately. In this paper I briefly describe the camera and the UKIDSS survey plans, highlighting the Ultra-Deep Survey in particular which plans to reach depths of K=23, H=24 and J=25 over an unprecedented 0.8 deg^2 region. This will map an enormous volume of the high-redshift Universe, providing the first detailed picture of large scale structure at $z \sim 3$.

1 The UKIRT Wide-Field Camera

WFCAM is a cryogenic near-infrared camera currently under construction at the UK Astronomy Technology Centre (ATC) in Edinburgh. The focal plane will hold four 2048$\times$2048 Rockwell HgCdTe arrays, spaced at 90% detector width. Using a 0.4 arcsec pixel size the instantaneous field-of-view is 0.2 square degrees, allowing a 0.8 square degree region to be tiled seamlessly in four pointings. A purpose-built f/9 secondary is combined with the existing tip-tilt hexapod to provide optimal image quality. The optical (instrumental) image quality is < 0.26 arcsec rms. To improve image sampling, and mitigate the effects of the large pixels, a 2×2 microstepping technique will be adopted as standard.

The large WFCAM focal plane has been achieved using a new forward-Cassegrain quasi-Schmidt design, which allows a high-degree of off-axis correction. This design also employs a cold-pupil stop for maximum K-band sensitivity. The instrument is due to be commissioned in the Spring of 2004. Further information on the camera and its progress towards completion can be obtained from the project website at http://www.roe.ac.uk/atc/projects/wfcam/.

2 The UKIDSS Surveys

Once commissioned, approximately 50% of all time on UKIRT will be dedicated to carrying out a major series of public surveys. These five surveys explore both low and high galactic latitudes to a variety of depths and will require 7 years to complete. The UKIRT board have approved this programme on a 2-year rolling basis. Originally a UK project, the raw and processed data will now be made available to the entire ESO community. The scope of these surveys is briefly outlined below, with further details in Table 1.

Table 1. Details of the five elements of UKIDSS. The quoted depths are total magnitude 5σ for a point source. The estimated number of nights for each survey includes an allowance for poor weather.

Survey	Filter	Area sq. degs	Mag. limit (Vega)	Nights
	Y		20.5	
Large Area Survey	J	4000	20.0	262
LAS	H		18.8	
	K		18.4	
	J		20.0	
Galactic Plane Survey	H	1800	19.1	186
GPS	K		19.0	
	H_2	300	...	
	J		19.7	
Galactic Clusters Survey	H	1600	18.8	84
GCS	K		18.7	
	J	35	22.5	
Deep Extragalactic Survey	H	5	22.0	118
DXS	K	35	21.0	
	J		25.0	
Ultra Deep Survey	H	0.77	24.0	296
UDS	K		23.0	

The Large Area Survey (LAS) aims to cover 4000 square degrees to a depth which will match the Sloan Digital Sky Survey (SDSS), reaching a depth of K=18.4 and matching depths in J, K and Y-bands. This combination of area, depth and multi-filter coverage make it a powerful tool for determining statistics on common objects (stars, galaxies, clusters of galaxies) and isolating extreme objects (low mass dwarfs and high redshift quasars). Multiple epoch observations in the J-band will allow proper motions to be determined. In addition we will use a new "Y" filter (Hillenbrand et al. 2002), covering 0.97 to 1.07 microns. This is specifically designed to detect extremely high redshift ($z = 7$) quasars, and to distinguish them from low mass L,M,T dwarfs.

The Galactic Plane Survey (GPS) will map a large fraction (1800 square degrees) of the galaxy, surveying a strip $\pm$ 5 degrees above and below the galactic plane along a length of 180 degrees. This will allow star formation regions to be mapped throughout the Milky Way and provide an order of magnitude increase in the known number of Young Stellar Objects (YSOs). We expect to map 10^{8-9} stars in total.

The Galactic Cluster Survey (GCS) aims to measure the sub-stellar mass function in approximately 10 galactic clusters, covering a total of 1400 square degrees. The mass limit reached is typically around 30 Jupiter masses. The aim is to determine the fundamental initial mass function (IMF) for stars as a function of environment and age.

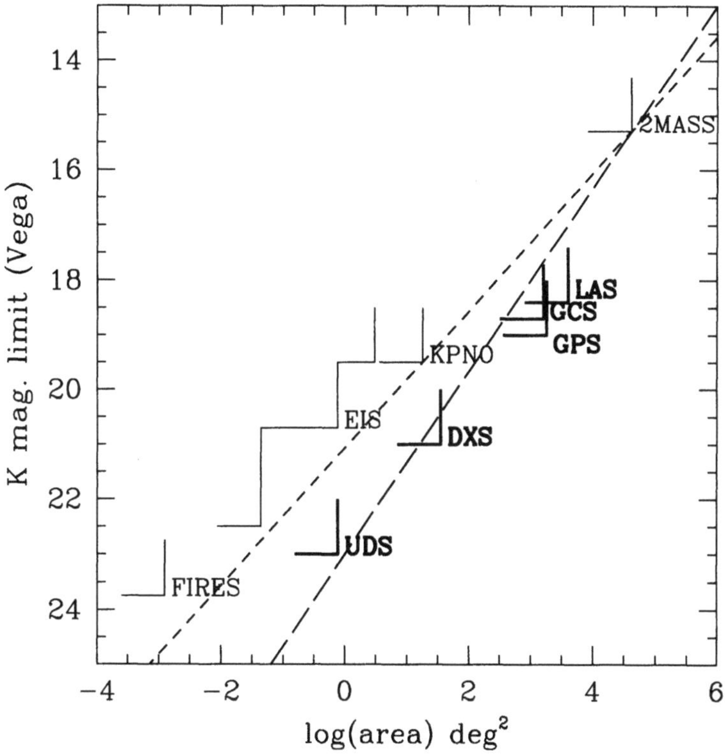

Fig. 1. Comparison of survey area and 5σ depth in K of existing and planned near–ir surveys. The most ambitious existing or planned surveys, besides UKIDSS, are plotted in light text. These are: 2MASS, the KPNO deep survey, the ESO Imaging Survey (EIS) and the very deep FIRES VLT image of HDF–S. The five UKIDSS elements are plotted dark. The short-dashed line, normalised to 2MASS, plots the relation $area \propto 10^{-0.8K}$, which is the trade–off between area and depth for a fixed amount of telescope time (for sky–limited observations). The position of a survey relative to this line is a relative measure of the total number of photons collected. The long–dashed line, again normalised to 2MASS, plots the relation $area \propto 10^{-0.6K}$. The position of a survey relative to this line is a relative measure of the volume of space surveyed.

The Deep Extragalactic Survey (DXS) survey aims to reach depths of K=21, H=22 and J=22.5 over 35 sq degrees, split over 4 fields well spaced in right ascension (XMM-LSS, Lockman Hole, ELAIS-N1 and VIRMOS-4). The scientific goal of the DXS is to obtain an extensive census of galaxies in the redshift range $1.0 < z < 1.5$, surveying a volume comparable to the 2df survey and the Sloan Digital Sky Survey (SDSS).

Finally, the Ultra-Deep Survey (UDS) aims to cover 0.8 sq degrees to a depth of K=23, H=24, J=25. This will give a hundred-fold increase in area compared to any previous survey to this depth. Crucially, we are mapping an enormous volume of the high-z Universe (30 × 30Mpc by 2 Gpc deep at 2 <

$z < 4$), which will provide the first detailed picture of large scale structure at $z \sim 3$. Understanding the evolution and clustering of galaxies at these redshifts, particularly as a function of mass, will provide key tests for the current paradigm of hierarchical structure formation.

3 The Formation of the Most Massive Galaxies

A key goal of the UDS, and to some extent the DXS, will be to address the major theme of this meeting: when were the massive elliptical galaxies formed? The ultimate test will be to determine the co-moving number density of the most massive galaxies at various epochs, particularly at $2 < z < 4$. This requires a combination of both depth and area which has previously been impossible to achieve. For a galaxy as massive as a local L^* elliptical, for example, even in the most "optimistic" PLE model one would expect only ~ 1 galaxy per square arcmin at $z > 2$. For the most massive elliptical galaxies (e.g. $4L^*$) the surface density falls to only 1 every 36 sq arcmin. With the unique capabilities of WFCAM we can unambiguously test whether these massive galaxies are already in place at $z = 3$, or whether they form at lower redshifts through the mergers of less massive systems (e.g. Lyman-break galaxies). Our goal is to directly measure the build-up of this population.

This large area must also be matched by depth. Assuming a passively-evolving model, an L^* elliptical galaxy formed at $z \simeq 5 - 10$ would have a total aperture magnitude of $K \simeq 21 - 22$ at $z = 3$ (assuming the models of Jimenez et

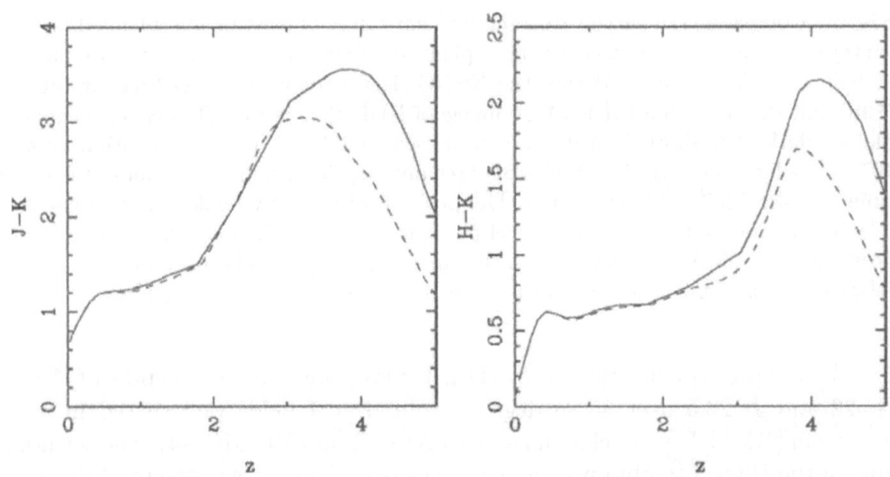

Fig. 2. Predicted $J-K$ and $H-K$ colours as a function of redshift, assuming passively evolving elliptical galaxies produced by the models Jimenez et al. (1999). The models assume a formation redshift $z_f = 10$. The model shown by the dashed line shows residual star formation to $z = 3$, with a more rapid burst of star formation for the solid line (terminating at $z = 5$).

al. 1999 and a standard Lambda-dominated cosmology). The requirement that we detect such an object at 5σ with WFCAM defines the depth requirements for the UDS.

Colour information is also vital for discriminating between objects, and for the determination of photometric redshifts, particularly using the 4000 break as it moves through the I, J, & H bands over the range $1 < z < 4$ (see Figure 2).

4 The Choice of Field for the UDS

The survey field chosen for the UDS is the Subaru/XMM Deep Survey field (02h, -05 dec), which has rapidly become the key extragalactic field over a ~ 1 sq deg region. Subaru has imaged this region in the B, R, i', z' bands to 5-sigma point source depths of B=28.2, R=27.5, i′=27.2, z′=26.3. Further information on the Subaru imaging can be obtained from http://www.naoj.org/staff/chris/SXDS/. Combined with the ultradeep J,H,K imaging data from WFCAM, this will enable photometric redshifts to be obtained for $\sim 10^6$ galaxies to fainter magnitudes and greater accuracy then ever previously attempted over such a wide area.

XMM-Newton has observed this field as part of the XMM Large-Scale Structure Survey, to a depth of 100ks in the central region and 50ks in six flanking fields. This is the widest contiguous hard X-ray survey ever conducted to this depth ($f_{2-10\mathrm{keV}} = 10^{-15}$ $\mathrm{erg\,s^{-1}cm^{-2}}$). The X-ray data will enable dozens of high-z X-ray clusters to be identified, in addition to locating several thousand AGN. Surveys to this depth and area are particularly powerful for determining the space density of quasars at high redshift (Haiman & Loeb 1999). We expect ~ 100 quasars to be identified at $z > 3$, for example, which will be isolated using their unusual optical/IR colours.

This field is also the target for the largest sub-mm survey ever undertaken, the SCUBA Half Degree Extragalactic Survey (SHADES). The major goal of this survey, which is currently underway, is to determine the $n(z)$ and clustering properties of several hundred sub-mm sources. Shorter wavelength sub-mm observations (250 and 350 μm) are planned with the Balloon-borne Large Aperture Submillimetre Telescope (BLAST), due for a long-duration flight in late 2004 (http://www.hep.upenn.edu/blast/). The field will also be observed with SIRTF as part of the SWIRE legacy programme, with observations using both IRAC (3.6, 4.5, 5.8, 8.0 μm) and MIPS (24, 70, 160μm), although we intend to apply for deeper GO observations.

Radio observations of the SXDS field were taken with the Very Large Array (VLA) during the period 10 August - 9 September 2002. A total of 64 hours of observation at 1.4 GHz in B-array were obtained, providing a uniform sensitivity of 60 uJy/beam (5 sigma) over the entire field.

5 UDS Plans for Year 1

The full UDS survey will require $5-7$ years to complete to the depths shown in Table 1, but we anticipate that the K-band imaging to a depth of $K = 22.5$

over the full area will be complete within the first year of operation, with complementary J-band imaging to a depth of $J = 24.0$. Combined with the Subaru data, this will immediately identify $\sim 10^4$ EROs and provide identifications for more than 95% of sub-mm and X-ray sources. Many of the key science goals may therefore be achieved on a relatively short timescale.

It is clear that the availability of wide-field near-infrared cameras will soon provide a major leap forward for observational astronomy. The brave decision by the UK community to use WFCAM on UKIRT to conduct a series of long-term public surveys will surely guarantee that this 4m telescope will continue to produce world-leading science for many years to come. We sincerely hope that the entire ESO community will work together productively to harvest this rich dataset to its full potential.

References

1. L.A. Hillenbrand, J.B. Foster, S.E. Persson, K. Matthews, PASP **114**, 708 (2002)
2. R. Jimenez, A.C.S. Friaca, J.S. Dunlop, R.J. Terlevich, J.A. Peacock, L.A. Nolan, MNRAS **305**, L16 (1999)
3. G. Kauffmann, S. Charlot, MNRAS **294**, 705 (1998)

Survey of Surveys: Past, Present & Future

Mara Salvato[1], Ralf Bender[1,2], and Alvio Renzini[3]

[1] MPE, Garching, Germany
[2] USM, München, Germany
[3] ESO, Garching, Germany

Abstract. To collect informations is often boring and time consuming task. For this reason, we thought that a collection of information about the known surveys may be useful especially for planning future work. We collected information available in literature on completed, work in progress and planed surveys. The list of surveys may not be complete but we are willing to update it (and distribute it) upon request[1].

1 Optical and Near-Infrared Ranges

We intend to give a visual impression of available, work-in-progress and planned imaging surveys, in the optical (Fig. 1) and near-infrared (Fig. 2). The plots are an updated version of those that are in the homepage of the NOAO Deep Survey. Surveys in red color are those for which HST observations are available, making them more effective. The values are collected from available information in the literature and are not homogeneous in terms of covered wavelength range, statistical significance and apertures. However, references and homepage for any displayed survey are also provided, where full information on wavelength coverage can be found, including Xray to radio (when available) and massive spectroscopic follow-up. The surveys are divided in completed, work in progress and planed.

1.1 Completed

ACS-UDF: *http://www.stsci.edu/hst/udf*,
(Bouwens, R. J., Illingworth, G. D., Thompson, R. I. et al.; 2004, astro-ph/0403167);
BTC40:
(Monier, E. M., Kennefick, J. D. et al., 2002, AJ, 124, 2971);
CADIS: *http://www.mpia-hd.mpg.de/GALAXIES/CADIS/welcome.html*,
(Meisenheimer, K., 1998, Astronomische Gesellschaft Meeting Abstracts, 14, 3);
CFDF: *http://virmos.bo.astro.it/hjmcc/cfdf/survey.html*,
(McCracken, H. J., Le Févre, O., Brodwin, M. et al.; 2001, A&A, 376, 756);
CNOC2: *http://www.astro.utoronto.ca/ cnoc/*,
(Lin, H,; Yee, H. K. C.; Carlberg, R. et al.; 1997, AAS, 29, 1400);
Deeprange: *http://www-int.stsci.edu/ postman/deeprange.html*,
(Postman, M.; Lauer, T. R.; Oegerle, W. et al.; 2002, ApJ, 579,93);

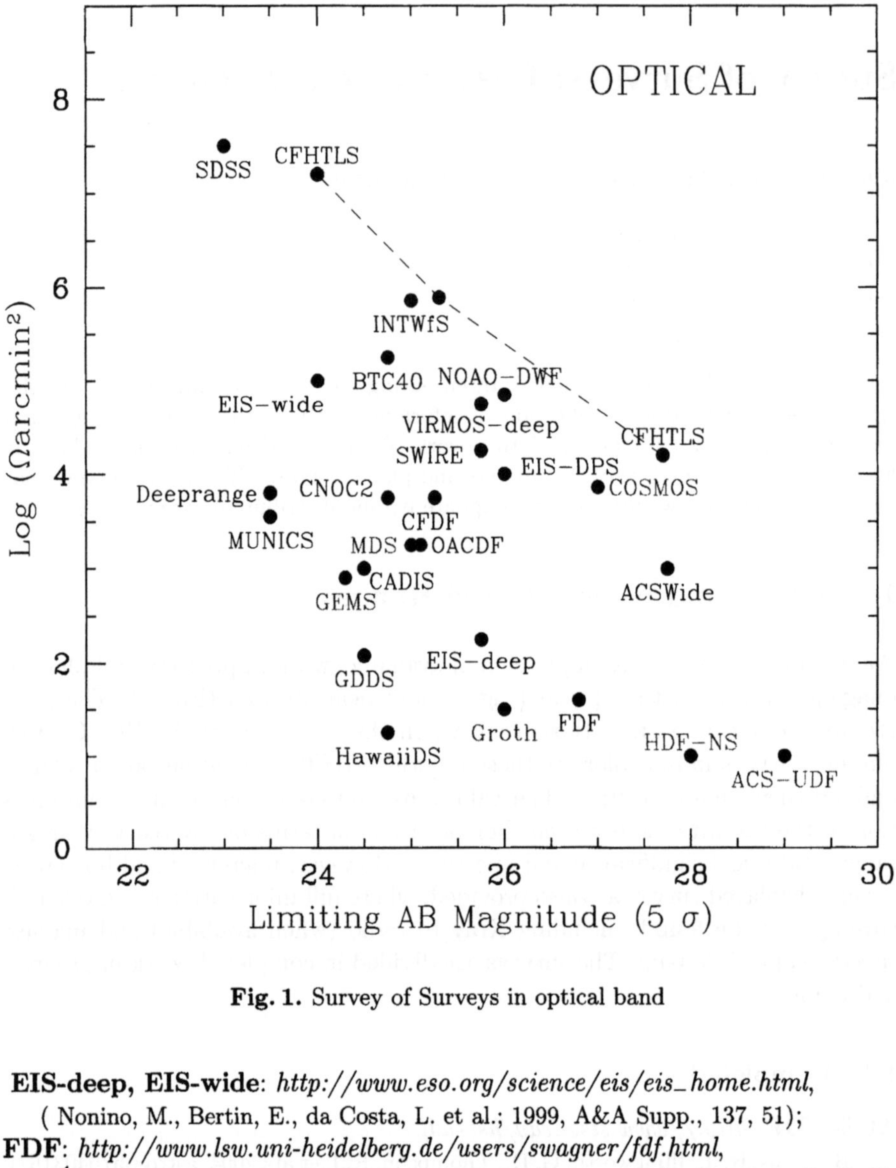

Fig. 1. Survey of Surveys in optical band

EIS-deep, EIS-wide: *http://www.eso.org/science/eis/eis_home.html*,
(Nonino, M., Bertin, E., da Costa, L. et al.; 1999, A&A Supp., 137, 51);

FDF: *http://www.lsw.uni-heidelberg.de/users/swagner/fdf.html*,
(J'ger, K.; Heidt, J.; Appenzeller, I. et al.,1999, Astronomische Gesellschaft, 15, 9);

GEMS: *http://www.mpia.de/GEMS/gems.htm*,
(Rix, H-W., Barden, M., Beckwith S. V. W.; 2004, astro-ph/0401427);

GDDS: *http://www.gemini.edu/project/announcements/press/2004-1.html*
(Abraham, R. G., Glazebrook, K. , McCarthy, P. J. et al.; 2004, astro-ph/0402436);

Grooth strip: *http://saci.ucolick.org/verdi/maps/gss.html* ,
(Groth, E. J., Kristian, J. A., Roger, L. et al.; 1994, AAS, 26, 1403);

Hawaii Deep survey:
(Cowie, L. L.; Gardner, J. P.; Hu, E. M. et al., 1994, ApJ, 434, 114);

[1] Please, email to mara@mpe.mpg.de

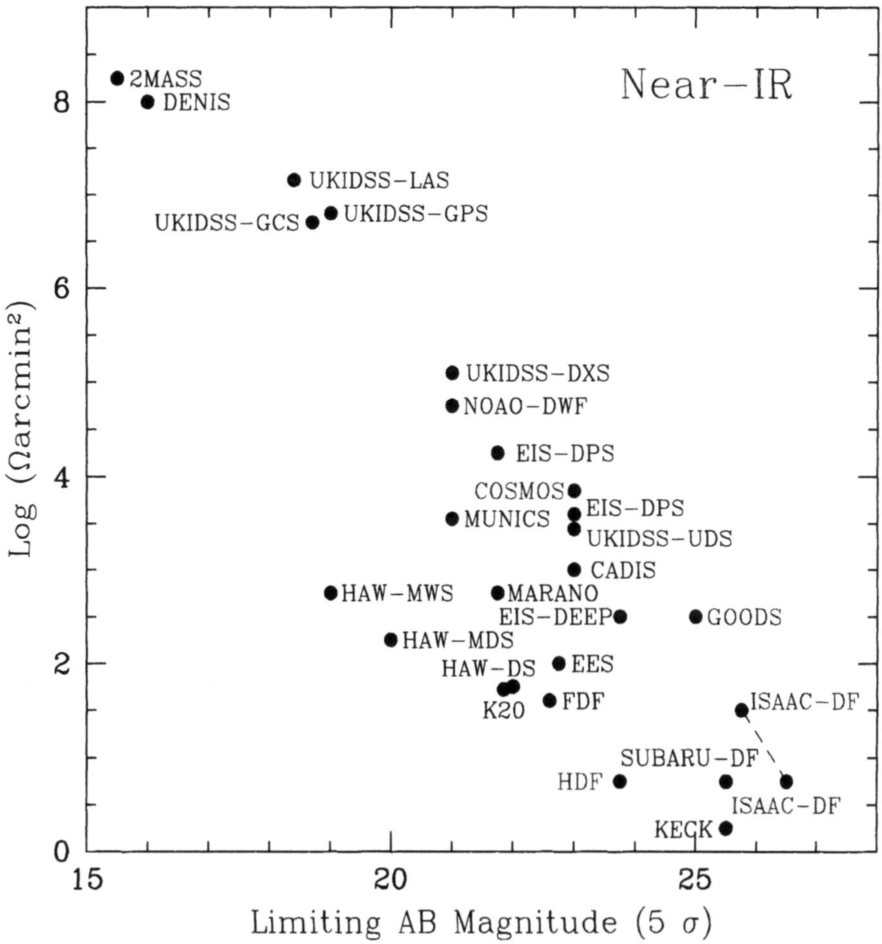

Fig. 2. Survey of Surveys in Nir-infrared band

HDF-N: *http://www.stsci.edu/ftp/science/hdf/hdf.html*,
(Ferguson, H.C.: 1995, STScI Newsletter 12, 2, 5);

HDF-S: *http://www.stsci.edu/ftp/science/hdfsouth/hdfs.html*,
(Casertano, S., de Mello, D., Dickinson, M. et al.; 2000, AJ, 120, 2747);

K20: *http://www.arcetri.astro.it/ k20/*,
(Cimatti, A., Daddi, E., Mignoli, M. et al. 2002, A&A, 392, 395);

MARANO: ,
(Marano, B., Zamorani, G., Zitelli, V.; 1988, MNRAS, 232, 111;

MDS: *http://astro.phys.cmu.edu/mds/*
(Griffiths, R. E., Casertano, S., Ratnatunga, K. U., et al. 1994, ApJ, 435, L19);

MUNICS: *http://capella.usm.uni-muenchen.de/ drory/munics/munics.html*
(Drory, N.; Feulner, G.; Hopp, U. et al.; 2001, ApSSS, 277, 579);

NOAO-DWF; *http://www.archive.noao.edu/ndwfs/*,
(Jannuzi, B. T. & Dey, A., 1999, ASP Conference Series, 191, 111);
SDSS: *http://www.sdss.org/*,
(Lenz, D. D.; Newberg, J.; Rosner, R. et al.; 1998, ApJ, 119, 1210;
Subaru deep survey: *http://zone.mtk.nao.ac.jp/ kashik/sds.html*,
(Maihara, T.,Iwamuro, F., Tanabe, H. et al., 2001, PASJ, 53, 25);
VIRMOS-deep: *http://www.astrsp-mrs.fr/virmos/vvds.htm*,
(McCracken, H. J., Radovich, M., Bertin, E. et al., 2003, A&A, 410, 17);

1.2 Work-in-Progress

CFHLS: *http://www.cfht.hawaii.edu/Science/CFHLS/*,
(Pain, R.; The Snls Collaboration; 2002, AAS, 34, 1169);
COSMOS: *http://www.astro.caltech.edu/cosmos/*,
(Scoville, N., 2004, HERE);
EIS-DPS: *http://www.eso.org/science/eis/eis_home.html*,
(Arnouts, S., Vandame, B., Benoist, C. et al.; 2001, A&A 379, 740);
GOODS (optical): *http://www.stsci.edu/science/goods/*,
GOODS (infrared): *http://www.eso.org/science/goods/*,
(Giavalisco, M.; Ferguson, H. C.; Koekemoer, A. M. et al.; 2004, ApJL, 600, 93);
INTWFS: http://www.ast.cam.ac.uk/ wfcsur/overview.php,
(McMahon, R. G.; Walton, N. A.; Irwin, M. J. et al., 2001, New Astron. Rev., 45, 97);
ISAAC-DF:
(Labbé, I., Franx, M., Rudnick, G. et al.;2003, AJ, 125, 1107);
OACDF: *http://www.na.astro.it/oacdf-bin/cdfcgi*,
Alcalá, J., Radovich, M., Silvotti, R. et al.; 2002, Proceedings of the SPIE, 4836, 406);

1.3 Planned

VST: *http://www.na.astro.it/oacweb/oacweb_ricerca/vst_site/index.html*;
VISTA: *http://www.roe.ac.uk/ifa/wfau/vsc/index.html*;

Radio Properties of EROs in the Phoenix Deep Survey

J. Afonso[1], A.M. Hopkins[2], M. Sullivan[3], B. Mobasher[4], A. Georgakakis[5], and L.E. Cram[6]

[1] CAAUL, Observatory of Lisbon, Tapada da Ajuda, 1349-018 Lisbon, Portugal
[2] Dept. of Physics & Astronomy, University of Pittsburgh, 3941 O'Hara St, Pittsburgh, PA 15260, USA
[3] Dept. of Astronomy and Astrophysics, University of Toronto, 60 St. George St, Toronto, Ontario M5S 3H8, Canada
[4] ESA, STScI, 3700 San Martin Drive, Baltimore, MD 21218, USA
[5] National Athens Observatory, I.Metaxa & Vas.Pavlou str., Athens 15236, Greece
[6] Australian Research Council, GPO Box 2702, Canberra, ACT 2601, Australia

Abstract. Insensitive to dust obscuration, radio wavelengths are ideal to study star-forming galaxies free of dust induced biases. Using data from the Phoenix Deep Survey, we have identified a sample of star-forming extremely red objects (EROs). Stacking of the radio images of the radio-undetected star-forming EROs revealed a significant radio detection. Using the expected median redshift, we estimate an average star-formation rate of $61\,\mathrm{M}_\odot\,\mathrm{yr}^{-1}$ for these galaxies.

1 Introduction

Identifying and studying dusty star-forming galaxies is of vital importance to understand galaxy evolution. Besides looking directly for the emission from dust, these objects can also be identified by their extreme optical to near-infrared (NIR) colours, forming a significant fraction of the so-called Extremely Red Objects (EROs) [3]. Radio wavelengths, unaffected by dust obscuration and being a measure of star formation rates in star-forming galaxies, can give an important insight to the characteristics of these dusty objects [1]. Performing an ERO search in an ultra-deep radio field can thus provide an important step in understanding some of the more obscured galaxies.

2 Star-Forming EROs in the PDS

The Phoenix Deep Survey is a multiwavelength survey based on deep radio (1.4 GHz) observations performed with the ATCA. With an area of over 4.5 square degree, and reaching flux densities of 60 μJy, the PDS is one of the deepest wide angle radio surveys already made [4]. Follow-up optical (*UBVRI*) data has recently been obtained, using the Wide-Field Imagers at the AAT, CTIO and ESO, reaching ~ 25 mag in each band. Near-infrared (K_s band) observations were made on the central 200 square arcmin of the PDS, using SOFI at the NTT (ESO) to a limiting magnitude of $K_s \sim 20$ [5]. Based on $3''$ aperture photometry,

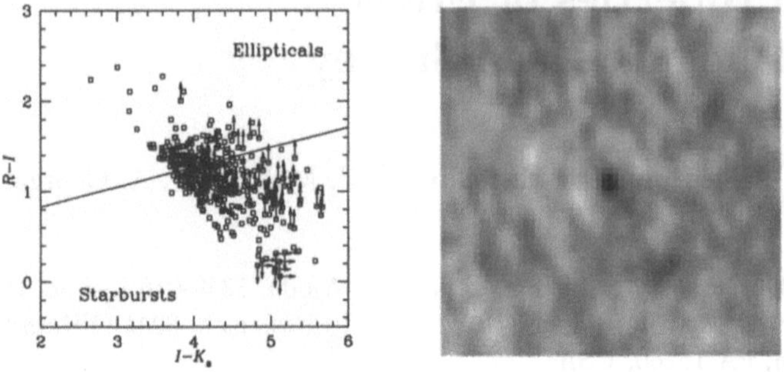

Fig. 1. Classification of EROs using the RIK_s colour-colour plot (left) and the stacked radio image of the radio-undetected star-forming EROs, revealing a 8.7 μJy detection.

a sample of 432 EROs ($R - K_s > 5$), were selected. Eighteen were directly identified in the radio catalogue, with radio fluxes between 65 and 991 μJy. The ERO population was separated into ellipticals and star-forming galaxies using the RIK_s magnitudes, following the recent work by Bergstrom et al. [2]. This method seems to perform well for EROs with redshifts $z < 1.4$ [2], which seems appropriate for this survey (a spectroscopic analysis of an ERO sample with similar K_s magnitude limits revealed a median redshift of $z \sim 1.2$, with a very small number of EROs having $z > 1.4$ [3]). A total of 214 objects fall in the star-forming region of the RIK_s colour-colour plot (Figure 1). After removing all ($> 3\sigma$) radio detections, the radio images at the locations of the star-forming EROs were stacked. A significant (6.7σ) detection of 8.7 μJy was revealed (Figure 1). Assuming a median redshift for the sample of $z \sim 1.2$, and assuming the radio emission is solely due to star-formation, this implies an average star-formation rate of 61 $M_\odot\,yr^{-1}$.

Improvement on the photometric analysis, by using the full photometric dataset to estimate photometric redshifts and classifications, is currently underway, and will result in a precise determination of the contribution of star-forming EROs to the star-formation rate in the Universe.

AMH acknowledges support provided by NASA through Hubble Fellowship grant HST-HF-01140.01-A awarded by STScI. JA acknowledges support from the Science and Technology Foundation (FCT, Portugal) through the fellowship BPD-5535-2001 and the research grant POCTI-FNU-43805-2001.

References

1. J. Afonso, B. Mobasher, B. Chan, L. Cram: ApJ, **559**, L101 (2001)
2. S. Bergström, T. Wiklind: A&A (2003) (in press; astro-ph/0310628)
3. A. Cimatti, et al.: A&A (2003) (in press; astro-ph/0310742)
4. A. M. Hopkins, J. Afonso, B. Chan, L. E. Cram, A. Georgakakis, B. Mobasher: AJ, **125**, 465 (2003)
5. M. Sullivan, A. M. Hopkins, J. Afonso, A. Georgakakis, B. Chan, L. E. Cram, B. Mobasher: ApJS, submitted.

Nuclei of Galaxy Bulges Through *HST* NIR Imaging

Marc Balcells[1], Alister W. Graham[2], and Reynier F. Peletier[3]

[1] Instituto de Astrofísica de Canarias, 38200 La Laguna, Tenerife, Spain
[2] Dept. of Astronomy, University of Florida, Gainesville, FL 32611-2055, USA
[3] Kapteyn Institute, Univ. of Groningen, Postbus 800, 9700 AV Groningen, The Netherlands

Abstract. We have analyzed the structure of bulges of early- to intermediate-type spirals, on nuclear and global scales, using *HST* NIR imaging together with ground-based NIR imaging. The goals are (a) to see whether results on bulge structure derived from ground-based, 1" FWHM seeing data remain valid when high-resolution *HST* data are used; (2) to compare nuclear properties of bulges to those of ellipticals; and (3) to provide bulge-disk scaling relations based on accurate profile decompositions employing the *HST* data. Details are presented in Balcells et al. (2003, 2004a, 2004b).

1 Nuclear Unresolved Sources

Most bulges show central excesses over the best-fit Sérsic $r^{1/n}$ model. Some of these central components are resolved but most are unresolved by *HST*. Resolved components have sizes of a few 100 pc, and may be inner disks or bars. Unresolved components (PS) have typical luminosities of $10^7 L_\odot$; their luminosities scale with bulge luminosity as $L_{PS} \sim (L_{Bulge})^{0.5}$. Similar unresolved sources are also found in ellipticals (Rest et al. 2000) and in dwarf ellipticals (Graham & Guzmán 2003), and appear to be a general property of galaxian spheroids (Balcells et al. 2003).

2 Nuclear Profile Slopes

Bulge profiles show nuclear slopes γ that range from $\gamma \sim 0$ for bulges with exponential profiles, to $\gamma \sim 1$ for bulges with higher Sérsic index (Balcells et al. 2004a). The distribution of γ is continuous; we disagree with Carollo & Stiavelli's (1998) claim that the γ distribution for bulges is bi-modal. Their result might arise from the later-type selection of their sample, and from their not performing bulge-disk decomposition. We find no evidence that bulges come in two distinct, "exponential" and "$r^{1/4}$" classes.

3 Global Scaling Relations

We re-examine bulge and disk global scaling relations in the light of our profile decompositions (Balcells et al. 2004b). Bulge and disk scaling relations are shown

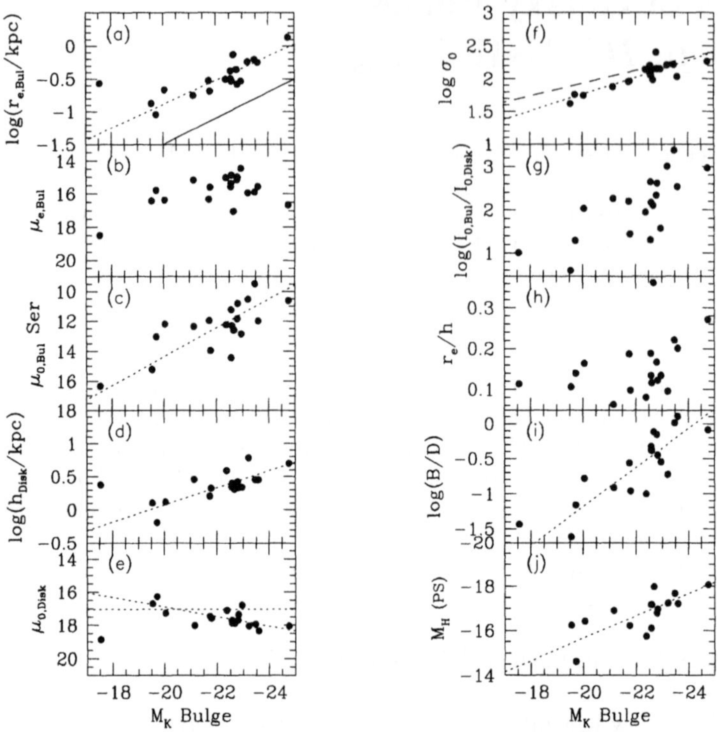

Fig. 1. Scaling relations for bulges and disks for our bulges sample, derived from the *HST*-plus-ground-based NIR profiles. Bulge luminosities are inferred from the total galaxy K-band luminosity and B/D from the profile decomposition. See Balcells et al. (2004b).

in Fig. 1. Courteau et al. (1996) proposed that the Hubble sequence is scale-free on the basis that the ratio of bulge-to-disk scale-lengths, r_e/h, does not vary with Hubble type T. We find that, while the ratio r_e/h does indeed not show a trend with T, it increases with bulge luminosity. The same occurs with other bulge-disk ratios, including even the B/D luminosity ratio, which only for late types (T>4) does show a clear trend with T, while it has a strong dependence on $M_{K,Bulge}$. While the Hubble sequence appears to be scale free, galaxies are not –the critical scale appears to be set by the bulge luminosity.

References

1. Balcells, M., Graham, A. W., Domínguez-Palmero, L,, Peletier, R. F. 2003, ApJ, 582, L79
2. Balcells, M., Graham, A. W., Peletier, R. F. 2004a, in preparation
3. Balcells, M., Graham, A. W., Peletier, R. F. 2004b, in preparation
4. Carollo. C. M., Stiavelli, M. 1998, AJ, 115, 2306
5. Courteau, S., de Jong, R. S., & Broeils, A. H. 1996, ApJ, 457, L73
6. Graham, A. W. 2001, MNRAS, 326, 543

The Bimodal Color-Magnitude Distribution of Galaxies from the SDSS

I.K. Baldry and K. Glazebrook

Dept. of Physics & Astronomy, Johns Hopkins University, Baltimore, MD 21218

We analyze the distribution in color versus absolute magnitude of a low-redshift sample of galaxies from the Sloan Digital Sky Survey [4]. For each magnitude bin, the color function can be approximated as the sum of two normal Gaussians, a bimodal function. Thus, the distribution can be deconvolved into two dominant components that have associated color-magnitude relations, dispersion-magnitude relations and luminosity functions. This differs from the classical approach of using cuts in morphology or color to define classes and instead allows for a natural overlap. The observed distribution and the deconvolved red and blue distributions are shown in Fig. 1. We describe some of the results and points to note about the figure, below. For full details, see [1].

(i) The color-magnitude relations are not well fit by straight lines. We use a straight line plus a tanh function, interpreted as a transition, to provide good fits. The color dispersion at the faint end of the red distribution is significantly higher than at the bright end. (ii) If we convert the galaxies' luminosities to stellar masses, the transitions occur around the same mass as that found using spectroscopic measurements, $3 \times 10^{10} \mathcal{M}_{\odot}$ [3]. This is an important confirmation of this transition mass using photometry, which uses apertures that scale with the size of the galaxy, as opposed to spectroscopy, which uses a $3''$ aperture. The transition involves a change in the properties of both distributions and a change over in dominance from one to the other. (iii) The significant overlap around M_r of -21.5 between the two distributions is at least partly due to a degeneracy between dust reddened late types and old stellar population early types. (iv) Other results suggest that galaxy environments affect the fraction of galaxies in each distribution but have little or no effect on the properties of galaxies within a distribution [2]. We suggest that major mergers can rapidly transform a galaxy from the blue to the red distribution but minor interactions are relatively unpredictable and cause stochastic or no significant changes to galaxies within each distribution.

Funding for the creation and distribution of the SDSS Archive has been provided by the Alfred P. Sloan Foundation, the Participating Institutions, the National Aeronautics and Space Administration, the National Science Foundation, the U.S. Department of Energy, the Japanese Monbukagakusho, and the Max Planck Society.

References

1. I. K. Baldry et al.: ApJ **600**, in press (2004) (astro-ph/0309710)
2. M. Balogh et al.: MNRAS, in press (2004) (astro-ph/0311379)
3. G. Kauffmann et al.: MNRAS **341**, 54 (2003)
4. D. G. York et al.: AJ **120**, 1579 (2000) (http://www.sdss.org/)

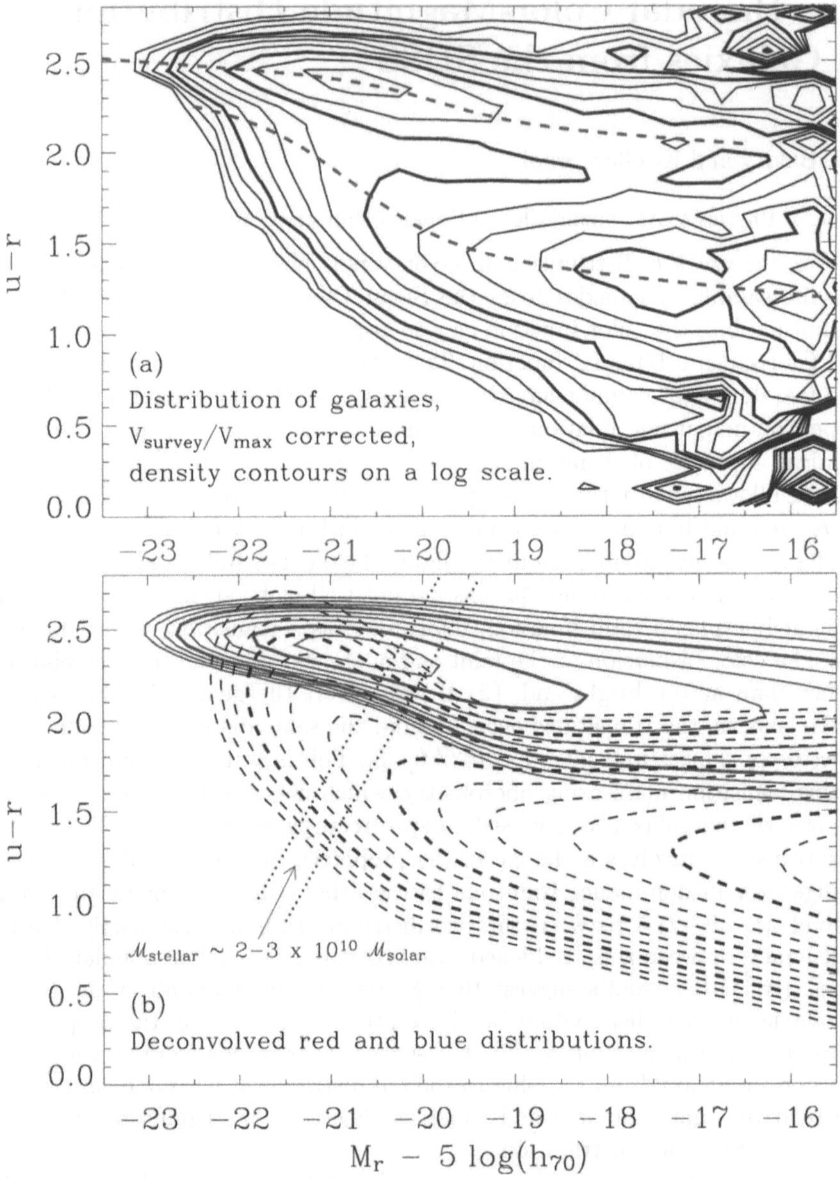

Fig. 1. Color-magnitude distributions. (**a**) Observed bimodal distribution, corrected for incompleteness. The contours are on a logarithmic scale in number density, doubling every two levels. The rapid changes at low luminosities are due to sample variance. The dashed lines represent the color-magnitude relations of the red and blue sequences. (**b**) Deconvolved and parameterized distributions. The solid contours represent the red distribution and the dashed contours represent the blue distribution. The dotted lines represent galaxies that have similar stellar masses, near the midpoints of the transitions.

Merging Clusters in the Core of Superclusters: A Multiwavelength View

S. Bardelli[1], E. Zucca[1], F. Gastaldello[2], F. Marini[1], S. Ettori[3], S. DeGrandi[4], T. Venturi[5], S. Giacintucci[1], S. Molendi[2]

[1] INAF-Osservatorio Astronomico di Bologna; [2] IASF-CNR, Milano; [3] ESO, Garching; [4] INAF-Osservatorio Astronomico di Merate; [5] IRA-CNR, Bologna

Abstract. The central regions of superclusters, being the observative counterpart of the intersections of the large scale filaments, are the formation loci of rich galaxy clusters. In these environments, the fraction of merging is enhanced and the probability of a major merging (the collision of similar mass clusters) event is high. In these conditions it is possible to study not only such mergings but also the related astrophysical consequences, like the formation of radio halos and relics and the effect on the emission of the galaxy population. Here we present our recent results on the A3558 complex and the A3528 cluster in the Shapley Concentration and on the A2061-A2067 pair in the Corona Borealis supercluster.

The A3558 complex This complex is formed by 3 clusters and two poor groups and is probably the most spectacular example of major merging. From a deep radio survey in this area (Fig.1 left) we found that the merging "switched off" the radio galaxies. Moreover, we detected a weak young radio-halo with an extended radio galaxy nearby (Fig.1 right). We showed that the number of electrons forming the halo is consistent with that lost by the radio galaxy. However, in order to justify the halo emission a reacceleration by the merging is needed [Giacintucci et al. 2003, A&A in press].

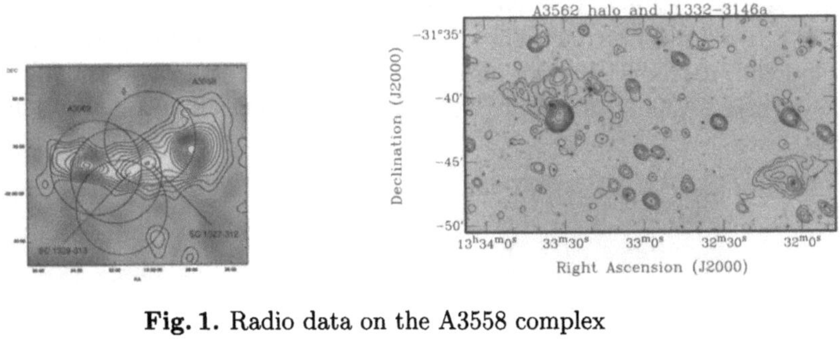

Fig. 1. Radio data on the A3558 complex

The A3528 cluster This cluster is actually formed by two twin subclusters, observed by us with XMM-Newton (Fig.2 upper left). We studied the temperature distribution, without detecting a shock front (Fig.2 upper right). However, clear signs of interaction such as dispersed hot gas (Fig.2 lower left) and an asymmetry in the total blue-band luminosities of the two components (Fig.2 lower right) suggest an off-axis merging [Gastaldello et al. 2003, A&A 411, 21].

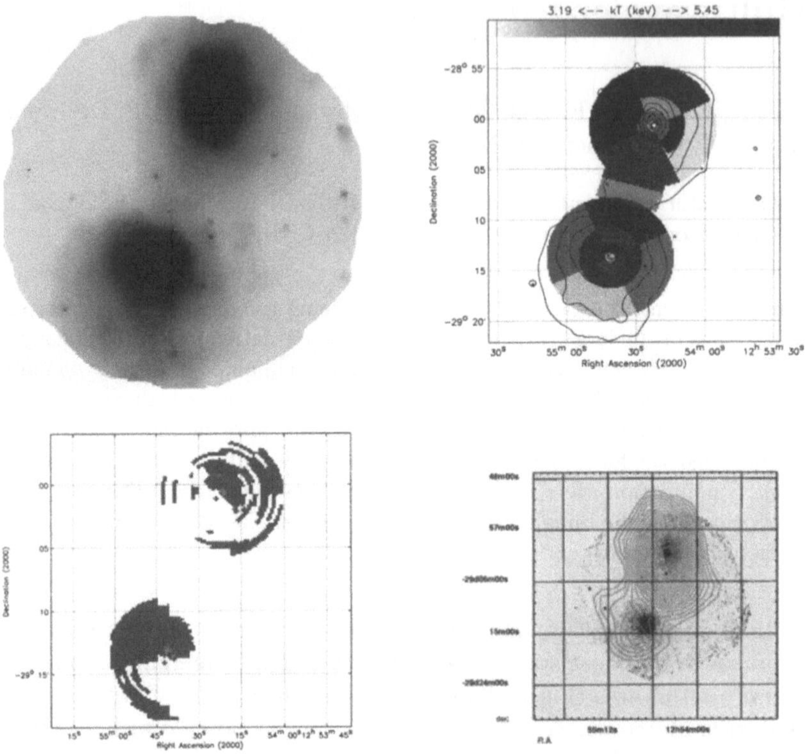

Fig. 2. XMM-Newton data on the A3528 cluster

The A2061-A2067 pair We observed with Beppo-SAX this pair (Fig.3 left): A2061 is on the bottom right of the image. From the temperature maps, we found that a group is falling inside A2061: this falling group probably caused a shock inside A2061 (in red in Fig.3 right) [Marini et al. 2003, A&A submitted].

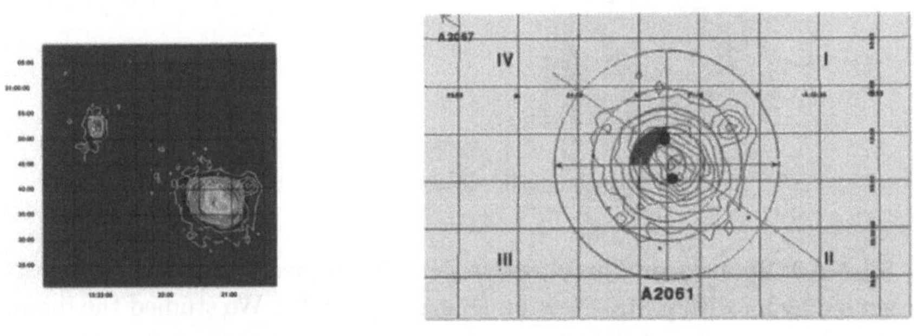

Fig. 3. Beppo-SAX data on the A2061-A2067 pair

Strange Hosts of Blue Compact Galaxies

Nils Bergvall[1], Thomas Marquart[1], Carina Persson[1], Erik Zackrisson[1], and Göran Östlin[2]

[1] Uppsala Astronomical Observatory, Box 515, SE-751 20, Uppsala, Sweden
[2] Stockholm University, AlbaNova University Centre, Stockholm Observatory, Department of Astronomy, SE-106 91 Stockholm, Sweden

1 Introduction

Luminous Blue Compact Galaxies (BCGs) have low metallicities, high gas consumption rates and are frequently involved in mergers. These are properties characteristic of young galaxies in the hierarchical formation scenario. Local BCGs can therefore be used as a complement to high redshift studies provided that the evolution of BCGs can be understood in more detail and be applied to early galaxy formation. Although much is known about the central starburst, few studies discuss the relation between the starburst and the progenitor, or the "host galaxy" of the starburst. Here we report of an effort to derive information about the host from the colours of the faint halo.

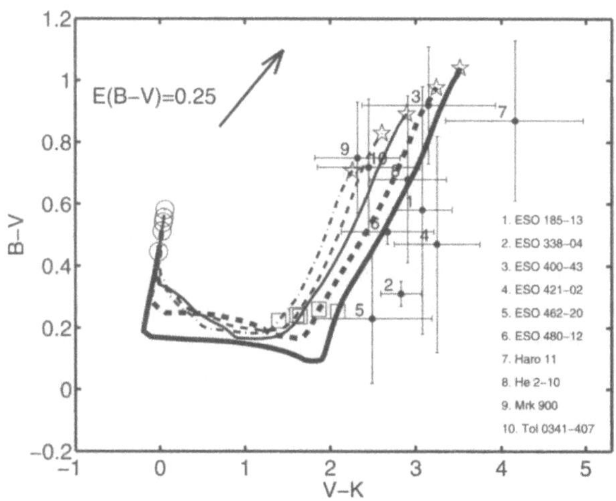

Fig. 1. Halo colours (corrected for galactic extinction but not for internal extinction; 1σ error bars) of 10 BCGs versus PÉGASE.2 evolutionary tracks. The evolutionary sequences include nebular emission and assume an exponentially decreasing star formation rate with an e-folding decay rate of 1 Gyr. The different lines correspond to constant metallicities of Z=0.001 (thin dash-dotted), 0.004 (thin dashed), 0.008 (thin solid), 0.020 (thick dashed), 0.040 (thick solid). Markers indicate ages of 0 yr (circle), 1 Gyr (square) and 14 Gyr (pentagram). The reddening vector is marked with an arrow.

2 The Data

We obtained surface photometry at ESO (IRAC2, FORS1 and SOFI) and NOT (ALFOSC and NOTCAM) in BVRIJHKs of 10 BCGs, mostly of high luminosity and low oxygen abundances, typically around 10–15% solar. The main objective was to constrain halo ages and metallicities from the colours, using spectral evolutionary models. Fig. 1 shows the observed B-V vs. V-K colours in comparison with evolutionary tracks from PÉGASE.2 [1].

We conclude that 1) The halos are older than 1Gyr and 2) the colours of more than half of the objects are not consistent with a metal poor stellar population. But what is most remarkable is that, although the scatter is large, about half of the objects show colours significantly redder than the most metal rich tracks. This conclusion is also confirmed by the spectral evolutionary model of Zackrisson et al. [2].

3 Possible Causes of the Red Excess

Problems with the instrumentation or reductions: The data have been obtained with three different near-IR cameras and with different observers. Comparisons with published data and careful checks of the background subtraction have been performed.

Nebular emission: The V-K colour of ionized gas of low metallicity is approximately 0 so an increased contribution from a nebular component would tend to make V-K bluer, i.e. in the opposite direction of the trend observed.

Hot dust: Emission from hot circumstellar dust should only appear in K. This would cause a red excess also in H-K that we do not observe. From our ISO observations of Haro 11 we have a direct check of the emission in the halo in the 4-12 μm bands. There is no signal at the level we need to explain the excess in this galaxy.

Dust extinction: As seen in Fig. 1, a normal reddening correction is running more or less in parallel with the evolutionary tracks and is thus incapable of causing the excess. From ISOPHOT data we can exclude cold dust in the halos of many of our BCGs.

Abnormal IMF: In order to increase the contribution of cool stars in the model stellar population we have tested a power-law IMF with a slope of 3.35, i.e. much steeper than the Salpeter. This diminished the excess with only 0.1 magnitudes. One possibility is that we have an excess of low mass stars, resulting from dynamical mass segregation coupled to the merger event.

References

1. M. Fioc, B. Rocca-Volmerange: astro-ph/9912179 (1999)
2. E. Zackrisson, N. Bergvall, K. Olofsson, A. Siebert: A&A 375, 814 (2001)

Estimating the Stellar Masses in 30000 Galaxies with Redshifts Below 1.0

Andrea Borch[1], Klaus Meisenheimer[1], and Christian Wolf[2]

[1] Max-Planck-Institut für Astronomie, Heidelberg, Germany
[2] Department of Physics, University of Oxford, U.K.

Abstract. We present a new method of constraining the star formation histories and hence the stellar M/L ratios by using the multicolor technique in COMBO-17 [7]. It estimates the stellar M/L ratio at the observed wavelength $\lambda = 815$ nm for each single galaxy on the basis of color information from 5 broadband and 12 medium band filters. The 815 nm mass estimations are in very good agreement to the masses derived from public J, H and K_s data from the ESO Imaging survey [4] in a tiny area (5 % of one of our fields). The stellar mass distribution in the range $0.1 < z < 0.3$ is compared to the result of Kauffmann et al. [5] for SDSS galaxies and shows a good agreement. Considering only galaxies with $M > 3 \times 10^9 M_\odot$, we find that the integrated stellar mass density in galaxies increased by a factor of about 1.5 since redshift 1.0.

The multicolor classification in COMBO-17 estimates the redshift and spectral energy distribution (SED) type simultaneously [7]. Different libraries for stars, galaxies and quasars are used for classifying the objects in a 16-dimensional color space. For the galaxy class, the observed Kinney et al. [6] library is fitted by PEGASE [3] spectra which can be directly translated into star formation histories and stellar M/L ratios. (See [1] for a former version of this library.)

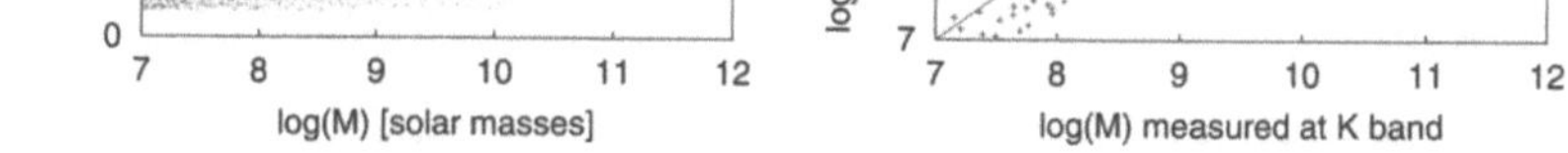

Fig. 1. Left: The COMBO-17 stellar mass estimations in 30000 galaxies as a function of redshift. **Right:** A comparison between the stellar mass estimations in the 815 nm and in the K_s filter for a subsample of 389 galaxies.

For the mass estimation we compare the flux in a filter with a observed central wavelength of 815 nm with the appropriate restframe flux of the PEGASE spectra. The results for 30000 galaxies as a function of redshift is shown in figure 1 (left). For a subsample of about 400 galaxies we compare these mass estimations to those obtained in J, H and K_s band ISAAC images. Therefore we use the public data from the ESO Imaging Survey in the GOODS [2] area which is located in one of our fields. Figure 1 (right) shows the mass estimations in the K_s filter in comparison to our $\lambda_{obs} = 815$ nm mass estimations. The agreement is within 0.5 dex. Both cases use the same multicolor classification in the optical regime. Figure 2 (left) shows the mass fraction in galaxies at $0.1 < z < 0.3$ in comparison to the Kauffmann et al. [5] result for the SDSS galaxies. The COMBO-17 result shows a good agreement. Figure 2 (right) shows the integrated stellar mass density in galaxies above $3 \times 10^9 M_\odot$ as a function of redshift. It seems to grow by roughly a factor of 1.5 since redshift 1.0. This is compared to the semianalytic models of C. Baugh (priv. comm.) and also of R. Somerville (priv. comm.) who provided us with a mock catalog. These models are consistent with the COMBO-17 result.

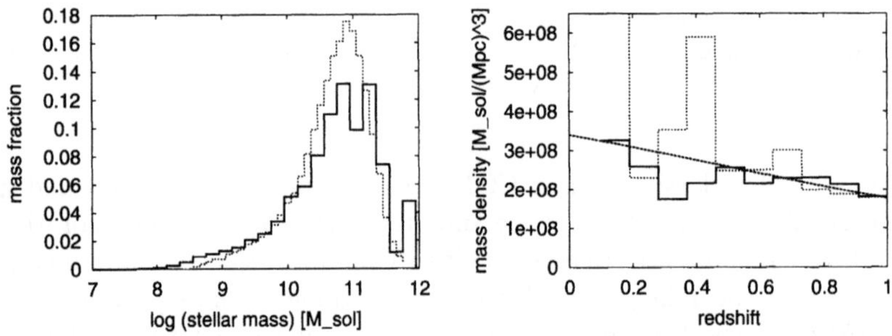

Fig. 2. Left: The fraction of stellar mass in galaxies. Solid histogram: COMBO-17, dotted histogram: Kauffmann et al. [5] **Right:** The stellar mass density as a function of redshift. As a comparison the results of the semianalytic models of C. Baugh (priv. comm.) and R. Somerville (priv. comm.) are plotted. Solid histogram: COMBO-17, dotted histogram: R. Somerville mock catalog, dashed line: C. Baugh et al., normalized to the COMBO-17 data.

References

1. A. Borch, K. Meisenheimer, C. Wolf, M. Gray: 2003, ApSS 284, 965
2. M. Dickinson: this conference
3. M. Fioc, B. Rocca-Volmerange: 1997, A&A 326, 950
4. Haziminaglou et al.: astro-ph/0201028
5. G. Kauffmann et al.: 2003, MNRAS 341, 33
6. A.L. Kinney, D. Calzetti et al.: 1996, ApJ 467, 38
7. C. Wolf, K. Meisenheimer, H.-J. Röser: 2001, A&A 365, 660

350μm Observations of Local IRAS Galaxies Using SHARC-II

Colin Borys

California Institute of Technology, Pasadena CA 91125, USA

1 SEDs and Redshifts of Sub-mm Bright Galaxies

The cosmologically significant population of dusty galaxies detected at sub-mm and mm wavelengths by SCUBA and MAMBO have played a central part in many of the discussions at this workshop. Obtaining their redshifts is critical in understanding their role in galaxy formation and evolution, but this is a challenging task. Nevertheless, Chapman et al. [1] have made significant process by obtaining spectroscopic redshifts of the subset of sub-mm galaxies detected in deep radio images that have optical counterparts. This exploits the well-known Far-Infrared/Radio correlation, and allows the high resolution radio maps to be used as a surrogate for finding the correct optical counterpart for which spectroscopy can then be attempted.

Because of the large investment of telescope time required to obtain and confirm the spectroscopic redshifts, and because the radio-detected sample consists only of the brighter sub-mm objects (only a few percent of the sub-mm source counts), some groups [2–4] have developed photometric techniques to estimate the redshifts of SCUBA/MAMBO detected galaxies. These are also potentially biased because the Spectral Energy Distribution (SED) basically follows a modified black-body curve, which results in a fundamental degeneracy between redshift and dust temperature [5]. However, by assuming that high-redshift galaxy SEDs resemble local ones, a library of templates derived from nearby galaxies can be used to essentially "fix" the temperature and break the degeneracy. One problem with this approach is that the SEDs of local infrared-bright systems are poorly constrained for all but a few of brightest local sources.

One way to address this issue is to obtain photometry at different wavelengths. However, the atmosphere is opaque to FIR emission except for the narrow windows at relatively long wavelengths that SCUBA and MAMBO have already exploited. Efficient observations at the shorter wavelength windows at 350 and 450μm have proved very difficult until now.

2 SHARC-II Observations of SLUGS Galaxies

The CSO recently commissioned the "CCD-style" SHARC-II camera [6] optimized for use at 350μm. The combination of 384 sensitive detectors and a dish with low surface error allow observations previously impractical with other cameras operating at similar wavelengths.

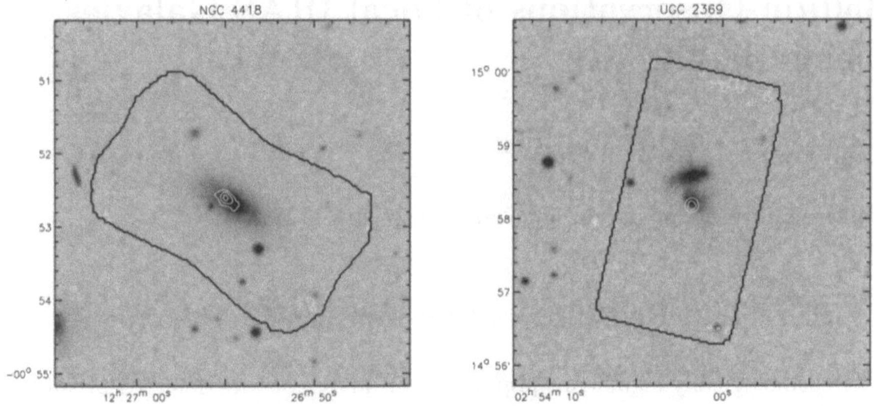

Fig. 1. Two IRAS galaxies imaged with SHARC-II. Overlaid on DSS images are SHARC-II Signal-to-Noise contours at 4 and 6 sigma/beam. The black outline denotes the area sampled by the SHARC-II array.

Using SHARC-II, we have been imaging the sample of 106 IRAS selected galaxies in the highly successful SCUBA Local Universe Galaxy Survey (SLUGS) [7]. These sources are lower luminosity analogues to the high-redshift SCUBA population. With observations at 60, 100 (IRAS) and 850μm (SCUBA), the SEDs were fit and dust parameters estimated. SCUBA was only able to detect 17 of these galaxies at 450μm, but from them it was recognized that the models required a component of cold dust to fit the the SED properly [8]. These results highlight the need for more short-wavelength observations. To date, we have imaged 60 of the 106 SLUGS sources with SHARC-II and have and detected every one. We are also conducting complementary observations of the Chapman et al. sample of SCUBA galaxies with spectroscopic redshifts. Using this spec-z to break the $T/(1+z)$ degeneracy, the combination of SCUBA and SHARC-II observations will allow a study of the physical dust properties as a function of redshift.

References

1. S.C. Chapman, A.W. Blain, R.J. Ivison, I.R. Smail: Nature **422**, 695 (2003)
2. D. Hughes et al.: MNRAS **335**, 871 (2002)
3. I. Aretxaga et al.:MNRAS **342**, 759 (2003)
4. T. Wiklind: ApJ **588**, 736 (2003)
5. A.W. Blain, V. E. Barnard, S.C. Chapman: MNRAS **338**, 733 (2003)
6. C.D. Dowell et al.: SPIE **4855**, 73 (2003)
7. L. Dunne, S.E. Eales, M. Edmunds, R. Ivison, P. Alexander, D. Clements: MNRAS **327**, 697 (2001)
8. L. Dunne, S.E. Eales: MNRAS **327**, 697 (2001)

Beyond the Lick Indices: The High-Resolution Spectral Synthesis of Stellar Populations

Alberto Buzzoni[1,2], Emanuele Bertone[3], Lino H. Rodríguez-Merino[3], and Miguel Chávez[3]

[1] INAF - Osservatorio Astronomico di Bologna, Italy
[2] INAF - Osservatorio Astronomico di Brera-Milano, Italy
[3] INAOE - Instituto Nacional de Astrofisica, Optica y Electronica, Puebla Mexico

Abstract. Up-to-date theoretical SED of stellar populations at 2 Å FWHM resolution is obtained by coupling a new library of synthetic spectra from the ATLAS and NextGen atmosphere database with the Buzzoni population synthesis code. We briefly outline the current status of the project, aimed at settling a new theoretical framework of narrow-band spectrophotometric indices to study stellar populations of galaxies.

1 A High-Resolution Approach to Population Synthesis

An exceptional effort has been carried out in the recent years in the computation of high-resolution stellar spectra from model atmospheres (e.g. [1], [2]) allowing us to tackle the fundamental properties of stars at unprecedented spectral detail ($\sim$ 0.1 Å resolution or even better). As far as the analysis of the integrated spectral energy distribution (SED) of stellar aggregates is concerned, this new generation of models has found a natural application in the domain of stellar population synthesis by matching synthetic spectra of stars and stellar isochrones. (e.g. [3] and [4]).

Such a remarkable advance calls however for a sensibly different strategy in the study of the integrated SED of galaxies and other stellar systems, as age and chemical properties now intimately "merge" at high spectral resolution with the dynamical piece of information of the system as a whole. In this sense, therefore, also important interpretative tools, like the Lick spectrophotometric indices [5], so extensively used in the population synthesis literature to assess the overall evolutionary properties of stellar populations in external galaxies ([6], [7]) need to be reconsidered on a new basis.

Here, we would like to briefly sketch our project aimed at settling new high-resolution spectrophotometric indices selectively sensitive to age, [Fe/H] and IMF in Simple Stellar Populations (SSPs). Our work relies on the match of a 2 Å FWHM theoretical grid of stellar spectra with the original code for evolutionary population synthesis by Buzzoni ([8], [9]). Our high-resolution spectra ([10], [11]), generated with the Kurucz SYNTHE code [1] and then broadened to 2 Å FWHM, are complemented, at low temperatures, with those based on the NextGen [2] model atmospheres.

As an illustrative example, in Fig. 1 we present our high-resolution synthesis of SSPs in the ultraviolet region around 2800 Å, where two outstanding absorption features of MgII and MgI are present.

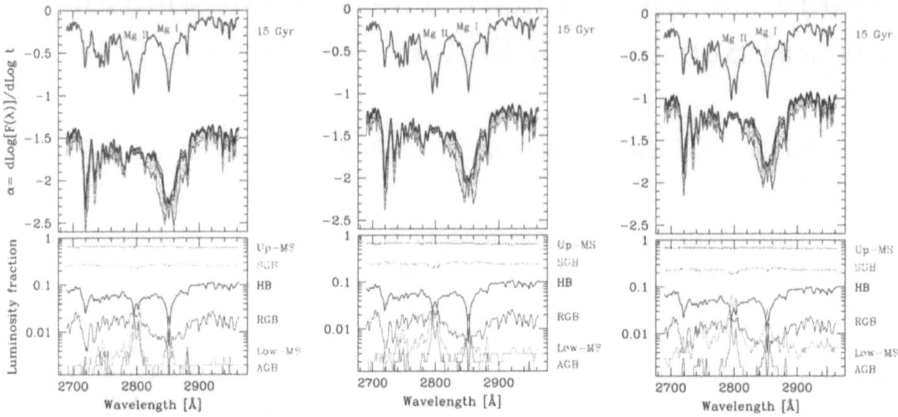

Fig. 1. SSP synthesis models in the MgII-MgI ultraviolet region around 2800 Å. Models assume solar metallicity, red Horizontal Branch (HB) morphology and IMF power-law index $s = 1.35$ (*left panel*), 2.35 (the Salpeter value; *middle panel*), and 3.35 (*right panel*). On the y axis of upper panels we report the rate of SSP luminosity evolution, $\alpha(\lambda) = d \log \text{flux}(\lambda)/d \log t$, evaluated numerically at different ages back to 1 Gyr and taking the 15 Gyr SSP model as a reference (see, as a guideline, its corresponding SED overplotted at the top in each panel, in arbitrary units). A "tuned" trend for α is evident in the MgII region about 2790-2810 Å, whith a steady "plateau" in all plots. A much "steeper" time evolution appears on the contrary for the MgI feature about 2850 Å ($\alpha \sim -2.1$ in this region for the Salpeter SSP) thus making the MgII/MgI line ratio a selective tracer of SSP age not depending on the IMF slope. The contribution of the stars in the different evolutionary phases for the 15 Gyr SSPs is displayed in the lower panels. One sees that most of the Mg features originate from the subgiant (SGB) and HB stars (see their "depressed" relative contribution, compared for example to the "emission" of low Main Sequence, MS, and asymptotic giant stars, AGB, given the lack of both MgII and MgI absorption lines in their spectra).

References

1. R.L. Kurucz: CDROM No. 18, SYNTHE spectrum synthesis program and line data (1993)
2. P.H. Hauschildt, F. Allard, E. Baron: ApJ, **512**, 377 (1999)
3. A. Vazdekis, E., Casuso, R.F. Peletier, J.E. Beckman: ApJS, **106**, 307 (1996)
4. G. Bruzual, S., Charlot: MNRAS, **344**, 1000 (2003)
5. G. Worthey, S.M. Faber, J.J. Gonzalez, D. Burstein: ApJS, **94**, 687 (1994)
6. G. Worthey, S.M. Faber, J.J. Gonzalez: ApJ, **398**, 69 (1992)
7. A. Buzzoni, L. Mantegazza, G. Gariboldi: AJ, **107**, 513 (1994)
8. A. Buzzoni: ApJS, **71**, 817 (1989)
9. A. Buzzoni: AJ, **123**, 1188 (2002)
10. E. Bertone, A. Buzzoni, L. Rodriguez-Merino, M. Chavez: In *Modelling of Stellar Atmospheres*, IAU Symp. no. 210, ed. by N.E. Piskunov, W.W. Weiss, D.F. Gray (ASP: San Francisco 2003) in press
11. L.H. Rodriguez-Merino, M. Chavez, E. Bertone, A. Buzzoni: In *Star Formation through Time*, ASP Conf. Series **297**, ed. by E. Perez, R. Gonzalez-Delgado, G. Tenorio-Tagle (ASP, San Francisco 2003) pp. 237-238

Constraining the Star Formation History with Photometric Redshifts

Peter Capak, Len L. Cowie, and Esther M. Hu

Institute for Astronomy, University of Hawaii,
2680 Woodlawn Dr., Honolulu, HI, 96822, USA

Abstract. Large scale structures can bias global measurements, such as the star formation history, due to over and under densities in the field at a given redshift. To overcome this problem large area redshift surveys are necessary. Photometric redshifts provide an economical method for estimating the distances to galaxies using the broad band photometry. To this end we are surveying 1.2 sq. degrees centered on the Hawaii survey fields in U, B, V, R, I, z', J, HK' to depths similar to the Great Observatories Origins Deep Survey (GOODS). Many of these fields have also been observed in the x-ray, sub-mm, and radio and will be observed with SIRTIF in the near future. Using, our data in the Hawaii Hubble Deep Field North (Hawaii HDF-N) and SSA22, we have obtained photometric redshifts accurate to dz/(1+z)=0.06. These can then be used to measure the global star formation history and the evolution of Active Galactic Nuclei (AGN).

1 Photometric Redshifts and Redshift Distributions

To achieve good results we first carefully calibrated our photometry to remove any systematics caused by seeing or filter systems [5]. We then applied small color offsets to optimize results for objects with known spectroscopic redshifts. Photometric redshifts were determined with the Bayesian photometric redshift code of Benitez et. al.[2]. The RMS scatter between the spectroscopic and photometric redshifts is $dz/(1+z) = 0.06$. Even in the case of catastrophic failures the redshift probability distribution usually includes the true redshift meaning the results are statistically accurate.

Shown in figure 1 are the redshift distributions for the Hawaii-HDF-N and SSA22. Note the over densities which have been spectroscopically confirmed and are especially strong in the radio and x-ray source population. Brodwin et. al.[4] see similar field to field variations in the Canada France Redshift Survey fields which include SSA22.

2 Evolution of the Luminosity Function

Photometric redshifts can be used to directly measure the optical luminosity function of galaxies in several bands. They are especially useful in measuring the faint end slope of the luminosity function at $z > 1.0$, which can not be constrained by spectroscopic redshifts. In addition a large enough sample can be obtained to measure the luminosity function's evolution in redshift bins as small

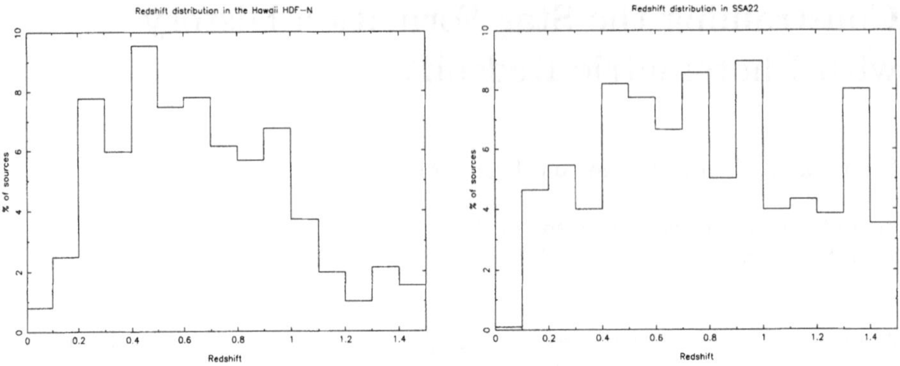

Fig. 1. On the left is the redshift distribution of the HDF-N on the right that of SSA22, all objects brighter than $R_{ab} = 26.5$ are shown. Note the difference in the distribution, especially the peaks which correspond to walls in the redshift distribution.

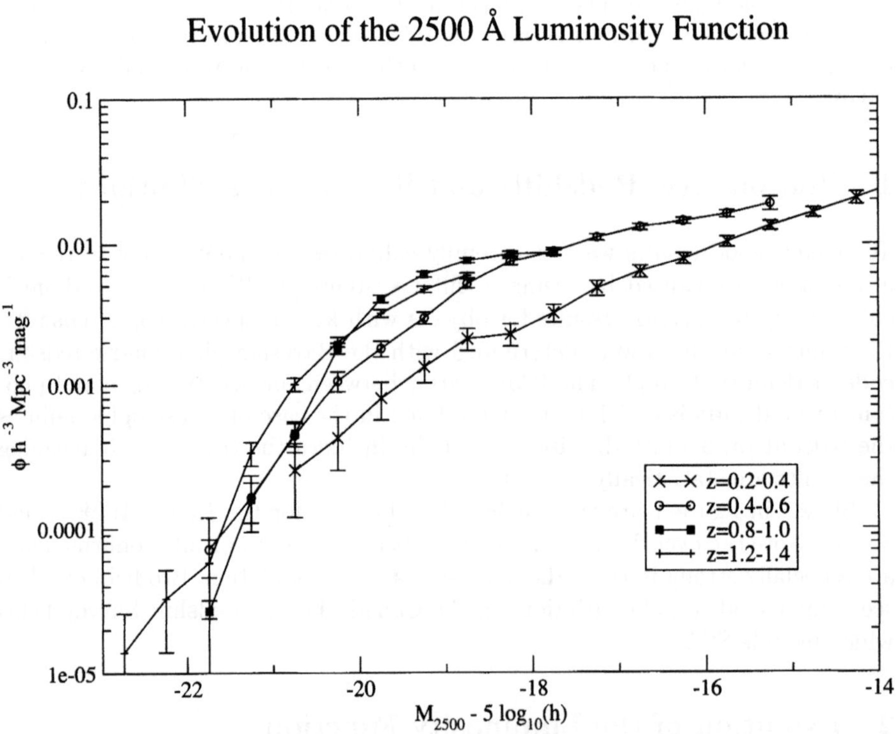

Fig. 2. Shown are 2500 Å luminosity functions in several redshift bins. These were calculated with the 1/Vmax method. Error bars reflect Possion variation and errors from photometric redshifts.

as $dz = 0.2$. We have used a Monte-Carlo method similar to Bolzonella et. al. [3] to account for the redshift errors using the redshift probability distribution. The 2500Å luminosity functions are shown in figure 2. The faint end slope of the 2500 Å luminosity function appears to be evolving from $\alpha = -1.47 \pm 0.03$ at $z = 0.3$ to $\alpha = -0.95 \pm 0.05$ at $z = 1.95$. The ultraviolet luminosity functions can be translated directly into a measure of the star formation rate by integrating the total ultraviolet flux [8]. Adopting the dust correction factors of 1.7 for $z < 1.5$ and 4.5 for $z > 1.5$ used by Steidel et. al. [9] we find that our results agree with those of other studies. These results will be presented in more detail in upcoming papers[6].

References

1. Barger, A.J, Cowie, L.L., Capak, P., Alexander, D.M., Bauer, F.E., Fernandez, E., Brandt, W.N., Garmire, G.P., Hornschemeier, A.E., 2003, AJ, 126, 632B
2. Benitez, N. 2000, ApJ, 536, 517B
3. Bolzonella, M., Pello, R. Maccagni, D., 2002, A&A, 395, 443B
4. Brodwin, M., Lilly, S.J., Porciani, C., McCracken, H.J., Le Fevre, O., Foucaud, S., Crampton, D., Mellier, Y.,2003, astro/ph, 0310038
5. Capak, P., Cowie, L.L., Hu, E.M., Barger, A.J., Dickinson, M., Fernandez, E., Giavalisco, M., Komiyama, Y., Kretchmer, C., McNally, C., Miyazaki, S., Okamura, S., Stern, D., 2004a, AJ, Accepted
6. Capak, P., Cowie, L.L, Hu, E.M., Barger, A.J., 2004b, in prep
7. Cowie, L.L., Barger, A.J., Fomalont, E.B., Capak, P., 2003, ApJL, submitted
8. Kennicutt, R.C., 1998, ARAA, 36, 189K
9. Steidel, C.C., Adelberger, K.L., Giavalisco, M., Dickinson, M., Pettini, M., 1999, ApJ, 519, 1S

Morphology and Redshifts of Extremely Red Galaxies in the GOODS/CDFS Deep ISAAC Field

K.I. Caputi, J.S. Dunlop, R.J. McLure, and N.D. Roche

Institute for Astronomy, University of Edinburgh, Royal Observatory, Edinburgh EH9 3HJ, Scotland.

Abstract. We present the photometric redshift distribution of a sample of 198 Extremely Red Galaxies (ERGs) with $K_s < 22$ and $(I_{775} - K_s) > 3.92$ (Vega), selected by Roche et al. [3] in 50.4 arcmin^2 of the Chandra Deep Field South (CDFS). The sample has been obtained using ISAAC-VLT and ACS-HST GOODS public data. We also show the results of a morphological study of the 72 brightest ERGs in the z band ($z < 25$, AB).

1 Photometric Redshifts

Figure 1 shows the redshift distribution of the ERGs in Roche et al.'s sample. The redshifts have been computed using the public code 'hyperz' [1] with photometry in seven bands: B, V, I_{775}, z (V1.0 ACS-HST release); J,H and K_s (ISAAC-VLT).

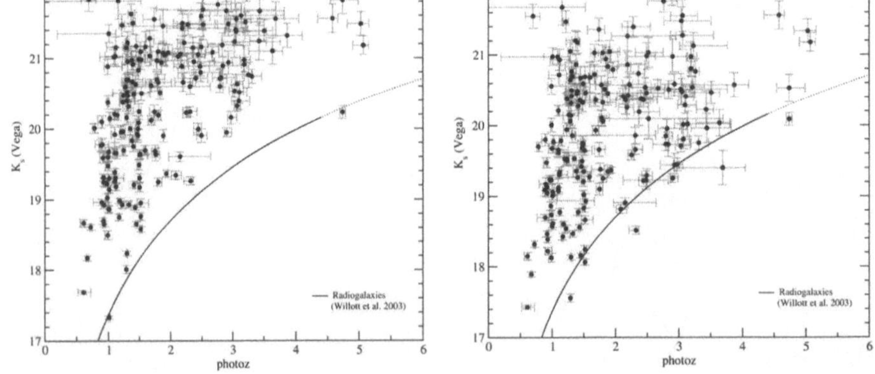

Fig. 1. Hubble diagram for the ERGs in Roche et al.'s sample. a) Original K_s magnitudes; b) dust-corrected K_s magnitudes.

Our results show that the ERGs span a wide variety of luminosities in the Hubble diagram up to the high mass limit observationally defined by massive radio galaxies [4].

2 Morphology

The radial dependence of the surface brightness distribution of regular galaxies is given by $I(r) \propto I_o \exp(-r/r_o)^{\beta}$, where I_o is the surface brightness at a characteristic distance r_o. Empirically, elliptical galaxies have $\beta \sim 0.25$, while pure disks have $\beta \sim 1$. We determined the best-fit β parameter of the 72 brightest ERGs in the z-band, using a 2D-modelling code developed by McLure et al. [2], on the 5-epoch GOODS stacked images. The results are shown in fig. 2.

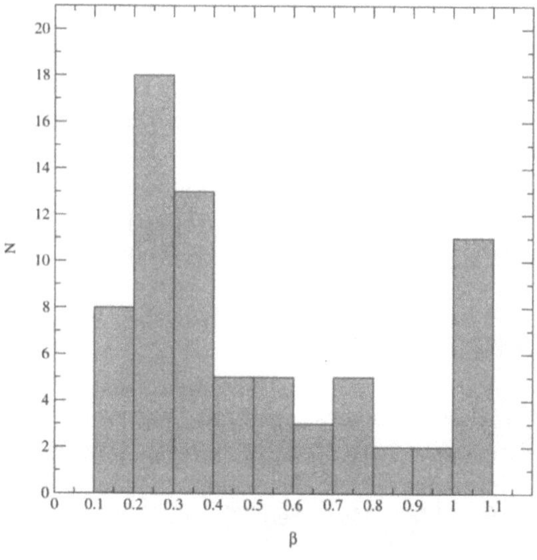

Fig. 2. Distribution of the β parameter for the 72 brightest ERGs in the z band.

We conclude that 61% and 24% of this subsample is composed of elliptical ($0 < \beta < 0.5$) and disky ($0.5 < \beta < 1.0$) galaxies, respectively. The remaining 15% are represented in the β interval $[1.0, 1.1]$ and are either irregular galaxies or a mix of disk and bulge.

References

1. Bolzonella M., Miralles J.M., Pelló R., 2000. A&A **363**, 476
2. McLure R.J., Dunlop J.S., Kukula M.J., 2000. MNRAS **318**, 693
3. Roche N.D., Dunlop J.S., Almaini O., 2003. MNRAS accepted, astro-ph/0303206
4. Willott C.J., Rawlings S., Jarvis M., Blundell K.M., 2003. MNRAS **339**, 173

Stellar Populations of Compact Group Galaxies

Paula Coelho[1], Cláudia Mendes de Oliveira[1,2], Jesus J. González[3], N. Visvanathan[4], and B. Barbuy[1]

[1] Universidade de Sao Paulo, IAG, Departamento de Astronomia, Rua do Matao 1226, Cidade Universitária, 05508-900, São Paulo, SP, Brazil
[2] Max-Planck-Institut für extraterrestrische Physik, Giessenbachstr. 85748 Garching, Germany
[3] Instituto de Astronomia, UNAM, Apdo Postal 70-264, 04510, Mexico D.F., Mexico
[4] Mount Stromlo and Siding Spring Obs., Inst. of Advanced Studies, ANU, Australia

1 Introduction

The study of the stellar populations of the elliptical galaxies in compact groups could give us important information on the merging history of the elliptical galaxies in the groups and could elucidate some of the questions regarding group evolution. We have initiated a program to study the stellar populations of the elliptical galaxies in compact groups. In the following, we describe our results for a sample of 12 galaxies. The kinematical properties of the same sample were previously studied by Mendes de Oliveira et al. (1993).

2 Line Strength Indices

Eight line-strength Lick/IDS indices, as defined by Worthey et al. (1994) were measured. The indices were corrected by velocity broadening and zero-point by the usual procedure. To derive the stellar populations parameters of the galaxies, the indices Hβ, Mg2, Mgb, Fe5270 and Fe5335 were compared to the Single Stellar Populations (SSP) models by Thomas, Maraston & Bender (2003, TMB03).

In Figure 1 we show the results in index planes for the galaxies in compact groups, those for the sample of galaxies in less dense environments studied by González (1993) and the SSP models superimposed. Although the statistics of the compact group galaxies is still small, it is possible to see that the locus of the compact group galaxies is characterized by old ages, solar or super-solar metallicities and alpha-enhancement. As a comparison, the galaxies from the sample of González, which inhabit less dense environments, seem to spam a wider parameter range, although these also span a larger range in mass.

3 Results and Conclusions

In this work we studied a sample of 12 compact group galaxies through their line-strength indices and compared them with SSP models.

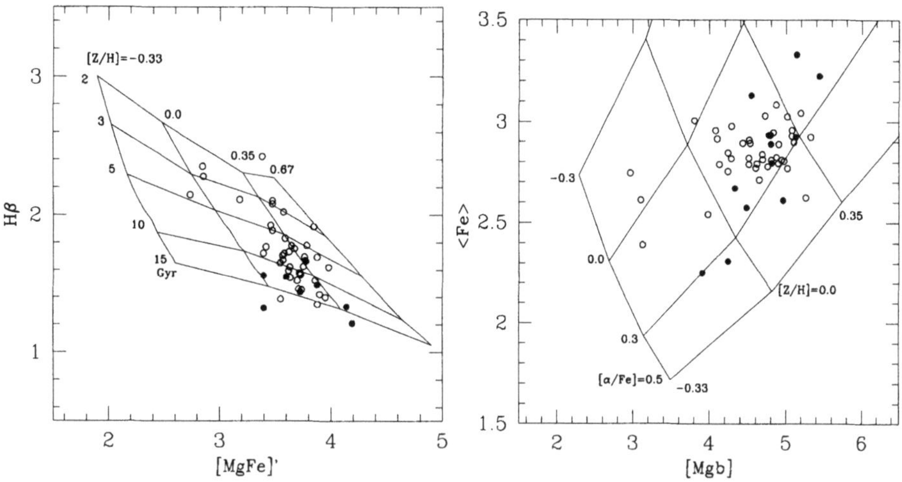

Fig. 1. Compact groups galaxies (filled circles), field galaxies (open circles) and SSP models (grid) in two index planes. The index $[MgFe]'$ is defined as $[MgFe]' = Mgb.(0.72Fe5270 + 0.28Fe5335)^{1/2}$ (TMB03). Labels are given to illustrate the loci occupied by different stellar population parameters.

Employing a Fortran code (kindly provided to us by D. Thomas, M.P.E.) that inspects correlations between Hβ and metallic indices to find the best fit (t, [Z/H], [α/Fe]) for each galaxy, we derived the parameters of the galaxies for which the Hβ errors were less than 0.25 and had no emission lines (eight galaxies). We have found that the luminosity weighted average [Z/H] ranges from solar to 0.5 dex and all galaxies show an overabundance of α-elements in respect to the solar-scale ([α/Fe] ranging from 0.1 to 0.4 dex). We find very old ages for the compact group galaxies (average age is 14 +/- 3 Gyr, with the exception of galaxy HCG 05B), for the eight galaxies in our sample with best S/N data.

We plan to increase the size of the sample in order to be able to make a definite statement about the mean ages of elliptical galaxies in compact groups.

References

1. González, J. 1993, Ph.D. Thesis, University of California at Santa Cruz
2. Mendes de Oliveira, C.; González, J. J.; Visvanathan, N.; Hickson, P. 1993, In: *Cosmic Velocity Fields, Proceedings of the 9th IAP Astrophysics Meeting*, Institut d'Astrophysique, Paris, p. 563
3. Thomas, D., Maraston, C., Bender, R. 2003, MNRAS 339, 897
4. Worthey, G., Faber, S., Gonzalez, J., Burstein, D. 1994, ApJS 94, 687

Nested Bars with Different Pattern Speeds: NGC 2950

Enrico Maria Corsini[1], Victor P. Debattista[2], and J.A.L. Aguerri[3]

[1] Dipartimento di Astronomia, Università di Padova, Padova, Italy
[2] Institut für Astronomie, ETH Hönggerberg, Zürich, Switzerland
[3] Instituto de Astrofísica de Canarias, La Laguna, Spain

Abstract. We studied the SB0 galaxy NGC 2950, which hosts two nested stellar bars. We used the Tremaine-Weinberg method to measure the pattern speed of the primary bar. We established directly and for the first time in a double-barred galaxy that the two nested bars are rotating with different pattern speeds.

Interest in secondary stellar bars, which are observed in $\sim 1/3$ of barred galaxies [1], is motivated by the hypothesis that they are a mechanism for driving gas to small radii to feed the supermassive black holes powering AGN's. However, the efficiency of such transport is uncertain because of our complete lack of knowledge of the relation between $\Omega_{\rm p}$ and $\Omega_{\rm s}$, the pattern speeds of the primary and secondary bars, respectively. A model-independent method for measuring pattern speeds is the Tremaine-Weinberg method ([2], TW hereafter). This gives the pattern speed Ω of a single bar as $\mathcal{X}\Omega \sin i = \mathcal{V}$, where $\mathcal{X}$ and $\mathcal{V}$ are luminosity-weighted average position and line-of-sight velocity measured parallel to the major axis of the galaxy disk, and i is the disk inclination. Long-slit spectra parallel to the disk major axis can measure all the quantities needed. If several parallel slits at different offsets Y relative to the major axis are available for a galaxy, $\Omega \sin i$ can be obtained as the slope of a plot of $\mathcal{V}$ versus $\mathcal{X}$. In a double-barred galaxy (hereafter S2B), for a slit passing through both bars, the TW equation is modified to $(\mathcal{X}_{\rm p}\Omega_{\rm p} + \mathcal{X}_{\rm s}\Omega_{\rm s}) \sin i = \mathcal{V}_{\rm p} + \mathcal{V}_{\rm s}$ provided the two bars are rigidly rotating through each other. It can can be solved for $\Omega_{\rm s}$ by first measuring $\Omega_{\rm p}$ with slits which avoid the secondary bar, and then modeling to obtain $\mathcal{X}_{\rm s}$ and/or $\mathcal{V}_{\rm s}$ from the observed $\mathcal{X} = \mathcal{X}_{\rm s} + \mathcal{X}_{\rm p}$ and $\mathcal{V} = \mathcal{V}_{\rm s} + \mathcal{V}_{\rm p}$. However, the two bars are not likely to rotate rigidly through each other when $\Omega_{\rm p} \neq \Omega_{\rm s}$ [3,4], requiring additional modeling to obtain $\Omega_{\rm s}$. Nonetheless, when $\Omega_{\rm p} = \Omega_{\rm s}$, the modified TW equation reduces to the original one and is satisfied exactly. Thus testing whether $\Omega_{\rm s} = \Omega_{\rm p}$ does not require additional assumptions.

We applied this technique to the S2B galaxy NGC 2950. We obtained surface photometry and medium-resolution spectra with the slit along the disk major axis and parallel to it. The value of $\Omega_{\rm p}$ we obtained is 11.2 ± 2.4 km s^{-1} arcsec^{-1} from the values of $\mathcal{X}$ and $\mathcal{V}$ for the slits at $|Y| > 3''$, which cross only the primary bar, and at $Y = 0''$. Then, from the circular velocity $V_{\rm c}$ and bar semimajor axis $a_{\rm p}$, we found that the corresponding ratio $\mathcal{R}_{\rm p} \equiv D_{\rm p}/a_{\rm p}$ of the corotation radius $D_{\rm p} = V_{\rm c}/\Omega_{\rm p}$ to the bar semimajor axis is $\mathcal{R}_{\rm p} = 1.0^{+0.3}_{-0.2}$, *i.e.* the primary bar of NGC 2950 is rapidly rotating as generally found for SB0's [5]. If this is the norm

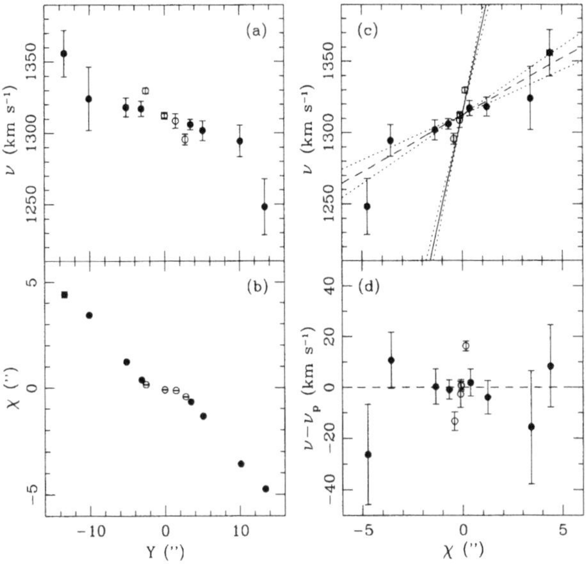

Fig. 1. *(a)* The kinematic integrals $\mathcal{V}$ as a function of the slit offset Y with respect to the major axis ($Y = 0''$). Open and filled circles correspond to slits at $|Y| < 3''$ and at $|Y| > 3''$, respectively. *(b)* The photometric integrals $\mathcal{X}$ as a function of Y. *(c)* $\mathcal{V}$ as a function of $\mathcal{X}$ with different straight-line fits. These were obtained including the slits at $Y = 0''$ and only the slits at $|Y| > 3''$ (dashed line) or the innermost slits at $|Y| < 3''$ (solid line). The different slope of the two straight lines strongly suggests that $\Omega_\mathrm{p} \neq \Omega_\mathrm{s}$. *(d)* The residuals from the dashed-line fit.

in S2B's, then it guarantees that primary bars are efficient at funnelling gas down to the radius of influence of secondary bars. Estimating Ω_s is model dependent, and we considered two extreme cases. First, we assumed that the secondary bar dominates at $|Y| < 3''$ and obtained $\Omega_{\mathrm{s},1} = 89.2 \pm 9.9$ km s^{-1} arcsec^{-1}. In the second case, by applying the modified TW equation we obtained $\Omega_{\mathrm{s},2} = -21.2 \pm 6.2$ km s^{-1} arcsec^{-1}, *i.e.* secondary bar is counter-rotating relative to the primary. Secondary bars may oscillate about an orientation perpendicular to the primary bar, possibly accounting for $\Omega_\mathrm{s} < 0$. This can be tested by repeating our measurements on a sample of S2B's. Nonetheless, we can confidently conclude that in NGC 2950 the two bars must have different pattern speeds [6].

References

1. Erwin, P. 2003, A&A, in press (astro-ph/0310806)
2. Tremaine, S., & Weinberg, M. D. 1984, ApJ, 282, L5
3. Louis, P. D., & Gerhard, O. E. 1988, MNRAS, 233, 337
4. Maciejewski, W., & Sparke, L. S. 2000, A&A, 313, 745
5. Aguerri, J. A. L., Debattista, V. P., & Corsini, E. M. 2003, MNRAS, 338, 465
6. Corsini, E. M., Debattista, V. P., & Aguerri, J. A. L. 2003, ApJ, 599, L29

Chemo-Photometric Models of Ring Galaxies

Anna Curir[1] and Paola Mazzei[2]

[1] INAF, Osservatorio Astronomico, Strada Osservatorio, 10025 Pino Torinese, Italy
[2] INAF, Osservatorio Astronomico, Vicolo dell'Osservatorio 5, 35122 Padova, Italy

1 Introduction

The peculiar ring galaxies are formed as a result of cosmic interaction. An intruder galaxy plunges through the center of a disk galaxy triggering radially expanding density waves on the disk. Here we present results of SPH simulations which focus on the role of different halos (live and analytical) and bulges on the morphology of the target galaxy and on the star formation rate. We find that inside a live triaxial halo arise unfavourable conditions to the development of a ring. Moreover, for same collision, different features appear if the target galaxy is embedded in an analytical halo or endowed with a massive bulge. The presence of a bulge noticeably affects the star formation rate and the dynamical evolution of the interacting system.

2 Simulations

Simulations performed are resumed in Table 1 where the unit mass is $5.97 \cdot 10^{10} m\odot$. In the live halo cases, we built up the target galaxy merging the bary-

Table 1. Simulations

Simulation	State of the halo	M_{halo}	M_{bulge}	intruder velocity
1	live	5	no bulge	120 Km/s
2	live	5	no bulge	720 km/s
3	live	1	no bulge	120 km/s
4	live	1	no bulge	720 km/s
5	Analytical	1	no bulge	120 km/s
6	Analytical	1	no bulge	720 km/s
7	no halo	0	1	120 km/s
8	no halo	0	1	720 km/s

onic disk in a dark matter (DM) triaxial halo far from virialization. In the no halo cases, we merge in the disk a king sphere with the same mass and radius equal to $0.5R_{disk}$. In the analytical halo cases, the gravitational potential is given by an isothermal sphere with the same mass as the disk, radius $3R_{disk}$ and core radius $0.1R_{disk}$. The intruder is a king sphere with the same mass and radius as the bulge. The intrusion is orthogonal to the disk plane.

3 Results

In all our live halo cases the dynamical friction of the halo on the intruder is very strong and provokes a diffusion of the intruder, so that the impact does not form a long lasting ring in the disk. The more interesting cases for the ring formation are the analytical halo and the no halo cases. In the last case, for the higher intrusion velocity, two order of rings connected by spokes appear in the gas. In the lower velocity case, we note the formation of two order of rings again but without spokes. The bulge-intruder interaction, i.e. the bulge oscillation, gives rise to a secondary shock propagating in the gas. This wave forms a second order ring which enhances the star formation in the disk. The ring-shaped morphology of the new stars is shown in the first phases of the interaction in the no halo cases, whereas it is appearing in the latest ones in the analytical halo cases (see Fig.1).

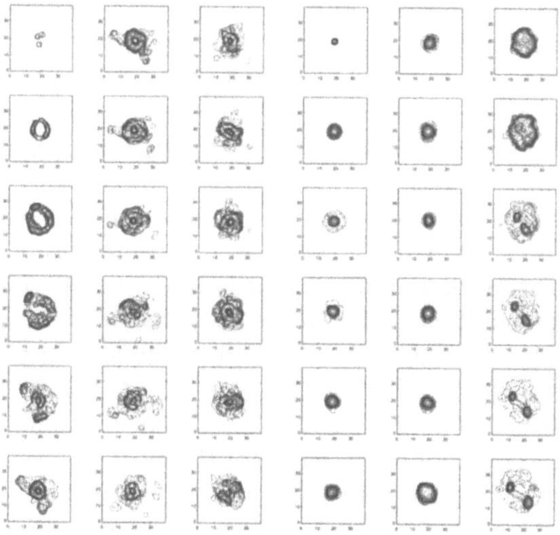

Fig. 1. B band isophotes of new stars formed in simulation 8 (left panel) and 6 (right panel); box size, spatial resolution and magnitude contrast are 40 kpc$x$40 kpc, 1 kpc and 3.7 mag respectively; time increases from top to bottom and left to right starting from $2x10^7$ yr to $3.55x10^8$ yr with time steps of $1.95x10^7$ yr; our chemo-photometric models are described in Curir and Mazzei: New Astr **1**, 20 (1999)

Star Formation in High Redshift Radio Galaxies

Carlos De Breuck[1] and Michiel Reuland[2]

[1] ESO, Karl Schwarzschild Straße 2, D-85748 Garching, Germany
[2] Sterrewacht Leiden, Postbus 9513, 2300 RA Leiden, The Netherlands

Abstract. We present evidence for star formation in distant radio galaxies based on deep FORS1/VLT spectropolarimetry and SCUBA/JCMT 850μm photometry.

High redshift radio galaxies are efficient tracers of massive stellar systems. Their powerful radio emission requires massive black holes ($> 10^9 M_\odot$), which are known to reside in massive ellipticals [e.g. 5]. An attempt to compare the stellar mass of radio galaxies with other galaxies can be made in the Hubble $K - z$ diagram. This shows that radio-selected galaxies have brighter $K-$band magnitudes than $K-$band selected galaxies [2], which can be interpreted in terms of radio galaxies having baryonic masses as high as $10^{12} M_\odot$, i.e. more than an order of magnitude higher than a typical HDF galaxy at $z \sim 3$ [7].

A serious complication is the presence of an AGN, which may also contribute to the *observed* $K-$band emission. A direct way to determine the contribution of scattered AGN emission is spectropolarimetry. [8] have found *rest-frame* UV polarizations P >10% in five out of 10 z ~2.5 radio galaxies, indicating scattered light from a burried AGN. However, at $z > 3$, 2 out of 3 radio galaxies with spectropolarimetry information have P <5% [4,8]. One of these (4C 41.17, z = 3.8) shows stellar photospheric absorption lines [3], providing direct evidence that the UV continuum is dominated by light from young, hot stars.

To verify whether the contribution of young stars to the UV continuum rises with redshift, we have observed four $z > 3$ radio galaxies with FORS1/VLT. All targets have been selected to have $22 < R < 23$ to ensure sufficient S/N to determine the polarization in integration times of 6 to 14 hours per target (De Breuck et al. in preparation). All four galaxies show $P \leq 8\%$, and two $P \leq 4\%$. In particular, we detect the CIII λ 1428 Å photosheric absorption line in one of the unpolarized sources, TN J2007−1316 at $z = 3.84$ (Fig. 1).

Another indication of star formation is submm dust emission. If the dust is heated by massive stars rather than directly by the AGN, this implies star formation rates as high as a few $1000 M_\odot yr^{-1}$. [1] and [6] have obtained SCUBA 850 μm photometry of 69 $z > 1$ radio galaxies. The detection rate and 850μm luminosity increase dramatically at $z > 3$, suggesting higher star formation rates.

Figure 2 shows that the UV polarization and dust emission appear anti-correlated. This can be interpreted as evidence that these these sources selected to have $R < 23$, require either an important scattered quasar component (hence high polarization), or (if unpolarized), a high contribution from a young stellar population, which can heat the dust in these objects. A more detailed analysis will be presented by Reuland et al. (2004).

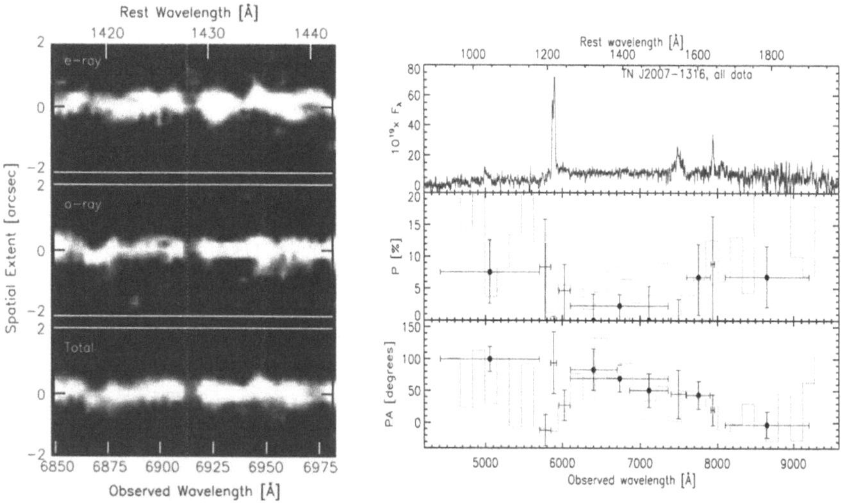

Fig. 1. VLT/FORS1 spectropolarimetry of TN J2007−1316: *Left:* Part of the summed extraordinary ray, ordinary ray, and total intensity 2-dimensional spectra. The dotted line indicates the wavelength of the CIII λ 1428 Å photospheric absorption line, assuming the redshift of the AGN. *Right:* Total intensity, percentage polarization, and polarization angle. The grey histograms show the data with a 150 Å binning.

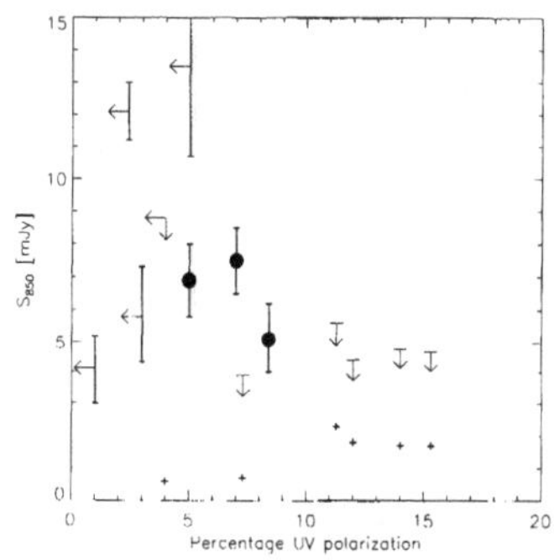

Fig. 2. Anti-correlation between UV continuum polarization and S_{850} for $z > 1$ radio galaxies. For non-detections in S_{850}, we show nominal detections (+) and 3σ limits.

References

1. Archibald et al. 2001, MNRAS. **323**, 417
2. De Breuck, C. et al. 2002, AJ. **123**, 637
3. Dey, A., et al. 1997, ApJ **490**, 698
4. Dey, A. 1999, “Most distant radio galaxies”, KNAW, Amsterdam, p. 19
5. Gebhardt, K. et al. 2000, ApJ. **539**, L13
6. Reuland et al. 2004, MNRAS, submitted
7. Rocca-Volmerange et al. 2004, A& A., in press, astro-ph/0311490
8. Vernet et al. 2001, A& A., **366**, 7

The Nature of UV-Selected Galaxies in the CDFS

Duília F. de Mello[1,2], Jonathan P. Gardner[1], T. Dahlen[3], C.J. Conselice[4], N.A. Grogin[5], and A.M. Koekemoer[3]

[1] NASA's GSFC, LASP code 681, Greenbelt, MD 20771, USA
[2] Catholic University of America, Washington, DC 20064, USA
[3] STScI, Baltimore, MD 21218, USA
[4] Caltech, Pasadena, CA 91125, USA
[5] CAS, Johns Hopkins University, Baltimore, MD 21218, USA

1 Introduction

The compilation of the cosmic star formation history suggests that no particular epoch of star formation is seen at redshifts z>1 (Madau et al. 1996). The rise of the volume-averaged star formation rate from z=0 to 1 has been attributed to a disk epoch when gaseous galactic disks were converting into stars. In order to search for these predictions deep high resolution images of large areas of the sky are required. Only recently, surveys with such requirements have become available. One of these surveys has been taken with the HST's Advanced Camera for Surveys (ACS) by the GOODS team (Great Observatories Origins Deep Survey). Treasury fields (Chandra Deep Field South and Hubble Deep Field North) have been visited by ACS during 5 epochs (see M. Giavalisco contribution in this proceedings for more details) from October 2002 until March 2003. In this contribution we present the results of the survey using the HST/WFPC2 (F300W) in combination with the HST/ACS multi-wavelength images. The main goal is to establish the nature of the star-forming galaxies at intermediate redshifts.

2 The Method and Results

We implemented an alteration of the WFPC2 pure parallel program and took near-UV images (filter F300W) of the GOODS fields that have been imaged at redder (B,V, i, z) wavelengths with ACS. The exposure times varied between 2 and 5 orbits depending on which filter ACS was observing with.

We used SExtractor v2.2.2 (Bertin & Arnouts 1996) to detect the objects in the U band images. We matched the U catalog with the ACS B band catalog (B_{435}) produced by the GOODS team. For two of the five epochs a total of 137 objects were identified adopting 1.5 arcsec as a maximum offset radius between the WFPC2 coordinates and ACS B band. Photometric redshifts were estimated by the GOODS team (Mobasher et al. 2003) using templates of the spectral energy distribution of E-Sa, Sbc, Scd, Im (Coleman et al. 1980), and two starbursts (Kinney et al. 1996). The majority of the objects in the U-band

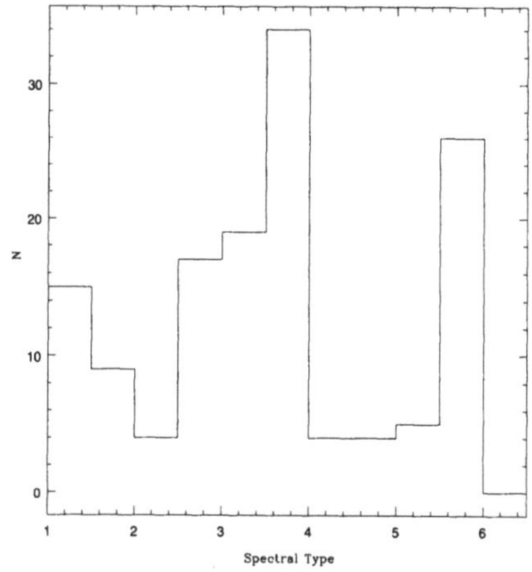

Fig. 1. Spectral type distribution. 1=Early-type, 2=Sbc, 3=Scd, 4=Im from Coleman et al. 1980 and 5,6=starburst from Kinney et al. 1996.

sample has z<1. The template fitting were used to classify galaxies as early-, late-types and starbursts. Fig.1 shows the spectral-type distribution.

We have compared the colors of the objects in the first epoch with the colors produced with the stellar evolutionary synthesis code Starburst99 (Leitherer et al. 1999) in order to date the stellar population.

Galaxy morphology of these objects was evaluated by measuring the light concentration (C), asymmetry (A) and clumpiness (S) of the ACS images in the rest-frame B band using the method by Conselice (2003).

The majority of the galaxies have: (i) spectral energy distribution of late-type galaxies; (ii) lower values of light concentration and higher asymmetries than typical early and late-type galaxies; (iii) colors typical of stellar populations with age < 1 Gyr. However, there are objects of all types, from dwarf ellipticals, early and late-types to peculiar objects which resemble mergers in progress. Our results suggest that galaxies of all types are contributing to the rise in the volume-averaged star formation rate.

References

1. Bertin, E., & Arnouts, S. 1996, A& A S.S. 117, 393
2. Coleman, G.D., Wu, C-C., & Wedman, D.W. 1980, ApJS, 43, 393
3. Conselice, C. 2003, ApJS, (astro-ph/0303065)
4. Kinney, A. et al. 1996, ApJ, 467, 38
5. Leitherer, C., et al. 1999, ApJS, 123, 3 (Starburst99)
6. Madau, P., Ferguson, H., Dickinson, M., Giavalisco, M.,Steidel, C., Fruchter, A. 1996, MNRAS, 283, 1388
7. Mobasher, B., et al. 2003, ApJ (astro-ph/0309068)

The Rest-Frame Optical Properties of Star-Forming Galaxies at $z \sim 2$

Dawn K. Erb[1], Charles C. Steidel[1], Alice E. Shapley[2], Max Pettini[3], and Kurt L. Adelberger[4]

[1] Department of Astronomy, California Institute of Technology, MS 105-24, Pasadena, CA 91125, USA
[2] Department of Astronomy, University of California at Berkeley, Berkeley, CA, USA
[3] Institute of Astronomy, Cambridge, UK
[4] Carnegie Observatories, Pasadena, CA, USA

Abstract. We present preliminary results from Hα spectroscopy and deep K-band imaging of UV-selected star-forming galaxies at $z \sim 2$. We see evidence for relatively mature, massive galaxies in the form of an increase in the mean velocity dispersion and the mean $\mathcal{R}-K_s$ color from $z \sim 3$ to $z \sim 2$, and in the approximately solar metallicities of a significant number of the $z \sim 2$ galaxies. There appears to be substantial diversity in the $z \sim 2$ sample, as other kinematic measurements suggest that some of the galaxies are still in the process of formation.

1 Introduction

It is increasingly evident that $z \sim 2$ is an important epoch in the evolution of the universe: galaxies at $z \sim 3$ are compact, rapidly star-forming, and morphologically disordered [7,8], whereas those at $z \sim 1$ have become the normal Hubble sequence galaxies of the universe today. QSO activity reaches a peak near $z \sim 2$ [6], and the median redshift of bright sub-millimeter galaxies is $z = 2.4$ [2]. Recent studies also suggest that most of the stellar mass in the universe today formed during this epoch [4].

We have obtained a sample of ~ 800 star-forming galaxies spectroscopically confirmed to be at $z \sim 2$, using $U_nG\mathcal{R}$ color selection criteria and the LRIS-B spectrograph on the Keck I telescope [1,9]. We have observed Hα emission in ~ 70 of these with the near-IR spectrograph NIRSPEC on the Keck II telescope, and obtained deep K-band images of ~ 250 of them using the Wide-field IR Camera (WIRC) on the Palomar 200-inch telescope.

2 Preliminary Results

Our initial results suggest that the $z \sim 2$ sample contains mature galaxies with substantial masses and significant star-formation histories. Several lines of evidence imply this. The mean Hα velocity dispersion of 61 galaxies in the sample is $\langle\sigma\rangle = 114 \pm 51$ km s^{-1}. In Fig. 1 we compare this with the mean velocity dispersion of the $z \sim 3$ LBGs (measured from the [OIII]λ5007 emission line), $\langle\sigma\rangle = 76 \pm 23$ km s^{-1}; we see an increase of 50% from $z \sim 3$ to $z \sim 2$, implying

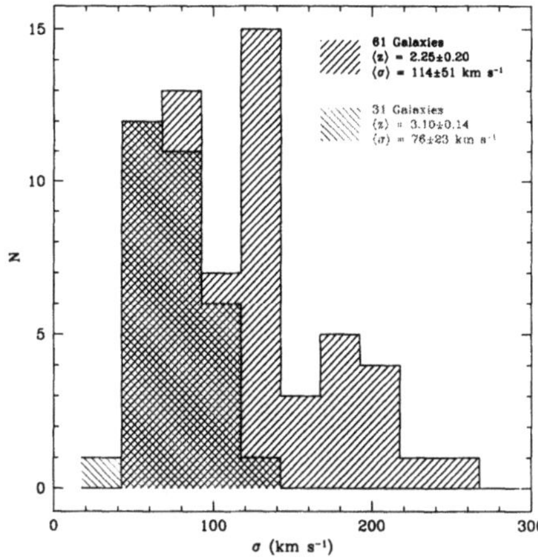

Fig. 1. A comparison of the one-dimensional velocity dispersions of galaxies at $z \sim 2$ and $z \sim 3$ suggests that the $z \sim 2$ galaxies have larger dynamical masses on average.

larger dynamical masses and suggesting substantial evolution between the two epochs. We have also compared the $\mathcal{R} - K_s$ colors of the $z \sim 2$ and $z \sim 3$ samples, and we find that the mean $\mathcal{R} - K_s$ color increases from 2.85 ± 0.59 at $z \sim 3$ to 3.29 ± 0.54 at $z \sim 2$, indicating an increase in age and/or dust content. Finally, we see significant [NII]λ6583 emission in many of the rest-frame optical spectra, which can be used to estimate the galaxies' metallicity [3]; we find metallicities of up to 1.5 $Z_{\odot}$.

Our sample also contains 13 galaxies with Hα spectra in the GOODS-N field. These were primarily selected because they appear to be extended in the high-resolution ACS images, and we placed slits along their major axes to see if they might be more likely to yield spatially resolved kinematic information. We find tilted emission lines for only 2 of the 13 galaxies, however, and see some evidence that the extended galaxies may still be in the process of formation and that the tilted lines, when observed, are due to mergers [5]. This, when combined with the above results, suggests significant diversity in the $z \sim 2$ sample.

References

1. Adelberger, K. L. et al. 2004, ApJ, submitted
2. Chapman, S. C. et al. 2003, Nature, 422, 695
3. Denicoló, G. et al. 2002, MNRAS, 330, 69
4. Dickinson, M. et al. 2003, ApJ, 587, 25
5. Erb, D. K. et al. 2004, ApJ, submitted
6. Fan, X. et al. 2001, AJ, 121, 54
7. Giavalisco, M. et al. 1996, ApJ, 562, 95
8. Shapley, A. E. et al. 2001, ApJ, 562, 95
9. Steidel, C. C. et al. 2004, ApJ, submitted

The Cosmic Evolution of Quasar Hosts

R. Falomo[1], J.K. Kotilainen[2], C. Pagani[3], R. Scarpa[4], and A. Treves[3]

[1] INAF - Observatory of Padova, Italy
[2] Tuorla Observatory, Turku, Finland,
[3] Insubria University, Como, Italy,
[4] ESO, Santiago, Chile

1 Introduction, Observations and Data Analysis

Ground based and HST imaging of quasars (see e.g. Dunlop et al 2003; Pagani et al 2003 and references therein) clearly indicate that at z $<$ 0.5 quasars are hosted in massive galaxies dominated by the spheroidal component. While radio loud quasars (RLQ) are exclusively hosted by ellipticals exceeding by 2-3 mag L*, radio quiet quasars (RQQ) are found both in ellipticals and in spirals. On the other hand there is mounting evidence that most if not all nearby massive spheroids have inactive supermassive BH in their centers (see e.g. Ferrarese 2002 for a review). Taken together these findings depict a scenario where nuclear activity may be a common phenomenon during the lifetime of a galaxy. Understanding how the properties of the galaxies hosting quasars change with the cosmic time is therefore a fundamental step to investigate the link between the evolution of the galaxies and the nuclear activity.

We have carried out a program to characterize the host galaxies of quasars in the redshift range $1 < z < 2$ using VLT capabilities. Here we present some results concerning the host galaxy luminosity. The full results for this program are given in Falomo et al (2003). We secured near-IR images of 17 quasars (both RLQ and RQQ with matched z-L distribution) using ISAAC in service mode during excellent seeing conditions (FWHM 0.4 arcsec). For each quasar we performed a detailed modeling of the luminosity profile using an iterative least-squares fit from which we derive the luminosity and of the host galaxies and of the nuclei. In this work we assume H_o=50 Km s^{-1} Mpc^{-1}Ω_o=0.

2 Results

We plot in Figure 1 the absolute magnitude of the host galaxies versus the redshift for the 16 resolved quasars. These host galaxies have luminosity ranging between M* and M*-2, (where M*(K) = -25.2). The average absolute K-band magnitude of the host galaxies of RLQ ($< M_K >$(host) = -27.55) is found $\sim$ 0.7 mag brighter than those of RQQ ($< M_K >$ (host) = -26.83).

In order to investigate the evolution of the host luminosity of quasars up to z = 2, we report in Fig. 1 the average luminosities of the host galaxies derived from various quasar samples (at z $<$ 2) whose data have been made homogeneous with ours. It turns out that the host galaxies luminosity of both types of quasars

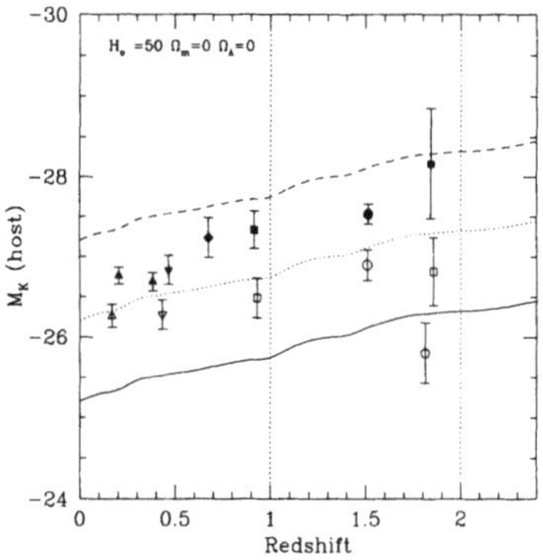

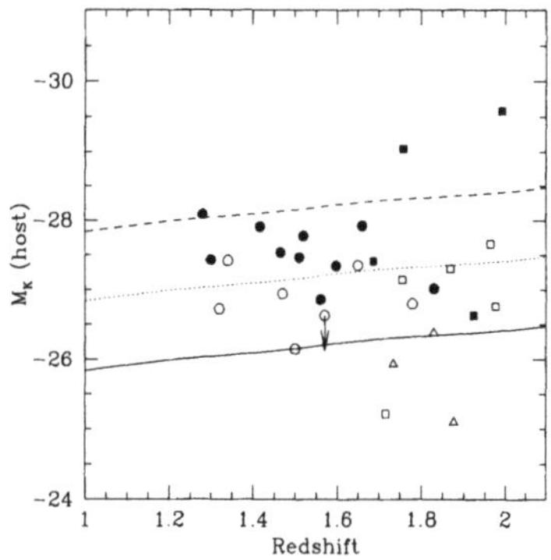

Fig. 1. Left Both RLQs (filled symbols) and RQQs (open symbols) appear to follow the standard passive evolution for luminous elliptical galaxies. The lines represent the expected behaviour of a massive elliptical (at M*, M*-1 and M*-2; solid,dotted and dashed line) undergoing passive stellar evolution. Squares by Kukula et al 2001. Triangles by Ridgway et al 2001. **Right** The evolution of quasar host luminosity from various samples. RLQ hosts are systematically more luminous than RQQ hosts. No drop of host luminosity (mass) is observed up to z=2.

appear to follow the expected behavior of massive inactive ellipticals undergoing passive evolution. The present data also suggest that the luminosity difference between RLQ and RQQ hosts does not significantly depend on the redshift. The evolution traced by quasar hosts up to z $\sim$2 appears similar to that of massive inactive spheroids (see e.g. Cimatti 2003) and disagrees with the expectations of the models of AGN and galaxy formation and evolution based on the hierarchical scenario. These models predict fainter (less massive) hosts at high redshift, which then merge and grow to form the massive spheroids observed in the present epoch. From this analysis we do not find any significant drop in the host mass. If this drop occurs, it must take place at epochs earlier than $z = 2$. Since the peak of quasar activity occurs at $z \sim 2.5$, it is of great importance to assess whether the trend exhibited by the quasar hosts at $z < 2$ is also followed by the higher redshift ($z = 2$ to 3) objects.

References

1. Cimatti A. 2003 astro-ph/0303023.
2. Dunlop, J.S., McLure, R.J., Kukula, M.J., et al 2003, MNRAS, 340, 1095
3. Ferrarese L., 2002 astro-ph/0203047
4. Falomo, R. Kotilainen, J.K., Pagani C., Scarpa R. & Treves, A. 2003, ApJ in press (astro-ph/0312234)
5. Kukula, M.J., Dunlop, J.S., McLure, R.J., et al MNRAS, 326, 1533
6. Pagani, C., Falomo, R., Treves, A., 2003, ApJ, in press (astro-ph/0307043)
7. Ridgway, S., Heckman, T., Calzetti, D., Lehnert, M. 2001, ApJ, 550, 122

On the Dearth of Low-Luminosity, High-Redshift Quasars

F. Fontanot[1], P. Monaco[1], G. Taffoni[2], S. Cristiani[2], and M. Nonino[2]

[1] Dipartimento di Astronomia, Università di Trieste - ITALY
[2] Osservatorio Astronomico di Trieste, INAF - ITALY

Abstract. The GOODS Team [2] presented a sample of 17 high-redshift QSO candidates from the GOODS fields with $3.5 < z < 5.2$. The number counts of their $z > 4$ quasar revealed a dearth of low-luminosity objects both with respect to naive Pure Luminosity Evolution (PLE) predictions and to models based on the statistic of major mergings of dark matter halos (DMHs) [5]. A Pure Density Evolution (PDE) prediction of the quasar luminosity function (LF) at $z \sim 2.7$ [1] appeared able to reproduce both the SDSS counts and the GOODS number counts.

We present 3 models of quasar formation, in a standard cosmological scenario, that follow the processes of DMH mergings and galaxy formation. It is often assumed that quasars are triggered by episodes of major merging of DMHs: in our first model we assume a constant ratio ε between black hole (BH) and DMH mass, and that BHs shine at their Eddington Luminosity for a duty-cycle time t_{duty}. Figure 1-left shows the resulting LF of quasars, which evolves differently from observations. However the timescale necessary for gas in halos to cool, lose angular momentum and accrete onto a seed BH is not negligible at high redshift, so the shining of the quasar will be delayed with respect to the merger of their DMHs. This delay will be regulated by feedback from star formation and from the quasar itself [4]. We implement this idea in the previous model by letting the quasar shine after a time t_{shine} [4] after the major merger of DMHs: this delay depends on the BH mass M_{BH} in the sense that greater masses correspond to shorter delays. As a result of the delay, the number of low-luminosity quasars decreases, but with this approach the suppression is typically too strong.

We present the results of a preliminary attempt of introducing quasars in a galaxy formation model. This model takes into account: (a) merger histories of DMHs [6]; (b) mergers of satellite galaxies [7] ; (c) hot virialized gas in hydrostatic equilibrium within the DMH ; (d) metal enrichment and the metallicity dependence of cooling; (e) formation of discs from the cooled gas; (f) disc instabilities and galaxy major merger (that bring cold gas into the bulge); (g) feedback from SNe. Cold gas in the bulge must lose angular momentum to fall into the inner part of the potential well and constitute a reservoir of gas ready to be accreted by a BH. According to [8], angular momentum loss is due to radiation drag from the radiation field, in which case $\mathrm{d}M_{BH}/\mathrm{d}t \propto SFR$. Taking into account the Eddington limit on the bolometric luminosity of a quasar and the corresponding Eddington time t_{ED} $\mathrm{d}M_{BH}/\mathrm{d}t = \min(f_{BH}SFR, M_{BH}/t_{ED})$. This model introduces two parameters: f_{BH}, which represents the efficiency of

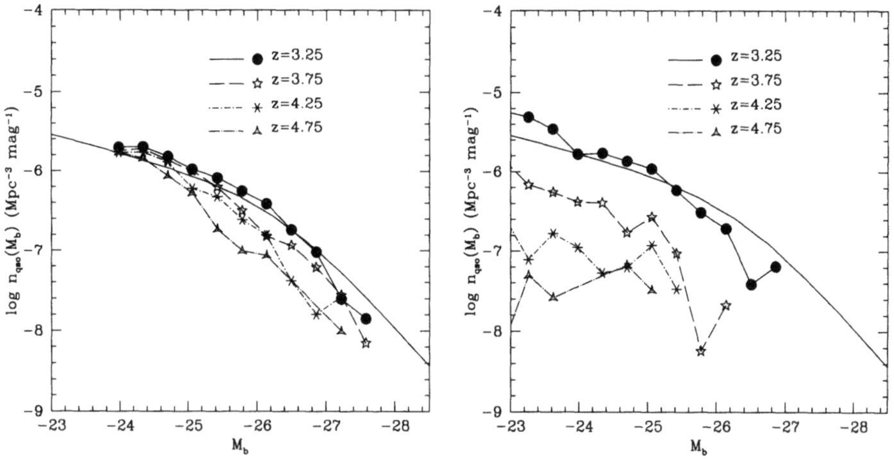

Fig. 1. Evolution of the LF for: **Left** the model based on major mergers of DMHs ($t_{duty} = 2 \times 10^7 yr$, $\varepsilon = 2 \times 10^{-4}$); **Right** galaxy formation model with quasars ($f_{BH} = 0.03$ and $M_{seed} = 10^4 M_{\odot}$). Different symbols refer to different redshifts as stated in figure. The PDE analytical fit of the observational LF (extrapolated from the 2QZ LF [1] with a PDE of a factor 3.5 per unit redshift after z=2.7 [3]) is shown at z=3.25

the accretion with respect to the SFR, and M_{seed}, which represents the mass of the seed BH that is put in each forming galaxy. The resulting LF is shown in Fig. 1-right. Remarkably, a density evolution of the LF at low luminosity is found, although this evolution is faster than observed.

Conclusions A simple direct connection between mergers of DMHs and the shining of quasars is not valid at $z > 3$,and this is mostly easily understood in terms of timescales needed to the gas to cool, lose angular momentum and accrete onto a BH. The inclusion of quasars in a galaxy formation model based on hierarchical clustering of DMHs is able to reproduce a PDE-like growth of low-luminosity quasars. This result is preliminary, as the model has not yet been widely tested. However, we can conclude that no sophisticated feedback recipes are needed to obtain a PDE-like evolution of the quasar LF at high redshift.

References

1. B.J. Boyle, T. Shanks, S.M. Croom, et al., 2000, MNRAS, 317, 1014
2. S. Cristiani, et al., 2003, ApJL, in press (astro-ph0309049)
3. X. Fan, et al., 2001, AJ, 121, 54
4. G.L. Granato, L. Silva, P. Monaco, et al., 2001, MNRAS, 324, 757
5. Z. Haiman, L. Hui, 2001, ApJ, 547, 27
6. P. Monaco, T. Theuns, G. Taffoni, et al., 2002, ApJ, 564, 8
7. G. Taffoni, L. Mayer, M. Colpi, F. Governato, 2003, MNRAS, 341, 434
8. M. Umemura, 2001, ApJ, 560, 29

The Hα-Based Evolution of Star-Forming Galaxies from z = 0.8 to Now

Jesús Gallego[1], Sergio Pascual[1], Jaime Zamorano[1], Roser Pelló[2], Alfonso Aragón-Salamanca[3], Pablo G. Pérez-González[4], Víctor Villar[1], and Cristina Díaz[1]

[1] Universidad Complutense de Madrid, 28040 Madrid, Spain
[2] Laboratoire d'Astrophysique de l'Obs. Midi-Pyrénées, France
[3] University of Nottingham, UK
[4] Steward Observatory, University of Arizona, USA

Abstract. We present a long-term project to study the characteristics and evolution of current star-forming galaxies at different redshifts. The selection of the samples is carried out in an homogeneous way by detecting Hα emission-line objects in deep images taken with narrow-band filters tuned to the corresponding wavelength for a given redshift. Target redshifts are z=0.24, z=0.4 and z=0.8. Here we present photometric redshifts, emission-line luminosities and multi-band photometry for a sample of emission-line galaxies at z=0.24, selected by their contrast in a narrow band filter centered in the 8200Å atmospheric window. The physical properties of these galaxies appear to be remarkably similar to those of local galaxies selected in the same way. Our results support the idea that the higher Star Formation Density measured at redshift z=0.24 is due to an increase of the density of bursting galaxies and not to an intrinsic change on overall galaxy properties.

1 The Survey

Galaxy evolution during the last half of the life of the Universe is now well established. The period since z∼1 until the current epoch has been characterized by clear changes in the overall properties of the Universe such as Star Formation Rate (hereafter SFR) density [5]. Simultaneously, the galaxy population of the Universe since z∼1 has experienced substantial changes.

Our goal is to build homogeneous samples of current star-forming galaxies covering the redshift range from z∼1 to the current epoch. The objects are selected by their flux excess in a narrow-band filter centered at the wavelength corresponding to Hαλ6563Å redshifted at the target redshift. At this moment, optical filters at 8200Å and 9200Å have been used to study the z=0.24 and z=0.4 universes. Work at 11810Å is in progress to study the z=0.8 universe.

2 Overall Properties of Hα Emitters at z = 0.24

Here we analyze the sample of candidates to emission-line galaxies used by Pascual et al. (2001; [6]) to estimate the Hα-based SFR density of the Universe at z=0.24. In that paper, we presented deep images taken with the 2.2m telescope of Calar Alto Observatory (Almería, Spain) on the ELAIS-N1 region [7] using a narrow band filter centered in 8200Å with a FWHM of 160Å and a broad-band filter to probe the continuum. This specific narrow-band filter was selected because at 8200Å there is a substantial gap in the sky-night OH emission lines. The broad-band was selected because it has almost the same effective wavelength.

Deep multi-band photometry for our sample was obtained from the INT Wide Field Survey (hereafter INT-WFS). Our field was imaged with five filters (*Ugriz*), ranging from 3664Å (U filter) to 8953Å (z filter). Limiting magnitudes were ~22.9 in the broad band and ~21.0 in the narrow band. There were 61 objects in the original sample. 16 of them were marginal detections, detected only in the narrow band filter. Stars were flagged using the stellarity parameter CLASS_STAR given by Sextractor [1].

In order to discriminate between genuine Hα emitters and potential sources of contamination, i.e. stars and other emission line objects ([OIII]5007 at z~0.64, Hβ at z~ 0.68 or [OII]3727 at z~ 1.2), photometric redshifts were computed through a standard minimization procedure, using the public code *hyperz* [2]. Based in both the stellarity and photometric redshift, 13 objects were finally classified as stars and 33 objects were confirmed as Hα emitters at z=0.24. This number is compatible with the one obtained evolving with a $(1+z)^3$ the local Hα luminosity function obtained by [4]. We have found 6 candidates to be galaxies at z~0.6 and 3 at ~1.2. The presence of these objects at these relatively bright fluxes implies a stronger evolution of the population of [OIII]5007 and [OII]3727 emitters than the corresponding to Hα.

We find a total extinction-corrected Hα luminosity density of $(4.6\pm0.3)\ 10^{39}$ erg s^{-1} Mpc^{-3} at z=0.24. Assuming a constant relation between the Hα luminosity and star formation rate, the SFR density in the volume covered is $(0.036\pm0.002)\ M_\odot\ yr^{-1}\ Mpc^{-3}$. This value is a factor of ~ 3.8 higher than the local SFR density and consistent with the strong increase in the SFR density from z=0 to z=1.

Hα emitters at z=0.24 exhibit properties similar to the local analogs, in distribution of Hα fluxes and equivalent widths. The downsizing scenario [3], in which rapid star-forming galaxies evolve smoothly in luminosity with decreasing redshift, seems to be a possible model to explain the evolution of these galaxies.

References

1. Bertin, E. & Arnouts, S. 1996, Astronomy & Astrophysics Supp. Series, 117, 393
2. Bolzonella, M., Miralles, J., & Pelló, R. 2000, Astronomy & Astrophysics, 363, 476
3. Cowie, L., Songaila, A., Hu, E., & Cohen, J. 1996, Astronomical Journal, 112, 839
4. Gallego, J., Zamorano, J., Aragón-Salamanca, A., & Rego, M. 1995, Astrophysical Journal Letters, 455, L1
5. Madau, P., Pozzetti, L., & Dickinson, M. 1998, Astrophysical Journal,, 498, 106
6. Pascual, S., Gallego, J., Aragón-Salamanca, A., & Zamorano, J. 2001, Astronomy & Astrophysics, 379, 798
7. Rowan-Robinson, M., & Elais Consortium. 1999, ESA SP-427: The Universe as Seen by ISO, 427, 1011

Multi-Wavelength Luminosity Functions of Galaxies

Jonathan P. Gardner

GSFC, Code 681, Greenbelt MD 20771

Multivariate or multi-wavelength luminosity functions will reveal the interplay between star formation, chemical evolution, and absorption and re-emission of dust within evolving galaxy populations. By using principal component analysis to reduce the dimensionality of the problem, I optimally extract the relevant photometric information from large galaxy catalogs. As a demonstration of the technique, I derive the multi-wavelength luminosity function for the galaxies in the released SDSS catalog, and compare the results with those obtained by traditional methods.

A luminosity function is essentially a histogram, and therefore must either be parameterized, or binned. The Schechter (1976) parameterization is successful at fitting most LFs measured in a single filter; however there is no equivalent parameterization for multi-variate LFs, and so they must be binned. If the variables are strongly correlated, however, appropriate binning is not straight-forward.

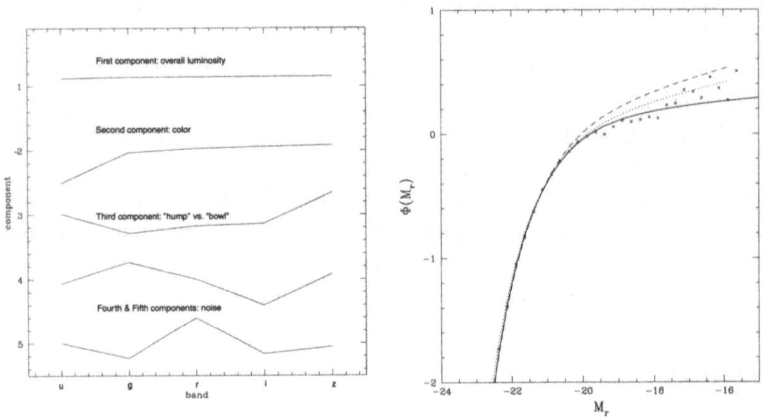

Fig. 1. The left panel shows principal component analysis of the absolute magnitudes of the SDSS early release dataset. The components are: (1) overall luminosity, comprising 90% of the information, (2) overall color, red or blue (dominated by the $u - g$ color), (3) shape of the spectrum, "hump" vs. "bowl", as discussed by Koo (1985), (4) and (5) noise. The right panel shows the "luminosity function" of the first principal component, transformed back to absolute r magnitude (solid line). For comparison, I plot the r-band LF from Blanton et al. (2001) (dashed line) and the LF determined through traditional methods (dotted line and points). At bright magnitudes the LFs are nearly identical, but they diverge at faint magnitudes, where the color-magnitude relation is more important.

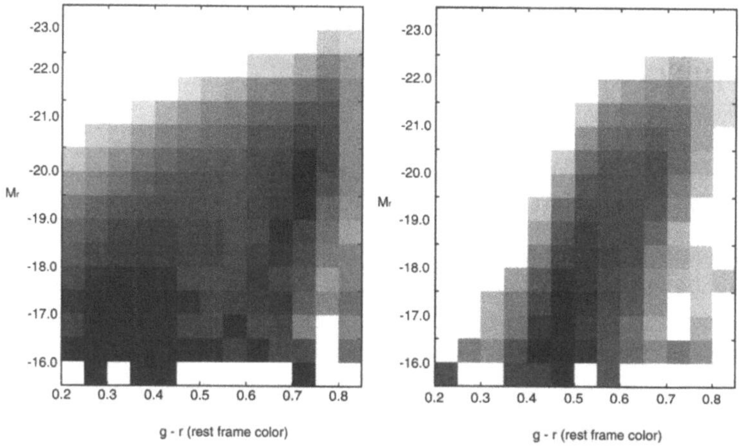

Fig. 2. The left panel shows the joint luminosity-color distribution $\Phi(M_r, g-r)$ for the SDSS galaxies, plotted on a log scale. The right panel shows the distribution of the 1st and 2nd principal components, converted back to M_r and $g-r$ color. In both panels, the color-luminosity relation is clearly evident. In the original data the E/S0 ridgeline is visible at $g-r \simeq 0.7$, however, this is not seen in the reconstruction. Although I have converted the second principal component back to $g-r$ rest frame color, this color component is dominated by $u-g$, as can be seen in the left panel of Figure 1. The color-luminosity relation is strong in this color, but the red galaxies do not have much effect. The E/S0 ridge is a subtlety that requires 3 or more principal components.

Principal Component Analysis (PCA) determines an orthogonal set of principal components, or axes, in a multi-dimensional space which maximize the variance for each axis, and is ideally suited for finding the appropriate dimensionality of the distribution. Components which do not carry much weight can be discarded, and the system can then be transformed back onto the original axes.

This is work in progress, and I have not yet included a proper treatment of errors. The photometric errors in the SDSS dataset are small in comparison to the intrinsic distribution, and to the bin size. However, the u-band photometry of the reddest galaxies have the highest noise, and this might have a significant effect on the second principal component. I have also not yet included normalization constraints, or ϕ^*. Finally, in order to apply this technique to other datasets such as the 2MASS survey, I will need to take into account the multiple selection functions, and different magnitude limits in different bands. Adding SIRTF and GALEX data to this analysis will enable a comparison of the ultraviolet light that escapes the galaxies with the light that is absorbed and re-emitted at longer wavelengths.

References

Blanton, M. R., et al. 2001, AJ, 121, 2358
Koo, D. C. 1985, AJ, 90, 418
Schechter, P. 1976, ApJ, 203, 297

Linking Star Formation and Environment in Supercluster Galaxies

Meghan Gray[1], Chris Wolf[2], Klaus Meisenheimer[3], Andy Taylor[1], Simon Dye[4], Andrea Borch[3], and Martina Kleinheinrich[3]

[1] Institute for Astronomy, Blackford Hill, Edinburgh, EH9 3HJ, UK
[2] Department of Physics, University of Oxford, Keble Road, Oxford OX1 3RH, UK
[3] Max-Planck-Institut für Astronomie, Königstuhl 17, D-69117, Heidelberg, Germany
[4] Astrophysics Group, Imperial College, London SW7 2BW, UK

1 The Dark Matter Environment

We investigate correlations between the location of galaxies in dense environments and their degree of star-formation activity. Using photometric redshifts and spectral classifications from the COMBO-17 survey [1] we are able to precisely isolate galaxies from a 4×4 h^{-1} Mpc region around the Abell 901/902 supercluster in the range $0.15 < z < 0.18$. We measure rest-frame photometric quantities from PEGASE template fits to the 17-band photometric data, and divide supercluster galaxies into two classes ('quiescent' and 'star-forming') based on deviation blueward of the $(U-V)_{\rm rest}$ vs. M_V colour-magnitude sequence.

We sample the weak gravitational lensing mass-map of [2] at the location of each galaxy in order to use mass as a direct measurement of environment for the first time. We observe:

- strong evidence for segregation by type, with the highest density regions populated almost exclusively by galaxies classified according to their rest-frame $U-V$ colours as quiescent (Fig. 1);
- a threshold surface mass density from lensing, $\kappa = 0.05$, above which star-formation activity is rapidly suppressed;
- that this abrupt transformation affects primarily the faint end of the star-forming galaxy population (down to $M^* + 6$) and occurs at a local surface number density that is an order-of-magnitude above the density of the field;
- that when only galaxies brighter than $M_V^* + 1$ are considered the trends with environment remain, but are more gradual and extend beyond $2h^{-1}$ Mpc radius.

2 The Multiwavelength Dataset

We are currently engaged in enlarging the existing multicolour dataset in order to assemble a comprehensive view of galaxy evolution as a function of environment. To the aforementioned weak lensing map, photometric redshifts, and SED classifications, we have added a deep 90 ks XMM-Newton X-ray observation of

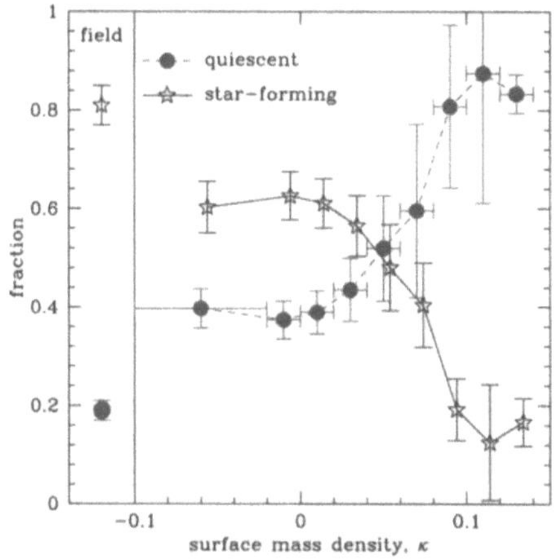

Fig. 1. The SF-κ relation for the A901 supercluster. A dramatic change in relative abundance of each galaxy type at a given density is seen as the proportion of star-forming galaxies declines and quiescent galaxies become more dominant in dense regions. The symbols at left show the corresponding fractions reached in the field, obtained from an identical redshift slice within two blank COMBO-17 fields.

the supercluster field in order to trace out the hot inter- and intracluster medium and to investigate if this correlates well with the distributions of dark matter and galaxies. We have also collected 2dF spectra (using 1200B and 600V gratings) for 285 galaxies within the supercluster, with which we will derive detailed star-formation histories and search for signatures for both ongoing and recently truncated star formation.

Our goal is to accurately characterize the environmental influences that may be acting on the infalling galaxies, whether by the influence of the cluster dark matter, hot X-ray gas, or high galaxy number density. We hope to disentangle which interactions (whether between the galaxy and the cluster gravitational field, the galaxy and the intracluster medium, or galaxy-galaxy interactions) are responsible for the observed transformation of photometric properties. Finally, we wish to investigate the equivalence of the spectral and photometric star-formation indicators. Preliminary results indicate a possible population of strong [OII] emitters whose colours place them firmly on the cluster colour-magnitude relation and so would be classified as 'quiescent' by photometry alone.

References

1. C. Wolf, K. Meisenheimer, H.-W. Rix, A. Borch, S. Dye, M. Kleinheinrich: A&A **401**, 73 (2003)
2. M. Gray, A. Taylor, K. Meisenheimer, S. Dye, C. Wolf, E. Thommes: ApJ **568**, 141 (2002)

A 1200 μm MAMBO Survey of the ELAIS N2 and Lockman Hole East Fields

T.R. Greve[1], R.J. Ivison[1], F. Bertoldi[2], J.A. Stevens[1], S.C. Chapman[3], I. Smail[4], and A.W. Blain[3]

[1] Royal Observatory Edinburgh, Blackford Hill, Edinburgh EH9 3HJ, UK
[2] Max-Planck Institut für Radioastronomie, Bonn, Germany
[3] Astronomy Department, Caltech, Pasadena, CA 91125, USA
[4] Department of Physics, University of Durham, South Road, Durham DH1 3LE, UK

Abstract. Using the MPIfR Max Planck Millimeter Bolometer array (MAMBO) on the IRAM 30m Telescope we have mapped the ELAIS N2 and Lockman Hole East Fields at 1200 μm to a rms noise level of $0.8-1.0$ mJy per $11''$ beam. The areas surveyed are 326 arcmin^2 in the ELAIS N2 field and 212 arcmin^2 in the Lockman Hole[1], and cover the 260 arcmin^2 previously observed by SCUBA [5].

The 1200 μm number counts derived from the survey are shown in Fig. 1a (Greve et al. in prep.). At flux levels $\lesssim$ 3.5 mJy the power-law slope of the number counts is about $\alpha \sim -1.6$, while at the brighter end there is evidence for a turn-over in the number counts, as is illustrated by the fact that the data are well matched by an integrated Schechter function with a knee at 3.5 mJy. At a redshift of 2.5, this corresponds to a far-IR luminosity of $10^{13}\,L_\odot$ assuming a modified black body law with $\beta = 1.5$ and $T_d = 40$ K. For comparison we have also plotted the 850 μm counts from the HDF-N SCUBA Supermap [1], scaled by a factor of $S_{850\mu\rm m}/S_{1200\mu\rm m} = 2.5$ which is expected for a starburst galaxy at $z = 2.5$ [2]. Even though this scaling-factor is highly uncertain, the agreement between the 1200 μm and scaled 850 μm counts in terms of the shape of the number counts is remarkably good.

Deep radio observations currently provide the most efficient way of determining the exact positions of (sub)-mm sources, and thus positively identifying them in the optical/NIR [3,6]. Using deep Very Large Array radio maps [4] we have searched for statistically robust radio counterparts within $6''$ of each of the MAMBO sources in our sample. We find that about two-thirds of the MAMBO sources have counterparts in the radio, which is comparable to the radio-identification fraction found for SCUBA sources [4]. The MAMBO source shown in Fig. 1b is associated with a very strong radio counterpart ($S_{1.4\,GHz}$ = 189 μJy) which lies on top of a compact optical/NIR galaxy. A Keck LRIS-B spectrum of this source reveals that it is a type II QSO at $z = 2.6$ (Ivison et al. in prep.). This source lies well within the SCUBA map yet is not included in the $\geq 3.0\sigma$ SCUBA catalogue [5]. While it is conceivable that a certain fraction of the 1200 μm sources might be at extremely high redshifts ($z > 8$) and

[1] The Lockman data are part of the MAMBO 1sq. deg. survey (Bertoldi et al. in prep.)

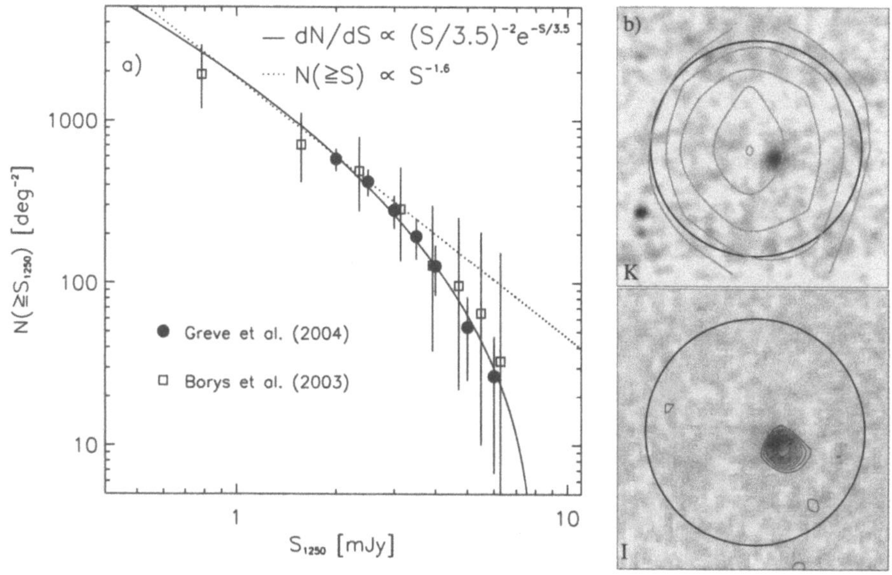

Fig. 1. a) Preliminary cumulative number counts at 1200 μm (filled circles) based on $\geq 3.5\sigma$ sources extracted from our MAMBO map of the ELAIS N2 and Lockman Hole East fields. 850 μm counts based on the HDF-North SCUBA Super-map are shown as squares (see Borys et al. (2003) for details). Note the 850 μm fluxes have been scaled by a factor of $1/2.5 = 0.4$. **b**) An example of a MAMBO source with a strong radio counterpart. *Top:* The 1200 μm-emission shown as gray contours: 3.5,4.0,4.5,5.0, 5.5$\times\sigma$ with $\sigma = 0.8$ mJy; *bottom:* Radio (1.4 GHz) contours (black) starting at 3σ and increasing in steps of $\sigma = 9.5\,\mu$Jy. The thick black circle is the 6″ search radius adopted.

thus can 'drop-out' at 850 μm if the dust is cold [2], it is clearly not the case here since it is detected in the I-band which is shortward of 912 Å for $z > 8$. Comparing the MAMBO and SCUBA maps we find that, although, a few MAMBO sources are not detected by SCUBA and vice versa, there is a fair overall correlation between the 1200 μm and 850 μm counts and galaxy positions, suggesting that both surveys are tracing the same high-redshift dusty population (Greve et al. in prep.). If this is the case, the faster mapping speed (about a factor of $\times 6$) and smaller beam size of IRAM 30m/MAMBO over that of JCMT/SCUBA make the former the facility of choice for wide-field extragalactic surveys.

References

1. C. Borys, et al.: MNRAS, **344**,385.
2. S. Eales, et al.: MNRAS, **344**, 169.
3. R.J. Ivison, et al.: MNRAS, **298**, 583 (1998).
4. R.J. Ivison, et al.: MNRAS, **337**, 1 (2002).
5. S.E. Scott, et al.: MNRAS, **331**, 817 (2002).
6. I. Smail, et al.: ApJ, **528**, 612 (2000).

Surface Density of Extremely Red Objects with R-J≥5

Angela Hempel[1], Tom M. Herbst[1], and Dave Thompson[2]

[1] Max-Planck Institute for Astronomy, Königstuhl 17, 69117 Heidelberg, Germany
[2] Palomar Observatories, Caltech, Pasadena, CA 91125, USA

Abstract. A catalogue of galaxies with extremely red colours (R-J≥5), covering 8.43 square degrees has been selected from optical and near-infrared data. Based on their bright near-infrared magnitude and their very red colour, these objects are likely to be very massive galaxies with little or no ongoing star formation. We have derived a surface density of 0.0396 ±0.0011arcmin^{-2} (Poisson uncertainties) for these Extremely Red Objects.

1 Introduction

Extremely Red Objects (EROs) were discovered in deep optical and near-infrared surveys ([2], [3]). Defined mostly by a single colour, primarily R-K, they do not represent a uniform class of objects, but may contain contributions from substellar objects, cool stars, galaxies or active galactic nuclei. The very red optical-to-near-infrared colour covers a wide range (R - K ≥ 5-7 or I - K ≥ 4-6). This allows for a variety interpretations regarding their nature. The first detections (R-K≥5) were classified as large galaxies at z> 6 undergoing a rapid luminous phase of star formation [2]. Optical spectroscopy proofed the same objects were elliptical galaxies at z∼0.8 dominated by an old stellar population [3]. Determining the nature of EROs and their abundance will help explain open issues regarding the number density evolution of elliptical galaxies and structure formation on large scales. Monolithic collapse models predict pure luminosity evolution (PLE) ([6], [9]) of early-type galaxies from the period of formation until the present-day, while hierarchical formation models predict a significant decline in the comoving density of ellipticals with z, as they should form through merging at z ≤2 [8],[11],[12]. Observational evidence has been found for both scenarios.

2 Data Set

The R-band observations were made by the *High-z Supernova Search Cosmology Project* ([10] with the Cerro Tololo Inter-American Observatory (CTIO) 4.0 meter telescope during observing runs in February 1996. The J-band data were taken during observing runs in October 1997, 1998 and May 1999 with the 3.5 m telescope on Calar Alto [7]. The J-band field of view covers about 6.8 x 6.8 arcmin. The final image is constructed from a mosaic of 4 such fields.

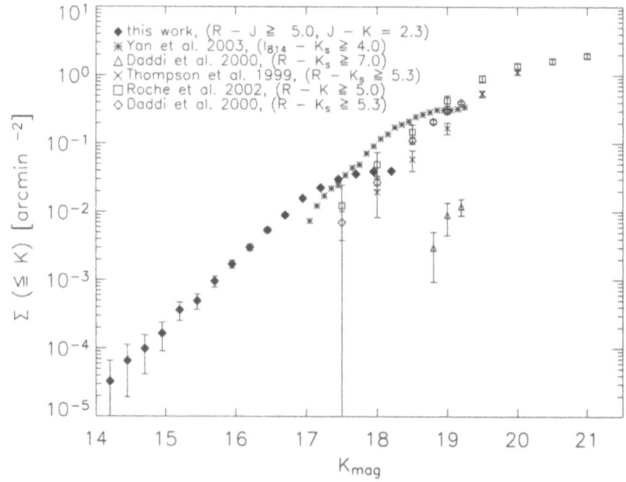

Fig. 1. Surface density of EROs compared with other surveys.

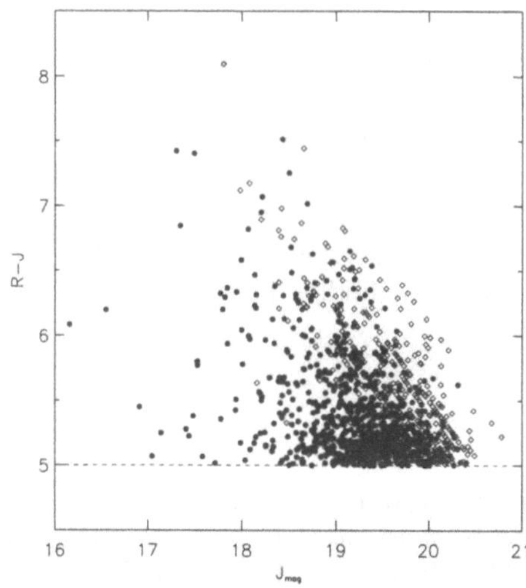

Fig. 2. Colour-magnitude diagram of all EROs. Objects with R and J-band detection (dots) and EROs without optical counterpart (diamonds) are shown.

Adding up the area of all 146 fields results in a total survey area of ~8.43 degree^2.

As criteria for classification as a galaxy we used the star/galaxy classifier from SExtractor, based on numerous isophotal areas of each object.

3 Summary

In an area of 8.43 degree^2 we found 1204 EROs with R-J$\geq$5^{mag} and J$\leq$20..21^{mag}, yielding a surface density of 0.0396 $\pm$ 0.0011 arcmin^{-2} (Fig.1). As seen in the colour-magnitude diagram (Fig.2), there is a large fraction, $\sim$ 35 percent, of EROs have no optical counterpart. Excluding these from our sample, the number of EROs decreases to 784, dropping the surface density to 0.026 $\pm$ 0.001 arcmin^{-2}. Still, this is by far the largest sample of EROs obtained to date. This surface density is comparable to that of bright (B$\leq$16.5) quasars with a ultraviolet excess [5] and high redshift quasars in general [13], and constitute a significant population.

Figure 1 compares our result to other ERO surveys, using a colour relation of J-K = 2.3. The only comparable survey by Daddi et al. [1] selecting objects with R-K$\geq$7, found 5 of objects within the completeness limit of their deep survey (K$\leq$19.2, 445 arcmin^2). The resulting surface density of 0.0011 arcmin^{-2} is smaller than our result by a factor of 2.4 or 3.5 respectively, depending on whether or not we include EROs with no R-band detection. This difference can be attributed both to the small number statistics of Daddi's result and field to field variations caused by cosmic variance.

The wide scatter in surface density around the average value of 0.0396 arcmin^{-2} for all 146 fields, includes also the result of Daddi for galaxies with R-K$\leq$7 [1]. This is supported also by the very inhomogeneous spatial distribution of our ERO candidates. Such overdense regions are indicators for clusters at higher redshift, found also for EROs with less strict colour thresholds (R-K$\geq$5) [1]. Proposed deep multi-colour observations with the VLT will help to confirm their cluster nature by finding fainter members and an estimate of their photometric redshift.

References

1. Daddi, E., Cimatti, A., Pozzetti, L.. et al., 2000, A&Ap, 361, 535-549
2. Elston, R., Rieke, G.H., & Rieke, M.J., 1988, ApJ, 331, L77-80
3. Elston, R., Rieke, M.J., & Rieke, G.H., 1989, ApJ, 341, 80-88
4. Fioc, M. & Rocca-Volmerange, B., 1997, A&A, 326, 950-962
5. Goldschmidt, P., Miller, L., La Franca, F. & Cristiani, S., 1992, MNRAS, 256, 65P-68P, *Short Communication*
6. He, P. & Zhang, Y.-Z., 1998, MNRAS, 298, 483-496
7. Herbst, T.M., Thompson, D., Frockenbrock, R., Rix, H.-W. & Beckwith, S.V.W, 1999, ApJ, 526, L17-L20
8. Kauffmann, G., White, S.D.M. & Guiderdoni, 1993, MNRAS, 264, 201-218
9. Pozzetti, L., Bruzual, G.A. & Zamorani ,G., 1996, MNRAS, 281, 953-969
10. Schmidt, B.P., Suntzeff, N.B. & Phillips, M.M. et al., 1998, ApJ, 507, 46-63
11. Somerville, R.S., Primack, J.R. & Faber, S.M., 2001, MNRAS, 320, 504-528
12. White, S.D.M. & Rees, M.J., 1978, MNRAS, 183, 341-358
13. Wolf, C. Meisenheimer, K. & Röser, H.-J. et al., 1999, A&Ap, 343, 399-406

The TOOT Survey: Typical Radio Sources at High Redshift

Gary J. Hill[1] and Steve Rawlings[2]

[1] McDonald Observatory, University of Texas at Austin, 1 University Station C1402, TX 78712, USA
[2] Astrophysics, Oxford University, Keble Road, Oxford OX1 3RH, UK

1 Introduction and Motivation for TOOT

The TexOx 1000 (TOOT) radio source redshift survey [1] is designed to find and study *typical* radio-loud active galaxies to high redshift. They are typical in the same sense that L* galaxies are typical of galaxies in the optical. Previous surveys have only included the most luminous, rare objects at and beyond the peak of activity at z~2, but in going a factor of 100 fainter than the 3C survey, and in assembling a large sample, TOOT probes for the first time the objects that dominate the radio luminosity density of the universe at high redshift.

The TOOT survey has made use of UK and Texas facilities to identify and measure the redshifts of a large sample drawn from the low-frequency 7C survey with $S_{151MHz}>0.1$ Jy. It is already the largest radio source (RS) redshift survey in existence that extends to high redshift and is unbiased by optical selection criteria. Previous low frequency surveys (3CRR [2], 6CE [3], and 7CRS [4]) have a total of 356 RS, but still suffer from small number statistics and the fact that even the 7CRS does not reach the radio luminosity function (RLF) break at z>1. As a result, there is still controversy about the form of the RLF evolution at high redshift, specifically whether there is a cut-off beyond z=2 [5–7]. The TOOT survey on the other hand probes low-enough L_{151} at high redshift, and is sufficiently large, to resolve the form of the radio source population (RSP) evolution. Even at high redshift, TOOT contains objects with L_{151} typical of 3CR galaxies at z~0.3, allowing direct comparison of properties as a function of redshift without the usual problems of the redshift-luminosity degeneracy. In particular, TOOT contains the radio equivalent of L* galaxies at the RLF peak at z ~ 2, which dominate the radio luminosity density of the universe.

The survey covers six 36-sq. deg. patches at 00, 02, 08, 12, 16, 18 hours RA, dubbed TOOT00, TOOT02, etc. Reference [1] provides a full description. So far, ~540 of the 1000 radio sources have redshifts with ~460 of those in well-defined, complete, sub-regions of the survey (TOOT00, TOOT08, TOOT12). The TOOT survey is the realisation of a survey proposed by Willott *et al.* [7], and it has several key aims:

- Probe the existence of a redshift cutoff for the high-luminosity sources at z>2. We will be able to finally get away from small number statistics and measure the magnitude of any high-z space density decline.

- Establish whether the low-luminosity population evolves at all. At present we cannot discriminate reliably between constant space density and a rise by an order-of-magnitude: TOOT will establish the evolution to ± 0.1 dex.
- Probe large scale structure. Steep spectrum RS are ideal sparse tracers of LSS, having high bias and being easy to observe over large volumes [8,9].

The preliminary redshift distribution [1] has a median redshift of $z \sim 1$. The evolution of the low-luminosity population is about a factor of 10 out to this redshift, and there is not a sharp cut off in the high redshift distribution of the high-luminosity sources.

2 Typical Radio Sources at $z \sim 3$

In order to illustrate the power of TOOT to investigate radio galaxy properties, we consider the properties of $z \sim 3$ TOOT RS [10] in comparison with 3C RS of similar power at $z \sim 0.5$. The TOOT and 7CRS RS in this redshift range are all reddened quasars with properties similar to low redshift broad-line radio galaxies. They also show significant sub-mm flux from SCUBA observations [11] indicating star formation rates of several 100 $M_{\odot}$/yr. Hence, the typical RS at $z \sim 3$ is dusty and undergoing significant star formation. In contrast, the typical RS of similar luminosity at $z \sim 0.5$ is hosted by what is normally considered a typical radio galaxy: an evolved elliptical galaxy with little star formation or dust The fact that TOOT probes the population that is most representative of RS in general, over the history of the Universe, shows that the radio galaxies studied at low redshift are in fact quite atypical, and are certainly rare objects when compared to the RSP as a whole. We speculate that high redshift RS are associated with major mergers of gas-rich systems, probably in dense environments that will evolve into rich clusters.

References

1. G.J. Hill & S. Rawlings, New Ast. Rev. **47**, 373 (2003)
2. R.A. Laing, J.M. Riley, M.S. Longair, *MNRAS* **204**, 151 (1983)
3. S. Rawlings, S.A. Eales, M. Lacy, *MNRAS* **322**, 523 (2001)
4. C.J. Willott, S. Rawlings, K.M. Blundell, M. Lacy, G.J. Hill, S.E. Scott, *MNRAS* **335**, 1120 (2002)
5. J.S. Dunlop, J.A. Peacock, *MNRAS* **247**, 19 (1990)
6. M.J. Jarvis, S. Rawlings, C.J. Willott, K.M. Blundell, S. Eales, M. Lacy, *MNRAS* **327**, 907 (2001)
7. C.J. Willott, S. Rawlings, K.M. Blundell, M. Lacy, S.A. Eales, *MNRAS* **322**, 536 (2001)
8. K. Brand, S. Rawlings, G.J. Hill, M. Lacy, E. Mitchell, J. Tufts *MNRAS* **344**, 283 (2003)
9. G.J. Hill, *New Era in Cosmology*, T. Shanks, N. Metcalf, Eds.,
10. S. Rawlings, G.J. Hill, C.J. Willott, *AJ* (2004), in preparation
11. S. Rawlings, C.J. Willott, G.J. Hill, Archibald, E.N., Dunlop, J.S., Hughes, D.H., *MNRAS* (2004), submitted

Deep Submillimetre Imaging of Distant AGN: Visualising the Formation of Cluster Ellipticals

Rob Ivison

Astronomy Technology Centre, Royal Observatory, Edinburgh EH9 3HJ, UK

Abstract. We[1] report deep submm mapping of 7 high-redshift radio galaxies (HzRGs). These data confirm not only the presence of spatially extended massive star-formation activity in the radio galaxies themselves, but also in companion objects previously undetected at any wavelength. The prevalence, orientation, and inferred masses of these companion galaxies suggest that we are witnessing the synchronous formation of the most luminous elliptical galaxies found today at the centres of rich galaxy clusters.

The most massive galaxies in the present-day Universe lie in the centres of rich clusters. They have old, coeval stellar populations suggesting that the bulk of their stars must have formed at early epochs in spectacular starbursts. Popular models of galaxy formation predict that these galaxies form in proto-clusters at high-density peaks in the early Universe, signposted by massive HzRGs.

Existing submm studies of HzRGs have shown that their star-formation rates are large enough to build a massive galaxy in < 1 Gyr, but provide no information on the spatial extent of this emission or on the star-formation activity in their environments. We have mapped 7 HzRGs with SCUBA, illustrating graphically the distribution of dust-reradiated emission in and around the HzRGs on scales from 5–160″, or 30–1000 kpc [1,2] (Fig. 1).

The dust emission from the central HzRG is resolved in at least 5 of the 7 sources – even with SCUBA's beam. This emission is sometimes in the form of semi-merged clumps (e.g. 8C 1909+722), sometimes in an smoother distribution (e.g. 4C 60.07), and is usually more extended than the radio emission.

The extent of the dust emission, 50–250 kpc, is equivalent to both the half-light radii (30–150 kpc) of brightest cluster galaxies in the local Universe or the predicted separation at the epoch when star formation peaks in major mergers. Higher resolution observations will discriminate between these possibilities.

Several fields also contain serendipitous detections of new, luminous submm galaxies. Several are, like the HzRGs, resolved at 850 μm. The source density is higher than found in blank fields by $\times 2$. Two fields each contain a >20- mJy companion. The largest blank-field survey contains no >15-mJy sources.

A surprising feature of the maps is the observed alignments between the HzRG radio axis and (1) the submm emission; (2) the brightest submm 'companions'.

[1] Rob Ivison, Jason Stevens, James Dunlop, Ian Smail, Wil Percival, David Hughes, Huub Röttgering, Wil van Breugel and Michel Reuland

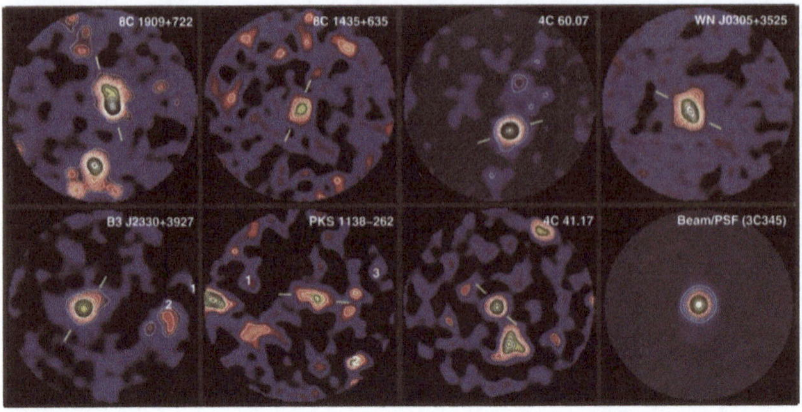

Fig. 1. Submm continuum emission from dust in and around 7 HzRGs.

We propose that the brightest companions trace the large-scale structure around the HzRG and that these directions thus contain the densest cross-sections of gas. By selecting the brightest radio sources at this epoch we will then have (inadvertently, but inevitably) selected sources which have produced jets aligned with the densest regions of gas, producing the brightest hot spots.

Analysis of the possible alignment effects indicates that the submm emission from the radio galaxies themselves, and that of the brightest companions, is tracing the large-scale structure around radio galaxies at $z \simeq 2$. Indeed, some companions have been shown to be at the same redshift as the radio galaxy [3].

What is the typical mass of these galaxies? Using standard assumptions we can convert the dust masses into total gas masses, and hence estimate the masses of their associated dark matter (DM) haloes to be in excess of $10^{12}\,M_\odot$.

Can we reconcile this result with hierarchical models of galaxy and structure formation? Simulations of the collapse of ΛCDM haloes do appear to be in accord with the observations, i.e. both data and theory suggest that the submm companions revealed by SCUBA are associated with $>10^{12}\,M_\odot$ DM halos.

The average number of companions, the distribution of companion incidence, and the inferred baryonic gas masses of the companions are all most consistent with the interpretation that the SCUBA sources uncovered in this study are the progenitors of massive present-day cluster ellipticals.

Since elliptical galaxies are known to contain massive black holes with mass proportional to that of the spheroid, it is reasonable to assume that the black hole and stellar mass grow coevally from the same gas reservoir. This in turn suggests that the companion objects should contain buried AGN. First results [4] show a high rate of correspondence between the submm companions and luminous X-ray sources, suggesting that this is indeed the case.

References

1. R. J. Ivison et al.: ApJ, **542**, 27 (2000)
2. J. A. Stevens et al.: Nature, **425**, 264 (2003)
3. I. Smail et al.: ApJ, **583**, 551 (2003)
4. I. Smail et al.: ApJ, astro-ph/0307560 (2003)

Quasar Host Galaxies of GEMS, First Results: 0.5<z<2.75

K. Jahnke[1], S.F. Sánchez[1], L. Wisotzki[1,2], M. Barden[3], S.V.W. Beckwith[4], E.F. Bell[3], A. Borch[3], J.A.R. Caldwell[4], B. Häußler[3], S. Jogee[4], D.H. McIntosh[5], K. Meisenheimer[3], C.Y. Peng[6], H.-W. Rix[3], R.S. Somerville[4], and C. Wolf[7]

[1] Astrophysikalisches Institut Potsdam, Germany
[2] Universität Potsdam, Germany
[3] Max-Planck-Institut für Astronomie, Germany
[4] Space Telescope Science Institute, USA
[5] University of Massachusetts, USA
[6] Steward Observatory, University of Arizona, USA
[7] University of Oxford, UK

Understanding the host galaxies (HGs) of quasars is the key to linking the evolution of galaxies and their central black holes. Within the GEMS project we are studying quasar HGs and their morphologies, luminosities, ages and environment up to redshift $z \sim 3$. GEMS – Galaxy Evolution from Morphologies and SEDs – is a large, two-filter (F606W, F850LP) imaging survey with the *HST* ACS, centered on the Chandra Deep Field South. For ~8 000 galaxies and ~130 AGN photometric redshifts are available in this field from the COMBO-17 project (Wolf et al., 2003, A&A, 401, 73). Full GEMS overview by Rix et al. (these proceedings). For the initial study we defined two quasar samples with $0.5 < z < 1.1$ and $1.8 < z < 2.75$ ('low' and 'high', 16 and 22 quasars, respectively), complete down to a flux limit of $R = 24$ and representing intermediately luminous AGN.

Host galaxy extraction: We estimate flux, colours, and morphologies of the underlying HGs with two-dimensional modelling using GALFIT (Peng et al., 2002, AJ, 124, 266), testing the modelling and determine error bars with extensive

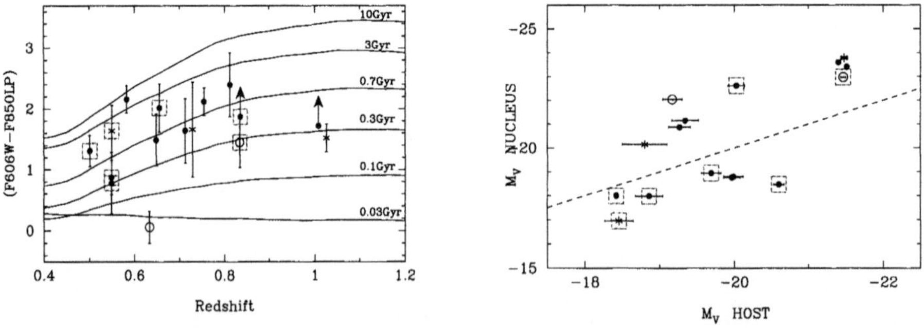

Fig. 1. Symbols mark the morphology of the 'low'-z HGs: 2/3 are spheroid-dominated (filled circles), 1/4 disk dominated (stars). The remaining (open circles) have uncertain morphologies. Squares mark interacting objects (45%), judged from visual inspection. In the right plots a line of equality is shown to guide the eye.

simulations of ~4000 artificial HGs. From these we can confidently detect HGs that have a flux > 5% of the total flux.

'Low'-z results: In the low redshift quasars 100% (88%) of the HGs are detected in the z (V) band, comprising 60% (30%) of the total flux. In Fig. 1 the resulting colours are shown. The average half-light radius of the HGs is $r_{1/2} = 1.9$ kpc, with an rms spread of 0.8 kpc. Overplotted are single age star formation burst models by Bruzual & Charlot (2003, MNRAS, 344, 1000; solar metallicity). The colors suggest young luminosity-weighted ages, indicating significant recent star formation despite the fact that 2/3 of the HGs show spheroidal morphologies. Converting the observed z band to absolute rest frame V band magnitudes, we recover the (cor)relation between nuclear and HG luminosity (Fig. 1) as described e.g. by McLeod & Rieke (1995, ApJ, 441, 96), and confirmed also at $z \sim 1$ (Sánchez & González-Serrano, 2003, A&A, 406, 435).

High-z results: Despite the well-known difficulties in extracting faint hosts, 8 of 22 HGs are at the current stage resolvable in both bands. To allow a statement on the 'unresolved' quasar hosts, we coadded the quasar images and successfully detected the 'mean' HG in the much higher S/N coadded V band image. The hosts make up ~40% (~25%) of the total flux in the V (z) band, albeit with a large spread, and ~6% in the 'average' quasar, respectively. A morphological classification of the HGs is possible for only one object, which is a spheroidal.

Resulting colours and derived star formation rates are shown in Fig. 2. We find a luminosity-weighted age range very similar to the lower redshift sample. In addition, the dust-uncorrected UV-derived star formation rates are $\gtrsim 10\ M_\odot/yr$ (error bars are photometric errors only).

Conclusions Our results are consistent with multicolour and spectral data showing strong signs of recent star formation for both moderately luminous quasar HGs (Jahnke, Kuhlbrodt, Wisotzki, subm. to MNRAS, astro-ph/0311123; Jahnke & Wisotzki, 2003, MNRAS in press, astro-ph/0308177) and less luminous Type 2 AGN hosts (Kauffmann et al., these proceedings) at low redshift. Taken together, this strongly supports a scenario in which star formation in the host and nuclear fuelling have a common cause, e.g. tidal interactions.

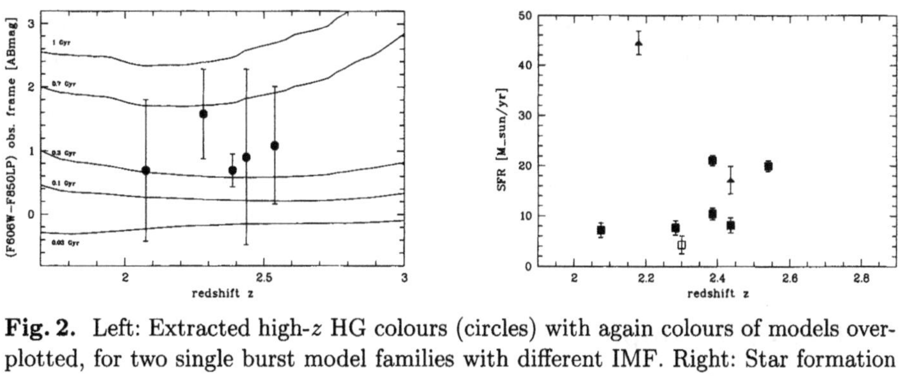

Fig. 2. Left: Extracted high-z HG colours (circles) with again colours of models overplotted, for two single burst model families with different IMF. Right: Star formation rates derived from rest-frame HG UV fluxes following Kennicutt (1998, ARAA, 36, 189). Squares are from V band. If not available, z band (triangles) is used.

A Sample of Radio Galaxies Spanning Three Decades in Radio Luminosity – an Overview

Matt J. Jarvis[1], Ross J. McLure[2], Chris J. Willott[3], Steve Rawlings[1], Gary J. Hill[4], Ewan Mitchell[1], and James Dunlop[2]

[1] Astrophysics, Department of Physics, Keble Road, Oxford, OX1 3RH, UK
[2] Institute for Astronomy, Royal Observatory, Edinburgh EH9 3HJ.
[3] National Research Council, 5071 West Saanich Rd., Victoria, B.C., V9E 2E7, Canada
[4] McDonald Observatory, University of Texas at Austin, 1 University Station C1402, Austin, TX 78712-0259, USA

Abstract. In this contribution we highlight a large programme that is being undertaken to establish the physical links between black-hole mass, host galaxy properties and cluster environments of a sample of radio galaxies spanning ~ 3 dex in radio luminosity at a single cosmic epoch. The sample of 41 radio galaxies is drawn from four complete radio samples and is comprised of all narrow-line radio galaxies in the redshift range $0.4 < z < 0.6$. This allows us to delineate the effects of cosmic evolution from genuine radio-power dependent characteristics.

1 Introduction

In the last few years the link between galaxy bulge masses, their stellar velocity dispersions and their central black-hole masses has been established by a number of authors (e.g. Magorrian et al. 1998; Gebhardt et al. 2000; Ferrarese & Merritt 2000). This provides strong evidence that the formation and subsequent evolution of galaxies at high redshift may be strongly dependent on the accretion activity of the central supermassive black hole. One possibility is that outflows from the central active nucleus may limit the mass of the black hole (e.g. Silk & Rees 1998). Black-hole driven feedback activity may also hold the key in preventing the overcooling of baryons in semi-analytic models of galaxy formation, a key unsolved problem in theoretical models (Balogh et al. 2001).

One way in which to investigate the effects of black-hole mass, host galaxy properties and environments on AGN activity is to carry out a study which delineates cosmic evolution from the intrinsic luminosity of the AGN. We are now in a position to achieve this with four complete samples of radio galaxies at progressively lower flux-density limits.

2 The ZP5 Sample

Our subsample of galaxies is drawn from the complete low-frequency selected 3CRR (Laing, Riley & Longair 1983), 6CE (Rawlings et al. 2001); 7CRS (Willott et al. 2002) and TOOT (Hill & Rawlings 2003) radio samples. The combination of these sample allows us to sample a factor of ~ 1000 radio luminosity over a small range in redshift, i.e. $0.4 < z < 0.6$. We select all narrow-line radio galaxies within this range, and exclude broad-line quasars, to allow us to make detailed investigations of the host galaxy properties, without the presence of a bright quasar nucleus.

3 Host Galaxy Properties

We obtained 44 orbits of Cycle 10 HST-WFPC2 time in order to study the host galaxies of our sample. The full results of this analysis will be presented in McLure et al. (2004), but to summarize. All of the galaxies are well fit by an elliptical $r^{1/4}$ law. They all lie on the Kormendy relation between surface brightness and scalelength defined by low-redshift radio galaxies, with an offset consistent with purely passive evolution of the stellar population. There is evidence for a correlation between radio luminosity and host galaxy luminosity.

4 Stellar Velocity Dispersion and the Fundamental Plane

We have also carried out spectroscopic observations with ISIS on the WHT in order to measure the stellar velocity dispersions of our sample. This will allow us to place the radio galaxies on the fundamental plane of ellipticals (Willott et al. in prep.).

5 Cluster Environments

The third part of the investigation utilizes three-band photometry to quantify the richness of the environments surrounding the radio galaxies and will lead to firm statements regarding the rôle that environment plays in determining the radio power, morphology (e.g. FRI or FRII) and linear size (Jarvis et al. in prep.). We are also undertaking a weak lensing analysis of a subset of our sample to allow us to estimate the actual mass in the clusters and to map out the dark matter structure (Wold et al. in prep.).

References

1. M.L. Balogh, F.R. Pearce, R.G. Bower, S.T. Kay, MNRAS, **326**, 1228 (2001)
2. G.J. Hill, S. Rawlings, NewAR, **47**, 373 (2003)
3. R.A. Laing, J.M. Riley, M.S. Longair, MNRAS, **204**, 151 (1983)
4. R.J. McLure, C.J. Willott, M.J. Jarvis, S. Rawlings, E. Mitchell, G.J. Hill, J.S. Dunlop, M. Wold, MNRAS, submitted, (2004)
5. S. Rawlings, S.A. Eales, M. Lacy, MNRAS, **322**, 523 (2001)
6. J. Silk & M.J. Rees, A&A, **331**, L4 (1988)
7. C.J. Willott, S. Rawlings, K.M. Blundell, M. Lacy, G.J. Hill, S.E. Scott, MNRAS, **335**, 112 (2002)

Probing the Redshift Desert with GOODS

C. Kretchmer[1], M. Dickinson[2], L. Moustakas[2], B. Mobasher[2], T. Dahlen[2], and GOODS Team

[1] Johns Hopkins University, Baltimore MD 21211, USA
[2] Space Telescope Science Institute, Baltimore MD 21211, USA

1 Introduction

A crucial step in understanding the evolution and development of present day galaxies is to study the redshift desert ($1 < z_{phot} < 2$) and map how the population has changed from the Lyman Break/SCUBA era to the epoch dominated by the Hubble sequence. Using deep ground-based near-infrared data (SOFI, ISAAAC) and ACS images of the GOODS/CDF-S region, we select a sample of luminous, high redshift galaxies to study their spectral and morphological evolution towards the Hubble sequence. To see a more complete picture of galaxies that dominate the stellar mass density in the high-z universe, we use an approximately volume-limited K_s selected sample with $1 < z_{phot} < 2$ and $M_V < -21.5$, which should be largely (if not entirely) complete out to $z_{phot} = 2$. By restricting the sample to the most optically luminous objects, we aim to trace the evolution of the most massive galaxies which are the likely precursors to present-day ellipticals and spirals. For morphological analysis, we restrict the sample selection to the GOODS area where we have deep, high resolution ACS imaging in the $B_{435}, V_{606}, i_{775}, z_{850}$ filters. The ACS z_{850} filter provides the rest-frame optical morphology at $z_{phot} < 1.2$ and UV rest-frame morphology at $1.2 < z_{phot} < 2$. Deep ISAAC imaging (0.4-0.5 arcsec FWHM) also gives some rest-frame optical morphological information for galaxies at $z_{phot} > 1.2$, when searching for bulges in large disk galaxies. Currently, ISAAC imaging covers $\sim$ 30% of the GOODS/CDF-S area, and is being extended to encompass the whole field.

The SOFI K_s sample ($\sim 360 arcmin^2$) consists of 735 sources, *while the ACS sample ($\sim 182 arcmin^2$) is limited to* 375 sources.

2 Morphology and EROs

We derive spectral type classifications by SED template fitting as a byproduct of the photometric redshift estimation. The spectral type distribution at $1 < z_{phot} < 1.5$ is skewed towards early type galaxies where 57 % of the sample is classified as E to Sbc. This substantial fraction have older stellar populations which dominate the rest-frame optical light. At higher redshift ($1.5 < z_{phot} < 2.0$) the distribution is bluer with 41% of the sample classified as E to Sbc. At $z_{phot} > 2$ most optically bright galaxies (the LBG population) have very blue, starburst-like colors. Only a small portion of the most luminous K_s selected galaxies at $1 < z_{phot} < 2$ have extremely blue SEDs characteristic of local starburst galaxies or $z_{phot} > 2$ Lyman Break galaxies. UV-bright starburst are certainly present at these redshifts, but do not contribute substantially to the bright end of the rest-frame optical luminosity function. Our $1 < z_{phot} < 2$ sample suggests a decline in the the specific star formation rates among the most luminous galaxies, and the build-up of older stellar populations. However, even out to $z_{phot} = 2$, a substantial fraction of the most luminous galaxies have redder SEDs like those of present-day spirals and ellipticals. These redder galaxies may be missed in samples selected by UV continuum, and will be extremely challenging targets for spectroscopy.

EROs, an important class of objects at $z_{phot} > 1$, are defined by their optical-near-infrared colors and are composed of two possible populations: actively star forming galaxies with large amounts of reddening or passive, old systems. Their relative contributions are important to unfolding galaxy evolution. 12 % of our sample meets the stringent color criterion of $R - K_s > 4.35$, most likely starburst or early type galaxies. 40 % of the bright galaxy population at $1 < z_{phot} < 2$ meets the less stringent color criterion of $R - K_s > 3.35$ These EROs are not so "extreme" after all, but simply the redder component of the bright galaxy population in our redshift range. Moustakas et al. (2003) find that 75% of the $R-K_s > 3.35$ EROs have relatively "ordinary" morphologies resembling elliptical or disk galaxies, with only 25 % showing irregular or disturbed morphologies. Most of the the $R - K_s > 3.35$ EROs are evidently luminous elliptical and disk galaxies with mature stellar populations and morphologies, without rapid star formation. They comprise a significant fraction of the $M_V < -21.5$ population at $1 < z_{phot} < 2$.

3 Conclusion

The galaxies in our $1 < z_{phot} < 2$, $M_V < -21.5$ sample span a wide range of morphology, and include some large galaxies with sizes of order ~ 42.3 kpc. We find bulge-dominated systems, galaxies with clear spiral structure and large disks, large LSB systems which may be relatively quiescent disks dimmed by redshift and k-correction, likely examples of merging and interacting galaxies, and peculiar/irregular objects.

The fraction of galaxies with blue spectral types (Scd and later), indicating star formation, increases at higher redshift. However, 57 % and 41 % of the sample at $1 < z_{phot} < 1.5$ and $1.5 < z_{phot} < 2$, respectively, have SEDs like those of present-day early-type galaxies through mid-type (Sbc) spirals. Only a small minority of the most optically luminous galaxies at $1 < z_{phot} < 2$ have spectral types as blue as Im or starburst galaxies.

40 % of galaxies with $M_V < -21.5$ and $1 < z_{phot} < 2$ meet the ERO criterion $R - K_s > 3.35$. These "moderate" EROs are therefore not rare and extreme, but quite common among the most luminous galaxies in this redshift range. Most (80 %) have relatively ordinary bulge- or disk-dominated morphology (Moustakas et al. 2003). 12 % of the sample meet the stricter $R - K_s > 4.35$ ERO definition.

A Search for Emission Line Galaxies at $z = 6.5$

Jaron Kurk[1], Andrea Cimatti[1], Sperello di Serego Alighieri[1], Joël Vernet[1], and Emanuele Daddi[2]

[1] INAF – Osservatorio Astrofisico di Arcetri, Largo E. Fermi 5, 50125, Firenze, Italy
[2] ESO, Karl-Schwarzschild-Str. 2, 85748, Garching bei München, Germany

Abstract. We are carrying out a search for Lyα emitting galaxies at $z = 6.5$ employing slitless spectroscopy at the VLT. In our 43 arcmin^2 field we find three isolated single emission lines with SEDs consistent with line emitting galaxies at $z = 6.5$.

1 Introduction

Since the discovery of $z \sim 3$ Lyman Break Galaxies (LBGs [1]), many efforts are being made to select the whole population of star-forming galaxies at the highest possible redshifts (see [2] for a review), and to constrain the star formation history of the universe (e.g. [3]). The detection of $z > 6$ galaxies allows to study the modes of early galaxy formation and the interplay between the first galaxies and the intergalactic medium (IGM). The redshift range $6 < z < 7$ is a very intriguing time during cosmic evolution, when hydrogen reionization is believed to be basically complete and the IGM starts to be polluted with metals [4,5].

Because a significant part of the bolometric luminosity of primeval star-forming objects escapes as Lyα emission, especially if dust is not yet ubiquitous, it seems profitable to search for the redshifted Lyα line of high redshift galaxies. The advent of ten meter class telescopes made this approach successful and efficient, and both narrow-band imaging and optical *serendipitous* spectroscopy found Lyα emitters at $3 < z < 6$ [2,6,7]. Thanks to the recently increased sensitivity of CCDs at $\lambda > 0.8\,\mu$m, the discovery of $z > 6$ galaxies has also become possible. Using narrow band filters sensitive to wavelengths in the range $9100 < \lambda < 9250$ Å, i.e. Lyα emission at $z \sim 6.5$ and subsequent spectroscopy, two groups have detected the first three galaxies known at $z > 6$. The first of these was found due to the strong lensing amplification (4.5) by the cluster Abell 370 [8]. The other two galaxies were found in a survey made with Subaru Suprime-Cam instrument which provides a huge field of view (814 arcmin^2 [9]).

2 Observations and Analysis

We have used a new approach to search for Lyα emitters at $z \sim 6.5$ based on slitless spectroscopy in combination with a narrow band (2.2%) filter and very deep imaging in a field with very low Galactic extinction (E_{B-V}=0.002). The z_{sp} filter employed has a central wavelength of 9135 Å and FWHM of 200 Å and is therefore sensitive to Lyα at $6.44 < z < 6.61$. The imaging was carried out through the z_{sp} filter (to identify the sources of emission lines detected in the spectroscopy) and the Bessel I filter (to determine the $I - z$ colour or spectral break over the Lyα line). The observations were carried out with the FORS2 instrument at the VLT during eight nights in October, November 2002 and February 2003. Total exposure times were 7.6, 6.33 and 4.0 hours for the slitless spectroscopy, z_{sp} and I band direct imaging, respectively. Observing conditions were excellent resulting in 0.7 arcsec seeing on the resulting combined images. The field was also imaged earlier in Bessel U and Gunn v band.

Three authors (JK, AC, SSA) have independently searched for emission lines on the spectroscopic frame, resulting in a combined list of 161 emission lines. Each emission line can originate from a counterpart on the direct image within a region of about $50'' \times 1''$. The emission lines are related to 345 possible counterparts in the z_{sp} image. Four of these have $I - z_{sp} > 1.5$, one of which is an [OIII] emitter at $z = 0.82$. This leaves three counterparts which may be Lyα emitters at $z \sim 6.5$. Furthermore, among the $\sim$ 4500 objects detected by SExtractor on the z_{sp} image, there are ten other objects with these colours, which are considered LBG candidates at $z > 6$.

3 Conclusions and Future Work

The number of possible Lyα emitters at $z = 6.5$ detected in our field is consistent with the number densities of $z > 5.5$ galaxies reported by other authors [2,6]. In particular, the recent Subaru survey by [9] resulted in the detection of 73 Lyα candidates at $z = 6.5$ (with $i' - z' > 1.3$) in an 814 arcmin^2 field. Based on this result, we expect to find in our available 43 arcmin^2 field three candidate Lyα emitters at $z = 6.5$.

Slitmask spectroscopic observations are needed to confirm the identity of the candidate Lyα emitters and the LBGs. These observations will provide higher signal to noise and higher spectral resolution to measure the characteristic asymmetric line profile of Lyα at high redshift. The larger wavelength coverage will allow the detection of emission lines from the presumed LBGs outside the z_{sp} filter. Time has been allocated to carry out these observations with FORS2.

References

1. Steidel C. et al. 1996, ApJ, 462, L17
2. Stern D. & Spinrad H. 1999, PASP, 111, 1475
3. Madau P. et al. 1996, MNRAS, 283, 1388
4. Gnedin N.Y. 2000, ApJ, 535, 530
5. Ciardi B. et al. 2003, MNRAS, 344, L7
6. Hu E.M. et al. 2000, in "The Hy-Redshift Universe: Galaxy Formation and Evolution at High Redshift", eds. A.J. Bunker and W.J.M. van Breugel, A.S.P. Conf. Series, astro-ph/9911477
7. Malhotra S. et al. 2001, ApJ, 561, 766
8. Hu E.M. et al. 2002, ApJ, 568, L75
9. Kodaira K. et al. 2003, PASJ, 55, L17

Clues on Galaxy and Cluster Formation from Their Scaling Relations

Barbara Lanzoni[1], Luca Ciotti[2], Alberto Cappi[1], Giuseppe Tormen[3], and Gianni Zamorani[1]

[1] INAF – Osservatorio Astronomico di Bologna, via Ranzani 1, 40127, Bologna, Italy
[2] Dip. di Astronomia, Università di Bologna, via Ranzani 1, 40127 Bologna, Italy
[3] Dip. di Astronomia, Università di Padova, vicolo dell'Osservatorio 5, 35122 Padova, Italy

Abstract. By means of high-resolution N-body simulations in a ΛCDM cosmology, we verify that scaling relations similar to those observed for nearby galaxy clusters are also defined by their dark matter hosts; the slopes, however, are not the same. We then show that the scaling relations of galaxy clusters can be explained as the result of the cosmological collapse of density fluctuations at the appropriate scales, plus a systematic trend of the M/L ratio with cluster mass. The empirical fact that the exponent of the Faber-Jackson relation of elliptical galaxies is significantly different (higher) than that of clusters, force us to conclude that the galaxy scaling laws might derive from the cosmological collapse of density fluctuations at the epoch when galactic scales became non-linear, plus modifications afterward due to early-time dissipative merging.

1 Observed and Simulated Scaling Relations

It has been shown [7,1] that nearby galaxy clusters define scaling relations involving their optical luminosity L, effective radius R_e and velocity dispersion σ. Reanalyzing the sample of [7], we obtain $L \propto \sigma^{2.18\pm0.52}$ for the Faber-Jackson (FJ) relation, $L \propto R_e^{1.55\pm0.19}$ for the Kormendy relation, and $L \propto R_e^{0.9\pm0.15}\sigma^{1.31\pm0.22}$ for a nearly edge-on view of the Fundamental Plane, where L is in $10^{12}L_\odot/h^2$, σ in 1000 km/s, R_e in Mpc/h.

Similar scaling laws are also followed by early-type galaxies, with the only significant difference of the FJ relation: in fact, $L \propto \sigma^4$ for luminous ellipticals [2].

For a sample of 13 massive ($M > 10^{14}M_\odot/h$) DM halos obtained from N-body cosmological simulations [4], we have computed the projected radius containing half the total mass (R_h) and the mean projected velocity dispersion within it (σ_h). Well defined scaling relations relating R_h, σ_h and the total mass M are found, and are interpreted as the result of the cosmological collapse of density fluctuations and the weak homology of DM halos [6]. The slopes of these relations, however, do not coincide with the observed ones: the simplest way to satisfy at once the theoretical and observational scaling laws is to assume that $M/L \propto L^{0.3}$, in good agreement with what inferred observationally [3].

Since a similar trend of M/L with L is also found for early-type galaxies, and the cosmological predictions for the DM halo scaling relations are the same

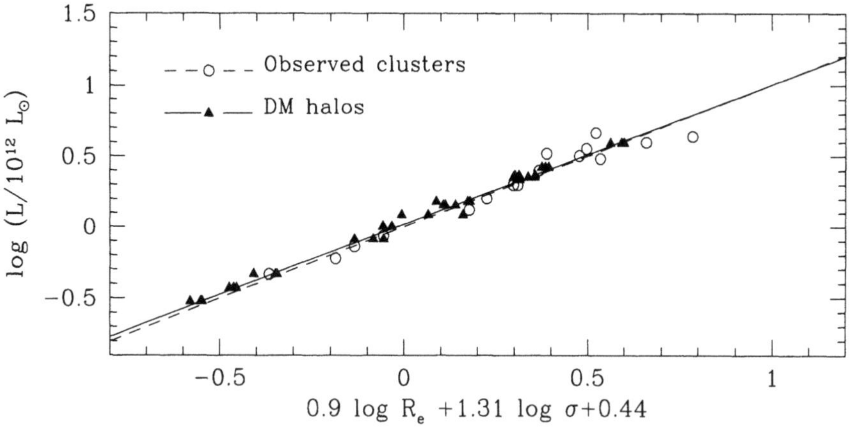

Fig. 1. Nearly edge-on view of the FP of galaxy clusters, and of their DM halos under the assumption $M/L \propto L^{0.3}$

at galactic and cluster scales, the higher slope of the galactic FJ relation implies that some additional mechanism played a role in the formation/evolution of ellipticals: indeed, the combined effect of dissipationless merging (which tends to increase the FJ exponent) and gas dissipation (which acts in the opposite direction) might be the solution [5].

2 Conclusions

We conclude that, while the scaling relations of galaxy clusters can be explained by the cosmological collapse of density fluctuations at the appropriate scale, plus a systematic trend the total M/L ratio with cluster luminosity, the scaling laws of elliptical galaxies seem to suggest that additional processes (like early-time dissipational merging) probably played an important role after the cosmological collapse of the density fluctuations at these scales. For a discussion of how M/L for clusters could increase with L, we refer to [5].

References

1. Adami C., Mazure A., Biviano A., Katgert P., Rhee G.: 1998a, A&A 331, 493
2. Bernardi M., et al.: 2003a, AJ 125, 1882
3. Girardi M., Manzato P., Mezzetti M., Giuricin G., Limboz F.: 2002, ApJ 569, 720
4. Lanzoni B., Cappi A., Ciotti L.: Mem. S.A.It. Suppl. Vol. 1, 145
5. Lanzoni B., Ciotti L., Cappi A., Tormen G., Zamorani G.: ApJ in press (astro-ph/0307141)
6. Navarro J.F., Frenk C.S., White S.D.M.: 1997, ApJ 490, 493 (NFW)
7. Schaeffer R., Maurogordato S., Cappi A., Bernardeau F.: 1993, MNRAS 263, L21

Tracing the Evolution of Massive Galaxies up to z ~ 3

M. Longhetti[1], P. Saracco[1], S. Cristiani[2], A. Fontana[3], E. Giallongo[3], M. Nonino[2], and E. Vanzella[4]

[1] INAF - Osservatorio Astronomico di Brera, Milano, Italy
[2] European Southern Observatory
[3] INAF - Osservatorio Astronomico di Roma, Italy
[4] INAF - Osservatorio Astronomico di Trieste, Italy

1 J-K > 3 Galaxies: Unveiling M> $10^{11}M_\odot$ Early-Type Galaxies at z ~ 3

Objects with unusual red near-IR color J-K>3 are extremely rare at magnitude brighter than K=20, while their surface density increases at fainter magnitudes. The nature of these sources has not yet been firmly established even if all the analysis performed so far conclude that they are galaxies at z > 2-3 ([1], [2], [3]). A detailed analysis based on a multi-band data set (from 0.3μm to 2.15μm) of three J-K>3 sources selected at Ks < 22 on the HDF-S ([4]) has shown that: i) the three galaxies are at redshift 2.5<z<3; ii) they have a stellar mass content of $M_{star} \sim 10^{11}M_\odot$; iii) they would populate the bright end ($L_{z=0} \sim L^*$) of the local luminosity function of galaxies even assuming they evolve passively; iv) they cannot follow an evolution significantly different from passive aging.

2 Depicting the Evolution of Massive Galaxies from z=3 to z=0

Thus, these J-K>3 selected are the z~3 counterpart of local massive early-type galaxies. Between z=3 and z=0 we expect to find the same population of galaxies at a different evolutionary status, namely with older stellar content. For instance, at z=1-1.5 we expect to find early-type galaxies ($t_{z=3}$ - $t_{z=1.5}$) ≈ 2 Gyr older. Such old early-type galaxies have been found among the EROs selected with R-K¿5.3 (e.g.[5]). In order to find their counterpart at the intermediate redshift z~2, we have identified a suitable color selection criterion by exploring the the expected colors evolution as derived by synthetic models ([6]). The results are displayed in Fig. 1: for short SF time scale ($\tau <$ 1Gyr galaxies at z~2 have colors J-K<3 and I-H>4. In fact, adopting these selection criteria on the HDF-S data, we identified 4 massive evolved galaxies at 1.5< z <2.5 which seem to be the link between the J-K>3 galaxies (z~ 3) and the EROs (z~1). An example of this passive evolution connection is proposed in the right panel of Fig. 1: 822 (upper panel) a J-K>3 galaxy selected in the HDF-S well described by a young SSP at z~2.9; 1203 (middle panel) one of the four massive ($M_{star} \geq 10^{11}$ $M_\odot$) galaxies selected in the HDF-S on the basis of the selection criteria above described, well

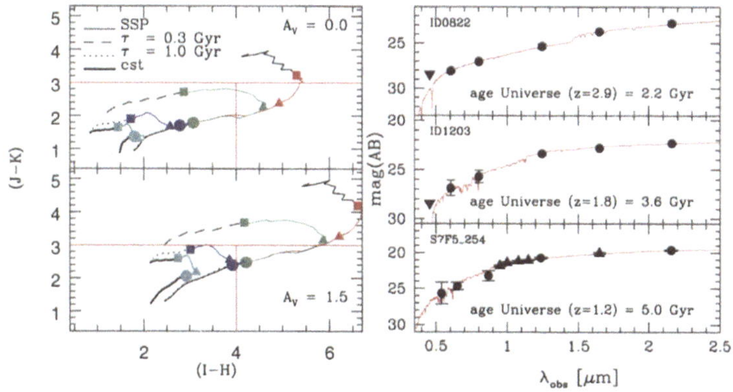

Fig. 1. *left panel*: each line represents the evolution with redshift of the J-K and I-H colors for different SFHs between z=1 (filled circles), z=2 (filled triangles) and z=3 (filled squares)(solar metallicity, Salpeter IMF, formation redshift z_f=6); *right panel*: possible passive evolution scenario (details in the text)

described by a 1.4 Gyr old SSP at z∼1.8; S7F5_254 (lower panel) the spectrum of a massive early-type galaxy at z=1.2 ([5]) superimposed on the template of a 3 Gyr old SSP.

The photometry, the stellar masses and the mean ages of the stellar populations of these three massive galaxies at different redshift are apparently those we expect from a massive galaxy fully assembled at z∼3 which evolves passively in time. This result is based, in fact, on a couple of objects and may be simply a coincidence. However, it is worthy to note that the number density of the three classes of objects represented in the figure and that of the local massive galaxies are well compatible with the depicted scenario. Actually, the number density of J-K>3 and K<22 galaxies (2.5<z<3, $M_{star} \geq 10^{11}$ $M_\odot$ galaxies) is $\rho = 1.2 \times 10^{-4}$ Mpc^{-3} ([4]) and that of J-K<3 and I-H>4 (1.5<z<2.5) calculated from the HDF-S data results to be the same within the errors. The number density of the early-type galaxies at 1.0<z<1.5 (R-K>5) is consistent with the previous value once the same stellar mass threshold ($M_{star} \geq 10^{11}$ $M_\odot$) is taken into account ([4]). The local density of $M_{star} \geq 10^{11}$ $M_\odot$ elliptical galaxies is $\rho \simeq 3 \times 10^{-4}$ Mpc^{-3} ([7]): at least 40% of the local population of bright massive early-type galaxies can be explained as the results of a simple passive evolution started from galaxies already assembled at z∼3.

References

1. M. Dickinson, C. Hanley, R. Elston, et al. 2000, ApJ, 531, 624
2. P.B. Hall, M. Sawicki, P. Martini, et al. 2001, ApJ, 121, 1840
3. T. Totani, Y. Yoshii, F. Iwamuro, et al. 2001, ApJ, 558, L87
4. P. Saracco, M. Longhetti, E. Giallongo, et al. 2003, astro-ph/0310131
5. P. Saracco, M. Longhetti, P. Severgnini, et al. 2003, A&A, 398, 127
6. M. Longhetti, P. Saracco, E. Giallongo, et al. 2004, in preparation
7. R.O. Marzke, L.N. Da Costa, P.S. Pellegrini, et al. 1998, ApJ, 503, 617

Morphological CAS Parameters of a Sample of Very Luminous Infrared Galaxies (VLIRGs) – Preliminary Results

Ray A. Lucas[1], Chris Conselice[2], Santiago Arribas[1], Howard Bushouse[1], Kirk D. Borne[3], and Luis Colina[4]

[1] Space Telescope Science Institute, 3700 San Martin Dr., Baltimore, MD 21218, USA
[2] Caltech, Department of Astronomy, 1201 E. California Blvd., MC 105-24 Pasadena, CA 91125, USA
[3] George Mason University, Department of Physics, 4400 University Dr., Fairfax, VA 22030, USA
[4] IEM/CSIC, Serrano 121 Madrid 28006, Spain

Abstract. We have observed a sample of 19 Very Luminous Infrared Galaxies (VLIRGs, $L_{\rm IR}$[8-1000μm] $10^{11} < L_{\odot} < 10^{12}$) with the Nordic Optical Telescope (NOT) + Andalusia Faint Object Spectrograph and Camera (ALFOSC) on La Palma, in B and I-bands. We discuss here the results of measuring preliminary morphological Concentration, Asymmetry, and Clumpiness (CAS) parameters for these galaxies. The results show VLIRGs as having similar Concentration to ULIRGs, but the Asymmetry and Clumpiness values show a considerably wider dispersion, which is unexpected. We are reanalyzing these preliminary results to assess their reliability.

1 Introduction

Morphological CAS parameters are useful in helping to trace the formation and evolution of galaxies by providing some quantitative measurements which can be related to star formation, a phenomenon which is closely linked to the history of a galaxy and which can help provide clues to the means by which it formed (Conselice 2003). Similarly, such parameters for a sample of galaxies can tell us more about the sample as a whole. We have observed 19 VLIRGs in B- and I-bands using NOT + ALFOSC on La Palma, and have run a modified version of the CAS code of Conselice on the VLIRGs in the sample.

2 Data and Method

The data were reduced in standard manner, with flat fields and biases applied, and the dithered science images were aligned using the IRAF task IMALIGN. Then, using the IRAF task IMEDIT, objects near the galaxy or galaxies which were assumed to be unrelated to the object(s) of interest were removed. The objects removed were primarily obvious stars and other galaxies which did not appear to be related by eye. After the removal of extra objects in the vicinity of the object(s) of interest, the images were convolved with a gaussian based on the factor required to normalize them to the FWHM of the seeing value of the Frei galaxies, using the IRAF task GAUSS. Then, the CAS software was run on the list of x,y pixel coordinates corresponding to the nuclei or bright spots within the galaxies of interest in each image. We compared the results to those of several samples from Conselice (2003), including two ULIRG samples from Borne et al. and Rowan-Robinson et al., the Frei galaxies local sample, starbursts, and dEs from both van Zee and Conselice. (See Conselice 2003 for these references.)

3 Results

We have plotted C vs A, A vs S, and C vs S. Although there is not the space to show them here, there is wider scatter in the initial VLIRG plots for all three parameters, C, A, and S, than for any of the other samples. Although C is generally more similar to other samples, the size and range of some A and S values appear much larger than expected, and we are investigating some sources of possible error or reasons for this which may include problems in cleaning the area of other objects around some of the galaxies.

4 Conclusions

VLIRGs seem a less homogeneous sample than the various others in Conselice (2003) in terms of morphologies and interaction stages, and our VLIRGs may represent the more diverse elements of the range, even though biased because we observed the ones in the sample which showed the most extended structure in the sky survey images. But a more traditional photometric analysis (Arribas et al., 2003, submitted) leads us to believe that VLIRGs as a class are basically very similar to ULIRGs, photometrically, although there may be some effects related to interaction or merger stage as well as to the masses and luminosities of the galaxies involved.

5 Future Work

We plan to try to identify and understand the reasons for higher than expected CAS values, especially for A and S. Eventually, we wish to artificially redshift these VLIRG images and place them in PSF-degraded GOODS images and then remeasure the CAS parameters to test for and measure their recoverability at higher redshifts. This is important for understanding what can be seen and measured for galaxies in deep survey images such as GOODS, the UDF, and etc., especially since VLIRGs are more numerous than ULIRGs.

References

1. S. Arribas, H. Bushouse, R. A. Lucas, L. Colina, K. D. Borne, AJ submitted (2003)
2. C. Conselice, ApJS 147, 1 (2003)

Stellar Populations in Spiral Galaxies

Lauren A. MacArthur[1], Stéphane Courteau[2], Eric F. Bell[3], and Jon Holtzman[4]

[1] University of British Columbia, Vancouver BC, V6T 1Z1, Canada
[2] Queen's University, Kingston ON, K7L 3N6, Canada
[3] Max-Planck-Institut für Astronomie, Königstuhl 17, D-69117 Heidelberg, Germany
[4] New Mexico State University, Las Cruces, NM 88003-8001, USA

1 Introduction

Color gradients in galaxies reveal information about the nature of their stellar populations via age and metallicity trends, and the amount and distribution of dust in their primary constituents. In this study (see [1]), we use a large catalog of galaxy colors to revisit the comparison of broad-band color gradients with stellar population models using a range of star formation histories (SFH) and basic assumptions about the dust distribution. We follow the approach developed by [2], exploring additional parameter ranges and using an extended data base. Trends in the measured local and global luminosity-weighted age and metallicities and their gradients with galaxy parameters are explored and interpreted in the context of existing models of galaxy formation and evolution.

2 Data and Age and Metallicity Determinations

Our data base is a combination of the optical and NIR radial surface brightness (SB) profiles of nearby, low inclination, spirals spirals from [2] & [3] (172 galaxies). For this color-based analysis, we require the galaxies to have measured SB profiles in at least two optical (BVR) and one NIR (H or K) band out to ≥ 1.5 NIR-band disk scale lengths, $h_{\rm NIR}$, with SB errors < 0.12 mag arcsec^{-2}. To increase the S/N of our color profiles, SB profiles were averaged into radial bins scaled by $h_{\rm NIR}$, and we required at least 3 data points per bin. Structural parameters were computed using our bulge-to-disk decomposition scheme [5].

In order to translate color information into constraints on the underlying ages and abundances of the stellar populations, color gradients are compared with stellar population synthesis (SPS) models. We convolve the latest single burst SPS models of [4] with an exponential SFH to produce model grids. Colors are fit to the grids using a maximum-likelihood method to measure luminosity-weighted average ages and metallicities for each radial bin. We explored the effects bursts of SF super-imposed on the underlying exponential SFH. Very recent bursts (~ 1 Gyr ago) have a significant effect on the underlying shape of the model grid, but older bursts only change the average age computed. Given the non-linear nature of the age and metallicities as a function of radius, gradients are computed for two regions in the disk that we refer to as "inner" (0–1.5 $h_{\rm NIR}$) and "outer" (1.5–3.5 $h_{\rm NIR}$). Using fixed-length baselines for the gradient measurements prevents discerning false trends with galaxy central SB [1].

3 Results

The main findings in this analysis are as follows:

- Relative age determinations are robust under the assumption that the underlying SFH of all disk galaxies are similar, and no major star bursts have occurred in the past 1–2 Gyr (justified based on the results of [6]).
- Dust reddening cannot be ruled out completely, but it is unlikely that the gradients are largely due to dust extinction.
- Age and metallicity correlate strongly with local SB: higher SB regions tend to be older and more metal-rich, but with large scatter. This indicates that the local potential plays a significant role in the SFH of spiral galaxies.
- Age and metallicity, measured at $1\,h_{\rm NIR}$, increase with earlier Hubble type (T), $M_{\rm NIR}$, $V_{\rm rot}$, and $\mu_{0,\rm NIR}$ but the trends flatten for $T < 4$, $M_{\rm NIR} < -22.5$ mag, $V_{\rm rot} > 120$ km s^{-1}, and $\mu_{0,\rm NIR} < 18.5$ mag arcsec^{-2}. Age also correlates weakly with scale length but with large scatter (decreasing with galaxy size).
- Age gradients are steeper (negative) in more luminous and bigger galaxies, but no correlation is seen with CSB. Metallicity gradients do not correlate strongly with any galaxy parameters.

These observations are consistent with a picture where higher SB regions of galaxies formed their stars earlier on than lower SB regions, or at a similar epoch but on shorter timescales, and the SFHs at a given SB level, which lead to age gradients, are modulated by the overall potential of the galaxy such that brighter/higher rotational velocity galaxies formed earlier. These trends reach saturation for the brightest and highest CSB galaxies.

An earlier formation time for more massive galaxies is clearly in conflict with hierarchical galaxy formation which predicts that more massive galaxies form late. This discrepancy could be remedied if there is a mechanism at work that prevents the gas in lower mass galaxies from cooling and forming stars at early times. Feedback processes are often invoked as a possible solution, but no prescription has been found thus far that agrees with all observational constraints (e.g. [6]; [7]). While the above trends in the stellar populations of spiral galaxies are not well reproduced by current semi-analytical models of galaxy evolution, they will undoubtedly serve as effective constraints for future models.

References

1. L.A. MacArthur, S. Courteau, E.F. Bell, J. Holtzman: ApJ. submitted (2003)
2. E.F. Bell, R.S. de Jong: MNRAS **213**, 497 (2000)
3. S. Courteau, J.A. Holtzman, L.A. MacArthur: in prep. (2004)
4. G. Bruzual, S. Charlot: MNRAS **344**, 1000 (2003)
5. L.A. MacArthur, S. Courteau, J.A. Holtzman: ApJ. **582**, 689 (2003)
6. G. Kauffmann, et al.: MNRAS **341**, 54 (2003)
7. E.F. Bell, C.M. Baugh, S. Cole, C.S. Frenk, C.G. Lacey: MNRAS **343**, 367 (2003)

Large Scale Structure in the Two Micron All Sky Survey

Ariyeh H. Maller, Daniel H. McIntosh, Neal Katz, and Martin D. Weinberg

Astronomy Department, University of Massachusetts, Amherst, MA 01003

1 The Two Micron All Sky Survey

The unprecedented sky coverage and photometric uniformity of the Two Micron All Sky Survey (2MASS) [1] provides a rich resource for obtaining an understanding of the galaxies populating the local Universe. Characterizing the large-scale clustering distribution is important for theoretical studies of structure formation. 2MASS offers an all-sky view at two microns, unbiased by young stellar light and minimally affected by dust. We use 2MASS to measure the clustering dipole in the K_s band and compare this to the velocity of the Local Group as measured by the CMB dipole [2]. We also measure the angular correlation function and from this infer the K-band three dimensional power spectrum [3].

2 The Clustering Dipole

Because both gravitational force and flux fall off as the square of the distance we expect the flux dipole to be proportional to the gravitational acceleration of the Local Group, if mass traces light. In linear theory the Local Group velocity (which can be measured relative to the CMB) is proportional to the acceleration. The direction of the 2MASS clustering dipole is 15 degrees from the direction of the Local Group velocity as measured against the CMB. This discrepancy reduces to a few degree if galaxies with $K_s < 8$ are removed (375 galaxies). From this we conclude that the assumption of a constant mass-to-light ratio is not valid for the brightest galaxies but is an excellent assumption when averaged over many thousands of galaxies [2]. From the magnitude of the 2MASS dipole, the K-band luminosity density, $\Omega_m = 0.27 \pm 0.04$ and $H_0 = 71 \pm 4$ km s^{-1} [5] we determine a K_s-band linear bias parameter of $b = 1.06 \pm 0.17$. Thus the K_s-band selected galaxies are an unbiased tracer of the underlying mass distribution.

3 The Angular Correlation Function

We measure the angular correlation function of 2MASS galaxies on scales of half an arcminute to fifteen degrees. The angular correlation function looks like it is a pure power law at small angles, but there are oscillations around the power law as can be seen in Fig. 1a. The best fit power law has an exponent of -0.76 and an amplitude at 1 degree of 0.11. However, this is not a good fit with $\chi^2/d.o.f. = 3.9$.

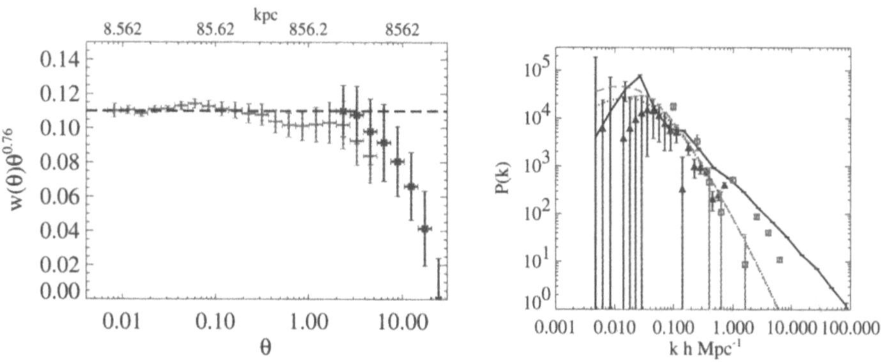

Fig. 1. a) The angular correlation function of 2MASS galaxies multiplied by $\theta^{0.76}$. Small oscillations around the power-law form are apparent and are statistically significant. The horizontal dashed line with amplitude $A = 0.11$ is the best fit power law. b) The power spectrum measured from 2MASS (solid line). For comparison the results from the APM survey (triangles) [7], and for the $r = 18 - 19$ magnitude bin of the SDSS survey (squares) [8] are shown. In all cases error bars are the square root of the diagonal elements of the inverse covariance matrix, $\sqrt{C^{-1}_{P,ii}}$. The dashed line is the best fit CDM type power spectrum to the linear regime $k < 0.2$. The dotted line is the power spectrum from the CMB as determined by [5].

The three dimensional power spectrum can be determined from the angular correlation function using Limber's Equation [6]. We use singular value decomposition [7] to do the inversion and find the power spectrum shown in Fig. 1b in the black. The results for SDSS [8] (red) and APM [7] (blue) are shown for comparison. A fit of a CDM type power spectrum to $k > 0.2h^{-1}$ Mpc^{-1} is shown by the dashed line. The best fit CDM parameters are $\Gamma_{eff} = 0.116 \pm 0.01$ and $\sigma_8 = 0.96 \pm 0.15$.

References

1. Skrutskie, M. F. et al.: 'The Two Micron All Sky Survey (2MASS): Overview and Status'. In: *ASSL Vol. 210: The Impact of Large Scale Near-IR Sky Surveys*, ed. by F. Garzon et al., (Kluwer Academic Publishing Company, Dordrecht 1997) pp. 25–32
2. A. H. Maller, D. H. McIntosh, N. Katz, M. D. Weinberg: ApJL, **598**, L1 (2003a)
3. A. H. Maller, D. H. McIntosh, N. Katz, M. D. Weinberg: ApJ, submitted, astro-ph/0304005
4. M. Rowan-Robinson, M. et al.: MNRAS, **314**, 375 (2000)
5. D. N. Spergel et al.: ApJS, **148**, 175 (2003)
6. D. N. Limber: ApJ, **117**, 134 (1953)
7. D. J. Eisenstein, M. Zaldarriaga: ApJ, **546**, 2 (2001)
8. S. Dodelson, et al.: ApJ, **572**, 140 (2002)

Internal Kinematics of AGN Hosts

Josefa Masegosa[1], Isabel Márquez[1], Florence Durret[2], and DEGAS Consortium

[1] Instituto de Astrofísica de Andalucía, CSIC, POBox 3004, 18080 Granada, Spain
[2] Institut d'Astrophysique de Paris,CNRS, 98bis Bld Arago, 75014 Paris, France

1 The Sample

In order to understand the fueling mechanism to power AGNs, a large effort has been done by the DEGAS (Dynamics and nuclear Engine of GAlaxies of Spiral type) consortium to collect the necessary observational material. We defined an AGN sample which is not contaminated by interacting galaxies and a control sample made of normal spirals with the same properties than the AGN sample, i.e. luminosity and redshift distribution, morphology and percentage of bars. In total we have analyzed a sample of 17 active galaxies and 16 normal spirals. All the galaxies are isolated with the following criteria: not having a companion within 0.4Mpc and cz< 500 km/s. We want to stress this aspect of our sample, since claimed differences between active and normal galaxies (Knapen et al. 2000) disappear when the samples are decontaminated of interacting systems.

2 Morphology

The analysis of the NIR morphologies of the galaxies of our sample (Márquez et al. 1999,2000) has shown no clear difference between AGN and normal spirals. Their disk and bulge structural parameters show the same properties, and bar characteristics are similar in both types of objects. The only difference, if any, may be found in the very central colours (down to a few hundreds central parsecs): the AGNs seem to be redder than normal spirals, probably due to the contamination by the light from the central engine. On the other hand, the hosts of active and non-active galaxies look similar in all respects.

3 Gaseous & Stellar Kinematics

The other approach we have taken in our analysis has been to study the kinematics of the ionized gas, through the Hα emission, and the stellar component by using the Calcium triplet, CaT (Márquez et al. 2003b). Active and non active spirals define similar Tully-Fisher relations, populating the region occupied by the early type spirals. A pure rotation model fails to account for large scale peculiarities in the rotation curves or small scale deviations indicative of non circular and/or decoupled disks. We cannot identify, however, a kinematical distinction between active and non active galaxies. The differences related to the mechanism driving AGNs have to be searched for in the very central nuclear regions.

4 Central Velocity Dispersion Drops

To search for such effects in the very central regions of AGNs, the CaT velocity dispersion for the eight galaxies with high spatial and spectral resolution data has been investigated (see Márquez et al. 2003a). In five of them we have detected drops in the velocity dispersion of the CaT, similar to those reported by Emsellem et al. (2001) using CO IR lines. For all of them archival HST images have been analyzed to search for the presence of central structure. In addition to this, we have also considered the inner morphology in nine galaxies with drops reported in the literature. Disk like structures have been detected in 12 out of the total number of 14 galaxies. It is very likely that these disk structures correspond to a nuclear disk decoupled from the general kinematics of the large scale galactic disk. The measure of the CaT equivalent width (CaT EW) shows that in the region where the drop has been detected, a higher CaT EW is measured, supporting the idea that these disks have been formed from cold material accreted into a circumnuclear disk which is responsible for the dip in the velocity dispersion. These results agree with numerical simulations by Wozniak et al. (2003), which predict a young stellar population born from dynamically cold gas accreted on to the central region.

5 Conclusions

The large scale properties of the active and non-active spiral galaxies in the DEGAS sample are equivalent in all respects. The differences have to be found in details related to the circumnuclear region barely resolved in our observations. Considering the growing evidence that all galaxies above a certain luminosity host central massive compact objects, our results are fully compatible with the idea that the AGN activity could take place recurrently in a given galaxy and the AGN would be seen when fuel is made available in the most internal regions.

References

1. E. Emsellem, D. Greusard et al. : Astron. & Astroph. **368**, 52 (2001)
2. J.H. Knapen, I. Shlosman, R.F. Peletier: Ap. J. **529**, 93 (2000)
3. I. Márquez, F. Durret et al. : Astron. & Astroph. Supp. **140**, 1 (1999)
4. I. Márquez, F. Durret et al. : Astron. & Astroph. **360**, 431 (2000)
5. I. Márquez, J. Masegosa et al. : Astron. & Astroph. **409**, 459 (2003a)
6. I. Márquez, F. Durret et al. : Astron. & Astroph. , in press astro-ph/0311616 (2003b)
7. H. Wozniak, F. Combes I. et al. : Astron. & Astroph. **409**, 469 (2003)

Probing Galaxy Evolution via Interactions and Mergers in the IRAS Revised Bright Galaxy Sample

Joseph M. Mazzarella[1], David B. Sanders[2], Catherine M. Ishida[2,3], Joseph B. Jensen[4], and Dong-Chan Kim[5]

[1] MS 100-22, California Institute of Technology & Jet Propulsion Laboratory, Pasadena, CA 91125; *mazz@ipac.caltech.edu*
[2] Institute for Astronomy, University of Hawaii, 2680 Woodlawn Drive, Honolulu, HI 96822; *sanders@ifa.hawaii.edu*
[3] Subaru Telescope, National Astronomical Observatory of Japan, 670 North Aohoku Place, Hilo, HI 96720; *cat@naoj.org*
[4] Gemini Observatory, 670 North Aohoku Place, Hilo, HI 96720; *jjensen@gemini.edu*
[5] School of Earth and Natural Sciences (BK21), Seoul National University, Seoul, Korea; *dckim@astro.snu.ac.kr*

1 Motivation

In order to interpret observations of luminous infrared/sub-mm galaxies and AGNs at high redshifts, and to study their evolution from the earliest epochs of star formation to the present, it is important to construct a comprehensive understanding of the properties of objects that are thought to be the closest analogs in the local Universe. These are the luminous infrared galaxies (LIRGs, $L_{ir} > 10^{11} L_{\odot}$) and ultraluminous infrared galaxies (ULIRGs, $L_{ir} > 10^{12} L_{\odot}$) identified in the IRAS survey. The IRAS Revised Bright Galaxy Sample (RBGS; [1]) is a complete sample of extragalactic objects with total 60 μm flux density > 5.24 Jy, covering the entire sky surveyed by IRAS at Galactic latitudes $|b| > 5$ degrees. The 629 objects in the RBGS have a median redshift of 0.0082, and their visual and near-infrared properties span a wide range from normal, isolated gas-rich spirals at low luminosities ($L_{ir} < 10^{10.5} L_{\odot}$) through an increasing fraction of interacting galaxy pairs and ongoing mergers powered by enhanced star formation and AGNs among the LIRGs and ULIRGs. There are 22 ULIRGs (3.5%) and 180 LIRGs (29%) in the RBGS, making this an ideal sample for detailed study and comparison with observations of high-redshift galaxies. A primary goal of this study is to delineate the transformations that occur in basic properties as galaxies evolve from separate disks through the interaction and merger sequence, and possibly into elliptical galaxies.

2 B, V, R, I, and K′ Imaging

We have obtained images in the R (0.6 μm) and K′ (2.2 μm) bands for a sample of over 150 LIRGs and ULIRGs in the RBGS with Galactic latitude $|b| > 10$

degrees and declination > −40 degrees. In addition, B (0.4 μm), V (0.5 μm), and I (0.8 μm) band images have been analyzed and color maps have been constructed for 56 LIRGs in the sample. Most of the observations were obtained using the University of Hawaii 88-inch telescope on Mauna Kea, and some of the R band images were acquired using the 60-inch telescope at Mt. Palomar.

The poster contained a collage of 56 LIRGs and 10 ULIRGs. The images were constructed by stacking data in the B, V and I bands to optimize sensitivity to faint tidal debris, and they were displayed using a common field of 100 kpc X 100 kpc to allow direct comparison of their environments. Samples of (B-I), (B-V) and (V-I) color maps were also shown, in addition to some R and K′ images with surface brightness profiles and initial ellipse fitting results. Due to the two page limit, none of these graphics can be included here, but some can be found in the contribution by Sanders et al. in this same volume.

3 Initial Results

We are in the process of analyzing nuclear and annular colors, surface brightness distributions and light concentrations, physical sizes and separations of components, and morphological measurements as a function of L_{ir} and other parameters. Initial results are as follows:

1) BVI "true color" images and color maps reveal the distribution of dust and young (blue) versus old (red) stellar populations. 2) Very deep coadded B, V and I ("open filter") images reveal morphological features not seen in previous surveys. 3) The ratio $L_{ir}/L_{K'}$ correlates with far-IR color temperature ($f_{60\mu m}/f_{100\mu m}$) and the degree of interaction; this is a more robust indicator than the traditional L_{ir}/L_B ratio, which is more highly influenced by dust extinction in the B band. 4) Among the interacting pairs with separated nuclei that have not yet merged below the resolution limit of the data, both galaxies tend to have comparable K′ luminosities; few pair component K′ flux ratios are above 3:1, indicating that major mergers among two (not more) galaxies are usually involved. 5) Combining the K′ data with IRAS data for individual galaxies that have been resolved using the HIRES algorithm in pairs with separations greater than 30 arcseconds [2], the total L_{ir} for the pairs, even when normalized by total $L_{K'}$, is often dominated by one of the galaxies in the pair.

In addition to presentation of full details and analysis results in a series of journal publications, the calibrated pixel data are being prepared for public accessibility in FITS format through the NASA/IPAC Extragalactic Database (NED, *http://nedwww.ipac.caltech.edu*) to facilitate investigations using the extensive multiwavelength data and distributed resources linked in NED.

References

1. D. B. Sanders, J. M. Mazzarella, D.-C. Kim, J. A. Surace, & B. T. Soifer, AJ, **126**, 1607 (2003)
2. J. A. Surace, D B. Sanders, & J. M. Mazzarella, AJ, submitted (2003)

Multiwavelength Maps of Simulations of Galaxy Formation

Paola Mazzei[1] and Anna Curir[2]

[1] INAF, Osservatorio Astronomico, Vicolo dell'Osservatorio 5, 35122 Padova, Italy
[2] INAF, Osservatorio Astronomico, Strada dell'Osservatorio 20, 10025 Pino Torinese, Torino, Italy

Abstract. We performed smoothed particle hydro-dynamical simulations of isolated collapsing triaxial systems initially composed of dark matter (DM) and gas with cooling, shocks, star formation and feedback self-consistently accounted for. The initial configuration is built up as expected in the CDM scenario. We analyze the role of the initial geometry, dynamical state and mass of the DM halo on the galaxy formation and evolution. We find intriguing connections between dark and luminous matter. Multiwavelength maps shed light into different galaxy properties.

Results

Our knowledge of the star formation rate, which leads galaxy evolution, i.e. luminosity, colors and dust enrichment, is still very poor. Smoothed particle hydrodynamic (SPH) simulations including chemo-photometric evolutionary population synthesis models of isolated collapsing systems endowed with triaxial dark matter halos, gas and the star formation switched on, are a useful tool to shed light on several open questions. This approach, which allows a greater particle resolution than a fully consistent cosmological scenario, gives us important insights into connections between dark and luminous matter. By constraining the systems with the same initial density, [1] pointed out that the star formation rate (SFR) depends both on the total mass of the system and on the dynamical state of the halo whose initial geometry influences the results further. We resume here the major points of our works, starting from [2] (see also [3]) showing multiwavelength maps. We find that the total mass of the system drives galaxy formation and evolution. In our framework spiral galaxies arise only in systems both less massive than $10^{12} m\odot$ and with a critical value of the baryonic-to-total mass ratio, around its cosmological value, 0.1. Stellar bars are common transient features in such models (see Fig 1). as a consequence of the halo evolution [4], [5]. For the same value of such a ratio, more massive systems host more luminous and redder galaxies as a consequence of the stronger burst of star formation arising into their halos earlier on (see Fig 1). Moreover the star formation does not switch on in the less massive DM halos so that the critical total mass to entail a luminous system is $10^{11} m\odot$. For a given value of M_{tot} higher values of the baryonic-to-total mass ratio provide galaxies with lower rotational support, so that no constraints on the total mass of the system hosting an elliptical galaxy have been derived.

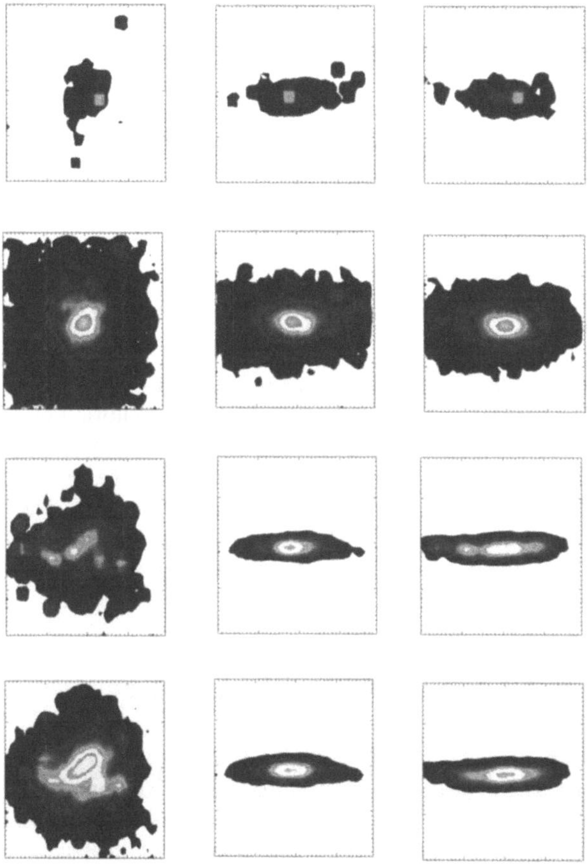

Fig. 1. Face-on (i.e.xy projection, left column), yz (middle column) and xz (right column) rest-frame maps in B and K from top to bottom respectively, for models with different M_{tot} at 15.5 Gyr: $M_{tot} = 2 \times 10^{12}\, m_{\odot}$ (upper panels) and with$M_{tot} = 2 \times 10^{11}\, m_{\odot}$ (lower panels). The box size is 40 *kpc x* 40 *kpc*, 1 *kpc* the spatial resolution and 3.7 mag the magnitude contrast achieved.

References

1. P. Mazzei, A. Curir: ApJ **591**, 784 (2003)
2. A. Curir, P. Mazzei: New Astr **1**, 20 (1999a)
3. P. Mazzei: 'Dark and luminous matter connections. Towards understanding galaxy formation' review paper In: *Recent Research Developments in Astronomy and Astrophysics, 2003*, ed. G.S Pandalai (Research Signpost) in press.
4. P. Mazzei, A. Curir: A&A **372**, 803 (2001)
5. A. Curir, P. Mazzei: A&A **352**, 103 (1999b)

Multiwavelength Study of the NEPR Sample. The 60μm Luminosity Function

Paola Mazzei[1], Daniela Bettoni[1], Antonio Della Valle[1,3], Hervé Aussel[2], Gianfranco De Zotti[1], and Alberto Franceschini[3]

[1] INAF, Osservatorio Astronomico, Vicolo dell'Osservatorio 5, 35122 Padova, Italy
[2] University of Hawaii, USA
[3] Universitá di Padova,Vicolo dell'Osservatorio 6, 35122 Padova, Italy

Abstract. We present the 60μm luminosity function (LF) of the deepest complete IRAS selected sample [1] so far available, in the North Ecliptic Polar Region (NEPR). The LF here is the first step in the multi-wavelength study of this sample which, thanks to our ISOCAM observations, provides a direct link between IRAS and ISOCAM surveys and the forthcoming deeper SIRTF cosmological surveys. The sample is unique in allowing a direct investigation of the evolution of the global Spectral Energy Distribution (SED) of dusty galaxies up to a redshift of about 0.3, where the global star formation rate is known to evolve very fast.

1 Redshift Distribution and 60 μm LF

The IRAS Deep Survey (IDS) sample was defined by Hacking and Houck [1] co-adding IRAS scans of the North Ecliptic Polar Region (NEPR), representing more than 20 hours of integration time. It comprises 98 galaxies with S(60 μm)> 50 mJy over an area of 6.25 sq. deg. and is still the deepest complete and statistically significant 60 μm selected sample available before the advent of the SIRTF observatory; for comparison the deepest ISOPHOT surveys, FIRBACK at 175μm [2], ELAIS and Lookman Hole at 90 μm, [3], [4], are all complete at 100 mJy. The IDS sample is also superior to those in the coverage of the SED of its member galaxies. In fact, in addition to 60 μm fluxes, most of them have also 100 μm fluxes and several have 25 μm fluxes. VLA observations for a good fraction of sample sources have been made by [5]. Moreover we have obtained ISOCAM observations in the LW3 filter (range 12-18 μm) of 94 of these sources [6]. Our spectroscopy runs at the TNG, Keck II, Subaru and Asiago telescopes, from 1999 to 2002, have yielded redshifts for all the optical counterparts of our ISOCAM sources, the best candidates of the IRAS sources. The data by [7], who measured the redshifts of 76 tentative optical counterparts of IDS sources, turned out indeed to be not the correct identifications, based on our ISOCAM positions. The redshift distribution of our sample is shown in Fig. 1 (left panel). Two sub-clusters, at z=0.055 and 0.080, are visible. IRAS fluxes at ISOCAM positions have been recently revised [8] showing that our sample is complete for S(60 μm)> 80 mJy. Our derived 60 μm LF, in Fig. 1 (right panel), is in good agreement with previous estimates [9].
Optical imaging has been planned for the whole sample; B, R and Ks images

have been already performed for a significant fraction of our sources. These observations, together with ISO, IRAS, VLA and 2MASS data when available, will allow a systematic study of the broadband spectral properties of the deepest IRAS-selected sample, on which numerous studies of the far-IR evolution of galaxies rely. The sample is unique in allowing a direct investigation of the evolution of the global spectral energy distribution, together with morphological information coming from optical imaging, of dusty galaxies up to a redshift of about 0.3, where the star-formation rate is known to evolve very fast. Chemo-photometric evolutionary population synthesis models developed by Mazzei et al. ([10], [11]) covering the full frequency range from UV up to 1 mm (i.e. including dust effects) coupled with chemo-dynamical models by [12] and [13], will been used to understand these data.

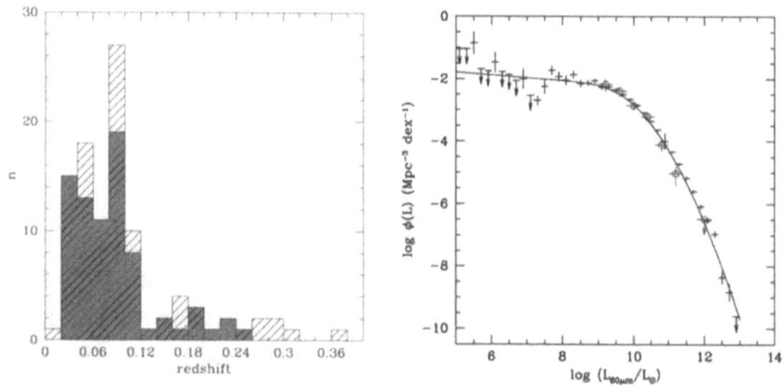

Fig. 1. Left panel compares redshift distribution of our sample with that by [7] (shaded); right panel shows LF from our data (circles) compared with that by [9] (crosses) and their corresponding best fit (continuous line).

References

1. Hacking, P., Houck, I.R.: ApJS **63**, 311 (1987)
2. Puget, J-L. et al.: A&A **345**, 29 (1999)
3. Rowan-Robinson, M. et al: *The Universe as Seen by ISO*, ESA SP-427
4. Rodighiero, G et al: MNRAS **343**, 1155 (2003)
5. Hacking, P., et al: ApJ **339**, 12 (1989)
6. Aussel et al.: A&AS **141** ,257 (2000)
7. Ashby M., et al: ApJ **456**, 428 (1996)
8. Mazzei P. et al. New Astr. **6**, 265 (2001)
9. Saunders et al.: MNRAS **242**, 318 (1990)
10. Mazzei et al: A&A **256**, 45 (1992)
11. Mazzei et al: ApJ **422**, 81 (1994)
12. A. Curir, P. Mazzei: New Astr **1**, 20 (1999)
13. Mazzei P. and Curir, A.: ApJ **591**, 784 (2003)

The Cosmological Evolution of Quasar Black-Hole Masses

Ross J. McLure and James S. Dunlop

Institute for Astronomy, University of Edinburgh, Edinburgh, EH9 3HJ, UK

1 Introduction

The strong correlation observed between black-hole and bulge mass in the local Universe (eg. Magorrian et al. 1998; Gebhardt et al. 2000) directly implies that the evolution of the supermassive black holes and their host galaxies are intimately related. In this proceedings we present one of the main results of our recent paper (McLure & Dunlop 2003) which analyses the evolution of quasar black-hole masses drawn from the Sloan Digital Sky Survey (SDSS) first data release (Schneider et al. 2003).

2 Sample and Analysis

Our sample consists of 12,286 broad-line (FWHM ≥ 2000 kms^{-1}) quasars with $z \leq 2.1$ drawn from the SDSS Quasar Catalog II (Schneider et al. 2003). The black-hole masses for each quasar have been determined using the calibrations of the virial mass estimator based on the $H\beta$ and MgII emission lines as derived by McLure & Jarvis (2002). The line-widths of either $H\beta$ or MgII were determined for this sample using an automated fitting algorithm.

3 The Activation Rate of Supermassive Black Holes at z=2

One of the main results presented in McLure & Dunlop (2003) is an estimate of the activation rate of supermassive black holes at $z \simeq 2$. Firstly, using the Boyle et al. (2000) quasar luminosity function the number density of $z = 2$ quasars is calculated as a function of luminosity. Secondly, our new SDSS results are then used to estimate the fraction of these quasars which have black-hole masses above three mass thresholds: $\geq 10^{8.5}\,M_{\odot}, \geq 10^{9}\,M_{\odot}$ & $\geq 10^{9.5}\,M_{\odot}$. As shown in Panel A of Fig 1, a direct comparison with the local mass function of dormant black holes then suggests that lower limits on the activation fraction of $z \simeq 2$ black holes are: ~ 0.005, ~ 0.02, and ~ 0.05 at black-hole masses of $\geq 10^{8.5}\,M_{\odot}, \geq 10^{9}\,M_{\odot}$ & $\geq 10^{9.5}\,M_{\odot}$ respectively. In panel B of Fig 1 we show the estimated number densities of active supermassive black holes corrected for increasing quasar lifetime as a function of black-hole mass as predicted by Yu & Tremaine (2002). Finally, in panel C of Fig 1 we further correct the estimated number densities of active $z \simeq 2$ black holes by a factor of two to conservatively account for geometric obscuration.

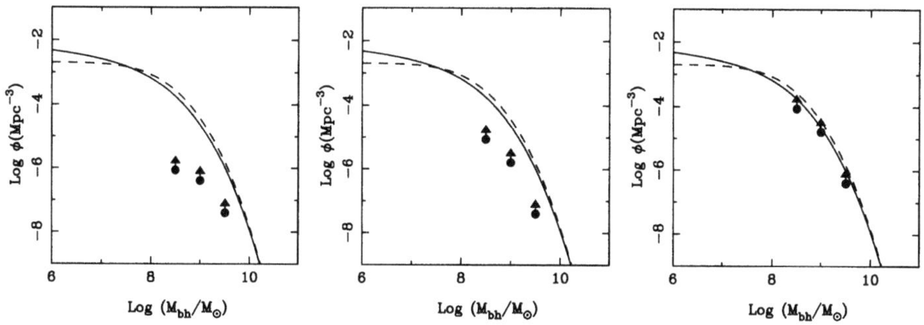

Fig. 1. In each panel the solid and dashed lines show the cumulative local (z=0) dormant black-hole mass functions as derived using the $M_{bh}-L_{bulge}$ and $M_{bh}-\sigma$ relations respectively. Panel A shows the estimated number densities of active supermassive black-holes at $z \simeq 2$ for three mass thresholds ($\geq 10^8\,M_\odot, \geq 10^9\,M_\odot$ & $\geq 10^{9.5}\,M_\odot$). Panel B shows the number densities adjusted to account for a possible relationship between mean quasar lifetime and black-hole mass (Yu & Tremaine 2002). Panel C shows the number densities boosted by a factor of two to account for geometric obscuration.

4 Conclusions

The activation fraction of supermassive black-holes at $z \simeq 2$ is apparently an increasing function of mass, with an activation rate of $f \simeq 0.005$ at $M_{bh} \simeq 10^{8.5}\,M_\odot$ rising to $f \simeq 0.05$ at $M_{bh} \simeq 10^{9.5}\,M_\odot$. This result is consistent with theoretical work which predicts quasar lifetimes to be an increasing function of black-hole mass. Correcting for this expectation we find that the shape of the active black-hole mass function at $z \simeq 2$ is consistent with that of the local dormant black-hole mass function, with a normalization a factor of $\simeq 5$ lower. Making a conservative correction of a factor of two to account for geometric obscuration, the direct implication of this result is that the fraction of black-holes with mass $\geq 10^{8.5}\,M_\odot$ which are in place and active at $z \simeq 2$ is ≥ 0.4. Furthermore, McLure & Dunlop (2003) find a fairly robust limit on the lifetime of quasars with black-hole masses $\geq 10^{9.5}\,M_\odot$ to be $t_Q > 2 \times 10^8$ years. Black holes of this mass appear to be prevented from growing to masses $> 10^{10}\,M_\odot$ by some physical mechanism, other than the Eddington limit, which prevents accretion at rates $\gtrsim 10\,M_\odot$ per year.

References

1. Boyle B.J., et al., 2000, MNRAS, 317, 1014
2. Gebhardt K., et al., 2000, ApJ, 539, L13
3. Magorrian J., et al., 1998, AJ, 115, 2285
4. McLure R.J., Jarvis M.J., 2002, MNRAS, 337, 109
5. McLure R.J., Dunlop J.S., 2003, MNRAS, in press, astro-ph/0310267
6. Schneider D.P., et al., 2003, ApJ, in press, astro-ph/0308443
7. Yu Q., Tremaine S., 2002, MNRAS, 335, 965

The Tully–Fisher Relation of Cluster Spirals at $z = 0.83$

Bo Milvang-Jensen[1,2] and Alfonso Aragón-Salamanca[2]

[1] School of Physics and Astronomy, University of Nottingham, UK
[2] Max–Planck Institut für extraterrestrische Physik, Garching, Germany

Abstract. We study Tully–Fisher relation for a sample of 7 cluster spiral galaxies at $z = 0.83$ and 19 field spirals at $z = 0.15$–0.90 based on VLT spectroscopy and HST photometry. No strong difference is detected between the cluster and the field galaxies, but we find some evidence that the cluster spirals are ~ 0.5–1 mag brighter in the rest-frame B–band than the field ones at a fixed rotation velocity. Although only a ~ 1.5–2σ result, if confirmed with larger samples this effect could be due to the cluster spirals experiencing a period of enhanced star formation while falling into the cluster. The Tully–Fisher residuals for the cluster spirals are found to correlate with the star formation rate, which supports the above hypothesis.

We study spiral galaxies in the high redshift cluster MS1054–03 at $z = 0.83$ and field spirals at similar redshifts. The aim is to shed light on the processes governing the evolution of galaxies in clusters, in particular the processes responsible for the morphological transformation of spiral galaxies into S0s. The main diagnostic used is the Tully–Fisher relation, see Fig. 1.

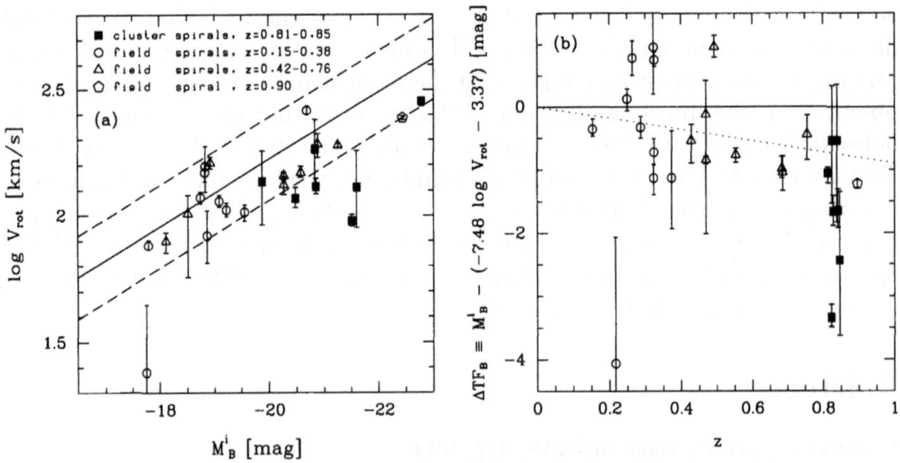

Fig. 1. (a) High redshift cluster and field B–band Tully–Fisher relation. The solid line is the local Tully–Fisher relation from [7] and the dashed lines mark the 3σ limits. (b) Residuals in absolute magnitude from the local Tully–Fisher relation versus redshift. The cluster spirals are marginally (1.5–2σ significance) brighter than the field spirals at a fixed rotation velocity. (Dotted line: fit to the field galaxies: $\Delta \mathrm{TF}_B = (-0.9 \pm 0.3)z$.) Figure adopted from [6], see also [5].

A difference in the Tully–Fisher residuals (Fig. 1b) is seen at 1.5–2σ significance between cluster and field spirals. If interpreted as a difference in absolute magnitude at a fixed rotation velocity, the mean difference is ~1 mag in the sense that the cluster sample is brighter than the field sample. This could indicate a brightening due to enhanced levels of star formation. This hypothesis is supported by the finding that the Tully–Fisher residuals for the cluster spirals correlate with the star formation rate, see Fig. 2.

We have also studied the rest-frame H–band Tully–Fisher relation using data from [3]. The cluster–field difference is as large as in the B–band. Using "standard" stellar population models, this is at variance with the above interpretation of the B–band results. However, recent models [4], which include a thermally pulsating AGB phase, predict a substantial brightening in the NIR bands for a population with age ~0.2–1 Gyr. The impact of this has still to be explored.

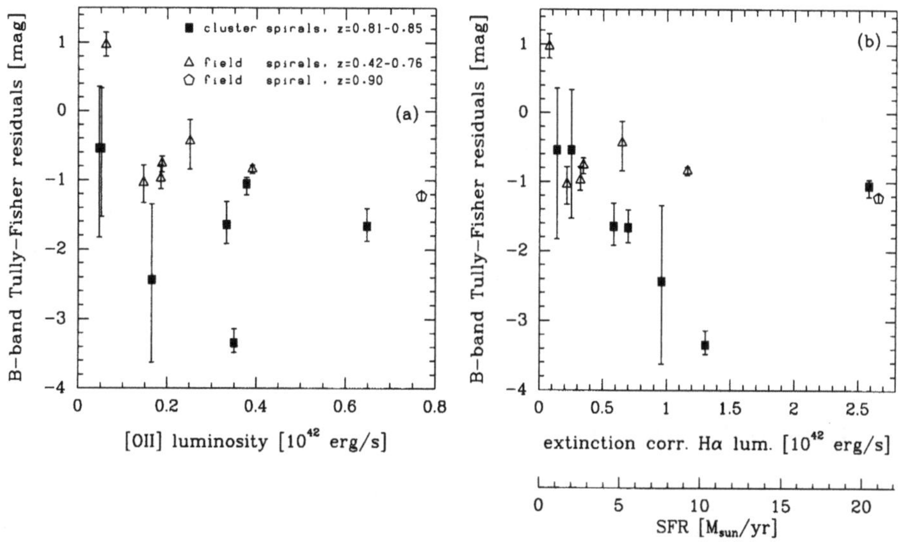

Fig. 2. B–band Tully–Fisher residuals (cf. Fig. 1) versus [OII] luminosity (panel a) and versus an estimate of the extinction corrected SFR (panel b; calculated from the [OII] luminosity, the [OII] EW and the absolute B–band magnitude using an empirical relation from [1,2]). The correlations for the cluster spirals (~2σ significance) suggests that the brightening is due to star formation. Figure adopted from [6].

References

1. A. Aragón-Salamanca et al.: AJ, submitted
2. A. Aragón-Salamanca et al.: in Star Formation Through Time, astro-ph/0210123
3. N. M. Förster Schreiber et al.: in prep.
4. C. Maraston: in prep.
5. B. Milvang-Jensen et al.: MNRAS **339**, L1 (2003)
6. B. Milvang-Jensen: PhD Thesis, University of Nottingham (2003)
7. M. J. Pierce, R. B. Tully: ApJ **387**, 47 (1992)

Chemical Evolution of Bulges of Spiral Galaxies: Environmental and Morphological Influence

Lorenzo Morelli[1,2], Emanuela Pompei[1], Alessandro Pizzella[1], Enrico Corsini[1], and Francesco Bertola[1]

[1] European Southern Observatory, 3107 Alonso de Cordova, Santiago, Chile
[2] Dipartimento di Astronomia, Università di Padova, Vicolo dell'Osservatorio 2, I-35122 Padova, Italy

1 Introduction

This project is aimed at studying the chemical evolution of bulges of spiral galaxies, with the goals to understand how much the properties of bulges differ from those of elliptical galaxies and to assess if the environment plays a significant role in the chemical evolution. To achieve these goals, we observed 11 spiral galaxies in field and 9 in the Fornax, Hydra and Pegasus cluster. A second run has been completed in September 2003 to complement the cluster sample with 10 additional galaxies. We measured the main Lick indexes Hβ, Mg$_2$, <Fe>, MgFe and the velocity dispersion as function of the galactocentric radius, in order to determine the age and chemical abundances for bulges of galaxies with morphological type ranging from S0 to Sc. Here we present first preliminary results.

2 Kinematic and Index Analysis

We obtained long slit spectroscopy at the ESO 3.6 telescope using EFOSC2 spectrograph with grism #9 and 1" slit. The wavelength range is 4700Å and 6700Å, with a instrumental FWHM=6.5Å. The spectra has been rebinned, in natural logarithmic scale, along the dispersion direction and summed line by line, along the spatial direction, in order to obtained at least a S/N 30-35 for each summed bin. For each galaxies we measured the kinematics for the stellar component, from the absorption features present in the spectral region ranging between 5000Å and 6100Å. Using FCQ (Bender et al. 1990, Bender et al. 1994) based on Fourier correlation quotient we derived the LOSVD along the slit to measure velocity (v), and the velocity dispersion (σ).
From the flux calibrated spectra, spatially rebinned with exactly the same radial steps as done for kinematic measurement, we measured line strength and the radial profile of the Lick indexes present in our wavelength range, following Faber et al. (1985) and Worthey et al. (1994). All the measurement were corrected for the velocity dispersion broadening using the results of kinematical analysis and calibrated to the Lick indexes using Lick template stars with spectral type ranging from GIII to KIII.

3 Element-σ Relation and Age

It is known that for early type galaxy exist a strong correlation between the Mg absorption feature and the central velocity dispersion (σ). Less evident are so far the correlation between σ and <Fe> or Hβ. In Fig. 1 we show the results for 7 galaxies out from a sample of 9 observed (2 galaxies had a S/N too low to be included in the analysis). We find a very well defined linear relation for all the indexes. <Fe> and Mg_2 slopes are consistent with the literature (Jorgensen 1999, Kuntschner 2000). The slope discrepancy in the Hβ fit relation could be due to the Hβ emission features contamination.
Galaxies (Fig. 1) cover a wide range in age and metallicity without showing any particular trend related to their morphological type.

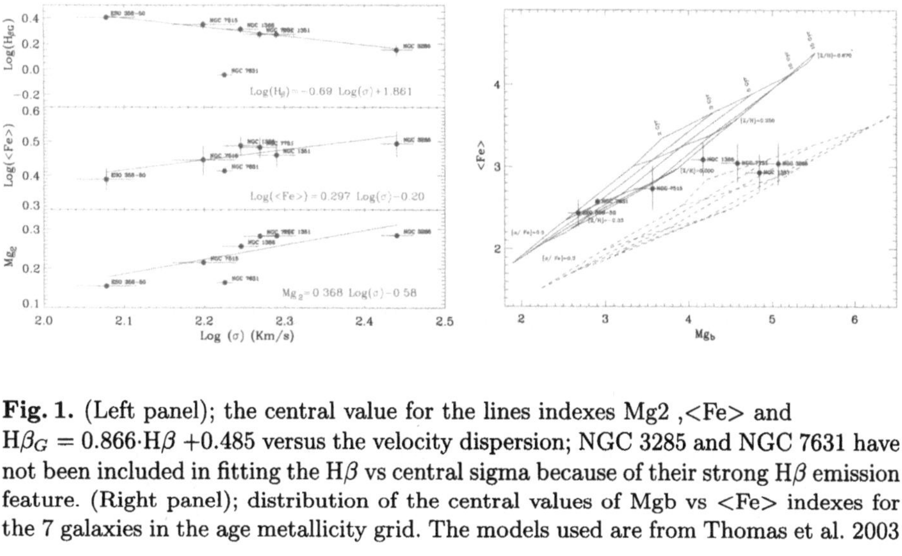

Fig. 1. (Left panel); the central value for the lines indexes Mg2 ,<Fe> and Hβ_G = 0.866·Hβ +0.485 versus the velocity dispersion; NGC 3285 and NGC 7631 have not been included in fitting the Hβ vs central sigma because of their strong Hβ emission feature. (Right panel); distribution of the central values of Mgb vs <Fe> indexes for the 7 galaxies in the age metallicity grid. The models used are from Thomas et al. 2003

We are increasing the cluster galaxies sample with a second run (September 2003). A more extended sample is crucial to have more detailed information on the chemical abundances in bulges of spiral galaxies. The whole sample will be compared with a similar sample of spiral galaxies in field and in group to understand if and how the environment can constrain the evolution of galaxies.

References

1. Bender, R. 1990, ApJ, 229, 441
2. Bender, R.et al, 1994, MNRAS, 269, 785B
3. Jorgensen, I., 1999, MNRAS, 306, 607J
4. Kuntschner, H., 2000, MNRAS, 315, 184K
5. Mehlert, D. et al. 2003, A&A, 407, 423M
6. Thomas, D. et al 2003, MNRAS, 343, 279T

Brightest Cluster Galaxy Formation in the Cluster C0037-2522: Flattening of the Dark Matter Cusp

Carlo Nipoti[1,2], Massimo Stiavelli[3], Luca Ciotti[2], Tommaso Treu[4], and Piero Rosati[5]

[1] Theoretical Physics, 1 Keble Road, Oxford, OX1 3NP, UK
[2] Dept. of Astronomy, University of Bologna, via Ranzani 1, 40127 Bologna, Italy
[3] Space Telescope Science Institute, 3700 San Martin Drive, Baltimore, MD 21218
[4] Dept. of Physics & Astronomy, UCLA, Box 951547, Los Angeles, CA 90095-1547
[5] ESO, Karl-Schwarzschild-Strasse 2, 85748 Garching, Germany

Abstract. The X-ray cluster C0337-2522 at redshift $z = 0.59$ hosts in its core a group of five elliptical galaxies. Using N-body simulations we show that a multiple merging event among the five galaxies is expected to take place in the next few Gyrs, forming a central brightest cluster galaxy. We also find indications that dynamical friction heating associated with this event is likely to modify the central slope of the cluster dark matter density profile.

1 Introduction

We have identified a group of five ellipticals (Es) located in the core of the X-ray cluster C0337-2522 at redshift $z = 0.59$ (ROSAT Deep Cluster Survey [12]). This system represents a strong candidate for the initial stages of brightest cluster galaxy (BCG) formation through galactic cannibalism [9]. Here, we explore how many of the five galaxies under consideration are expected to merge before $z = 0$ and we study the properties of the merger remnant. We address these questions by using N-body simulations, exploring several initial conditions compatible with the imaging and kinematic information from our ESO-VLT data. The details of observations and numerical simulations are given in [8].

Recent observational studies [13,14] of a few galaxy clusters find central dark matter (DM) density distributions ($\rho_{\rm DM} \propto r^{-\beta}$) flatter ($\beta \sim 0.5$) than predicted by cold dark matter simulations [6,7] ($\beta \sim 1 - 1.5$). A possible interpretation of this discrepancy is that dynamical friction heating [1] is effective in flattening the DM cusp. We investigate this hypothesis using our simulations, where the initial DM profile is cuspy ($\beta = 1$), galaxies are deformable, and initial conditions correspond to an observed cluster.

2 Results

In all the simulations 3 to 5 galaxies merge before $z = 0$. The merger remnant is similar in its main structural and dynamical properties to a real BCG (but with no evidence of the diffuse luminous halo typical of cD galaxies). In particular, it satisfies the Faber-Jackson [3] relation and the K-band Fundamental

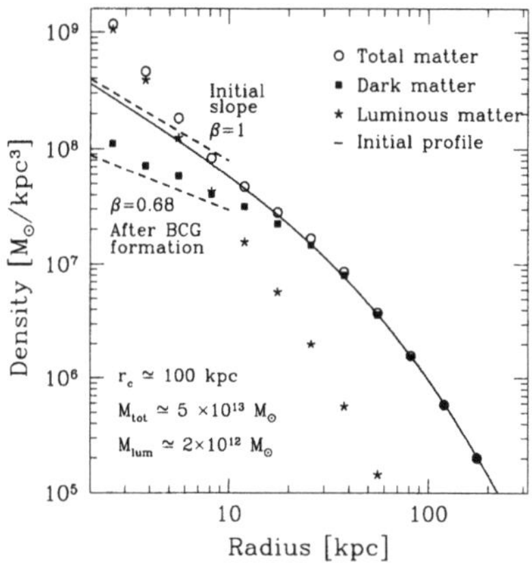

Fig. 1. Final DM, stellar, and total density profiles for a representative simulation. In this case, the best–fitting inner slope of the final DM profile is $\beta = 0.68$. The solid curve is the initial ($\beta = 1$) cluster DM profile.

Plane [10], under the hypothesis that the five Es are placed on these scaling relations, and that the mass–to–light ratio remains unchanged in the merging process. However other features of Es, such as the $M_{\rm BH}$-σ_0 relation [5,4], and the metallicity gradient [11], are hardly reproduced by this multiple dissipationless merging scenario. As regards the properties of the cluster DM halo, we find final profiles flatter than the initial $\rho_{\rm DM} \propto r^{-1}$ profile (see Fig. 1). We fit the cluster DM density profiles with the formula $\rho_{\rm DM} = \rho_{\rm DM,0}(r/r_{\rm c})^{-\beta}[1 + (r/r_{\rm c})]^{-4+\beta}$, with β, $r_{\rm c}$, and $\rho_{\rm DM,0}$ free parameters. For the final DM profiles in our simulations $\langle \beta \rangle \simeq 0.66 \pm 0.15$ to be compared with $\beta = 1$ of the initial profile. Our results (in accordance with recent simulations using rigid galaxy models [2]) point towards a major role of dynamical friction associated with BCG formation in determining the cluster DM profile on small scales.

References

1. G. Bertin, T. Liseikina, F. Pegoraro: A&A **405**, 73 (2003)
2. A. El-Zant, Y. Hoffman, J. Primack, F. Combes, I. Shlosman: preprint (2003, astro-ph/0309412)
3. S.M. Faber, R.E. Jackson: ApJ **204**, 668 (1976)
4. L. Ferrarese, D. Merritt: ApJ **539**, L9 (2000)
5. K. Gebhardt et al.: ApJ **539**, L13 (2000)
6. J.F. Navarro, C.S. Frenk, S.D.M. White: ApJ **462**, 563 (1996)
7. B. Moore, F. Governato, T. Quinn, J. Stadel, G. Lake: ApJ **499**, L5 (1998)
8. C. Nipoti, M. Stiavelli, L. Ciotti, T. Treu, P. Rosati: MNRAS **344**, 748 (2003)
9. J.P. Ostriker, S.D. Tremaine: ApJ **202**, L13 (1975)
10. M.A. Pahre, S.G. Djorgovski, R.R. de Carvalho: AJ **116**, 1591 (1998)
11. R. Peletier: PhD thesis, Univ. Groningen (1989)
12. P. Rosati, R. della Ceca, C. Norman, R. Giacconi: ApJ **492**, L21 (1998)
13. D.J. Sand, T. Treu, R.S. Ellis: ApJ **574**, L129 (2002)
14. D.J. Sand, T. Treu, G.P. Smith, R.S. Ellis: preprint (2003, astro-ph/0309465)

The FORS Deep Field Spectroscopic Survey of High-Redshift Galaxies

Stefan Noll, Dörte Mehlert, Immo Appenzeller, and Christian Tapken

Landessternwarte Königstuhl, 69117 Heidelberg, Germany

1 The Project

The evolution of the properties of galaxy populations with redshift provides important constraints on current models of galaxy formation and evolution. In order to investigate this evolution we used the FORS instruments at the VLT to obtain high-S/N galaxy spectra at redshifts up to 5.0 in the FORS Deep Field (FDF). A detailed discussion of the project can be found in [4].

2 The Sample

Our targets were selected on the basis of photometric redshifts [1], which were derived using the deep photometric data in seven filter bands from U to K, presented in the FDF photometric catalogue [2]. The high-redshift candidates were usually brighter than 24.5 mag in I.

The multi-object spectroscopy (MOS) observations were carried out using the FORS instruments at the VLT. All spectra were obtained with the low-resolution grism 150 I and a slit width of $1''$ corresponding to a resolution element of 18 – 23 Å (FWHM). Typical total integration times for the high-redshift objects were 10 h, leading to $\langle \mathrm{S}/N \rangle \approx 13$ for $z > 2$.

Including objects which coincided with the slits by chance, reliable redshifts could be derived for 341 FDF objects. 98 objects are starburst galaxies and QSOs at $z > 2$. The spectroscopic redshift distribution indicates significant redshift clustering in the whole redshift range, which can be attributed to the large-scale structure in direction to the FDF.

3 Results

We analysed the spectral properties of the high-redshift FDF galaxies using the individual as well as composite spectra for different redshift ranges, luminosities, Lyα emission strengths, and continuum slopes. Significant evolutionary effects were found comparing the redshift ranges $2 < z < 3$ and $3 < z < 4$ (Fig. 1). Most conspicuous are the decrease of the frequency and strength of Lyα emission, the increase of the C IV absorption [3], the enhanced dust reddening of the UV continuum, and the increase of the intrinsic UV luminosity (and consequently the star-formation rate) of the most luminous galaxies with decreasing redshift.

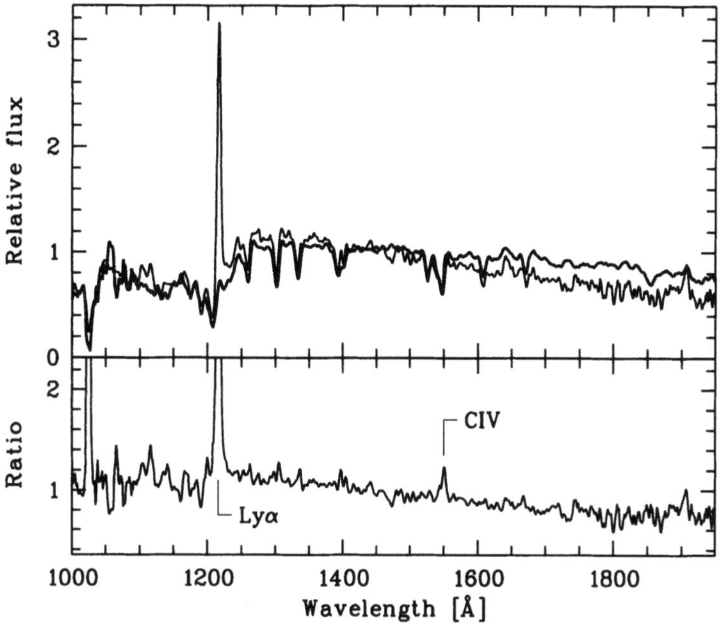

Fig. 1. Comparison of FDF composite spectra covering the redshift intervals $2 < z < 3$ (42 galaxies; *thick line*) and $3 < z < 4$ (22 galaxies; *thin line*), respectively. The lower panel gives the ratio of the $3 < z < 4$ and the $2 < z < 3$ composite spectrum

The majority of the mutual relations between Lyα, C IV, the low-ionisation interstellar absorption lines, and the continuum slope β turned out to be independent of redshift and in accordance with the results of [5], based on composite spectra covering the redshift range $2.5 < z < 3.5$. In contrast, the relation between β and the Lyα emission shows an increase of the average reddening for fixed Lyα emission strengths with decreasing redshift, which implies an evolution of the average properties of the galaxies' interstellar medium.

We interpret the evolutionary effects between $z \sim 3.3$ and 2.3 in terms of an increase of the metallicity, which is traced by C IV. Moreover, we find with increasing cosmic age a predominance of more massive starbursts, characterised by higher star-formation rates and stronger obscuration of the starburst cores by dusty gas clouds.

References

1. Bender, R., Appenzeller, I., Böhm, A., et al. 2001, in Deep Fields, ed. S. Cristiani, A. Renzini, & R.E. Williams, ESO Astrophysics Symposia (Springer), 96
2. Heidt, J., Appenzeller, I., Gabasch, A., et al. 2003b, A&A, 398, 49
3. Mehlert, D., Noll, S., Appenzeller, I., et al. 2002, A&A, 393, 809
4. Noll, S., Mehlert, D., Appenzeller, I., et al. 2003, A&A, submitted
5. Shapley, A.E., Steidel, C.C., Pettini, M., et al. 2003, ApJ, 588, 65

The Tully-Fisher Relation for Compact Group Galaxies

Cláudia Mendes de Oliveira[1], Philippe Amram[2], Henri Plana[3], and Chantal Balkowski[4]

[1] Universidade de São Paulo, Dept. de Astronomia, 05508-900, São Paulo, Brazil, and Max-Planck-Institut für Extraterrestrische Physik, Giessenbachstrasse, D-85748 Garching bei München, Germany
[2] Observatoire Astronomique Marseille-Provence & Laboratoire d'Astrophysique de Marseille, 2 Place Le Verrier, 13248 Marseille Cedex 04, France
[3] Universidade Estadual de Santa Cruz, Dept. Ciências Exatas e Tecnológicas, Rodovia Ilhéus-Itabuna, km. 16, 45650-000, Ilhéus, BA, Brazil
[4] Observatoire de Paris, GEPI, CNRS and Université Paris 7, 5 Place Jules Janssen, F-92195 Meudon Cedex, France

Abstract. Compact groups of galaxies are systems with high densities and low velocity dispersions, where galaxy interactions should play a major role in the dynamical evolution of the group members. They provide us with a unique laboratory in which to study galaxies that have likely undergone a substantial amount of dynamical evolution. In this contribution we determine the B-band Tully-Fisher (T-F) relation for 25 compact group (CG) galaxies, using $V_{\max}$ derived from 2-D velocity maps. Our main result is that CG galaxies lie on the T-F relation defined by more isolated systems. We also find that 75% of the CG galaxies studied (22 out of 29) have highly peculiar velocity fields. Nevertheless, a careful choice of inclination, position angle and kinematic center, obtained from the velocity field, and an average of the velocities over a large sector of the galaxy enabled the determination of fairly well-behaved rotation curves for the galaxies. We call attention to the importance of having 2-D velocity maps in order to obtain the correct internal kinematics of galaxies which live in such dense environments.

Introduction, Data and Main Result

It has become clear in the last few years that a number of the distant star-forming galaxies live in small dense groups. Numerous studies have also shown that a large fraction of ultra-luminous infrared galaxies are in interacting and merging systems which may be linked with the compact groups in our local universe. Understanding the physics and the kinematic properties of local interacting systems may then give us important clues about the evolution of distant galaxies.

The data used in the determination of the T-F relation and more details about this work are described in Mendes de Oliveira et al. (2003). In summary, our dataset consists of Hα emission-line velocity maps obtained with a Fabry-Perot system, from which rotation curves were obtained.

Fig. 1 shows our main result: galaxies in compact groups, as a whole, follow the T-F relation. This result is in contrast with an earlier result by Rubin et al. (1991), who found an offset of the T-F relation for most galaxies in CG in

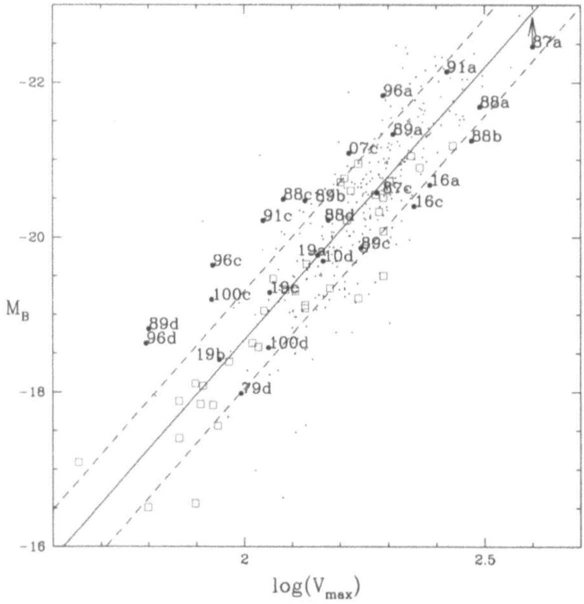

Fig. 1. The T-F relation in the B band for CG galaxies (black dots with numbers), for the Sb and Sc galaxies in the lower density environments studied by Courteau et al. (1997, small dots), and for Sb-Sd galaxies in the Ursa Major cluster studied by Verheijen (2001, open squares). The solid line represents a linear-squares fit to the Courteau's data ($M_B = -7.05 \log (V_{max}) - 4.57$). The broken lines indicate the $\pm$ one sigma dispersion (rms=0.63) for their data. The dispersion for the HCG galaxies around the solid line is rms=0.82 for the whole sample of 25 galaxies and rms=0.65 (similar to that for the control sample of Courteau 1997) if the outliers HCG 89d, 91c, 96c, 96d are not considered .

the sense that galaxies in CGs have lower velocities for a given luminosity. The disagreement is explained by the differences in the amplitudes of the derived rotation curves, V_{max}, mainly due to a different choice of galaxy parameters (center, inclination and position angle) and a less extended rotation curve in the case of curves obtained from long-slit spectroscopy. Generally, a choice of parameters guided by photometry alone (when no kinematic maps are available) will result in a lower V_{max} of a galaxy, affecting the T-F relation in the direction of the Rubin et al.'s result. This suggests that results based on rotation curves obtained from long-slit spectroscopy of distant interacting galaxies must be interpreted with caution.

References

1. Courteau S.: AJ, **114**, 2402 (1997)
2. Mendes de Oliveira, C., Amram, P., Plana, H., Balkowski, C.: AJ, in press.
3. Rubin V.C., Hunter D.A., Ford W.K.Jr.: ApJS, **76**, 153 (1991)
4. Verheijen M.A.W: ApJ, **563**, 694 (2001)

Extended HI Structures Around Early-Type Galaxies: Relics of Their Formation?

T.A. Oosterloo[1], E.M. Sadler[2], R. Morganti[1], and J.M. van der Hulst[3]

[1] Netherlands Foundation for Research in Astronomy, Dwingeloo, NL
[2] School of Physics, University of Sydney, NSW 2006, Australia
[3] Kapteyn Astronomical Institute, RuG, Landleven 12, 9747 AD, Groningen, NL

Abstract. Our systematic H I survey of early-type galaxies has revealed several objects with large, regular disks of neutral hydrogen of up to 200 kpc in size. Given their size and regular kinematics these disks must be quite old, in many cases well over 5×10^9 yr. They may represent key structures that trace the origin and evolution of these galaxies.

Hierarchical accretion and merging of small clumps appears to be a good description of the formation of early-type galaxies. In many early-type galaxies, (in particular field galaxies) signatures of such recent assembly can be observed. Optical fine structure – shells, dust lanes, "X" structures – revealed through unsharp masking of optical images, suggests a recent accretion event. Interestingly, these fine structures are correlated with the presence of an intermediate age (1-2 Gyr) population of stars (e.g. Schweizer et al. 1990). Moreover, the comparison of detailed stellar population models with high signal-to-noise optical spectra indicates the presence of an intermediate age population in several early-type galaxies (e.g. Trager et al. 2000). A small but significant fraction of early-type galaxies contains neutral hydrogen. The origin of this H I is generally thought to be external and due to recent accretion of companions. Thus, the H I appears to fit into the picture of continued accretion of small clumps and the associated intermittent episodes of modest star formation in some early-type galaxies. However, there appears to be a subclass of H I-rich early-type galaxies that does not fit into this picture. In some early-type galaxies the H I may not be due to a recent accretion but instead has been part of the galaxy for at least half a Hubble time.

Large Regular Gas Disks in Early-Type Galaxies

Our systematic H I survey of all early-type galaxies south of $\delta < -25°$ with $V < 6000$ km s^{-1} (based on the Parkes All Sky H I Survey: HIPASS, Barnes et al. 2001) has revealed several early-type galaxies that have very large (up to 200 kpc in size), regular disks of low column density H I. The largest of these disks have H I masses up to $10^{10}\ M_\odot$. Given their size and regular appearance, these disks must be quite old, in many cases well over 5×10^9 yr. Hence, they are

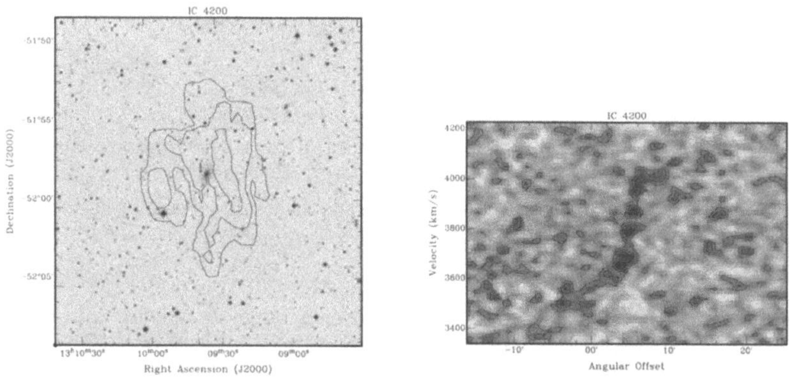

Fig. 1. Total intensity (left) and position-velocity plot along the major axis (right) for the early-type galaxy IC 4200. The H I disk is about 200 kpc in size. Contour levels are: $2, 4, 8, 16 \times 10^{19}$ cm^{-2} (left) and -3, 3, 6, 8 mJy beam^{-1} (right).

not related to recent accretions. The H I column density in these disks peaks at only 10^{20} atoms cm^{-2}. Thus, despite the large H I reservoir, no significant star formation is occurring and the H I disks will evolve only very slowly. In Fig. 1 the case of IC 4200 is shown as an example.

As part of this project, it is also interesting to report the discovery, in the vicinity of the field elliptical NGC 1490, of very small, isolated H II regions that are located about 100 kpc from from any galaxy visible in the optical (see Oosterloo et al. 2003). These H II regions reside at the edge of an H I cloud that, together with 3 other H I objects, forms an arc of 500 kpc long that half encircles NGC 1490. The total H I mass of the H I clouds is almost 10^{10} M$_{\odot}$. Only one of the clouds has a faint, low surface brightness optical counterpart. This is the H I-richest object known that is also detected optically: $M_{\rm HI}/L_V \sim 200$. The Hα luminosities of the H II complexes are $3-5 \times 10^{38}$ erg s^{-1} (i.e. 30-50 times Orion). The star formation rate estimated from these luminosities is very modest: around 0.005 $M_{\odot}$ yr^{-1} per region. The ionized gas has relatively high metallicities. Therefore, the gas clouds are most likely tidal remnants related to the large gas-rich merger that formed the field elliptical NGC 1490. Apparently, in such gaseous tidal remnants, isolated star formation on a very small scale can occur in the form of H II regions that are located far away from any optical identifiable galaxy.

References

1. Barnes et al. 2001, MNRAS 322, 486
2. Oosterloo et al. 2003, in *Recycling Intergalactic and Interstellar Matter*, eds. P.-A. Duc, J. Braine, and E. Brinks, in press (astro-ph/0310632)
3. Schweizer F. et al. 1990, ApJL 364, L33
4. Trager S.C., Faber S., Worthey G., Gonzáles J.J. 2000, AJ 120, 165

Analysing Galaxy Clustering in Future Surveys

Will J. Percival[1], Licia Verde[2], and John A. Peacock[1]

[1] Institute for Astronomy, University of Edinburgh, Royal Observatory, Blackford Hill, Edinburgh EH9 3HJ, UK
[2] Department of Physics & Astronomy, University of Pennsylvania, 209 South 33rd Street Philadelphia, PA 19104, USA

Abstract. Future deep galaxy surveys will cover a range in redshift over which the galaxy-bias and density field evolve significantly. In this brief conference proceedings contribution, we describe how models for these effects can be factored into a clustering analysis and can then be constrained by the data. Even for recent low-redshift surveys such as the 2dF galaxy redshift survey (2dFGRS), luminosity-dependent galaxy-bias can change the recovered clustering signal from that of the underlying density field [5].

Analysis Method

The key idea is to correct standard methods [2] by multiplying galaxies by both a weight and by the reciprocal of the expected offset $b(\mathbf{r}, L)$ from a fixed density field. This factor allows for an offset as a function of galaxy type (here parameterised by luminosity L), and as a function of distance $\mathbf{r}$. The weighted galaxy fluctuation field is defined as

$$F(\mathbf{r}) \equiv \frac{1}{N} \int dL \, \frac{w(\mathbf{r}, L)}{b(\mathbf{r}, L)} \left[n_g(\mathbf{r}, L) - \alpha n_s(\mathbf{r}, L)\right] \tag{1}$$

where $n_g(\mathbf{r}, L) = \sum_j \delta(\mathbf{r} - \mathbf{r}_j)\delta(L - L_j)$, with $\mathbf{r}_j$ being the location of the jth galaxy of luminosity L_j, and $n_s(\mathbf{r}, L)$ defined similarly for a synthetic catalogue with the same radial and angular sampling, but with no correlations. $w(\mathbf{r}, L)$ is the weight and N is the normalization

$$N = \left\{ \int d^3r \left[\int dL \, \bar{n}(\mathbf{r}, L) w(\mathbf{r}, L) \right]^2 \right\}^{1/2} . \tag{2}$$

$\bar{n}$ is the expected galaxy density. Following the standard procedure [2], in Fourier space we have that

$$\langle |F(\mathbf{k})|^2 \rangle = \int \frac{d^3k'}{(2\pi)^3} P(\mathbf{k}') |G(\mathbf{k} - \mathbf{k}')|^2 + P_{\text{shot}}, \tag{3}$$

with

$$G(\mathbf{k}) = \frac{1}{N} \int d^3r \int dL \, \bar{n}(\mathbf{r}, L) w(\mathbf{r}, L) e^{i\mathbf{k}\cdot\mathbf{r}}, \tag{4}$$

and

$$P_{\text{shot}} = \frac{1+\alpha}{N^2} \int d^3r \int dL \, \bar{n}(\mathbf{r}, L) \frac{w^2(\mathbf{r}, L)}{b^2(\mathbf{r}, L)}. \tag{5}$$

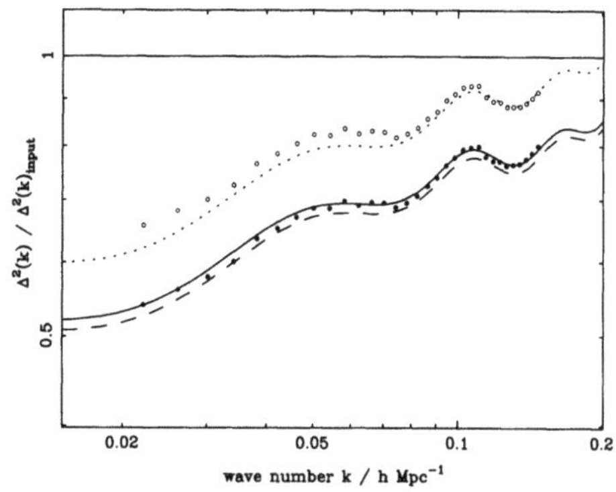

Fig. 1. the average recovered power spectrum from 1000 mock catalogues designed to mimic the part-complete 2dFGRS sample [4] (solid circles), weighted by optimal k-dependent weights, compared with the expected power (solid line). All the data are ratioed to the input power. Open circles give the same data, analysed with the same weights, but ignoring luminosity-dependent bias. The offset is caused by the change in normalization N and corresponds, on average, to a change in the mean galaxy bias [3]. Ignoring luminosity-dependent bias in the theory suggests that these data should be modeled by the dotted line (which has been normalized to the data). The dashed line shows the dotted line normalized as for the solid line: comparing these lines highlights the small systematic shape change caused by luminosity-dependent bias on the window shape (fixed N): the major systematic comes from the change in N and results in the offset between dotted line and open circles.

Ignoring differences in bias between galaxies of different luminosity reduces $G(\mathbf{k})$ and $P_{\rm shot}$ to the standard expressions [2]. Optimal weights to recover the matter power spectrum using this revised method in the absence of a window can be calculated and are dependent on k and the galaxy bias [5].

We highlight the usefulness of the revised method by considering the effect of luminosity-dependent bias on the recovered galaxy power spectrum from the 2dFGRS [1,4] in Fig. 1. The effect of a varying bias changes the weights $w(\mathbf{r}, L)$, producing a varying normalization constant N (Eq. 2) and altering the form of the expected window $G(\mathbf{k})$ (Eq. 4). These effects are demonstrated in Fig. 1 using log-normal mock catalogues [5] and convolved model power spectra. Only the second of these effects was an issue for the initial 2dFGRS analysis [4] which used k-independent weights, and was shown to be negligible [5].

References

1. Colless M., et al., MNRAS, **328**, 1039 (2001)
2. Feldman H.A., Kaiser N., Peacock J.A., ApJ, **426**, 23 (1994)
3. Lahav O., MNRAS, **333**, 961 (2002)
4. Percival W.J., et al., MNRAS, **327**, 1297 (2001)
5. Percival W.J., Verde L., Peacock J.A., MNRAS accepted, astro-ph/0306511, (2003)

The Relation Between Bulge Velocity Dispersion and Disk Circular Velocity in Galaxies

A. Pizzella, E. Dalla Bontà, E.M. Corsini, L. Coccato, and F. Bertola

Dipartimento di Astronomia, Università di Padova, Padova, Italy

Abstract. We investigate the relation between the central velocity dispersion, σ_c, and the circular velocity, V_{circ}, in galaxies, which could be interpreted as a relation between the masses of central black holes and dark matter halos. We consider an observationally homogeneous sample of 52 high surface brightness (HSB) spiral galaxies and 11 low surface brightness (LSB) spiral galaxies in addition to HSB literature data. We performed a straight line regression analysis in a linear scale, finding a good fit, also for low σ_c galaxies, always rejected in the previous studies. LSB galaxies, for the first time considered for this purpose, seem to behave differently, showing either higher values of V_{circ} or lower values of σ_c with respect to their HSB counterparts.

Studying the interplay between ionized-gas and stellar kinematics allows to address issues concerning the dynamical structure of disk galaxies, and to constrain the processes leading to their formation and evolution. In this regard it has been shown that masses of the central supermassive black holes, $M_\bullet$ [2,3], and dark matter halos, M_{DM} [1], are correlated with the stellar velocity dispersion, σ_c, of the host galaxy.

We derived the bulge velocity dispersion, σ_c (not corrected to an aperture of size $r_e/8$, as discussed in [4]) and the disk circular velocity, V_{circ} (from the flat part of the ionized-gas rotation curve), using the stellar and ionized-gas kinematics measured for 52 S0/a-Sc HSB spiral galaxies [5,6,4] and for 11 Sa-Sc LSB spiral galaxies [7]. Recently, Ferrarese [1] and Baes et al. [8] derived the $\sigma_c - V_{circ}$ relationship by fitting a power law to their dataset of 28 spiral galaxies, rejecting 4 galaxies with $\sigma_c < 70$ km s^{-1} because these have been considered as outliers. Enlarging the sample with the addition of the 20 elliptical galaxies of Kronawitter et al. [9] (who derived V_{circ} from mass models) the relation keeps its validity. From our sample we selected 6 HSB galaxies and 9 LSB with $R(V_{circ}) > R_{25}$ (to ensure a reliable determination of V_{circ}). We performed a straight line regression analysis in a linear scale (and not a power law as in previous studies) for the total sample of 30 HSB galaxies and 20 ellipticals. We find that the data are well fitted. Moreover also galaxies with $\sigma_c < 70$ km s^{-1} follow the same linear law and have been included in the sample (Fig. 1a). LSB galaxies are not fitted by the same law as the HSB and ellipticals. They have higher values of V_{circ} or lower values of σ_c with respect to HSBs. In Fig.1b we plot the $M_\bullet - M_{DM}$ relation derived from the $\sigma_c - V_{circ}$ relation of Fig.1a. The collapse of baryon matter has been claimed to induce a further concentration in the dark matter distribution, and a deepening of the overall gravitational well in the central regions. If this is the case, the finding that at a given V_{circ}

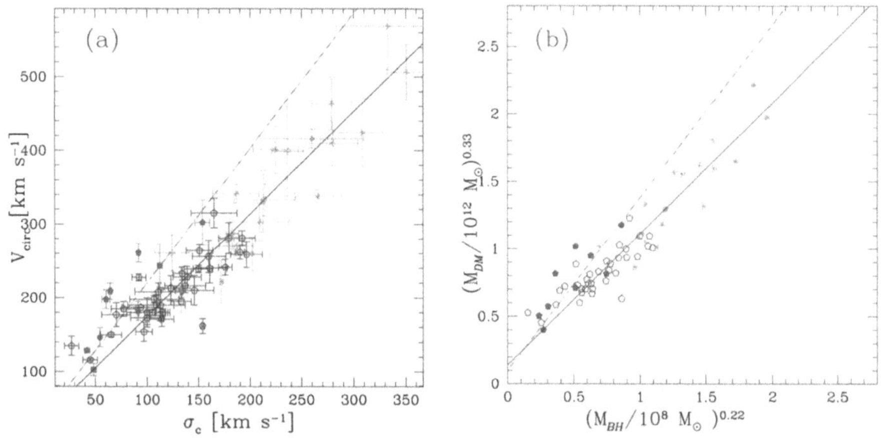

Fig. 1. (**a**): $\sigma_c - V_{circ}$ relation for the HSB spiral galaxies (empty symbols) from [1,8] and this work, for the LSB spiral galaxies (full symbols) from this work, and for the ellipticals from [9] (asterisks). The *continuous line* corresponds to the linear fit with all HSBs and ellipticals plotted. The *dashed line* corresponds to the linear preliminary fit with all LSBs plotted. (**b**): The $M_\bullet - M_{DM}$ relation for the galaxies of the left panel. The *dashed line* corresponds to the $M_\bullet - M_{DM}$ relation for LSB galaxies with $R(V_{circ}) > R_{25}$ only.

(which corresponds to a given M_{DM}, e.g. [10]) the central σ_c of LSBs is smaller than in their HSB counterparts indicates that in LSBs the baryon collapse did not produce a relevant DM central collapse and the dark halo radial density is therefore similar to the primordial one.

References

1. Ferrarese, L. 2002, ApJ, 578, 90
2. Ferrarese, L., & Merritt, D. 2000, ApJ, 539, L9
3. Gebhardt, K., Bender, R., Bower, G., et al. 2000, ApJ, 539, L13
4. Pizzella, A., Corsini E. M., Vega Beltràn, J. C. et al. 2003, A&A, submitted
5. Corsini, E. M., Pizzella, A., Sarzi, M., Cinzano, P., Vega Beltràn, J.C., Funes, J. G., Bertola, F., Persic, M., & Salucci, P. 1999, A& A, 342, 671
6. Vega Beltràn, J. C., Pizzella, A., Corsini, E. M., Funes, J. G., Zeilinger, W. W., Beckman, J. E., & Bertola, F. 2001, A&A, 374, 394
7. Pizzella, A., Corsini E. M., Magorrian, J. et al. 2003, in preparation
8. M., Buyle, P., Hau, G. K. T., & Dejonghe, H. 2003, MNRAS, 341, L44
9. Kronawitter, A., Saglia, R. P., Gerhard, O., & Bender, R. 2000, A&AS, 144, 53
10. Bullock, J. S., Dekel, A., Kolatt, T. S., Primack, J. R., & Somerville, R. S. 2001, ApJ, 550, 21

The RASS-SDSS Galaxy Cluster Survey

Paola Popesso, Hans Böhringer, and Wolfgang Voges

Max-Planck-Institut fur extraterrestrische Physik, 85748 Garching, Germany

Abstract. Solid observational evidences indicate a strong dependence of the galaxy formation and evolution on the environment. In order to study in particular the interaction between the intracluster medium and the evolution of cluster galaxies, we have created a large database of clusters of galaxies based on the largest available X-ray and optical surveys: the ROSAT All Sky Survey (RASS), and the Sloan Digital Sky Survey (SDSS). We analyzed the correlation between the total optical and the X-ray cluster luminosity. The resulting correlation of L_X and L_{op} shows a logarithmic slope of 0.6, a value close to the self-similar correlation. We analysed also the cluster mass to light ratio, by finding a significant dependence of M/L on the cluster mass with a logarithmic slope ranging from 0.27 in the i and r bands to 0.22 in the z band.

1 The RASS-SDSS Galaxy Cluster Catalog

For a detailed comparison of the appearance of cluster of galaxies in X-ray and in the optical, we have compiled a large database of X-ray and optical properties of a sample of clusters based on the largest available X-ray and optical surveys: the ROSAT All Sky Survey (RASS) and the Sloan Digital Sky Survey (SDSS). The X-ray-selected galaxy clusters of this RASS-SDSS catalog cover a wide range of masses, from groups of $10^{12.5}$ $M_\odot$ to massive clusters of 10^{15} $M_\odot$ in the redshift range from 0.002 to 0.45. The RASS-SDSS sample comprises 114 clusters. For each system we have uniformly determined the X-ray and optical properties . For a subsample of 53 clusters we also compiled the temperature and the iron abundance from the literature.

1.1 Correlating X-Ray and Optical Cluster Properties

The cluster optical luminosity is calculated, after statistical background subtraction, as the sum of all the galaxies within the N magnitude bins. A tight relation exists between L_{op} and L_X. In the $log(L_X)$-$log(L_{op})$ plane, the $L_{op}-L_X$ relation is a line with slope 0.44±0.03 and with a scatter of 0.3.

Given the $L_X - M$ relation of Reiprich & Böhringer 2001 and our new $L_X - L_{op}$ relation , also the mass should increase faster than L_{op}. To check this, we used the subsample of clusters with temperature, and with the $M-T$ relation of Finoguenov et al. 2001 we estimated the mass and analysed the $M - L$ and the $M/L - M$ relations (M and L_{op} are calculated both within r_{500}). M/L shows a clear dependence on the cluster mass with a slope from 0.27 in the i and r bands to 0.22 in the z band.

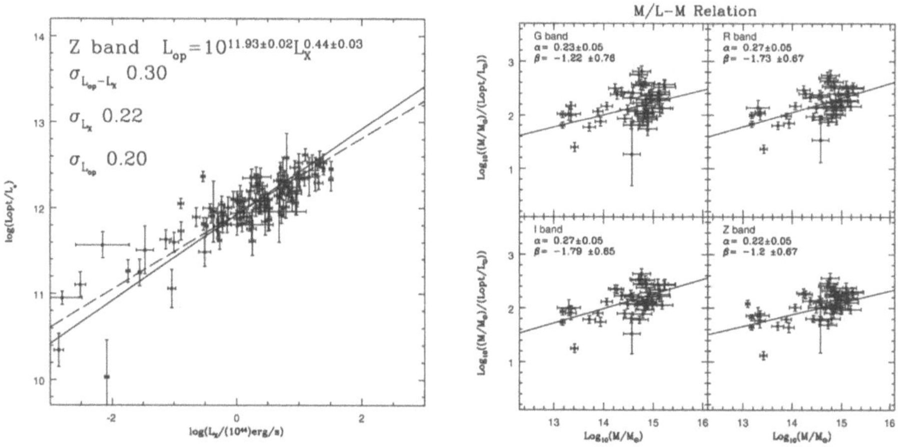

Fig. 1. A tight relation exists between L_X and L_{op} (left panel). The right panel shows the cluster M/L as a function of the M_{500} mass. The mass is estimated from the temperature with the $M_{500} - T_{ICM}$ relation of Finoguenov et al. 2001. The mass increases faster than the optical luminosity as predicted by our new $L_{op} - L_X$ relation.As a consequence of the $M_{500} - L_{op}$ relation, the M/L ratio is clearly an increasing function of the cluster mass. This is in agreement with the prediction based on our new $L_X - L_{op}$ relation.

Funding for the creation and distribution of the SDSS Archive has been provided by the Alfred P. Sloan Foundation, the Participating Institutions, the National Aeronautics and Space Administration, the National Science Foundation, the U.S. Department of Energy, the Japanese Monbukagakusho, and the Max Planck Society. The SDSS Web site is http://www.sdss.org/. The SDSS is managed by the Astrophysical Research Consortium (ARC) for the Participating Institutions. The Participating Institutions are The University of Chicago, Fermilab, the Institute for Advanced Study, the Japan Participation Group, The Johns Hopkins University, Los Alamos National Laboratory, the Max-Planck-Institute for Astronomy (MPIA), the Max-Planck-Institute for Astrophysics (MPA), New Mexico State University, University of Pittsburgh, Princeton University, the United States Naval Observatory, and the University of Washington.

References

1. Finoguenov, A. and Reiprich, T. H. and Böhringer, H.,2001,A&A,368,749
2. Reiprich, T. H. and Böhringer, H., 2002, ApJ,567,716
3. York, D.G., Adelman, J., Anderson, J.E., et al 2000, AJ, 120, 1579

Probing the Faint Radio Population Through Multi-Wavelength Information

Isabella Prandoni[1], Paola Parma[1], Mark H. Wieringa[2], Loretta Gregorini[3], Hans R. de Ruiter[4], Giampaolo Vettolani[1], and Ron D. Ekers[2]

[1] Istituto di Radioastronomia del CNR, Bologna, Italy
[2] Australia Telescope National Facility, Sydney, Australia
[3] Università di Bologna, Bologna, Italy
[4] Osservatorio Astronomico di Bologna, INAF, Bologna, Italy

1 Introduction

Deep 1.4 GHz counts show an upturn below a few milliJansky (mJy), corresponding to a more rapid increase in the number of faint sources. This change of slope is thought to be due to the emergence of a new population of radio sources, which does not show up at higher flux densities. Today we know that the faint radio population is a mixture of different classes of objects (AGNs, star-forming galaxies, normal ellipticals and spirals), but the relative fractions are still uncertain and very little is known about the luminosity properties and the redshift distribution of such sources. Understanding whether the dominant triggering process is star formation or nuclear activity has important implications on the study of the star-formation/black-hole-accretion history with radio-selected samples.

2 The ATESP Survey and its Follow-Up

In order to probe the origin of the radio emission in mJy and sub-mJy radio sources we are collecting multiwavelength information for the ATESP radio sources. The ATESP 1.4 GHz radio sample consists of a catalogue of 2967 objects complete down to $\sim$ 0.5 mJy and distributed in a region of 26 sq. degr ([3],[4]). In this work we present preliminary results obtained from the analysis of two sub-regions: (a) a 3 sq. degr area (ATESP-EIS region) where optical identification and spectroscopy are complete down to $I = 19$; (b) a 1 sq. degr. area where 5 GHz imaging down to $\sim$ 0.2 mJy (3σ limit) has been obtained with the ATCA, and very deep UBVRIJHK multicolor data will be soon available.

3 Evidences from the Optical Data

From the analysis of the ATESP-EIS region we find that, as for local samples, star-forming galaxies are typically characterized by low radio-to-optical ratios ($10 < R < 100$), while early-type objects and AGNs can reach $R >> 100$ ([5]). This means that evolutionary effects, if any, play a similar role on both the optical and the radio properties of the ATESP sources. If this behaviour holds also at deeper magnitudes (or higher redshifts) than probed in this analysis, we expect that the ATESP sample would not be sensitive to the low-R (star-forming) radio population at redshifts larger than $z \sim 0.5$. On the other hand, sub-mJy samples like the ATESP can be very helpful to probe the evolutionary properties of $z >> 0.5$ low-luminosity AGNs ($P \sim 10^{23-24}$ W/Hz).

4 Evidences from the Radio Data

A complementary way to constrain the origin of the radio emission in the faint radio sources is the analysis of their radio spectral properties. This approach is especially meaningful when high resolution images are available and radio source structures can be inferred.
In the 1 sq. degr. area observed at 5 GHz the spectral index $\alpha_{1.4}^{5}$ (where $S \sim \nu^{\alpha}$) measured for the ATESP sources shows a significant flattening with decreasing flux. At mJy level the radio spectra are steep, with a narrow dispersion around the expected value for standard synchrotron radio emission ($\alpha = -0.7$). At fainter fluxes, however, almost half (46%) of the radio sources show flat spectra ($\alpha > -0.5$) with a significant fraction (25%) of inverted-spectrum sources ($\alpha > 0$).
This result supports previous works suggesting a flattening of the spectra going to sub-mJy flux densities (see e.g.[2]). We notice that an increasing flattening going to sub-mJy fluxes is consistent with the finding of many flat and/or inverted spectral index sources at μJy flux densities ([1],[6]).
Flat spectra in radio sources usually indicate the presence of a self-absorbed nuclear core, but can also be produced on larger scales by thermal emission from stars. In order to better assess the origin of the radio emission in this sample, we are currently investigating the radio morphology of the sources. Nevertheless, hints toward a nuclear origin in our radio sources come from the optical information available. Most of the flat-spectrum sources have high radio-to-optical ratios ($R > 1000$), typically associated to classical powerful radio galaxies and quasars; while flat-spectrum sources with low R values are preferentially identified with early-type galaxies, where the radio emission is probably triggered by low-luminosity AGNs.
Further important information will come from the multicolor optical data (see § 2), which will allow us to identify $\sim 70\% - 80\%$ of the ATESP sources in this 1 sq. degr. area and to immediately determine their spectral types and (photometric) redshifts.

References

1. E.B. Fomalont, R.A. Windhorst, J.A. Kristian, et al.; AJ, **102**, 1258 (1991)
2. C. Gruppioni, G. Zamorani, H.R. de Ruiter, et al.; MNRAS, **286**, 470 (1997)
3. I. Prandoni, L. Gregorini, P. Parma, at al.; A&AS, **146**, 31 (2000)
4. I. Prandoni, L. Gregorini, P. Parma, et al.; A&AS, **146**, 41 (2000)
5. I. Prandoni, L. Gregorini, P. Parma, et al.; A&A, **369**, 787 (2001)
6. R.A. Windhorst, E.B. Fomalont, R.B. Partridge, et al.; ApJ, **405**, 498 (1993)

The VIRMOS U–Band Imaging Survey: The Deep and Wide Fields

Mario Radovich[1], Magda Arnaboldi[2], Vincenzo Ripepi[1], Henry Joe McCracken[3], Yannick Mellier[3], Emmanuel Bertin[3], Sebastien Foucaud[4], Stephen Gwyn[4], and Olivier Le Fèvre[4]

[1] INAF/Osservatorio Astronomico di Capodimonte, via Moiariello 16, 80131 Napoli, Italy
[2] INAF/Osservatorio Astronomico di Pino Torinese, Via Osservatorio 20, 10025 Torino, Italy
[3] Observatoire de Paris, LERMA, 61 Avenue de l'Observatoire, 75014 Paris, France
[4] Laboratoire d'Astrophysique de Marseille, Traverse du Siphon, 13376 Marseille Cedex 12, France

Abstract. The VIRMOS preparatory imaging survey (Le Fèvre et al. 2003) was designed to provide a photometric coverage in the *UBVRI* bands of one deep and three wide areas to be be observed with VIMOS. The *U*–band data for the deep and one wide area are here presented.

1 The Data

A total area of 13 sq. degr. was observed in th *U*–band using the WFI camera at the ESO 2.2m telescope at La Silla, Chile. The survey was started in 1999; as until 2001 no *U*–band filter was available at WFI, a filter borrowed from the Loiano observatory was initially used. This filter, whose field of view does not cover the whole WFI area, was substituted by the ESO U/360 filter as soon as it became available. One deep field was observed with total exposure time $\sim$ 22 hrs, covering $\sim$ 1 sq. degr. Three wide fields (F10, F14, F22) were observed with exposure time $\sim$ 5000 s covering 4 sq. degr. each. Of these, only the F22 field was entirely observed with the ESO filter.

2 Data Processing and Quality Assessment Tests

The prereduction of the images was done using the *MSCRED* package in *IRAF*. Astrometric and photometric calibration were done using the *WIFIX* package (http://www.na.astro.it/~radovich). A global astrometric solution was computed over the whole field (see McCracken et al. 2003, Radovich et al. 2003 for more details). The resampling and coaddition were finally done using the *SWarp* tool developed by E. Bertin (http://terapix.iap.fr/soft). The result is a seamless image from which catalogs are extracted. Quality assessment tests were performed checking the colors of sources classified as point-like (Fig. 1) and magnitude-number counts of galaxies (Fig. 2). Work is in progress for the determination of the *U*–band luminosity function and the analysis of the *U*–band morphology of nearby galaxies compared to the *I*–band.

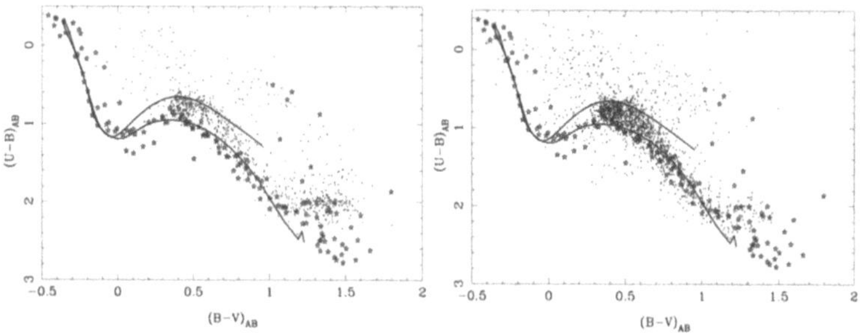

Fig. 1. Colors of bright ($U_{AB} < 22.5$) point-like sources (dots) from the deep F02 pointing (left) and the wide F22 pointing (right). Also displayed are: the library from Pickles (1998) (***stars***); the Kurucz model atmospheres (Kurucz 1979) with $\log g =$ 4.0, [M/H]=0 (lower curve) and [M/H]=-5.0 (upper curve). In both fields disk (solar metallicity) and halo (low metallicity) stellar populations are visible. The blue ($U - B <$ 1) sources outside the stellar locus are likely AGN candidates.

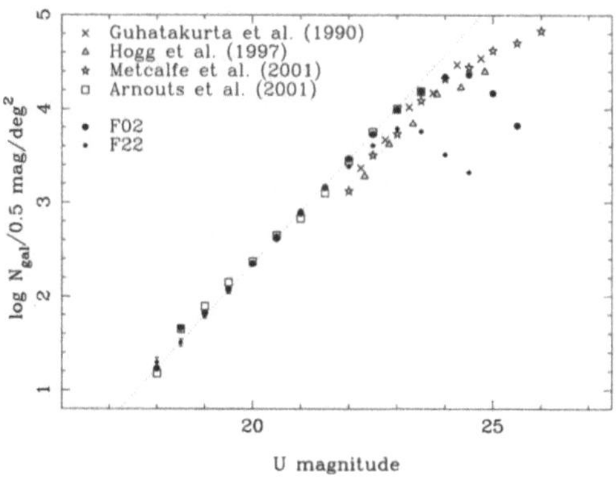

Fig. 2. U broad-band magnitudes vs. number counts of galaxies in the deep and wide fields, compared with literature data. The 50% completeness magnitudes are $U_{AB} =$ 25.4 mag (F02) and $U_{AB} = 23.8$ mag (F22).

References

1. Kurucz, R.L. 1979, ApJS, 40, 1
2. Le Fèvre, O., Mellier, Y., McCracken, H.J., et al., 2003, astro-ph/0306252
3. McCracken, H.J., Radovich, M., Bertin, E., et al., 2003, A&A, 410, 17
4. Pickles, A.J., 1998, PASP, 110, 863
5. Radovich, M., Arnaboldi, M., Ripepi, V., et al., 2003, A&A, submitted

UV – Optical Colors of the Host Galaxies of $z \sim 2-3$ Radio-Quiet Quasars

Susan E. Ridgway[1], Timothy M. Heckman[1], Daniela Calzetti[2], and Matthew Lehnert[3]

[1] Johns Hopkins University, Dept. of Physics and Astronomy, 3400 N. Charles St., Baltimore MD 21218

[2] Space Telescope Science Institute, 3700 San Martin Dr., Baltimore, MD 21218

[3] MPE, Postfach 1603, D-85740 Garching, Germany

1 Introduction

Understanding the links between the formation and evolution of the AGN population and the formation and evolution of galaxies requires investigation of the relationship between these populations at high redshift, where the quasar luminosity function is reaching a maximum. We have made a deep space-based near-infrared (NIR) imaging study of the properties of the host galaxies of $z \sim 2-3$ radio-quiet quasar (RQQ) hosts and compared them to those of the other known high-z galaxies, the Lyman break galaxies (LBGs) and the hosts of radio-loud objects. We found that the RQQ hosts have rest-frame optical magnitudes and sizes that are similar to the LBGs, and are much fainter than most radio-loud hosts at these epochs. We have also obtained rest-frame ultraviolet (UV) imaging of this sample, and use this to study the UV–optical colors of the hosts. These allow us to compare the star-forming properties of the quasar hosts with those of the LBGs.

2 Imaging and Results

The NIR observations were made in the F165M and F160W filters with HST's NICMOS camera, while the optical observations were made in the F606W filter with WFPC2. We interleaved observations of nearby stars with those of the quasars, and used these to characterize the point spread function (PSF). We make a simple 2D PSF subtraction, varying the centering and scaling of the PSF versus the quasar and assuming that the residual host flux decreases monotonically from the nuclear region.

At 1.6μm observed (rest-frame B and V), we detected extended flux around the quasar nucleus in all 5 objects [4]. The 5 radio-quiet quasars at redshifts of 1.8 – 2.7 have host galaxies with rest-frame absolute magnitudes of M_V −20 to −23, about 0.2 to 4 L_*. In Figure 1 we show the absolute rest frame V magnitudes for these RQQ hosts vs. those of various other samples of host galaxies: HDF Lyman break galaxies and radio-loud galaxies at similar redshifts, and the hosts of RLQs and RQQs at low-z. The dashed lines represent Bruzual & Charlot models for passive evolution of a galaxy formed at high-z. We determine half-light radii for

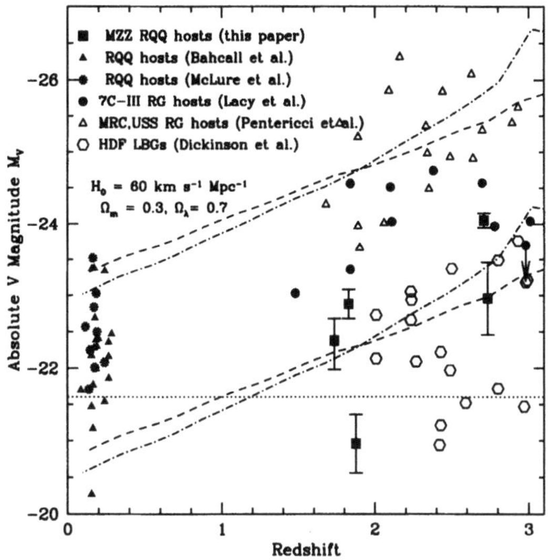

Fig. 1. The total absolute rest-frame V magnitudes versus redshift for our quasar hosts (squares) and other samples of host galaxies

the hosts of < 4 kpc, which are quite consistent with the half-light radii observed in LBGs at these z, while much more compact than the hosts associated with low-z quasar hosts or high-z radio galaxies ($\sim$ 8 - 10 kpcs).

Detection of the extended emission in the rest-frame UV is more difficult since the nucleus in the rest-frame UV is a much larger percentage of the total emission. Nonetheless, we were able to detect extended emission around the nucleus in 4 of the 5 cases. There are discrete off-nuclear components in some objects that are much bluer than the extended envelope;‘ however, we are interested in the global colors of these galaxies in order to compare them to other objects such as the LBGs.

We derive the rest-frame optical-UV colors with aperture photometry, parametrized as the spectral index α_R^H, where $S_\nu \propto \nu^{-\alpha}$. i We compare these colors to the LBGs using results from [1], and find that though the color distribution of our sample of objects overlaps that of the LBGs, the average color of the host galaxies may be redder than those of the LBGs. One object, MZZ 11408, is a relaxed galaxy both in morphology and in color.

3 Discussion

We present here the results of a rest-frame optical and UV imaging survey of a sample of 5 RQQ at $z \sim 2-3$, the peak of the quasar luminosity function. It now seems fairly conclusive from this work and many others that RQQs at high z are associated with L_* galaxies, in contrast to the bright, many times L_*

galaxies associated with radio-loud objects at these redshifts [4], [3], [5]. These results are roughly consistent with the hierarchical assembly theories of galaxy formation ([2]). However, our rest-frame UV imaging suggests that these hosts are not as actively star-forming as the LBGs at similar redshifts.

References

1. Dickinson, M. 2000, Philosophical Transactions of the Royal Society of London, 358, 2001.
2. Kauffmann, G. & Haehnelt, M. 2000, MNRAS, 311, 576
3. Kukula, M. et al. 2001, MNRAS, 326, 1533.
4. Ridgway, S.E., Heckman, T.M., Calzetti, D. & Lehnert, M. 2001, ApJ, 550, 122
5. Rix, H.-W. et al. 2001, ASP Conf. Ser. 237: Gravitational Lensing: Recent Progress and Future Goals, 169

Galaxies of 10^{12} $M_\odot$ at $z \geq 4$ Along the K-z Sequence

Brigitte Rocca-Volmerange[1,2]

[1] Institut d'Astrophysique de Paris, 98bis Bd Arago, F-75014 Paris
[2] Université de Paris-Sud, France

Abstract. Stellar masses of galaxies are derived from the K-band (2.2 μm) photometry by taking into account evolution and cosmological effects with the code PÉGASE. The evolution scenarios are robust: they give reference spectra for valid photometric redshifts and fit the deepest galaxy counts with the standard cosmological model. Baryonic masses of the progenitor gaseous clouds are then derived. The baryonic maximum value $M_{\mathrm{bar,max}} = 10^{12}$ $M_\odot$ fits the radio galaxy K-z sequence from z=0 to 4. Since this cut value is observed at all z in the Hubble diagram, it favors a physical origin: i. e. the fragmentation limit of 10^{12} $M_\odot$ predicted by Rees & Ostriker, 1977. One consequence is the discovery of high mass ($\simeq 10^{12}$ $M_\odot$) galaxies at $z \geq 4$, constraining the mass accumulation time-scales (Rocca-Volmerange et al, 2003).

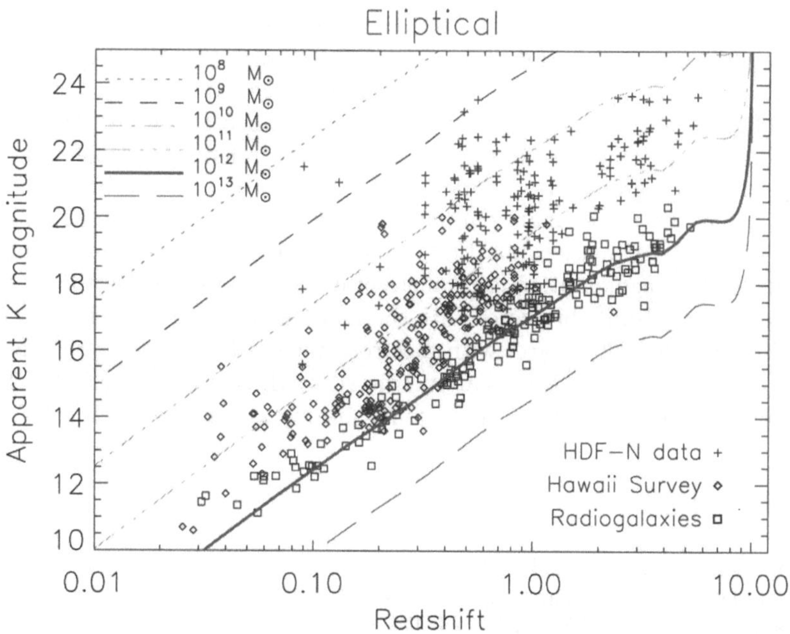

Fig. 1. The 6 K-z sequences predicted by elliptical galaxy models of baryonic masses $10^{8\ to\ 13} M_\odot$ compared to observations. The K-z sequence traced by the most powerful distant radio galaxies is fitted by $10^{12} M_\odot$ from z=0 to $\simeq$4. The IMF is from Salpeter.

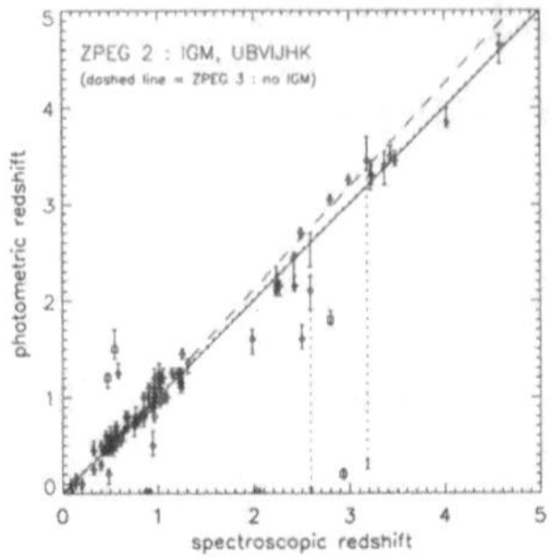

Fig. 2. Photometric versus spectroscopic redshifts of the HDF-N survey computed with similar evolution scenarios of PÉGASE.2 (ZPEG.2) (Le Borgne & Rocca-V., 2002).

Measuring the masses of distant galaxies imparts essential clues for galaxy evolution models. While dynamical measurements become less efficient at high z due to the $(1 + z)^{4}$ luminosity fading, the stellar mass estimates are potentially fruitful. The K-band photometry is favored because it is dominated by slowly evolving stars. The code PÉGASE.2 (http://www.iap.fr/pegase) is used with evolution scenarios (including inflow, star formation law, winds), typical of spectral types. The baryonic mass of the cloud is the initial mass of the progenitor from which the gas inflows to form the galaxy. Figure 1 compares the K-z sequences predicted for various baryonic masses with observed radio galaxy samples (De Breuck et al, 2002) and deep field samples. The most luminous galaxies and powerful radio galaxy hosts trace the K-z sequence, well fitted by the $M_{bar,max} = 10^{12} M_{\odot}$ model of elliptical galaxies (Rocca-Volmerange et al, 2003). The spiral models cannot fit the sequence at high z. At $z < 0.3$, the predictions of elliptical and spiral models are indistinguishable. The sensitivity to other parameters (cosmological, IMF, formation redshift) confirms that M_{bar} appears as the main parameter of the Hubble K-band diagram. The validity of the stellar mass calibration is due to i) the upper K-z mass limit corresponds to the fragmentation critic mass of models by Rees & Ostriker, (1977) ii) photometric redshifts predicted with the same template spectra (Le Borgne & Rocca-Volmerange, 2002) agree with spectroscopic redshifts on $0 \leq z \leq 4$ (Figure 2), iii) the deepest galaxy counts are predicted by these scenarios (Fioc & R.V., 1999). What is then the link of high stellar masses to massive black holes?

References

1. De Breuck, C., et al. 2002, AJ, 123, 637
2. Le Borgne, D., Rocca-Volmerange, B., 2002, A & A, 386, 446
3. Rees, M., Ostriker, J, 1977, MNRAS, 179, 541
4. Rocca-Volmerange, B., et al., astroph/0311490
5. Fioc, M., Rocca-Volmerange, B., 1999, A & A, 344, 393

Averaging the Universe: The Cosmic Color and Stellar Mass Density to $z \sim 3$

Gregory Rudnick[1], Hans-Walter Rix[2], Marijn Franx[3], Ivo Labbé[3], and the FIRES Collaboration

[1] MPI für Astrophysik, Karl-Schwarzschild-Str. 1, Garching, D-85741,Germany
[2] MPI für Astronomie, Königstuhl 17, Heidelberg, D-69117, Germany
[3] Leiden Observatory, PO BOX 9513, 2300 RA Leiden, Netherlands

One key element of the study of galaxy evolution is the measurement of the buildup of stellar mass in the Universe. While tremendous progress has been made in measuring the star formation rate (SFR) of the Universe as a function of redshift [4][6] it is also desirable to measure the stellar mass buildup directly. We have performed such a measurement using the very deep near infrared (NIR) data taken in the Hubble Deep Field South (HDF-S) as part of the Faint Infrared Extragalactic Survey (FIRES) [3]. Our deep NIR data, coupled with the extremely deep optical data from the Hubble Space Telescope (HST), allow us to select galaxies with $L > 0.5L_*$ in the K-band, which gives us access to the rest-frame optical out to $z \sim 4$. With our seven-band filter set $(0.3 - 2.2\mu m)$ we derived accurate photometric redshifts with $\langle\ |z_{spec} - z_{phot}|\ /\ (1 + z_{spec})\ \rangle = 0.05$ at $z_{spec} > 1.3$ using the technique from [5]. In addition the data on the HDF-S, we have also used the FIRES data on the MS1054-03 field [2].

We derived rest-frame optical luminosities for all of the galaxies in our sample and measured the rest-frame optical luminosity density $j_\lambda^{\rm rest}$ of galaxies (with $\mathrm{L}_\mathrm{V}^{rest}$ greater than a fixed threshold) in three equal volume bins. $j_\mathrm{V}^{\rm rest}$ is constant to within a factor of ~ 2 over the whole redshift range.

Using the $j_\lambda^{\rm rest}$ values in the U, B, and V-bands we calculated the cosmic mean color for all galaxies. The cosmic color bluens in all bands with increasing redshift. While the $(U-B)_{rest}$ and $(B-V)_{rest}$ colors at $z = 0.1$ are 0.16 and 0.75 respectively they are -0.39 and 0.29 at $z = 2.8$. We interpret bluing of the cosmic color primarily as a decrease in the mean stellar age of galaxies as a function of increasing lookback time. The cosmic colors vary much more smoothly with redshift than the $j_\lambda^{\rm rest}$ estimates, largely due to their lower sensitivity to field-to-field variance. Using the cosmic colors we modeled the mean properties of the Universe with a set of simple smoothly changing star formation histories (SFHs). Because the SFHs of individual galaxies can be - and at high redshift often are - quite bursty, large systematic errors can be introduced when modeling the individual colors with smooth or two component models. The mean cosmic SFH on the other hand is much smoother than that of the individual SFHs and a model with a smooth SFH can be applied in a more consistent fashion.

Using the stellar mass-to-light ratios $\mathrm{M/L}_\mathrm{V}^*$ derived from the models, we derived the evolution of the cosmic stellar mass density ρ_* from $z = 0.1$ to $z = 3.2$ (see Figure 1). We compared our measurements in the HDF-S and

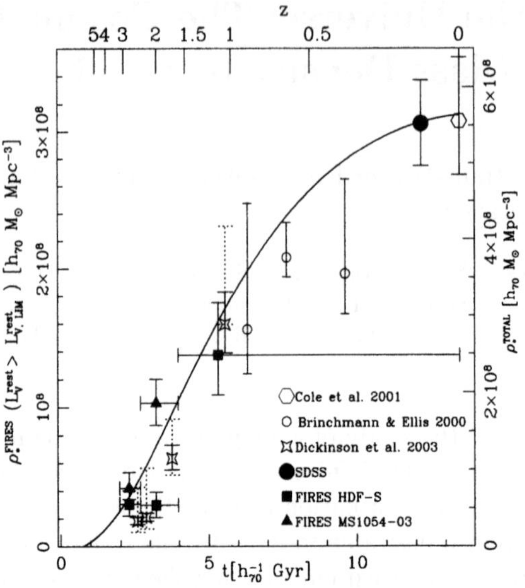

Fig. 1. The build-up of ρ_* as a function of redshift. The solid points are for galaxies with $L_V^{rest} > L_{V,\mathrm{thresh}}^{rest}\ h_{70}^{-2}\ L_\odot$. The FIRES data on the HDF-S are represented by the solid squares, the FIRES MS1054-03 data by the solid triangles, and the SDSS data by the solid circle. The y-axis scale on the left side corresponds to the ρ_* values for these points. The hollow points show the total ρ_* measurements from the one-component models in the HDF-N (D03; hollow stars; calculated assuming solar metallicity), the CFRS (Brinchmann & Ellis 2000; hollow circles), and the 2dFGRS + 2MASS (Cole et al. 2001; hollow hexagon). The y-axis scale on the right hand side corresponds to the ρ_* estimates for these points. The relative scaling of the two axes was adjusted so that our SDSS ρ_* estimate was at the same height as the total ρ_* estimate of [1]. The solid curve is an integral of the SFR(z) from [1].

MS1054-03 fields to those from other authors as well as to the integral of the SFR(z) curve from [1]. There is a global trend apparent in which ρ_* at $z \sim 3$ was $\sim 5-10$ times lower than at $z = 0.1$. The field-to-field variance is obviously large and a more accurate determination will require observations of similar depth over many different fields.

References

1. Cole, S. et al. 2001, MNRAS, 326, 255
2. Förster Schreiber, N. et al. in preparation
3. Labbé, I. et al. 2003, AJ, 125, 1107
4. Madau, P., Ferguson, H. C., Dickinson, M. E., Giavalisco, M., Steidel, C. C., & Fruchter, A. 1996, MNRAS, 283, 1388
5. Rudnick, G., Franx, M., Rix, H.-W., Moorwood, A., Kuijken, K., van Starkenburg, L., van der Werf, P., Röttgering, H., Labbé, I. 2001, AJ, 122, 2205
6. Steidel, C. C., Adqelberger, K. L., Giavalisco, M., Dickinson, M., & Pettini, M. 1999, ApJ, 519, 1

The TESIS Project: Revealing Massive Early-Type Galaxies at $z > 1$

P. Saracco[1], M. Longhetti[1], P. Severgnini[1], R. Della Ceca[1], V. Braito[1], R. Bender[2], N. Drory[3], G. Feulner[2], U. Hopp[2], F. Mannucci[4], and C. Maraston[5]

[1] INAF - Osservatorio Astronomico di Brera, Via Brera 28, 20121 Milano, Italy
[2] Universitäts-Sternwarte München, Scheiner Str. 1, 81679 München, Germany
[3] University of Texas at Austin, Austin, Texas 78712
[4] IRA-CNR, Largo E. Fermi 5, 50125 Firenze, Italy
[5] Max-Plank-Institut für extraterrestrische Physik, Garching bei München, Germany

1 Probing the Formation of Early-Type Galaxies

How and when present-day massive early-type galaxies built up and what type of evolution has characterized their growth (star formation and/or merging) still remain open issues. The different competing scenarios of galaxy formation predict much different properties of early-type galaxies at $z > 1$. The "monolithic" collapse predicts that massive spheroids formed at high redshift ($z > 2.5 - 3$) and that their comoving density is constant at $z < 2.5 - 3$ since they evolve only in luminosity. On the contrary, in the hierarchical scenario massive spheroids are built up through subsequent mergers reaching their final masses at $z < 1.5$ ([3], [5]). As a consequence, massive systems are very rare at $z > 1$, their comoving density decreases from $z = 0$ to $z \sim 1.5$ and they should experience their last burst of star formation at $z < 1.5$, concurrent with the merging event(s) of their formation. These opposed predicted properties of early-types at $z > 1$ can be probed observationally once a well defined sample of massive early-types at $z > 1$ is available. We are constructing such a sample through a dedicated near-IR very low resolution ($\lambda/\Delta\lambda \simeq 50$) spectroscopic survey (TNG EROs Spectroscopic Identification Survey, TESIS, [6]) of a complete sample of 30 bright (K<18.5) Extremely Red Objects (EROs).

The sample has been selected over two fields (~360 arcmin^2) of the Munich Near-IR Cluster Survey (MUNICS, [2]) covered by B, V, R, I, J and K-band observations. The red optical-to-near-IR colors (R-K$\geq$5.3) allow to select $z > 1$ evolved stellar systems; the bright K magnitudes assure the selection of massive galaxies and the near-IR spectra allow the detection of the 4000 break at $z > 1$. The survey is carried out at the 3.6m Italian Telescopio Nazionale Galileo (TNG) with the prism disperser Amici designed to cover the full spectral range (8000-25000 Å) in a single shot, thus being very efficient in detecting continuum breaks and in describing spectral shapes.

In parallel with the near-IR spectroscopic follow-up we obtained 150 ks of XMM-Newton observations for the two selected fields (75 ks each) to study the nature of the X-ray emitting EROs. The results relevant to the first set of XMM observations are presented by Severgnini et al. (see contribution in this volume).

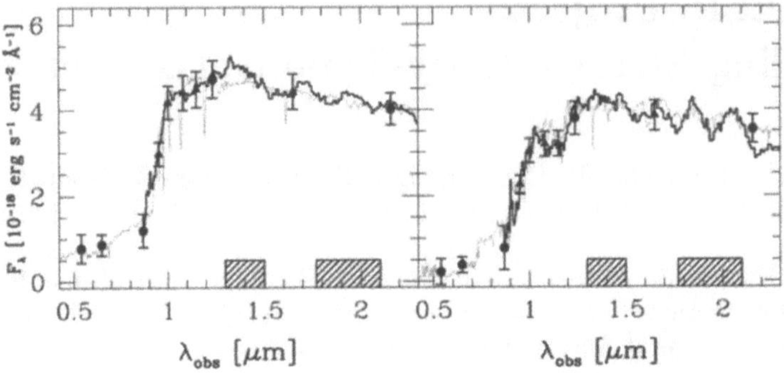

Fig. 1. Smoothed Amici spectra (black histogram) of two out of the 7 early-type galaxies superimposed on synthetic spectra ($\tau < 1$ Gyr and $A_V < 0.5$ mag; thick gray line). The circles are the photometric data in the V, R, I, J and K' bands and the filled triangles are additional photometric points derived from the observed spectra (see [6]). Atmospheric windows with opacity larger than 80% are marked by the shaded areas.

2 The Density of $\mathcal{M}_{star}>10^{11}M_\odot$ Early-Type Galaxies at $z > 1$

We classified 10 out of the 13 EROs (40% of the sample) observed so far: 7 early-type galaxies and 3 starbursts. The 7 early-type galaxies have apparent magnitudes $16.7 \leq K \leq 18.3$ and are at redshift $1.2 < z < 1.45$. Their rest-frame K-band absolute magnitudes are in the range $-26.9 \leq M_K \leq -25.7$, i.e. their luminosities are $L > 5L^*$ (we considered M^*_K=-24.2 from [1], H_0=70 Km s^{-1} Mpc^{-1}, $\Omega_m = 0.3$ and $\Omega_\lambda = 0.7$). In the hypothesis of passive evolution from $z \sim 1.5$ to $z = 0$ ($\Delta K \simeq 1$ mag), their resultant luminosities would be $L_{z=0} > 2L^*$. Thus, these early-types are the $z > 1$ counterpart of the ellipticals populating the bright end of the local luminosity function. Assuming a mass-to-light ratio $\mathcal{M}/L_K$=0.5 $[M/L]_\odot$, we derived a stellar mass well in excess to $2 \times 10^{11} M_\odot$ in all of them.

These 7 early-types account for a comoving density of about 2.6×10^{-5} Mpc^{-3}. This is a lower limit to the density of $\mathcal{M}_{star}>10^{11}M_\odot$ early-types at $z > 1$ since only 40% of the sample of EROs has been observed. Since, the number density of local $L_{z=0} > 2L^*$ is 7×10^{-5} Mpc^{-3} (we used $\phi^*_{E/S0} = 1.5 \times 10^{-3}$ Mpc^{-3}, from [4]), the comoving density of massive ellipticals cannot vary more than a factor ~ 2 from $z \sim 1.5$ to $z = 0$. The completion of the survey will allow us to further refine this value thus constraining severely the evolution of massive ellipticals.

References

1. Cole S., Norberg P., Baugh C., et al. 2001, MNRAS 326, 255
2. Drory N., Feulner G., Bender R., et al. 2001, MNRAS 325, 550
3. Kauffmann G., Charlot S. 1998, MNRAS 297, L23
4. Marzke R. O., Da Costa L. N., Pellegrini P. S., et al. 1998, ApJ 503, 617
5. Moustakas L. A., Somerville R. S. 2002, ApJ 577, 1
6. Saracco P., Longhetti M., Severgnini P., et al. 2003, A&A 398, 127

The TESIS Project: Are Type 2 QSO Hidden in X-Ray Emitting EROs?

P. Severgnini[1], R. Della Ceca[1], V. Braito[1], P. Saracco[1], M. Longhetti[1], R. Bender[2], N. Drory[3], G. Feulner[2], U. Hopp[2], F. Mannucci[4], and C. Maraston[5]

[1] INAF-Osservatorio Astronomico di Brera (OAB), via Brera 28, 20121 Milano, Italy
[2] Universitäts-Sternwarte München, Scheiner Str. 1, 81679 München, Germany
[3] University of Texas at Austin, Austin, Texas 78712
[4] IRA-CNR, Largo E. Fermi 5, 50125 Firenze, Italy
[5] Max-Plank-Institut für Extraterrestrische Physik, Garching bei München, Germany

1 Introduction

X-ray selected EROs are, on average, the hardest X-ray sources in medium and deep X-ray fields. This coupled with their extremely red colors (R-K>5) suggest that they represent one of the most promising population where looking for high-luminosity ($L_X > 10^{44}$ erg s^{-1}) and X-ray obscured ($N_H > 10^{22}$ cm^{-2}) type2 AGNs, the so called QSO2 (e.g. [5]; [4]; Mignoli et al. submitted to A&A). These latter are predicted in large density by the synthesis model of the Cosmic X-ray background ([3]) even if only few observational evidences have been found so far (e.g. [1] and references therein; Caccianiga et al. A&A accepted).

With the aim at studying the properties of X-ray emitting EROs our group has recently gained 75 ksec (AO-2 period) and 78 ksec (AO-3 period) of XMM-Newton observations centered on two MUNICS (MUNICS Near-IR Cluster Survey, [2]) fields of about 180 arcmin^2 each. For these fields photometric observations in the B, V, R, I, J, K' bands are available (Drory et al. 2001) down to limiting magnitudes of R$\sim$24.5 mag and K'$\sim$19.5 mag. All the EROs present in these 2 fields having a K' magnitude brighter than 18.5 are already under investigation by the TESIS (TNG EROs Spectroscopic Identification Survey, Saracco et al. this conference and references therein) project.

Here we present the analysis of the XMM-Newton data for the S2F1 field and we discuss the X-ray properties of the 6 X-ray emitting EROs detected down to $\sim 10^{-15}$ erg cm^2 s^{-1}. We assume H_0=65 km s^{-1} Mpc^{-1} and q_0=0.

2 X–Ray Data Analysis

The field S2F1 was observed by XMM-Newton on February 11, 2003. The data have been cleaned and processed using the Science Analysis System version 5.3.3. The net exposure time after data cleaning is about 41 ksec. We have cross–correlated the astrometrically corrected X–ray source catalogue with the TESIS-EROs catalogue. We found 5 secure X-ray emitting EROs and one more X-ray source which is more likely associated with an ERO counterpart. The detailed analysis of the 6 sources is presented in Severgnini et al. (in preparation). On the basis of their photometric redshifts all of the 6 X-ray emitting EROs have observed 2-10 keV luminosities ($L_{(2-10\ keV)}$ $>10^{42}$ erg s^{-1}) and X-ray-to-optical flux ratios ($Log(F_{(2-10\ keV)})/Log(F_R)$ >-1) suggesting the presence of AGN. For three out of the 6 EROs we have been able to perform a complete X-ray spectral analysis. The X-ray data are well fitted by a single power-law model (Fig. 1).

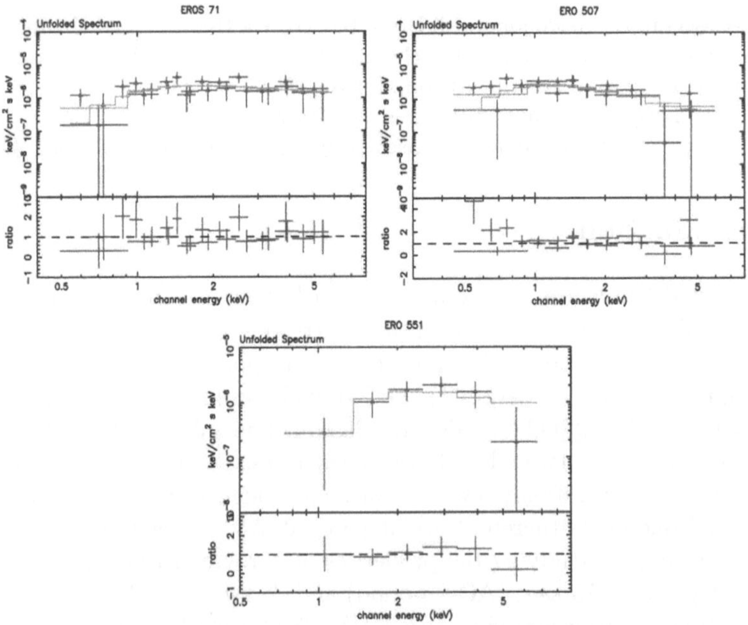

Fig. 1. *XMM-Newton* spectra in energy units (solid points) and best–fit model (continuous line). Ratios between data and the best–fit model values as a function of energy are reported in the lower panels.

A pure thermal component is rejected at more than 97% confidence level and the addition of a thermal component to the power-law model is not statistically required. These results, combined with the best fit parameters obtained, clearly indicate the presence of high luminosity (unabsorbed $L_{(2-10\ keV)}$ $>10^{44}$ erg s^{-1}), obscured (N_H $>10^{22}$ cm^{-2}) AGN, i.e. QSO2 candidates.

References

1. Della Ceca et al. 2003, A&A 406, 555
2. Drory et al. 2001, MNRAS 325, 550
3. Gilli et al. 2001, A&A 366, 407
4. Stevens et al. 2003, MNRAS 342,
5. Vignali et al. 2002, astro-ph/0209415

Radio Observations of the Subaru/XMM-Newton Deep Survey Field

Chris Simpson[1], Steve Rawlings[2], Rob Ivison[3], and the SXDS Team

[1] Department of Physics, University of Durham, Durham DH1 3LE, UK
[2] Astrophysics, University of Oxford, Oxford OX1 3RH, UK
[3] Astronomy Technology Centre, Royal Observatory, Edinburgh EH9 3HJ, UK

1 Introduction

The Subaru/XMM-Newton Deep Survey[1] (SXDS) comprises deep optical and X-ray observations of an area of sky larger than a square degree. With a transverse size of 100 Mpc at $z = 1.5$, the SXDS can detect and measure large-scale structure which will strongly affect the global results from smaller surveys. This region of sky is also home to a 0.8-deg^2 ultra-deep near-infrared imaging survey[2], a 0.25-deg^2 submillimetre survey[3], and also lies within the XMM-LSS region which will be observed by SIRTF as part of the SWIRE Legacy Program[4], and by GALEX as part of its Deep Imaging Survey and Medium–Deep Spectroscopic Survey. Extragalactic radio sources are able to probe large-scale structure and inform on the accretion and star formation histories of the Universe, and we have therefore undertaken deep radio observations to complete an unparalleled multi-wavelength dataset. This paper introduces the radio data in this field.

2 Observations

The SXDS field was observed with the Very Large Array at 1.4 GHz in A, B, and C configurations, to ensure sensitivity to structures on a large range of angular scales, and to obtain accurate total flux measurements even for extended sources. Observations in each configuration were made at the same 14 field centres, which were arranged on a triangular grid and separated by half the primary beam size. The final mosaicked image is shown in Fig. 1.

The rms sensitivity of the final map is 14 μJy/beam in the central 0.5 deg^2, and is less than 20 μJy/beam over the central 1 deg^2, which represents a dynamic range of several thousand, since the brightest radio source in the field is 90 mJy. Our 5σ limit of 70 μJy corresponds to a star formation rate of $\sim 80\,M_\odot\,yr^{-1}$ at $z = 0.5$, and is well below the break in the radio luminosity function, even at $z = 10$. There are approximately 1000 radio sources detected in the field, and the source number counts agree well with those from similar fields, notably the VLA–VIRMOS[5] and Phoenix[6] Deep Fields.

Further observations have been made with the Giant Metrewave Radio Telescope at 240, 327, and 610 MHz, reaching sensitivities of ~ 1 mJy/beam. These

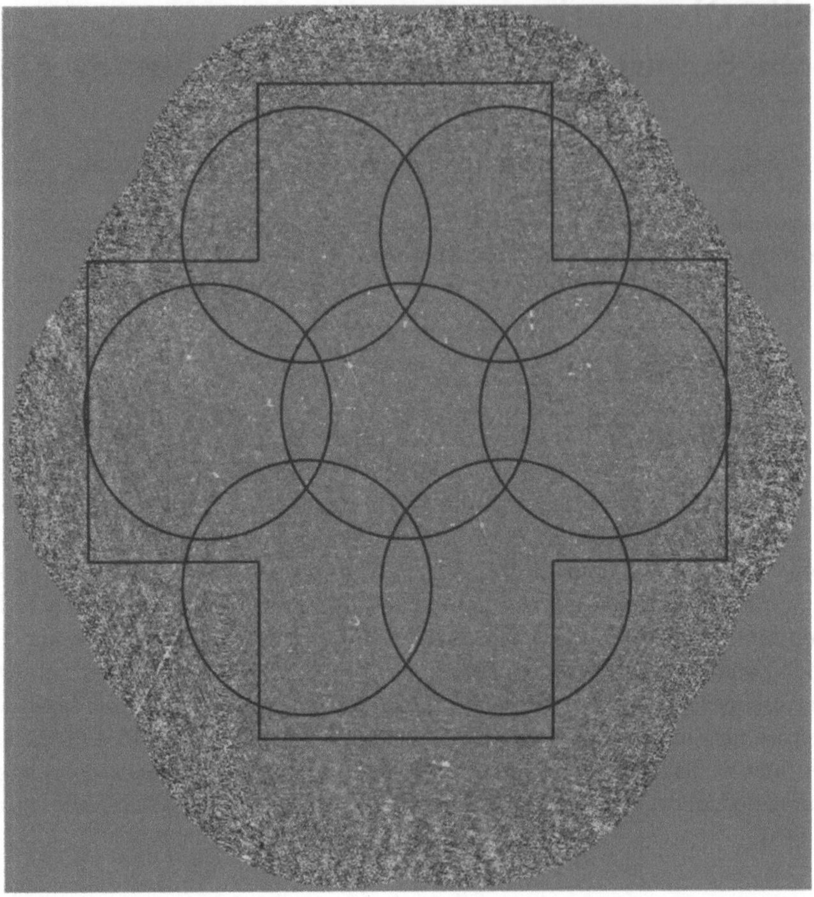

Fig. 1. VLA image of the SXDS field. The outlined cross indicates the region covered by the deep Suprime-Cam images, while the circles indicate the 7 XMM-Newton pointings.

data will provide important spectral index information and enable the construction of a low-frequency selected sample which is a less biased measure of radio luminosity.

References

1. K. Sekiguchi, et al.: these proceedings
2. O. Almaini, et al.: these proceedings
3. J. Dunlop, et al.: http://www.roe.ac.uk/ifa/shades
4. C. Lonsdale, et al.: PASP **115**, 897
5. M. Bondi, et al.: A&A **403**, 857
6. A. Hopkins, et al.: these proceedings

SED Diagnostics of Submillimetre Galaxies

Toshinobu Takagi[1], Nobuo Arimoto[2], and Hitoshi Hanami[3]

[1] University of Kent, Canterbury, Kent CT2 7NR, UK
[2] National Astronomical Observatory, 2-21-1, Osawa, Mitaka, Tokyo 181-8588, Japan
[3] Iwate University, Morioka 020-8550, Japan

Abstract. We apply an evolutionary SED model of starbursts to submm galaxies, in order to predict the present-day colour, magnitude, and size of submm galaxies. The reproduced colour-magnitude and size-magnitude relation of present-day descendants of submm galaxies suggest that they are progenitor of L^* elliptical galaxies on average, if the IMF of submm galaxies is slightly flatter than the Salpeter IMF.

1 Introduction

Submillimetre (submm) galaxies are thought to be the most plausible candidate of progenitors of elliptical galaxies at high redshifts. In this paper, we present a new clue to the evolutionary link between submm galaxies and elliptical galaxies, by using an evolutionary model of spectral energy distribution (SED) for starbursts [1].

2 Evolutionary SED Model of Starbursts

We approximated the star formation history of submm galaxies with the gas infall model, and solved the chemical evolution. The starburst regions were assumed to have centrally concentrated stars and homogeneously distributed dust. With this geometry, we solved the equations of radiative transfer, considering the temperature fluctuation of very small grains. The main parameters of this SED model are the starburst age and the size of the starburst region.

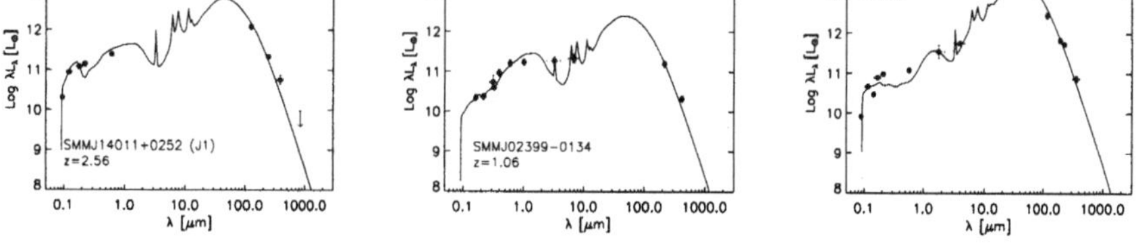

Fig. 1. Example of the result of SED fitting

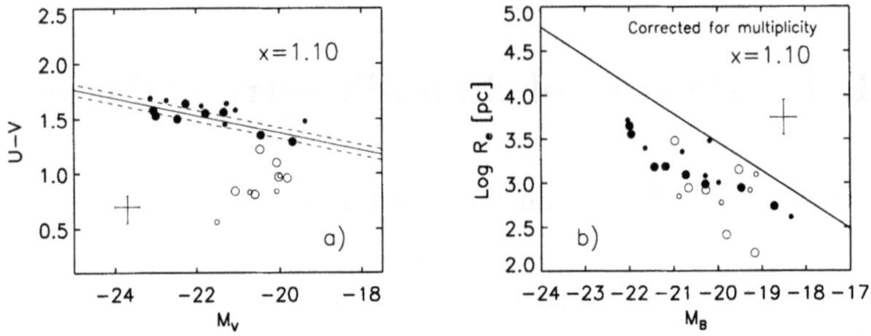

Fig. 2. The reproduced scaling relations of the present-day descendants of submm galaxies. Solid lines indicate the observed scaling relations of elliptical galaxies. (a) The colour-magnitude relation. Solid and open circles indicate older submm galaxies with age of $t/t_0 > 1.0$, where t_0 is the evolutionary time-scale of starbursts, and younger ones ($t/t_0 \leq 1.0$), respectively. Larger and smaller circles indicate the samples with the spectroscopic redshift and those with the photometric redshift, respectively. (b) The size-magnitude relation. The symbols are the same as (a). The derived sizes are corrected for the multiplicity of starburst regions in massive submm galaxies.

3 Results and Conclusions

We analysed 30 submm galaxies, of which 18 have spectroscopic redshifts. From the SED fitting, we derived the intrinsic properties of starbursts in submm galaxies, such as the starburst age, the stellar mass, and the mean stellar metallicity.

Based on the results of the SED fitting, we derived the colour-magnitude and the size-magnitude relation of present-day descendants of submm galaxies. We assumed that the effect of star formation after the observed epoch is not significant. This assumption is safe specifically for older submm galaxies.

We found that the present-day colours and magnitudes of older submm galaxies are consistent with the colour-magnitude relation of elliptical galaxies (Fig.2a), if the initial mass function (IMF) of submm galaxies has the power-law index of $x = 1.10$ ($x = 1.35$ for the Salpeter IMF). The mean present-day magnitude of submm galaxies corresponds to that of L^* elliptical galaxies. From the comparison of the size-magnitude relation, we found that massive submm galaxies tend to be smaller than elliptical galaxies with the similar mass. This suggests that massive submm galaxies may consist of multiple starbursts, which could eventually merge into more diffuse stellar systems. Once the effect of multiplicity is corrected, the present-day size-magnitude relation of submm galaxies becomes similar to that of elliptical galaxies (Fig.2b). These reproduced scaling relations suggest that submm galaxies are quite likely to evolve into elliptical galaxies after the starburst.

References

1. T. Takagi, N. Arimoto, H. Hanami: MNRAS **340**, 813

How Much Light Comes from Bulges and Disks?

Lidia Tasca and Simon D.M. White

Max Planck Institut für Astrophysik, Karl Schwarzschild Strasse 1,
D-85741 Garching bei München, Germany

Abstract. A complete sample of roughly 1800 morphologically classified galaxies in the Sloan Digital Sky Survey (SDSS) is used in order to study the amount of light in bulges and disks, after a detailed two–dimensional bulge to disk decomposition is performed. We conclude that at low redshift $(55 \pm 2)\%$ of the total light resides in the disk component, for a cosmological model with $\Omega_0 = 0.3$, $\Omega_\Lambda = 0.7$ and $h = 0.7$.

A detailed bulge to disk decomposition is performed on a complete magnitude limited sample of 1875 morphologically classified galaxies in the Sloan Digital Sky Survey (SDSS) [2]. We present our first results in the Sloan i band for a homogeneous sample, limited to bright galaxies with $r < 15.9$ after Galactic reddening correction.

The surface photometry of each galaxy component is modelled with a different parametric function: a Sérsic profile for the bulge, where $n = 4$ corresponds to the classical de Vaucouleurs profile, and an exponential profile $(n = 1)$ for the disk.

In order to compute the amount of total light which comes from the two different photometric components, we analyse the fraction of light in the disk component for 10 luminosity bins containing equal numbers of objects.

Only few very luminous galaxies actually have real thin disks detectable by the fitting algorithm, which is using the two components to improve fits to ellipsoidal galaxies where there are isophote twists or significant variation of ellipticity with radius.

In order to circumvent this problem we use in our analysis only the galaxies in each absolute magnitude bin which have a disk axial ratio (b/a) less than 0.5, which we assume to have real disks. We then assume that the same number of disk systems have $b/a > 0.5$, as expected for a uniform inclination distribution.

For the edge–on systems the fraction of the light in the disk component for each magnitude bin is estimated using the bulge to total light ratio (B/T), one of the structural parameters of our model. The light coming from disks with $b/a > 0.5$ is computed taking into account that dust attenuation of the disk light increases with inclination, whereas bulge attenuation is little affected. In order to correct for this effect, we adopt a standard empirical description for the extinction as function of the inclination [4].

In each absolute magnitude bin the disk contribution to the luminosity density is given by the total luminosity density in that bin, according to the Blanton et al. (2003) luminosity function, multiplied by the fraction of light in disks. The

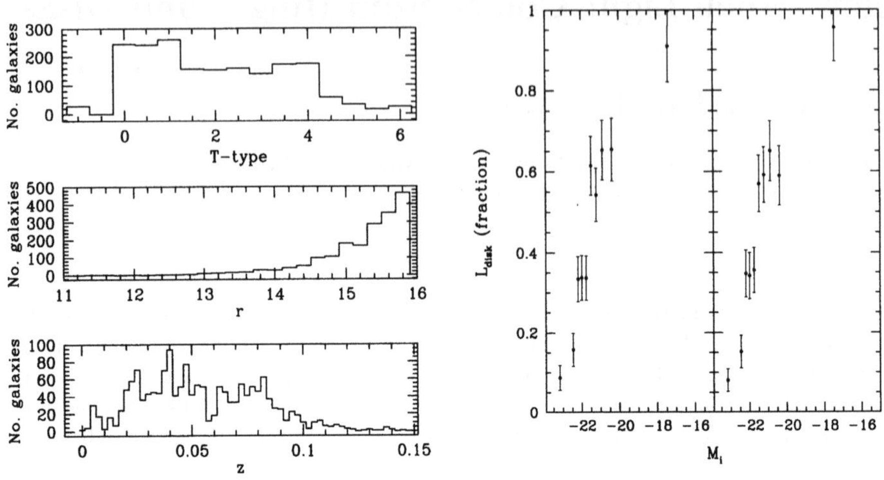

Fig. 1. (Left panel) The distribution of galaxies versus morphological type **(upper panel)** and against Petrosian magnitude in the r band after extinction correction **(middle panel)** are shown for the 1862 galaxies in the sample. The redshift distribution **(lower panel)** is given for the 1550 galaxies in the sample for which the redshift is measured. **(Right panel)** Luminosity fraction in disks for galaxies in the Sloan i band modelled with two different sets of parametric functions: de Vaucoulours profile for the bulge and exponential for the disk **(left panel)** or Sérsic profile for the bulge plus exponential for the disk **(right panel)**

sum over all absolute magnitude bins provides the total disk contribution to the luminosity density. A comparison with the total luminosity density in the i band brings us to the main result that roughly $(55 \pm 2)\%$ of the total light in the local Universe is in the disk component [3], independent of the profile model.

These preliminary results have to be supplemented by analysis in other bands. The next step will be to perform the same study at higher redshift, when the structures are less evolved, in order to try to discriminate between competing models of galaxy formation.

References

1. Blanton, M. R. et al. (2003), ApJ, 592, 819
2. Stoughton, C. , et al. (2002). *Sloan Digital Sky Survey: Early Data Release.* AJ, 123:485–548.
3. Tasca, L., White, S.D.M. (2004), in preparation.
4. Tully, R. B., Pierce, M. J., Huang, J., Saunders, W., Verheijen, M. A. W., & Witchalls, P. L. 1998, AJ, 115, 2264

Final Analysis of ELAIS 15 μm Observations

Mattia Vaccari[1,2,3]

[1] Dipartimento di Astronomia, Università di Padova, Italia
[2] CISAS "G. Colombo", Università di Padova, Italia
[3] Astrophysics Group, Imperial College, London, United Kingdom

Abstract. Reduction of ELAIS Fields

1 ELAIS

The European Large Area ISO Survey [10, ELAIS] was the largest Open Time project carried out with the Infrared Space Observatory [3, ISO], mapping areas of about 12 deg^2 at 15 and 90 μm and smaller areas at 7 and 175 μm with the ISOCAM [1] and ISOPHOT [8] cameras. Thanks to an extensive multi-wavelength coverage [11], the ELAIS fields have now arguably become the best studied sky areas of their size, and natural targets of on-going or planned large-area surveys with the most powerful ground and space-based facilities. Thus the need of reducing such data with the uttermost care and provide the community with an agreed-upon legacy from the ISO mission.

2 Data Reduction

ELAIS 15 μm observations were reduced using the LARI Method [5,6,12], a new data reduction method implementing a mathematical model describing the behaviour of ISOCAM LW detector. A fitting procedure models the transients attributable to changes in incident flux and the features on both short and long time scales produced by cosmic ray impacts ("glitches") on the time history, modeling them as discontinuities in the charge release towards the detector's contacts. The signal as a function of time is thus processed independently for each pixel, determining both the background and signal excess with respect to the background.

3 Map and Catalogue Construction

Flux excesses thus determined along the pixel time histories are then projected onto a sky map, on which source extraction is finally performed. Cross-correlation of source lists with each other and with optical astrometric catalogues allows to combine adjacent or superposed fields into mosaic maps Flux determination is then carried out by carefully modeling the effects due to mapping, noise and the data reduction method itself. The final catalogue contains about 2000 sources detected above a S/N threshold of 5, covering the 0.5 - 100 mJy flux range. Figure 1 shows a zoomed-in view of a S/N map and the catalogue flux distribution.

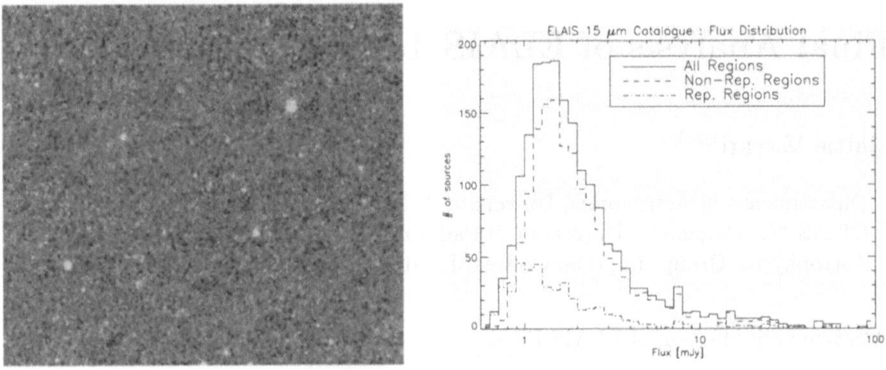

Fig. 1. Small portion of a S/N map (left) and catalogue flux distribution (right)

4 Simulations

Extensive simulations were carried out in order to carefully assess the performance of our data reduction method. Comparison of injected quantities with detected ones for simulated sources allowed to rigorously quantify the reliability, completeness, astrometric and photometric accuracy of our catalogue.

5 On-Going Work

15 μm data reduction has been completed [12], and a band-merged catalogue [11] has been made available. Extragalactic source number counts [7] and several multi-wavelength observational projects are underway, including identifications at optical [2,4], X-ray [9] and NIR [13] wavelengths, aiming at clarifying the nature of 15 μm sources and paving the way for the ambitious upcoming SIRTF extragalactic surveys such as GOODS as SWIRE.

References

1. Cesarsky C.J. et al., 1996, A&A, 315, L32
2. Gonzalez-Solares E. et al., 2003, in preparation
3. Kessler M.F. et al., 1996, A&A, 315, L27
4. La Franca F. et al., 2003, in preparation
5. Lari C. et al., 2001, MNRAS, 325, 1173
6. Lari C. et al., 2003, ESA SP-511, 349
7. Lari C. et al., 2004, in preparation
8. Lemke D. et al., 1996, A&A, 315, L64
9. Manners J. et al., 2003, in preparation
10. Oliver S. et al., 2000, MNRAS, 316, 749
11. Rowan-Robinson M. et al., 2003, MNRAS, submitted, astro-ph/0308283
12. Vaccari et al. 2003, in preparation
13. Väisänen P. et al., 2003, in preparation

The Relation Between Stellar Populations and Light Profiles in Early–Type Galaxies

Alexandre Vazdekis[1], Ignacio Trujillo[2], and Yoshihiko Yamada[3]

[1] Instituto de Astrofísica de Canarias, E-38200 La Laguna, Tenerife, Spain
[2] Max–Planck–Institut für Astronomie, Königstuhl 17, D–69117 Heidelberg, Germany
[3] National Astronomical Observatory of Japan, 2-21-1 Osawa, Mitaka, Tokyo 181-8588, Japan

1 Introduction and Galaxy Measurements

Early-type (E/SO) galaxies follow tight relations such as the Fundamental Plane, the Color-Magnitude Relation or the Mg_2–σ relation (Bender et al. 1992). However little is known about the relation between their morphological shapes and stellar populations (Conselice 2003), most likely because these two fields rely on methods that have grown up separately. Here we explore possible correlations between the Sérsic index n, and relevant stellar population parameters.

We use the mean luminosity weighted ages, metallicities ([M/H]) and abundance ratios ([Mg/Fe]) derived by Vazdekis et al (2001) and Yamada et al. (2003a,b in prep.) for a sample of Virgo and field early-type galaxies. Flux-calibrated spectra of extremely high signal–to–noise (S/N > 100) from galaxy central regions ($r_e/10$) were used. Relevant line-strengths, were measured and compared to those predicted by the model of Vazdekis (1999). Ages were determined on the basis of the $H\gamma_\sigma$ age indicator of Vazdekis & Arimoto (1999), which is free from metallicity effects.

The Sérsic shape parameters for most of Virgo galaxies were taken from Caon et al. (1993), or by fitting a Sérsic model to published (Peletier et al. 1990) or derived light profiles (HST archive images). We also use the concentration index of Trujillo et al. (2001) ($C_{re}(1/3)$), which evaluates the fraction of light enclosed within $r_e/3$ divided by the light within r_e, and is monotonically related to n.

2 Results and Discussion

No correlation can be found between galaxy ages and n as most of our galaxies are rather old (Vazdekis et al. 2001). Fig. 1 (left panel) shows a weak correlation (as shown by the Spearman test r_s) between the total metallicity and the shape parameters. However, Fig. 1 shows for the first time a strong positive correlation between [Mg/Fe] abundance ratio and the shape parameter indices. This result is expected since [Mg/Fe] correlates with σ (Trager et al. 2000), and σ correlates with $C_{re}(1/3)$ (Graham et al. 2001). Nonetheless, it is surprising to see that the [Mg/Fe] provides the strongest correlation among other stellar population parameters, reaching a Spearman coefficient 0.8. Finally, similar correlation values

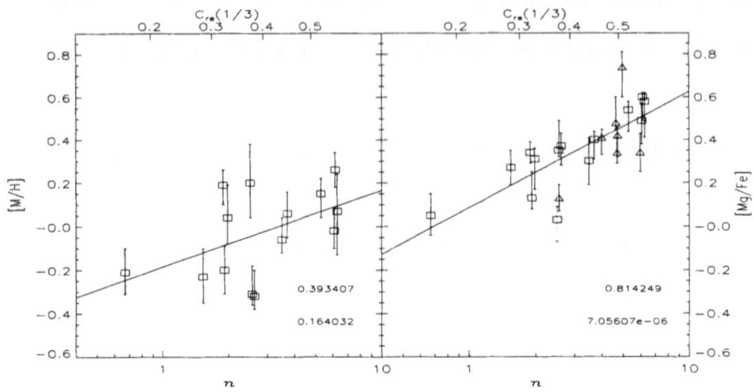

Fig. 1. Plot of the shape parameters, n and $C_{re}(1/3)$, versus [M/H], and [Mg/Fe]. Squares mean Virgo galaxies and triangles mean field galaxies. Spearman rank-order correlation coefficient and its significance value are written in each panel.

are obtained for the already known, $C_{re}(1/3) - \sigma$ and [Mg/Fe] $- \sigma$ relations on the basis of the same galaxies, probing the robustness of the new correlation.

Mg is ejected to the ISM by SN II on short time-scales ($\leq 10^6 - 10^7$ yr). If star formation stops before SN Ia release Fe (~1 Gyr), the [Mg/Fe] should be larger than 0. The central σ – [Mg/Fe] correlation suggests that the innermost regions of more massive galaxies stopped their star formation on short time-scales. Next, the fact that n correlates with [Mg/Fe] suggests that early-type galaxies settled up their global structure (and not only the innermost regions) on time-scales according to their [Mg/Fe] (an scenario favored if no [Mg/Fe] radial gradients were present, Mehlert et al. 2003). Since we do not find a correlation between n and galaxy mean ages but [Mg/Fe], suggests that the formation time-scale is the main process linked to galaxy global structure, rather than burst peppering as a function of time. Since our galaxies are old and the time scales of most massive galaxies very short, the new correlation suggests that the global structure was already at place at high z, without experiencing a significant time evolution.

References

1. Bender, R., Burstein, D., Faber, S.M. 1992, ApJ, 399, 462
2. Caon, N., Capaccioli, M. & D'Onofrio, M., 1993, MNRAS, 265, 1013
3. Conselice, C. 2003, ApJSS, 147, 1
4. Graham, A., Trujillo, I., Caon, N. 2001, AJ, 122, 1707
5. Mehlert, D., Thomas, D., Saglia, R., Bender, R., Wegner, G. 2003, A&A, 407, 423
6. Peletier, R., Davies, R., Illingworth, G., Davis, L., Cawson, M., 1990, AJ, 100, 1091
7. Trager, S., Faber, S., Worthey, G., González, J. 2000, AJ, 120, 165
8. Trujillo, I., Graham, A., Caon, N. 2001, MNRAS, 326, 869
9. Vazdekis, A. 1999, ApJ, 513, 224
10. Vazdekis, A. et al. 2001, 551, L127
11. Vazdekis, A., Arimoto, N. 1999, ApJ, 525, 144

The Submm Proprieties of Extremely Red Objects

Tracy Webb[1], Mark Brodwin[2], Steve Eales[3], and Simon Lilly[4]

[1] Leiden Observatory, Leiden, The Netherlands
[2] Dept. Astron. & Astrophys., Univ. of Toronto, Canada
[3] Dept. Phys. & Astron., Cardiff University, Wales
[4] ETH Zürich, Switzerland

1 Introduction and Data

A basic but crucial step toward our understanding of the evolution of the Universe and the objects within it is a complete accounting of the systems which produce the far-infrared background light. Once identified, the relationship between these populations and their role in the formation of galaxies must be determined. In this paper we focus on two equally interesting high-redshift populations: the submillimeter (submm) bright galaxies and the Extremely Red Objects (EROs).

We present the results of a study of the submm properties of EROs in the Canada-UK Deep Submillimeter Survey fields [2,7]. We select EROs from K-imaging from the CFH Telescope to a depth of $K \sim 20.7$ [8] and RIz' imaging from the Canada-France-Deep-Fields survey [4,1]. Our sample consists of 241 EROs selected using the $(I-K) > 4.0$ color criteria and 164 EROs selected by $(R-K) > 5.3$. The 850μm imaging reaches a 3σ depth of 3 mJy and covers 108 square arcminutes over two fields.

2 Results: The Submillimeter Flux of EROs

We measure the 850μm flux at each ERO location on the submm image and take a noise weighted mean of these measurements, after correcting for source confusion. In Figure 1 we show that the mean 850μm flux of K-selected galaxies increases to redder colours. Limiting the measurement to the EROs, we find the average 850μm flux for $(I-K) > 4.0$ EROs is 0.40$\pm$0.07 mJy and 0.56$\pm$0.09 mJy for those selected by $(R-K) > 5.3$. We estimate that EROs produce 7-11% of the far-infrared background at 850μm [6,3]. This is a much lower contribution than found by[9] and we claim this is primarily due to a difference in ERO number density and the effect of confusion correction. We note that these measurements are dominated by a minority of bright objects which are associated with the discrete submm-luminous systems uncovered in blank-field SCUBA surveys and thus, this contribution to the far-infrared background is not new, but was already accounted for from the deep SCUBA surveys.

Since any submm signal is diluted by the presence of early-type systems, we attempt to isolate the dusty star-forming galaxies through their location on a

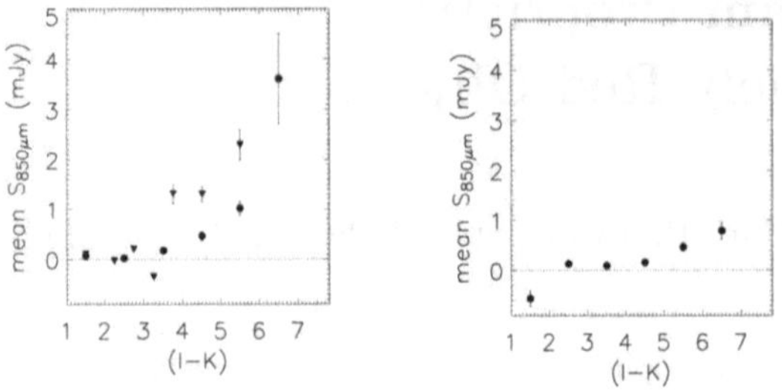

Fig. 1. The mean 850μm flux as a function of near-infrared colour. Left: $(I-K)$ colour. Right: $(R-K)$ colour. The solid circles correspond to this work and the solid triangles to the measurements of [9].

near-infrared color-color diagram, following [5]. All of the EROs identified with discrete submillimeter sources lie on the expected star-burst side of the diagram within their uncertainties. Repeating our analysis of the mean flux using only those galaxies with colors consistent with a starburst yields the following mean fluxes: 0.67±0.14 mJy for $(I-K) > 4.0$ EROs and 1.18±0.17 mJy for $(R-K) > 5.3$ EROs.

We note that 50% of the EROs which are the most likely counterpart to a discrete submm source have at least one other apparent ERO companion. These companions are close enough that the two or more EROs are confused within the JCMT beam at 850μm. A measurement of the mean 850μm flux as a function of the nearest neighbour distance for EROs reveals an increase in mean flux as the distance between nearest-neighbours decreases. It is not possible to determine from these data if (1) A single ERO is extremely luminous at 850μm and exists in an over-dense region of the universe, containing enhanced numbers of EROs, or (2) all of the EROs in these groups are producing submm emission at a relatively equal level and, due to confusion, appear to be one luminous source. Regardless, small groups of EROs appear to mark, on average, regions of enhanced star-formation and a reasonable explanation is that interactions between the EROs are driving up the star-formation rates in these galaxies.

References

1. Brodwin, M., et al., 2004, submitted to ApJ
2. Eales, S. et al., 2000, AJ, 120, 2244
3. Fixsen, D.J., Dwek, E., Mather, J.C., Bennet, C.L., & Schafer, R.A., 1998, ApJ, 508, 123
4. McCracken, H.J., et al., 2001, A&A, 376, 756
5. Pozzetti, L. & Mannucci, F., 2000, MNRAS, 317, L17
6. Puget, J.-L., et al., 1996, A&A, 308, 5
7. Webb, T.M.A. et al., 2003, ApJ, 587, 41
8. Webb, T.M.A. et al., 2003, ApJ, in press
9. Wehner, E.H., Barger, A.J., Kneib, J.-P., 2002, ApJ, 577, L83

The Luminosity Function of AGN at $z \sim 5 \ldots 1$

Christian Wolf[1], Lutz Wisotzki[2], Andrea Borch[3], Simon Dye[4], Martina Kleinheinrich[3], and Klaus Meisenheimer[3]

[1] Department of Physics, University of Oxford, Oxford, OX1 3RH, U.K.
[2] Astrophysikalisches Institut Potsdam, D-14482 Potsdam, Germany
[3] Max-Planck-Institut für Astronomie, D-69117 Heidelberg, Germany
[4] Imperial College, Prince Consort Road, London, U.K.

We present a determination of the optical/UV AGN luminosity function and its evolution, based on a large sample of faint ($R < 24$) QSOs identified in the COMBO-17 survey. Using multi-band photometry in 17 filters within $350\,\mathrm{nm} < \lambda_{\mathrm{obs}} < 930\,\mathrm{nm}$, we could simultaneously determine photometric redshifts with an accuracy of $\sigma_z \simeq 0.03$ and obtain spectral energy distributions. We select a sample covering the redshift range across $1.2 < z < 4.8$. Even at $z \simeq 3$, the sample reaches below luminosities corresponding to $M_B = -23$, conventionally employed to distinguish between Seyfert galaxies and quasars.

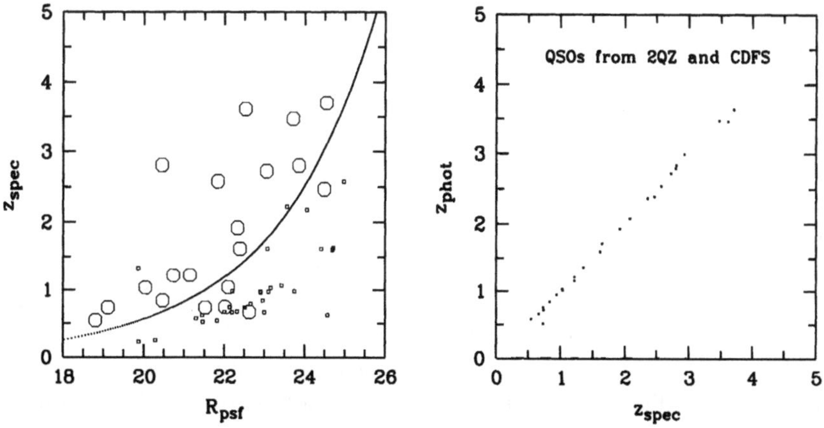

Fig. 1. *Left:* The purely optical COMBO-17 selection of broad-line AGN is basically complete at luminosities $M_{B,\mathrm{Vega}} < -21.5$ (solid line) for $z < 4$, and hence covers almost all of the type-1 luminosity emitted. Shown are all type-1 AGN identified in X-ray follow-up of the CDFS 1 Msec imaging (Szokoly et al., 2003). Large circles mark those found in COMBO-17, and small dots those only through X-rays and spectroscopy. The template based classification of COMBO-17 recognizes type-1 AGN, when the nuclear light contributes significantly to the whole spectrum. At low AGN luminosities the SED is dominated by stellar light from the host galaxy.
Right: Spectroscopic vs. photometric redshifts for 27 QSOs from 2QZ and Chandra X-ray follow-up. The mean redshift error is $\sigma_z/(1+z) \approx 0.015$. There is one outlier below the z range of our analysis with $(z_{spec}, z_{phot}) = (0.733, 0.51)$.

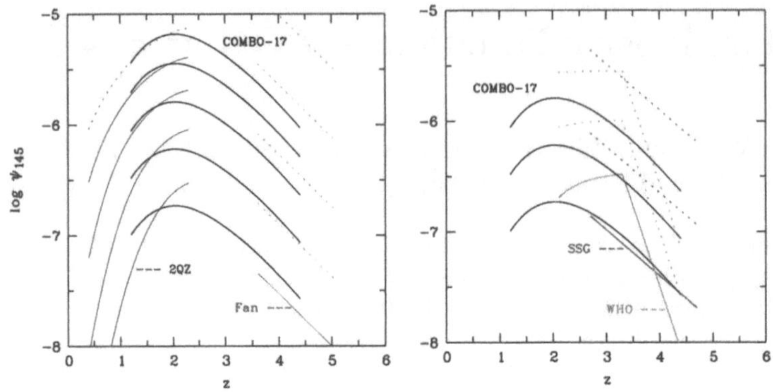

Fig. 2. Best-fit parametric models of quasar space density in comparison. Solid lines are constrained by observations, dashed lines are extrapolations to fainter luminosities. *Left:* COMBO-17 with 2QZ and Fan et al., at luminosities of $M_{145,Vega} < [-28 \ldots -24]$. The apparent disagreement of COMBO-17 with 2QZ at brightest luminosities should not be taken seriously as the curves show fits and not actual LF data.
Right: COMBO-17 with Warren, Hewett & Osmer (1994) and Schmidt, Schneider & Gunn (1995), at luminosities of $M_{145} < [-28, -27, -26]$.

We clearly detect a broad plateau-like maximum of quasar activity around $z \simeq 2$ and map out the smooth turnover between $z \simeq 1$ and $z \simeq 4$. The shape of the LF is characterised by some mild curvature, but no sharp 'break' is present within the range of luminosities covered. Using only the COMBO-17 data, the evolving LF can be adequately described by either a pure density evolution (PDE) or a pure luminosity evolution (PLE) model. However, the absence of a strong L^*-like feature in the shape of the LF inhibits a robust distinction between these modes.

We present a robust estimate for the integrated UV luminosity generation by AGN as a function of redshift. We find that the LF continues to rise even at the lowest luminosities probed by our survey, but that the slope is sufficiently shallow that the contribution of low-luminosity AGN to the UV luminosity density is negligible. Although our sample reaches much fainter flux levels than previous data sets, our results on space densities and LF slopes are completely consistent with extrapolations from recent major surveys such as SDSS (Fan et al., 2001) and 2QZ (Boyle et al., 2000). See Wolf et al. (2003) for further details.

References

1. Boyle, B. J., Shanks, T., Croom, S. M., et al., 2000, MNRAS **317**, 1014
2. Fan, X., Strauss, M. A., Schneider, D. P., et al., 2001, AJ **121**, 54
3. Schmidt, M., Schneider, D. P., Gunn, J. E., 1995, AJ **110**, 68
4. Szokoly, G., et al., 2003, ApJS, submitted, astro-ph/0312324
5. Warren, S. J., Hewett, P. C., Osmer, P. S., 1994, ApJ **421**, 412
6. Wolf, C., Wisotzki, L., et al., 2003, A&A **401**, 73

A Public Redshift Catalogue of the Chandra Deep Field South from COMBO-17

Christian Wolf[1], Martina Kleinheinrich[2], Klaus Meisenheimer[2], Andrea Borch[2], Simon Dye[3], Meghan Gray[4], and Lutz Wisotzki[5]

[1] Department of Physics, University of Oxford, Oxford, OX1 3RH, U.K.
[2] Max-Planck-Institut für Astronomie, D-69117 Heidelberg, Germany
[3] Imperial College, Prince Consort Road, London, U.K.
[4] Department of Astronomy, Blackford Hill, Edinburgh EH9 3HJ, U.K.
[5] Astrophysikalisches Institut Potsdam, D-14482 Potsdam, Germany

We have imaged the Chandra Deep Field South (CDFS) among other fields for the COMBO-17 survey. We used the Wild Field Imager at the MPG/ESO-2.2-m telescope at La Silla, Chile, to observe 17 passbands from 350 to 930 nm (UBVRI plus 12 medium-bands). Such a photometric dataset allows quite reliable multi-colour classification and accurate photometric redshift estimation. The resulting object catalogue for the CDFS contains $\sim$ 100 QSOs, $\sim$ 1000 stars and $\sim$ 10000 galaxies at $R < 24$. A further 50000 objects fainter than that are included in the catalogue, but have much less reliable redshifts. This catalogue was drawn from a deep R-band image with $R_{\mathrm{lim},5\sigma} = 26.0$ and $0.75''$ coadded PSF. It covers an area of $31.5' \times 30'$ centered on $\alpha_{2000} = 03^{\mathrm{h}}32^{\mathrm{m}}25^{\mathrm{s}}$ and $\delta_{2000} = -27°48'50''$.

We publish this catalogue with astrometry, photometry, classification, photometric redshifts and restframe luminosities alongside a reference paper with all details (Wolf et al., 2004). From January 2004 on, it is available for the benefit of the community interested in the CDFS at the COMBO-17 website (http://www.mpia.de/COMBO/index.html) and later from CDS.

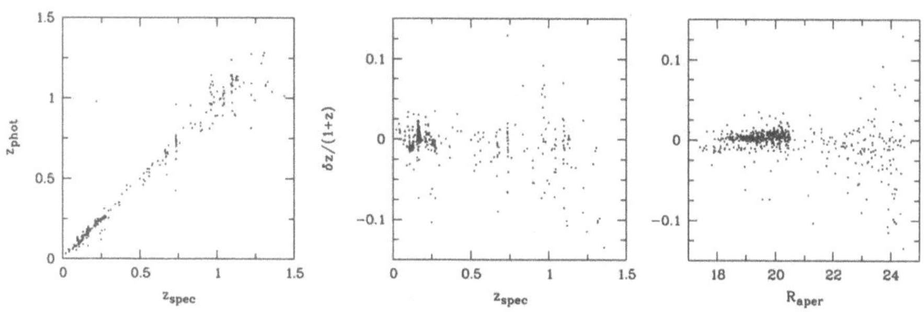

Fig. 1. Photometric vs. spectroscopic redshifts for 552 galaxies (from 2dF and VLT). *Left:* Redshift comparison z_{phot} vs. z_{spec}. Only few redshift outliers can be seen, which are mostly fainter than $R = 23$. *Center:* Redshift error $\delta z/(1 + z)$ vs. redshift z_{spec}. *Right:* Redshift error $\delta z/(1 + z)$ vs. aperture R-band magnitude. On the whole, the 1-sigma redshift error is < 0.01 for $R < 21$ and increases to ~ 0.06 at $R = 24$.

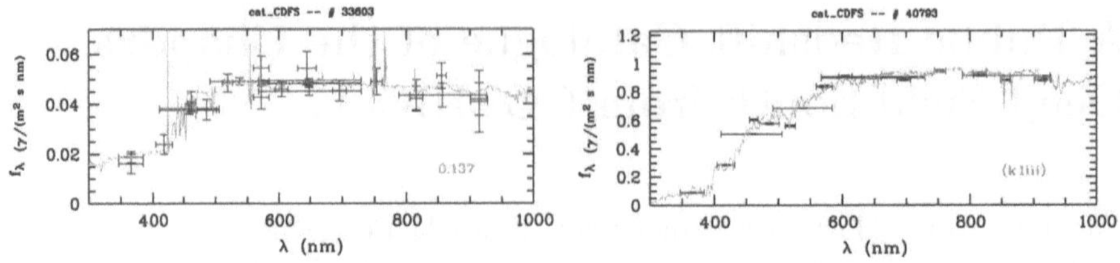

Fig. 2. Photometric COMBO-17 spectrum of two example objects in the CDFS, a galaxy with $R \sim 23$ at $z \sim 0.14$ (left) and a K-type star of $R \sim 20$ (right). The error bars resemble the width of the filters horizontally and the flux errors vertically. The line is the best fitting template SED.

The sample of stars is supposedly complete to $R = 22$, although M stars can be identified down to $R = 24$. Galaxy redshifts are $> 90\%$ complete down to $R \approx 23.5$ at all redshifts $z < 1.2$. QSOs and Seyfert-1 AGN are complete to a luminosity of $M_B = -21.5$ and across redshifts from $z = 0.5$ to $z = 5$. Below this luminosity limit the contribution from the nuclear activity to the 17-filter spectrum of the entire active galaxy is too little to be recognized. In that case the stellar continuum is too dominant and better matched with normal galaxy templates (see also Wolf, Wisotzki et al., these proceedings).

The quality of photometric classification and redshift estimation can be described by a simple picture with the magnitude range split into quality domains:

1. the *quality saturation domain* at $R < 22$, where systematic errors from calibration and insufficient match between templates and true object SEDs dominate the performance while observed photon statistics are irrelevant.
2. the *quality transition domain* at $R = 22 \ldots 24.5$, where photon statistics take over degrading the performance with decreasing signal.
3. the *quality breakdown domain* at $R > 24.5$, where the classification is undecidable using the given photons and a-priori knowledge completely dominates the interpretation of objects.

We also assess the classification and redshift performance by spectroscopy (from 2dFGRS and proprietary 2dF observations as well as from ESO/GOODS, Franx et al., priv. comm., and Szokoly et al., 2003). At present, we have redshifts available for 552 galaxies, 27 QSOs and 50 Seyfert galaxies. The most abundant problem objects are galaxies at $z > 1$ which are not well estimated due to the lack of NIR imaging to enclose the 4000Å-break.

Galaxy redshift errors are $\sigma_z/(1+z) < 0.01$ in the quality saturation domain. In the transition domain they increase with the photometric noise and lead to a redshift noise of $\sigma_z/(1+z) \approx 0.06$ at $R = 24$. QSO redshift errors are $\sigma_z/(1+z) \approx 0.015$ at all magnitudes to $R = 24.5$. Whenever QSOs are recognized their features are multiple and strong enough to constrain their redshift quite well. These results are consistently found in both the Monte-Carlo simulations of the survey and the spectroscopic checks.

Measuring the Ages of z∼1 Galaxies

Marsha J. Wolf, Niv Drory, Karl Gebhardt, and Gary J. Hill

University of Texas at Austin, Department of Astronomy, Austin TX 78712, USA

1 Introduction

One approach to studying the epoch of galaxy formation is to infer formation redshifts from the ages of galaxies at known redshifts. When observed with optical instruments, galaxies lying at redshifts of z∼1 reveal their restframe UV spectra, which contain some features that show promise for breaking age-metallicity degeneracies inherent in techniques used for estimating galaxy ages. Early-type galaxies around the z∼1 epoch are of additional interest because they lie only a few Gyrs after the probable peak when galaxy mergers occurred. The spectra of young galaxies, such as these merger products, with ages of 1-5 Gyr are much more distinguishable than those of older galaxies, which results in more accurate age estimates.

We take this approach by targeting the many early-type galaxies at z∼1 identified in recent and ongoing surveys and estimating ages by fitting their spectra and photometry to stellar population synthesis models. Because significant differences exist in the models, we combine results from Bruzual-Charlot 2003 [1], Maraston [2], and Worthey [3] models in the analysis. To help calibrate the technique, local galaxies and extragalactic globular clusters with independent age estimates have been added to the sample. The data are assembled from a number of sources. High redshift galaxies are from the DEEP surveys [4,5], moderate redshift galaxies are from the MUNICS survey [6], local galaxies are from the Storchi-Bergmann, et al. database of UV-optical spectra of nearby quiescent galaxies [7], and extragalactic globular clusters are from the Santos, et al. database [8].

2 Analysis and Results

To make use of the maximum amount of information, models are fit to the entire galaxy spectrum and also to the photometry when available. Best fitting models for each object are determined for the spectra and photometry separately via S/N-weighted χ^2 minimization.

This project began by using data from MUNICS, which includes B,V,R,I,J,K photometry and optical spectra. We hoped that adding such a wide photometric coverage would make the age estimates more robust. However, we found the simple stellar population (SSP) models to be bluer than the typical early-type galaxies, causing most of the estimated spectroscopic ages to be older than the

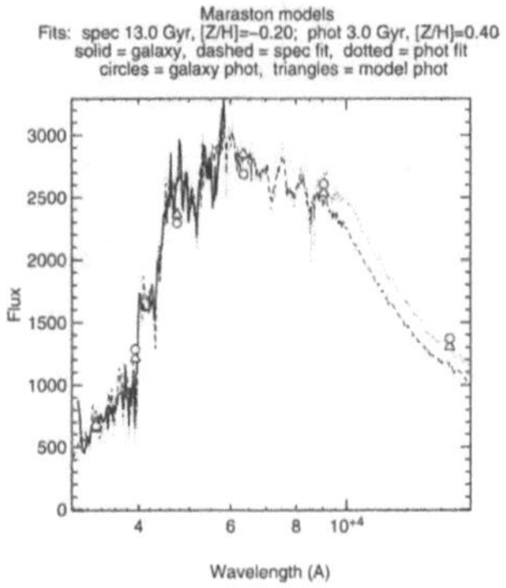

Fig. 1. Example of model fits to data.

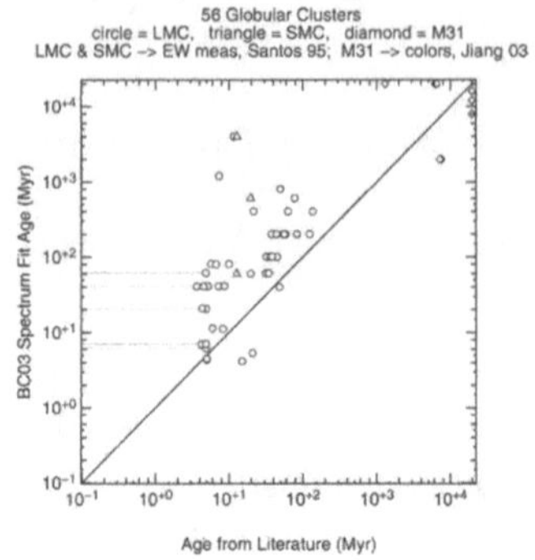

Fig. 2. Extragalactic globular clusters.

photometric ages. This discrepancy makes combining the spectral ages and photometric ages into a total galaxy age difficult. This difference is likely due to the model prescriptions of asymptotic giant branch star inclusion/exclusion. An example of model fits to a MUNICS galaxy at z=0.367 is shown in Figure 1.

We also fit to globular clusters, since these should be closer to single age stellar populations. Our spectral ages are compared to age estimates from the literature in Figure 2.

3 Conclusions

Good fits to optical spectra of early-type galaxies can be obtained with SSPs, but the resulting ages do not agree with those derived from wider range photometry of the galaxies. The models are bluer than the galaxies. The 3 sets of SSPs used do not agree on ages of individual galaxies, but come closer on globular cluster ages. Galaxy metallicities from these fits agree rather well. Globular cluster ages from SSP model fits agree fairly well with those derived from line indices and colors. However, for early-type galaxies multiple age stellar population models will be required to correctly estimate ages and metallicities.

References

1. G. Bruzual, S. Charlot: MNRAS **344**, 1000 (2003)
2. C. Maraston: MNRAS **300**, 872 (1998)
3. G. Worthey: ApJS **95**, 107 (1994)
4. D. Koo: in *Highlights of Astronomy* **11A**, ed. J. Andersen (Dordrecht: Kluwer, 1998), p. 468
5. M. Davis, et al.: Proceedings of SPIE **4834**, 161 (2003)
6. N. Drory, et al.: MNRAS **325**, 550 (2001)
7. T. Storchi-Bergmann, D. Calzetti, A. Kinney: *A Database of UV-Optical Spectra of Nearby Quiescent and Active Galaxies*, http://www.stsci.edu
8. J.F.C. Santos, D. Alloin, E. Bica, C. Bonatto: in *IAU Symposium 207*, ed. by D. Geisler, E.K. Grebel, D. Minniti (San Francisco: ASP 2002) p. 727

Understanding Current Star Formation Processes in Galaxies at Different Redshifts

Jaime Zamorano[1], Pablo G. Pérez-González[2], Jesús Gallego[1], Armando Gil de Paz[2], Almudena Alonso-Herrero[3], Alfonso Aragón-Salamanca[4], Sergio Pascual[1], Cristina Díaz[1], and Víctor Villar[1]

[1] Universidad Complutense de Madrid, 28040 Madrid, Spain
[2] IPAC, California Institute of Technology, USA
[3] Steward Observatory, University of Arizona, USA
[4] University of Nottingham, UK

Abstract. We have studied in detail the properties of local active star-forming galaxies from the UCM survey, and in particular their star-formation histories. We have quantified the relative importance of the current episode of star formation in comparison to the underlying stellar populations. We have also determined the total stellar mass function and burst mass function for the UCM sample using a M/L calculated for each galaxy. Integrating this mass function we obtained a value for the stellar mass density of current star-forming galaxies in the local Universe.

Stellar Populations of UCM Survey Galaxies

The main goal of our work [1] is studying the main properties of the stellar populations embedded in a statistically complete sample of local active star-forming galaxies: the Universidad Complutense de Madrid (UCM) Survey of emission-line galaxies[1]. This sample contains 191 local star-forming galaxies at an average redshift of 0.026. The survey was carried out using an objective-prism technique centered at the wavelength of the Hα nebular emission-line (a reliable tracer of recent star formation).

We have compiled an extensive multi-wavelength (from the optical to the near infrared) photometric (in narrow and broad band filters) and spectroscopic dataset for all the galaxies in the sample. The analysis of this dataset has been centered on the characterization of the morphology and star-formation processes of the objects, on the comparison of the global properties of the sample with those of other galaxy samples at several redshifts, and on the study of the stellar content of the UCM galaxies. In the latter research topic, a special attention has been paid to the most recent star-forming bursts, which are the main characteristic of our objects. We have also developed (and tested with the dataset for the UCM sample) a powerful stellar population synthesis tool (which is able to obtain robust estimates of the stellar mass, a very important parameter in the study of the evolution of galaxies), and a technique to calculate luminosity and stellar mass functions ([2] [3]). These techniques may be easily used in the characterization of samples of galaxies at any redshift.

[1] http://t-rex.fis.ucm.es/UCM_Survey/index.html

Main Results

The main results obtained in this work point out that the UCM Survey galaxies span a broad range in properties (among others, the total stellar mass and the star formation rate per stellar mass unit) between those of galaxies completely dominated by current/recent star formation (e.g., extreme dwarf HII galaxies or starburst galaxies) and those of 'normal' quiescent spirals. The UCM objects present intermediate-late Hubble type spiral morphologies and a recent instantaneous burst of star formation (occurred about 5 Myr ago) with subsolar metallicity. These bursts involved 5% of the total stellar mass of the UCM galaxies. An 'average' UCM galaxy has a total stellar mass of 10^{10} solar masses, i.e., about a factor of 7-10 lower than a typical local spiral galaxy.

The location of the star-forming knots, their intensity and concentration, and the relationship of these properties with those of the host galaxy have been studied using Hα imaging. We estimate that the amount of Hα emission that arises from the diffuse ionized gas is about 15%-30% of the total Hα flux for a typical UCM galaxy. This percentage seems to be independent of the Hubble type. Conversely, we find that an 'average' UCM galaxy harbors a star formation event with 30% of its Hα luminosity arising from a nuclear component [4].

We have determined the total stellar mass function and burst mass function for the UCM sample from the K band magnitudes. Instead of using the mean M/L ratio for the whole sample, the masses have been obtained using the M/L calculated for each galaxy. Integrating this mass function we obtained a value for the stellar mass density of the local Universe [5]. We estimate that $13 \pm 3\%$ of the total baryon mass density in the form of stars in the local Universe is associated with star-forming galaxies. Because our galaxies are selected based on star formation (via Hα emission), our data are directly comparable to high redshift survey galaxies, commonly dominated by intense burst of star formation (see, for instance, [6]).

References

1. P. Pérez-González: Poblaciones estelares en Galaxias con Formacion Estelar Activa en el Universo Local, PhD Thesis, Universidad Complutense de Madrid, (2003)
2. Pérez-González, P. G., Gil de Paz, A., Zamorano, J., Gallego, J., Alonso-Herrero, A., Aragón-Salamanca, A. Monthly Notices of the Royal Astronomical Society, **338**, 508 (2003)
3. Pérez-González, P. G., Gil de Paz, A., Zamorano, J., Gallego, J., Alonso-Herrero, A., Aragón-Salamanca, A. Monthly Notices of the Royal Astronomical Society, **338**, 525 (2003)
4. Pérez-González, P. G., Zamorano, J., Gallego, J., Aragón-Salamanca, A., Gil de Paz, A. Astrophysical Journal, **591**, 827 (2003)
5. Pérez-González, P. G., Gallego, J., Zamorano, J., Alonso-Herrero, A., de Paz, A. G., Aragón-Salamanca, A. Astrophysical Journal, **587**, L27 (2003)
6. Calura, F., Matteucci, F. Astrophysical Journal, **596**, 734 (2003)

Witnessing the Formation of a Brightest Cluster Galaxy at $z = 4.1$

Andrew W. Zirm, Roderik A. Overzier, George K. Miley, and the ACS/GTO Team

Leiden University, 2333 RA Leiden, The Netherlands

1 Introduction

Powerful radio sources mark the locations of massive galaxies. The K-band 'Hubble' diagram shows that radio hosts form the bright envelope of near-infrared sources at every observed redshift [1]. Within the context of biased galaxy formation, massive galaxies mark the peaks in the overall density distribution [2,3]. We have exploited this correspondence to identify galaxy overdensities in the high-redshift Universe.

2 A Proto-Cluster at $z = 4.1$

The radio galaxy TN J1338-1942 ($z = 4.1$) was imaged using a narrow-band filter to target the redshift Lyman-α emission line [4]. Of 50 line-emitting galaxies found, 33 have been confirmed spectroscopically to lie at the same redshift as the radio galaxy. To follow-up these ground-based observations, multicolor images were taken using the Advanced Camera for Surveys on-board the *Hubble Space Telescope* [5,6]. Using the very accurate colors determined from these imaging data, a large candidate population of star-forming galaxies have been identified in addition to the line-emitting population [7]. This discovery greatly strengthens the argument that 1338 marks the center of an overdense region. This protocluster is gravitationally destined to become a rich galaxy cluster in which the radio galaxy is the dominant, central, brightest cluster galaxy (BCG).

3 The Progenitor of a Present-Day BCG

The high spatial-resolution ACS imaging reveals the complex morphology of the radio galaxy in the rest-frame far-ultraviolet. The r band is completely dominated by the Lyman-α emission-line. In this poster contribution we focus on a particular feature of the emission-line morphology of this galaxy.

A large (50 kpc), low surface-brightness Lyman-α 'halo' surrounds the radio galaxy[4]. The spatial-resolution of the ACS data is only sensitive to the highest surface-brightness features within this halo. These include the radio galaxy itself. But, unexpectedly, also include a cone-shaped feature emanating from the center of the galaxy and perpendicular to its major axis. The apex of the 'cone' does not coincide with the peak of the K-band light or with the ostensible radio

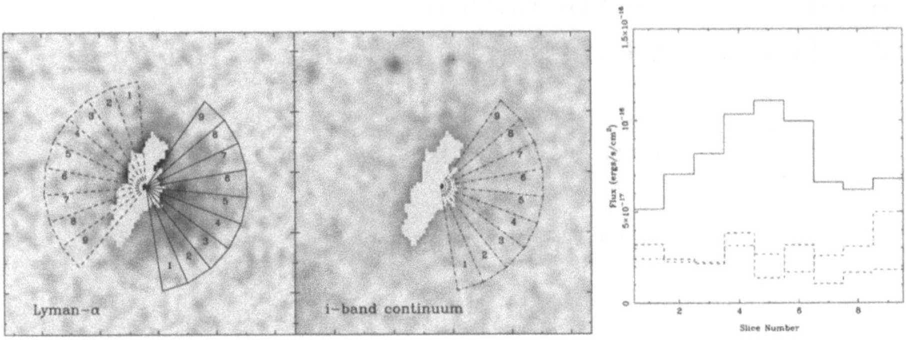

Fig. 1. Continuum-subtracted r-band (left) and i-band images of the radio galaxy. The galaxy has been masked-out of both images to isolate the 'wind' feature. The photometric points are plotted in the histogram on the right with line styles corresponding to the wedges overlaid on the images. There is a clear excess of line-emission on one side of the galaxy, and very little continuum.

core [1]. While we cannot rule out the possibility that an AGN is powering the line-emission, the evidence suggests an origin only indirectly related to the AGN.

We may then interpret the extended line-emission as an ionized outflow due perhaps to prodigious star-formation. Assuming that the wind can be described as a bubble expanding into a uniform medium the energy injection rate is given by Heckman, Armus & Miley 1990. For a fiducial shock velocity of 100 km/s and ambient density of 10 cm^{-3} and using the measured extent of the wind feature from the ACS Lyman-α image ($\sim$ 10 kpc) we derive an energy injection rate of 3×10^{44} ergs/s. This energy injection rate in turn implies a star-formation rate of $\sim 1000\,\mathrm{M_{\odot}yr^{-1}}$. If we instead estimate the SFR using the Lyman-α line luminosity to infer a ionizing photon production rate, we get a very robust lower limit of $\sim 1.2\,\mathrm{M_{\odot}yr^{-1}}$.

Regardless of the true nature of the outflow, whether starburst or AGN driven, it is clear that these phenomena can be very energetically important, and may contribute to the "pre-heating" of the intracluster gas which is inferred from thermodynamic studies of rich clusters of galaxies in the lower redshift Universe [9].

References

1. C. De Breuck et al., 2002, AJ, 123, 637
2. S.D.M. White and M.J. Rees, 1978, MNRAS, 183, 341
3. N. Kaiser, 1984, ApJ, 284, L9
4. B.P. Venemans et al., 2002, ApJ, 569, L11
5. H. Ford et al., 1998, SPIE, 3356, 234
6. J. Blakeslee et al., 2003, ASP Conf. Ser. 295, 257
7. G.K. Miley et al., 2003, Nature, in press
8. T.M. Heckman, L. Armus and G.K. Miley, 1990, ApJS, 74, 833
9. P. Tozzi and C. Norman, 2001, ApJ, 546, 63

Author Index

ESO ASTROPHYSICS SYMPOSIA
European Southern Observatory

Series Editor: Bruno Leibundgut

D. L. Clements, I. Pérez-Fournon (Eds.),
Quasar Hosts
Proceedings, 1996. XVII, 336 pages. 1997.

L. N. da Costa, A. Renzini (Eds.),
Galaxy Scaling Relations: Origins, Evolution and Applications
Proceedings, 1996. XX, 404 pages. 1997.

L. Kaper, A. W. Fullerton (Eds.),
Cyclical Variability in Stellar Winds
Proceedings, 1997. XXII, 415 pages. 1998.

R. Morganti, W. J. Couch (Eds.),
Looking Deep in the Southern Sky
Proceedings, 1997. XXIII, 336 pages. 1999.

J. R. Walsh, M. R. Rosa (Eds.),
Chemical Evolution from Zero to High Redshift
Proceedings, 1998. XVIII, 312 pages. 1999.

J. Bergeron, A. Renzini (Eds.),
From Extrasolar Planets to Cosmology: The VLT Opening Symposium
Proceedings, 1999. XXVIII, 575 pages. 2000.

A. Weiss, T. G. Abel, V. Hill (Eds.),
The First Stars
Proceedings, 1999. XIII, 355 pages. 2000.

A. Fitzsimmons, D. Jewitt, R. M. West (Eds.),
Minor Bodies in the Outer Solar System
Proceedings, 1998. XV, 192 pages. 2000.

L. Kaper, E. P. J. van den Heuvel, P. A. Woudt (Eds.), **Black Holes in Binaries and Galactic Nuclei: Diagnostics, Demography and Formation**
Proceedings, 1999. XXIII, 378 pages. 2001.

G. Setti, J.-P. Swings (Eds.), **Quasars, AGNs and Related Research Across 2000**
Proceedings, 2000. XVII, 220 pages. 2001.

A. J. Banday, S. Zaroubi, M. Bartelmann (Eds.),
Mining the Sky
Proceedings, 2000. XV 705 pages. 2001.

E. Costa, F. Frontera, J. Hjorth (Eds.),
Gamma-Ray Bursts in the Afterglow Era
Proceedings, 2000. XIX, 459 pages. 2001.

S. Cristiani, A. Renzini, R. E. Williams (Eds.),
Deep Fields
Proceedings, 2000. XXVI, 379 pages. 2001.

J. F. Alves, M. J. McCaughrean (Eds.),
The Origins of Stars and Planets: The VLT View
Proceedings, 2001. XXVII, 515 pages. 2002.

J. Bergeron, G. Monnet (Eds.),
Scientific Drivers tor ESO Future VLT/NLTI Instrumentation
Proceedings, 2001. XVII, 356 pages. 2002.

M. Gilfanov, R. Sunyaev, E. Churazov (Eds.),
Lighthouses of the Universe: The Most Luminous Celestial Objects and Their Use for Cosmology
Proceedings, 2001. XIV, 618 pages. 2002.

R. Bender, A. Renzini (Eds.), **The Mass of Galaxies at Low and High Redshift**
Proceedings, 2001. XXII, 363 pages. 2003.

W. Hillebrandt, B. Leibundgut (Eds.),
From Twilight to Highlight: The Physics of Supernovae
Proceedings, 2002. XVII, 414 pages. 2003.

P. A. Shaver, L. DiLella, A. Giménez (Eds.),
Astronomy, Cosmology and Fundamental Physics
Proceedings, 2002. XXI, 501 pages. 2003.

M. Kissler-Patig (Ed.),
Extragalactic Globular Cluster Systems
Proceedings, 2002. XVI, 356 pages. 2003.

P.J. Quinn, K.M. Górski (Eds.),
Toward an International Virtual Observatory
Proceedings, 2002. XXII, 341 pages. 2004.

W. Brandner, M. Kasper (Eds.),
Science with Adaptive Optics
Proceedings, 2003. XX, 387 pages. 2005.

A. Merloni, S. Nayakshin, R.A. Sunyaev (Eds.)
Growing Black Holes: Accretion in a Cosmological Context
Proceedings 2004. XIV, 506 pages. 2005

A. Renzini, R. Bender (Eds.)
Multiwavelength Mapping of Galaxy Formation and Evolution
Proceedings 2003. XXV, 487 pages. 2005

MIX
Papier aus verantwortungsvollen Quellen
Paper from responsible sources
FSC® C105338

If you have any concerns about our products,
you can contact us on
ProductSafety@springernature.com

In case Publisher is established outside the EU,
the EU authorized representative is:
Springer Nature Customer Service Center GmbH
Europaplatz 3, 69115 Heidelberg, Germany

Printed by Libri Plureos GmbH
in Hamburg, Germany